Aspects of Ecology and Zoogeography of Recent and Fossil Ostracoda

Proceedings of the 6th International Symposium
on Ostracods, Saalfelden (Salzburg),
July 30 - August 8, 1976

Edited by

Heinz Löffler and Dan Danielopol

Dr W. Junk b.v. Publishers, The Hague 1977

ISBN 90 6193 581 4

Cover design Max Velthuijs

© Dr W. Junk bv Publishers 1977

No part of this book may be reproduced and/or published in any form, by print, photoprint, microfilm or any other means without written permission from the publishers.

CONTENTS

Foreword IX

List of participants X

I. Introduction

I. G. Sohn, Zoogeography of Ostracodologists 3

II. General aspects and morphology

R. H. Benson, On the use of Ostracodes for tracing changes in oceanic circulation 15
M. C. Keen, Tertiary Ostracod provinces in Western Europe 23
R. L. Kaesler & P. S. Mulvany, Approaches to the diversity of assemblages of ostracoda 33
M. V. Angel, Some speculation on the signifiance of carapace length in planktonic halocyprid Ostracods 45
A. Rosenfeld & B. Vesper, The variability of the sieve pores in recent and fossil species of *Cyprideis torosa* (Jones 1850), as an indicator for salinity and paleosalinity 55
R. C. Whatley & J. M. Stephens, Precocious sexual dimorphism in fossil and recent Ostracoda 69
V. H. Valicenti, Some Hemicytherinae from the Tertiary of Patagonia (Argentina), their morphological relationship and stratigraphical distribution 93
A. Liebau, Carapace ornamentation of the Ostracode Cytheracea: principles of evolution and functional significance 107
G. Carbonnel & M. Jacobzone, L'ornamentation chez Loxoconchidae: un moyen de prédiction (paléo)écologique 121
E. Kristan-Tollmann, On the development of the muscle-scar patterns in Triassic Ostracoda 133

III. Ecology and zoogeography

R. F. Maddocks, Zoogeography of Macrocyprididae (Ostracoda) 147
L. S. Kornicker, Diversity of benthic myodocopid Ostracodes 159

W. A. van den Bold, Distribution of marine podocopid Ostrocoda in the Gulf of Mexico and the Caribbean 175
G. Hartmann-Schröder & G. Hartmann, Preliminary notes on the distribution of Polychaeta and Ostracoda on the Australian coast 187
U. Skaumal, Preliminay account of the ecology of Ostracodes of the rocky shore of Helgoland 197
D. Keyser, Ecology and zoogeography of recent brackish-water Ostracoda (Crustacea) from South-West Florida 207
A. Sokac, Ostrascoda from bottom cores off the Coast of Montenegro . . 223
R. L. Kaesler, P. S. Mulvany & L. S. Kornicker, Delimitation of the Antarctic convergence by cluster analysis and ordination of benthic myodocopid Ostracoda 235
J. H. Baker, Life history patterns of the myodocopid Ostracod *Euphilomedes producta* Poulsen, 1962 245
J. H. Martens, Distribution patterns of pelagic Ostracoda of the Peru current system (Crustacae: Ostracoda-Halocypridae) 255
A. Moguilevsky & A. Gooday, Some observations on the vertical distribution and stomach contents of *Gigantocypris mülleri* Skogsberg 1920 (Ostracoda, Myodocopina) 263
J. W. Neale, Ostracodes from the rice-fields of Sri Lanka (Ceylon) . . . 271
P. De Deckker, The distribution of the 'giant' Ostracods (family: Cyprididae Baird, 1845) endemic tot Australia 285
D. L. Danielopol, On the origin and diversity of European freshwater interstitial Ostracods 295

IV. Paleoecology and paleozoogeography

F. M. Swain, Paleoecological implications of Holocene and late Pleistocene Ostracoda, Lake Lahonton basin, Nevada 309
H. Löffler, 'Fossil meromixis' in Kleinsee (Carinthia) indicated by Ostracods 321
F. J. Gunther & A. S. Hunt, Paleoecology of marine and freshwater Holocene Ostracods from Lake Champlein, Nearctic, North America . . . 327
S. P. Cameron & R. F. Lundin, Environmental interpretation of the Ostracode succession in late Quaternary sediments of pluvial Lake Cochise, Southeastern Arizona 335
I. Purper, Some Ostracods from the Upper Amazon Basin, Brazil. Environment and age 353
J. P. Peypouquet, Les Ostracodes, indicateurs paléoclimatiques et paléogéographiques de quaternaire terminal (Holocene) sur le plateau continental Sénégalais 369
N. Krstić, The Ostracod genus Tyrrhenocythere 395
P. Carbonel, La conquête des milieux de Plateforme Continentale par l'ensemble Carinocythereis antiquata/carinata depuis le Miocène Moyen. . 407
O. Ducasse, La faune d'Ostracodes des depots tertiaires du plateau continental dans la partie centrale du golfe de Gascogne: Intérêt paléogéographique – Relations avec le continent 417

ASPECTS OF ECOLOGY AND ZOOGEOGRAPHY
OF RECENT AND FOSSIL OSTRACODA

K. Ishizaki, Distribution of Ostracoda in the East China Sea – A justification for the existence of the paleo-Kuroshio current in the late Cenozoic . . 425

P. Donze, Distribution comparée des populations marines d'Ostracodes en regard des modèles paléogéographiques récents concernant l'Ouest de la Téthys au Berriasien 441

I. Y. Neustrueva, Ostracod biofacies in Paleozoic and Mesozoic lake sediments of the USSR 451

G. Becker, Thuringian Ostracods from the Famennian of the Cantabrian Mountains (Upper Devonian, N. Spain) 459

S. M. Warshauer & R. Smosna, Paleoecologic controls of the Ostracod communities in the Tonoloway limestone of the central Apalachians (Silurian, Pridoli) 475

L. Ferguson, Some paleoecological and taxonomic problems in connection with growth series of the Ostracod genera Bairdia and Paraparchites from a Scottish Lower Carboniferous shale 487

V. Abstracts of the Discussion sessions

M. V. Angel, Discussion on sampling methods 491
J. W. Neale, Discussion on Ostracod terminology 495
I. G. Sohn, Discussion on Paleozoic Ostracoda 499

General index 501

FOREWORD

The sixth International Symposium on Ostracoda (Saalfelden, Salzburg, 30.7.1976–8.8.1976) was devoted mainly to the ecology and zoogeography of both recent and fossil representatives of this group. In contrast to other crustaceans – especially copepoda – ecological and zoogeographical orientated research on ostracoda sofar has received only little attention. Therefore it seemed worthwhile to stimulate an international debate about these subjects and the results are now presented, which leave no doubt that our knowledge is increasing rapidly. Of the 101 participants 51 contributed papers and it is hoped that their papers will be a stimulus for further work.

The editors are most grateful to the Federal Ministry for Science and Research and to the Federal County of Salzburg for their generous grant to help with the organization of the Symposium and to the mayor of Saalfelden for an unforgettable folkloric evening. The Austrian Oilcompany (ÖMV, Dr. A. Kröll) and the Creditanstalt Bankverein contributed to the organization of the excursions and provided a poster for the Symposium. Special mention must be made of the most generous help with the printing of the programme, abstracts and field excursions by the Facultas publishing company (Dr. A. Soritsch). The editors would like to express their thanks for the hospitality offered by the Bundeserziehungsanstalt Saalfelden (Direktor Dr. H. Leitgeb), to Prof. Dr. A. Tollmann and Dr. E. Kristan-Tollmann, Dr. T. Cernajsek, Dr. F. Steininger and Dr. P. Herrmann for preparing and conducting the field excursions and to Dr. E. Schultze, Dr. R. Schmidt, M. Bobek and Miss I. Gradl for their continuous efforts with the daily organization work.

We thank Dr. R. Whatley and Dr. H. J. Oertli for their editorial advice as well as Miss S. Powell for her help with the preparation of the circulars, the excursion guides and the translation into English of several papers.

H. Löffler & D. Danielopol,
Wien

LIST OF PARTICIPANTS

Adamczak, F., Stockholm, Sweden
Ahmad, M., Hull, United Kingdom
Al-Furaih, A., Riyad, Saudi Arabia
Al-Sheikly, S., Glasgow, United Kingdom
Andersson, A., Lund, Sweden
Angel, M. V., Wormley, Godalming, Surrey, United Kingdom
Baker, J., Houston, Texas, USA
Becker, G., Frankfurt, BRD
Benson, R., Washington D.C., USA
Berdan, R., Washington D.C., USA
Blaszyk, J., Warszawa, Poland
Bobek, M., Wien, Austria
Bodergat, A., Villeurbanne, France
van den Bold, W., Baton Rouge, USA
Bonaduce, G., Napoli, Italy
Braun, W., Saskatchewan, Saskatoon/Sask., Canada
Cameron, S., Phoenix, Arizona, USA
Carbonnel, G., Villeurbanne, France
Carbonel, P., Talence, France
Cernajsek, T., Wien, Austria
Colin, J. P., Begles, France
Damotte, R., Rueille-Malmaison, France
Danielopol D., Wien, Austria
Dealtry, J., Howden, North Humberside, United Kingdom
De Decker, P., Louvain-la-Neuve, Belgium
Donze, P., Villeurbanne, France
Ducasse, O., Talence, France
Ferguson, L. Sackville, New Brunswick, Canada
Gerry, E., Tel Aviv, Israel
Gottwald, J., Göttingen, BRD
Gramann, F., Hannover, BRD
Grekoff, N., Boulogne, France
Grigg, U., Halifax, Canada
Groos-Uffenorde, H., Göttingen, BRD
Guillaume, M., Paris, France
van Harten, D., Amsterdam, The Netherlands

Hartmann, G., Hamburg, BRD
Hartmann-Schröder, G., Hamburg, BRD
Haskins, C., Llarnhos Llandudno, Gwynedd, N. Wales, United Kingdom
Herrmann, P., Wien, Austria
Honigstein, A., Jerusalem, Israel
Ishizaki, K., Sendai, Japan
Jäger, P., Salzburg, Austria
Kaesler, R., Lawrence, Kansas, USA
Keen, M., Glasgow, United Kingdom
Kempf, E. K., Köln, BRD
Keyser, D., Hamburg, BRD
Kileny, T., London, United Kingdom
Kornicker, L., Washington D.C., USA
Kristan-Tollmann, E., Wien, Austria
Krömmelbein, K., Kiel, BRD
Krstić, N., Beograd, Yugoslavia
Le Fèvre, J., Pau, France
Liebau, A., Tübingen, BRD
Löffler, H., Wien, Austria
Maddocks, R., Houston, Texas, USA
Martens, J., Hamburg, BRD
Mertens, E., Dortmund, BRD
Michelsen, O., København, Denmark
Moguilevsky, A., Aberystwyth, Cardiganshire, United Kingdom
van Morkhoven, F., Houston, Texas, USA
Neale, J., Hull, United Kingdom
Oertli, H., Pau, France
Paik, K., Kiel, BRD
Pinto, I., Porto Alegre, RS Brasil
Plachy, H., Wien, Austria
Pugliese, N., Trieste, Italy
Purper, I., Porto Alegre, RS Brasil
Schallreuter, R., Hamburg, BRD
Scharf, B., Mainz, BRD
Schmidt, R., Wien, Austria
Schultze, E., Wien, Austria
Siddiqui, Q., Halifax, Canada
Siveter, D., Leicester, United Kingdom
Sivhed, U., Lund, Sweden
Skaumal, U., Hamburg, BRD
Sohn, I. G., Washington D.C., USA
Sokac, A., Zagreb, Yugoslavia
Swain, F., Minneapolis, Minnesota, USA
Sylvester-Bradley, P., Leicester, United Kingdom
Szczechura, J., Warszawa, Poland
Tambareau, Y., Toulouse, France

Uffenorde, H., Göttingen, BRD
Valicenti, V., Wales, Aberystwyth, Cardiganshire, United Kingdom
Ware, M., Waun Fawr, Aberystwyth, Dyfed Sy, United Kingdom
Warshauer, S., Morgantown, West Virginia, USA
Whatley, R., Aberystwyth, Cardiganshire, United Kingdom

Not attending, but paper submitted:
Godday, A. J., Wormley, Godalming, Surrey, United Kingdom
Gunther, F. J., Roselle, N.J. USA
Hunt, A. S., Vermont, USA
Jacobzone, M., Villeurbanne, France
Kotzian, S. B., Porto Alegre, RS-Brasil
Lundin, R. F., Phoenix, Arizona, USA
Mulvany, P. S., Lawrence, Kansas, USA
Neustrueva, I. Y., Leningrad, USSR
Peypouquet, J. P., Talence, France
Rosenveld, A., Kiel, BRD
Sanguinetti, Y. T., Porto Alegre, RS-Brasil
Smosna, R., Morgantown, West Virginia, USA
Stephens, J. M., Aberystwyth, Cardiganshire, United Kingdom
Vespers, B., Hamburg, BRD

I. INTRODUCTION

Sixth Intern. Ostracod Symposium, Saalfelden

ZOOGEOGRAPHY OF OSTRACODOLOGISTS

I. G. SOHN

Abstract

An ostracodologist is defined as an individual who has published at least one original paper on Ostracoda since January 1, 1966, except for those in the People's Republic of China, for whom, because of the Cultural Revolution, the time is extended to January 1, 1956. The card index of my reprint collection was used to obtain the names and titles of papers on Ostracoda published during the period 1966 through December 1975, and the countries in which the ostracodologists live were obtained either frommy mailing list or from the publications.

On the basis of my card index, 2295 papers, more than 40% of all the papers on Ostracoda in my records, were published during the past decade; 566 ostracodologists are distributed in 45 countries. Nine countries have approximately 68% of all the ostracodologists: Russia 113; USA 95; West Germany 59; France 39; India 35; England 29; and East Germany, Italy, and Poland 18 each. The remaining 32% are distributed in 36 countries: Rumania 13; Canada, and The Netherlands 9 each; Argentina, and Austria 8 each; Czechoslovakia, Spain and Sweden 7 each; and the remaining 24 countries 1 to 6 each. Living ostracodes are being studied by 133 ostracodologists; Tertiary by 176; Mesozoic by 134; and Paleozoic by 124. Many of these, however, also study ostracodes of other ages. These data should be useful in planning the revision of the ostracode volume of the Treatise, in training future ostracodologists, and in ordering reprints.

Introduction

This meeting marks the 13th anniversary of the first International Symposium on Ostracoda in Napels, June 10–19, 1963. At that meeting (Figure 1), 28 participants represented exactly 12 countries in Europe, Asia, North America, and Australia. The Hamburg Meeting in 1974, the 5th, had 84 participants representing 21 countries from all the inhabited continents (Dr. M. A. A. Bassiouni, University of Cairo, Egypt, was in Hamburg, although not listed). You can see for yourself the number of participants and countries represented at this meeting. For the purpose of this study, I define an ostracodologist as an individual who has published at least one paper on Ostracoda within the past 10 years (except for individuals in the People's Republic of China; because of the Cultural Revolution, I used the past 20 years). This paper shows the geographical location of ostracodologists and reviews the progress of ostracode studies during the past decade.

Previous research

Wise (1962) and Wise & Gerry (1967) listed more than 400 names and addresses of '. . . those who have written papers on ostracodes, those who are currently engaged in research relative to ostracodes and those who use ostracodes exclusively

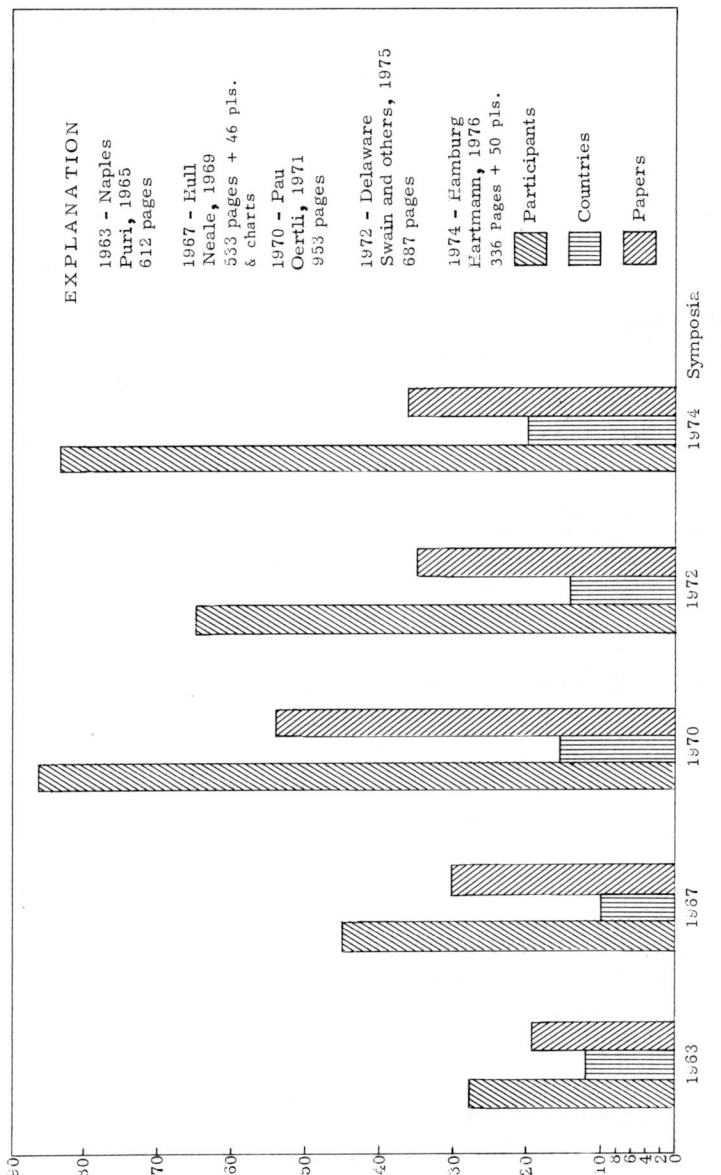

Fig. 1. Number of participants, countries represented, number of papers, and size of transactions of the earlier symposia.

in their work' (Wise, 1962, p. 269), and 'The Ostracodologist' nos. 10, 1967, through 21, 1974, lists additional names as well as changes in addresses.

Methods used

I used the card index for my reprint collection to obtain the names of the authors who had published on ostracodes during the period 1966 through 1975. The addresses of the individuals were obtained either from my mailing list, or from the reprints. In multiple-author papers containing information not related to ostracodes, only the authors dealing with Ostracoda were used.

In September 1975, I separated from my card index all those references dated 1966 and later, as the first step in obtaining the data for this paper. Approximately 40% of all the references in my index were published in the past decade. This is a conservative figure because not all the 1975 publications were processed, and also because the possibility of my missing papers published after 1966 is greater than it is for those published before that year. The number of papers pertaining to ostracodes published from the beginning of 1966 through 1975, as recorded by me up to to February 21, 1976, is by actual count 2,284. No wonder Malz (1975) titled his review paper 'The ostracodal deluge'.

As a check on the coverage of my card file of ostracode publications, I obtained on Sept. 11, 1975, through the U.S. Geological Survey Library a search by GEO-REF, operated by the American Geological Institute, the number of postings pertaining to ostracodes in its data bank for each of the years 1966–75. On the basis of the Sept. 1975 data, I conclude that my coverage is 85% or better.

Zoogeography of ostracodologists

Howe (1962) sketched the beginnings of ostracodology; Neale (1965) reviewed in detail what is known about the early British ostracodologists and also dealt with the later workers on living and post-Paleozoic fossil ostracodes; and I (Sohn, 1960, p. 2, 3, Fig. 1) reviewed the study of late Paleozoic ostracodes in North America and included some phylogenetic data by enumerating the teachers who trained Paleozoic ostracodologists in the United States. The present study shows the geographic location by countries of ostracodologists who published during the past decade (table 1). because ostracodologists work with material collected in various parts of the world, the table does not show the zoogeography of the animals studied.

The USSR has the largest number of ostracodologists because it is a confederation of 15 autonomous republics (Felber, 1974, p. 162). All the republics except the Kirghiz S.S.R. have one or more ostracodologists (table 2). Like all the data in this paper, these are minimum figures because my records may be incomplete, and also because individuals who have not published during the past decade were not included.

Table 1. Geographic distribution of ostracodologists and numbers of individuals publishing on various stratigraphic divisions during 1966-1975

Country	Holocene	Tertiary	Mesozoic	Paleozoic	Total
U.S.S.R. (all autonomous republics)	10	34 (+4)	24 (+4)	45 (+6)	113
U.S.A.	47 (+6)	21 (+6)	7 (+5)	20 (+3)	95
German Federal Republic	9 (+1)	14	24 (+4)	12 (+1)	59
France	5 (+1)	13 (+3)	18	3 (+1)	39
India	7 (+1)	21	5 (+2)	3 (+1)	36
England	7 (+4)	5 (+3)	12 (+1)	5 (+2)	29
Italy	5	11	1	1 (+2)	18
Poland	3 (+1)	3	5	7 (+1)	18
German Democratic Republic	– (+1)	2	10	5 (+2)	17
Romania	1	12	–	–	13
Canada	2 (+1)	2 (+1)	1 (+1)	4	9
The Netherlands	1 (+4)	3	3 (+1)	2	9
Austria	4	1 (+1)	3	–	8
Argentina	2 (+1)	1 (+1)	2 (+1)	3	8
Czechoslovakia	– (+2)	3 (+1)	1	3	7
Spain	1	4	1	1	7
Sweden	2	–	2	3	7
Denmark	4	–	2	– (+1)	6
People's Republic of China	–	1 (+1)	4	1 (+1)	6
Yugoslavia	1	5	–	–	6
Australia	2	– (+1)	–	3	5
Belgium	3	2 (+1)	–	– (+1)	5
Hungary	2	4	–	–	6
Brazil	3	–	2 (+1)	–	5
Egypt	– (+1)	2	– (+1)	3	5
New Zealand	1	3	– (+1)	–	4
Israel	1 (+1)	–	2	–	3
Japan	2	1	–	– (+1)	3
Bulgaria	1	1	–	–	2
Pakistan	–	1	1	–	2
South Africa	–	–	2	–	2
Turkey	–	2	–	–	2
Algeria	–	1	–	–	1
Chile	1	–	–	–	1
Gabon	–	1	–	–	1
Guam	1	–	–	–	1
Iran	–	1	–	–	1
Ireland	1	–	–	–	1
Jordan	1	–	–	–	1
Luxembourg	1	–	–	–	1
Mexico	1	–	–	–	1
Mongolian People's Republic	–	–	1	–	1
Portugal	–	–	1	–	1
Taiwan	1	–	–	–	1
Venezuela	–	1	–	–	1
Total	133	176	134	124	567

(+1) Number of individuals listed in other columns who have also published on ostracodes of this age.

Table 2. Geographic distribution of ostracodologists in the Soviet Republics and numbers of individuals publishing on various stratigraphic divisions during 1966-1975

Russian Republic Soviet Federated	Holo-cene	Terti-ary	Meso-zoic	Paleo-zoic	Total
Socialist	7	11	15	33	66
Ukrainian S.S.R.	1	8	2	7	18
Azerbaijan S.S.R.	1	3	1	–	5
Byelorussian S.S.R.	1	2	1	–	4
Kazakh S.S.R.	–	2	1	–	3
Tadzhikistan S.S.R.	–	–	3	–	3
Turkmen S.S.R.	–	3	–	–	3
Estonian S.S.R.	–	–	–	2	2
Georgian S.S.R.	–	2	–	–	2
Lithuanian S.S.R.	–	–	–	2	2
Uzbek S.S.R.	–	1	1	–	2
Armenian S.S.R.	–	1	–	–	1
Latvian S.S.R.	–	–	–	1	1
Moldavian S.S.R.	–	1	–	–	1
Total	10	34	24	45	113

Ages of the ostracodes studied

The ostracodes that were dealt with in the studies by the 567 individuals (table 1) were distributed stratigraphically as follows: living, 133, Teriary, 176; Mesozoic, 134, Paleozoic, 124. Because some students have interests that cross stratigraphic boundaries, the above figures include 46 individuals who published on fossil ostracodes adn also on living forms, and 68 who published on fossil forms from stratigraphic divisions other than the primary divison listed for them. These data indicate that 114 individuals, or 20% of the ostracodologists were not restricted to any one of the categories used.

Factors influencing the distribution of ostracodologists

Initially, the early studies of ostracodes were made because of intellectual curiosity. Academia contributed research by the faculty and by students (as dissertations) during the early part of the 20th century, and State and Federal governments published reports. The fossil fuels industry became a contributing factor shortly before the 1940's and became the major factor during the post-World War II period. Environmentalists began to contribute during the past decade, and will continue, together with the fuels industry, to be the major supporters of ostracodologists.

State of the art

Data processing and scanning electron microscopy (SEM) are the two major technological advances during the past decade that have been adopted by ostracodo-

Table 4. Attempts at ordinal refinement

Ivanova, 1960	Moore, 1961	Henningsmoen, 1965	Grundel, 1967	Grundel, 1969	Kozur, 1972	Kornicker & Sohn (1976)
Order Bradoriida	Order Archaeocopida new				Suborder Bradoriamorphes new	
Order Leperditiida	Order Leperditocopida				Order Bradoria	
Order Ostracoda					Superorder Podocopamorphes	
	Order Palaeocopida	Order Palaeocopa		Order Beyrichiida	Order Beyrichiida	
Suborder Paleocopida		Group Beyrichiida		Suborder Kirkbyocopina new		
		Suborder Beyrichiomorpha new			Suborder Beyrichiomorpha	
		Suborder Hollinomorpha new			Suborder Hollinomorpha	
					Suborder Binodocopina	
Order Podocopa	Order Podocopida		Order Podocopida	Order Podocopida	Order Podocopida	
	Suborder Podocopina				?Suborder Kirkbyocopina emend.	
			Suborder Cytherocopina new	Suborder Cytherocopina	Suborder Cytherocopina	
	Suborder Metacopina		Suborder Metacopina	Suborder Metacopina	Suborder Cypridocopina	

		Bairdiocopina new	Bairdiocopina	Bairdiomorpha new
Order Platycopina		?Group Kloedenellida new Suborder Kloedenellocopina	Order Platycopida Suborder Kloedenellocopina Superorder Platycopina	Suborder Bairdiomorpha new Order Platycopida
Order Myodocopa	Suborder Myodocopina	?Group Leperditiida		Superorder Myodocopa Suborder Myodocopamorphes new Order Leperditiida Order Myodocopida Suborder Myodocopina Order Halocypriformes Suborder uncertain Entomoconchacea	Order Myodocopida Suborder Myodocopina Order Halocyprida Suborder Halocypridina
	Suborder Cladocopina		Suborder Entomozocopina new	Order Cladocopida Suborder Entomozocopina emend.	Order Cladocopina

9

logists. The adage 'A good picture is worth a thousand words' has been validated time and again in publications from all the inhabited continents by excellent scanning electron micrographs. The state of the art of using SEM, however, is still in a preliminary stage, because considerable additional study is required in order to interpret correctly many of the fantastically highly magnified images now available. The increase of population studies (beta-taxonomy) is noteworthy. The papers dealing with Ostracoda during the past decade represent both applied and taxonomic studies. Papers in the International Symposia on Ostracoda have demonstrated the use of Ostracoda in the study of ancient and modern ecology and zoogeography; they, as well as other studies have applied Ostracoda to stratigraphy and paleogeography.

You all know of areas where descriptive work (alpha-taxonomy) is necessary. I have no precise data, but on the basis of my personal experience, I estimate that there are more undescribed ostracode species than described. The Russian 'Osnovy Paleontologii' and the 'American Treatise on Invertebrate Paleontology', both published in the early 1960's, laid the groundwork for ordinal classification (gamma-taxonomy), and several attempts have been made during the past decade to improve the classification of Ostracoda. Malz (1975) sketched the major advances during the past two decades.

Table 3. Categories assigned to the Crustacea and to the Ostracoda

	Crustacea	Ostracoda
Latreille, 1804 (1802)	Class	Order in subclass Entomostraca
Pearse, 1949	Superclass	Subclass in class Eucrustacea
Bassler & Kellett, 1934	Class	Superorder
Ivavona, 1960		Order in subclass Ostracodoidea
Sherov, 1966	Superclass	Class
Manton, 1969	Superclass	Class

Table 3 shows the history of classification of the group known as Ostracoda. The elevation of Ostracoda to class category has taken the pressure off the top, thus allowing more ordinal categories. Attempt at reclassification of ordinal and superordinal categories are shown in Table 4. I expect a refinement of the classification, based on modern methods, which will include the application of numerical taxonomy and the Henningian method within the next decades.

Acknowledgements

I wish to thank all my colleagues, both those who are here and those who are not here, for sending me reprints of their publications, without which this study could not have been made.

References

Bassler, R. S. & Kellett, Betty. 1934. Bibliographic index of Paleozoic Ostracoda: Geol. Soc. America Spec. Paper 1, 500 pp.
Felber, J. E. 1974. The American's tourist Manual for the U.S.S.R., 192 pp. International Intertrade Index, Newark, New Jersey.
Gründel, Joachim. 1967. Zur Grossgliederung der Ordnung Podocopida G. W. Müller, 1894 (Ostracoda). N. Jb. Geol. Paläont. Mh. 6: 321-332.
Gründel, J. 1969. Neue Taxonomische Einheiten der Unterclasse Ostracoda (Crustacea). N. Jb. Geol. Paläont. Mh. 9: 353-361.
Hartmann, Gerd (ed.). 1976. International Symposium on evolution of post-Paleozoic Ostracoda. Naturw. Ver. Hamburg, Abh. u. Ver., (NF), 18/19, Suppl., 336 pp.
Henningsmoen, Gunnar. 1965. On certain features of Palaeocope ostracodes. Geol. Foeren. Stockh., Foerh. 86: 329-394.
Howe, H. V. 1962. Ostracod taxonomy. 366 pp., Louisiana State University Press, Baton Rouge.
Ivanova, V. A. 1960. O proischozhdenii i filogenii Ostracodoidea. Paleont. Zhurnal 3: 21-27.
Kozur, Heinz. 1972. Einige Bemerkungen zur Systematik der Ostracoden und beschreibung neuer Platycopida aus der Trias Umgarns und der Slowakei. Geol. Paläont. Mitt. Innsbruck 2, 10: 1-27.
Kornicker L. S. & Sohn, I. G. 1976. Phylogeny, ontogeny, and morphology of living and fossil Thaumatocypridacea (Myodocopa: Ostracoda): Smithsonian Contrib. Zoology, no. 219, 124 pp.
Latreille, P. A. 1802. Histoire naturelle, générale et particulière des crustacés et des insectes. Ouvrage faisant suite à l'Histoire Naturelle générale et particulière, composée par Leclerc de Buffon, et redigée par C. S. Sonnini, membre de plusieurs Sociétés savantes: Paris, v. 3, p. 17; v. 4, pp. 232-254.
Latreille, P. A. 1804. Tableau méthodique des crustacés. In: Nouveau dictionnaire d'histoire naturelle appliqué aux arts. Par une société naturalistes et d'agriculteurs 1803-04, v. 24: Paris, p. 124 of third section in volume which consists of 258 pp.
Malz, Heinz. 1975. Die Ostracoden-Sintflut– ein Forschungsbericht. The ostracodal deluge – a research report. Paläont. Zeitschr. 49, 4: 461-476.
Manton, S. M. 1969. Introduction to classification of Arthropoda. Pp. R3-R15, in: Moore, R. C. (ed.), Treatise on invertebrate paleontology, Part R, Arthropoda 4 (except Ostracoda) Myriopoda-Hexapoda, v. 1. New York and Lawrence, Kans., Geol. Soc. America and Univ. Kansas Press, 389 pp.
Moore, R. C. (ed.). 1961. Treatise on invertebrate paleontology, Part Q, Arthropoda 3, Crustacea, Ostracoda: New York and Lawrence, Kans., Geol. Soc. America and Univ. Kansas Press, 442 pp.
Neale, J. W. 1965. Some factors influencing the distribution of Recent British Ostracoda. In: Puri, H. S. (ed.), Ostracodes as ecological and palaeoecological indicators. Publicazioni della Stazione Zoologica di Napoli, v. 33, Supplement, 1964, 247-307.
Neale, J. W. (ed.). 1969. The taxonomy, morphology, and ecology of Recent Ostracoda. Oliver and Boyd, Edinburgh, 553 pp.
Oertli, H. J. (ed.). 1971. Colloquium on the paleoecology of ostracodes. Bull. Centre Recherches Pau-SNPA, v. 5, Suppl., 953 pp.
Pearse, A. S. 1949. Zoological names. A list of phyla, classes and orders. Section F: Durham, N. C., Am. Assoc. Adv. Sci., 24 pp.
Puri, H. S. (ed.). 1965. Ostracods as ecological and palaeoecological indicators. Publicazioni della Stazione Zoologica di Napoli, v. 33, Supplement, 1964, 612 pp.
Sherov, A. G. 1966. Basic Arthropodan stock. New York, Pergamon Press, 271 pp.
Sohn, I. G. 1960. Paleozoic species of Bairdia and related genera. U.S. Geol. Survey Prof. Paper 303-A, 105 pp. [1961].
Swain, F. M., Kornicker, L. A. & Lunddin, R. F. (eds.).1975. Biology and paleobiology of Ostracoda. Bull. American Paleontology 65, 282: 687 pp.
Wise, C. D. 1962. Directory of ostracode workers. Micropaleontology 8, 2: 269-274.

Wise, C. D. & Gerry, Ephraim. 1967. Directory of ostracode workers. Micropaleontology 13, 3: 381-384.

Author's address:

Room E-214, National Museum Nat. History, Washington D.C., 20560, USA

II. GENERAL ASPECTS AND MORPHOLOGY

Sixth Intern. Ostracod Symposium, Saalfelden

ON THE USE OF OSTRACODES FOR TRACING CHANGES IN OCEANIC CIRCULATION

RICHARD H. BENSON

Abstract

This report summarizes some of the problems, points of view, and results of a series of studies on the phylogenetic and environmental histories of Neogene ostracodes from the deeper waters of Tethys and the Atlantic. With the 'devolution' of Tethys and the 'evolution' of the Atlantic, changes in basic water-mass structure have separated the deep and shallow faunas whose individual species have developed contrasting architectural properties. Through studies of the gradual formation of new skeletal structures, or the recognition of uniquely adaptive carapace architecture, and with a causally interrelated water-mass or ocean structure model, it is believed that the events in the threshold regions that led to the 'crisis' conditions of Tethys in the Late Miocene can be reconstructed.

Introduction

Ostracode species and faunas change greatly in composition and structure from those which inhabit the shallow seas and continental shelves to those which inhabit the deeper ocean floor. From their fossil record, it is possible to postulate some important changes in water mass distribution and the placement of currents that bring about balance between them over the past 150 million years. Several major geologic events have altered this balance, which is formed between those regions that produce warm dense ocean water and those that produce cold water of equal or greater density. The opening of the Atlantic and the closing of Tethys brought about a major shift in the World Ocean deep current system, when a dominantly latitudinal ocean of low latitude was replaced by a meridional one reaching to polar regions. This not only altered the temperatures of the deep bottom waters, but it also caused major changes in the faunas. These deeper faunas were subsequently isolated from the descendants of the older Tethyan faunas by a sharply grading thermocline that today forms one of the worlds most significant ecologic barriers.

Since the last meeting in Hamburg of the friends of the ostracodes, I have been concerned with trying to demonstrate the usefulness of the ostracode record of the deep-sea in its relationship to two major events in World Ocean history. The first event is the formation of the psychrosphere (or the present lower cold watermass system of the two-layer ocean circulation model), and the second is the final destruction of Tethys as a major ocean system. The Early Tertiary of the South Atlantic is the time and site of the controlling mechanism of major changes in the Atlantic. The Western Mediterranean and southern Spain is the major area of interest for the second event that may have continued through a series of greater and smaller crises over much of the last 40 million years.

This is a brief report of some of the problems, some of the important questions, and some of the results of these enquiries as they appear now. In my oral presentation I have endeavored to emphasize both the questions and the differing hypothetical positions being taken in order to attract interest in the historical problems, but even more importantly to demonstrate that there is a great interest in ostracodes or at least the history of ostracodes, outside of our working groups.

The potential of ostracodes as recorders of important events of the past has been an abiding faith of most of us for some time; however, it is both reassuring and helpful to know about those aspects of our work that are attractive to others, especially with major and exciting historical problems.

In this written report, I can only touch on some of the main points of the papers mentioned and illustrated in the oral presentation.

A changing view of the Ostracodes

If at Hull in 1968 (in Neale 1969, p. 238) I said, in the general discussions, that I did not believe in the importance of the SEM for basic morphology studies of ostracodes, both myself and my critics will look back on this statement with some chagrin. The SEM has changed the view of ostracodes, but not just because it makes better illustrations of carapaces than did light photomicrography. Perhaps in the case of the disproof of the presence of the bilamellar shell structure, the SEM did show the shell structure better than light microscope before it. But more importantly, the SEM has produced views of the carapace that have attracted the eye to the general architecture of the carapace. We can now see the subtle replacement of structures through time, the effects of stress near the adductor muscle scars, the underlying patterns conservative of reticulation in the histological framework of the carapace, as well as consistent and homologically traceable patterns of pore-canal patterns (Benson, 1975). Through these changing views of structure, it is now possible to connect and relate features of ostracode morphology that appeared to be different in the past for no reason. Or perhaps it would be better to say that it is now possible to see how different ostracodes are similar, whereas in the past we could only see the differences.

This advance in view is basic to the increasing confidence we can have in the ostracodes as recorders of history. As they have responded to change, we can now begin to recognise this change in terms of adaptation or architectural response, and not just as alterations in the species memberships of faunas or the appearance of new taxa. Hopefully this recognition will allow us to not only improve our arrangements of the more ornate taxa, but it will allow us to place them in some kind of successive order.

Perhaps this point of view is still optimistic, yet it is nonetheless necessary. As in mathematics, the hypothesis of development is as important as the description of points of progress. Without a calculus of historical progression, we can never hope to pass the stage or the definition of single events. All too often these 'single events' remain in our literature as disconnected taxa without such a history.

Ostracode biodynamics

The study of patterns of change in ostracodes as the products of structural response to stress leads directly from the linking of developmental stages, or at least the recognition of the linkage, to larger systematic relationships. If an ostracode has lived in warm shallow water where its eyes and eye tubercules are well developed, its shell is massive and obviously heavily calcified relative to the mass of the rest of him, and one can trace its descendants, which are blind and have fragile shells with more complex shell ornament; it is reasonable to assume that either the ostracode has moved across a thermal gradient or the gradient has moved across the ostracode. In either case an event (an historical happening) is identifiable. With knowledge of the relative reaction rates of both ostracode change and water-mass change, it is normally easier to postulate that the ostracode would not have moved into a zone of grater environmental pressure unless compelled to do so; that is to say, it is easier for it to move a great distance laterally to a similar environment than to invade a new one by crossing a thermal barrier even though this barrier may be only a few hundreds of meters braod.

I would expect you to say that this ecologic or adaptive relationship is too commonly accepted to bear repeating here. I have no doubt that it is; however, the need for biodynamics begins with the establishment of the fact of change, which can only happen when structural linkage is established. This linkage is sterile in direction until several of related changes can be shown to be occurring simultaneously within the same morphology, with one pattern of change remaining much less active than the others. This is similar to the inertial divisions in a watch mechanism. The cogs move at different rates but are linked structurally to each other. In many ways a fossil series of descendant ostracode valves is like a motion picture film of a watch mechanism. The problem is the question identification of those reactant subsets of organ patterns that are related to the skeletal function of the carapace so that these reflect the rate of reaction of the design to changing stress. The rearrangement in skeletal structure is always to provide sufficient strength, but to new demands.

In the example at the beginning of this argument, one could find that a change in size and decrease in relative wall thickness would be accompanied by movement of reticulate muri to form new ridges, a strengthening of free margins, a decrease in the massiveness of individual structural members as they moved away from the center of form, and many other predictable changes in the position or size of structural members. I have tried to demonstrate this sort of development with the ostracode *Oblitacythereis*, a new genus in the Neogene of the Mediterranean, which lived on the frontier between oceanic and warmer sea conditions (Benson, n.d.a.). Its importance as an example of the biodynamical point of view, that is the linking of a progression of interacting structural changes, can be compared with its confusion by others with similar but unrelated genera.

A historical problem and a historical synthesis

It is sometimes stated that the easiest way to resolve a problem is to conceive of a larger one in which the first is a functional part. I am not sure whether this is an intelligent strategy, or an obvious necessity. However, it would seem that the understanding of the 'devolution' of Tethys (its destruction, crises and ultimate extinction) leads quickly to the need to imagine the consequent patterns of change in World Ocean circulation, and a new water-mass formation, in order to regress to the problem of a single event of late Tethys history. The story of the search (Benson, n.d.b.) for oceanic ostracodes in Tethys begins with a problem that remains in part unresolved. And yet it leads to a synthesis that seems at present to clarify the general role of Tethys in its relationship to the Atlantic. The climax of this relationship may have occurred when Tethys was completely severed from the Atlantic during the late Miocene (Benson, n.d.a., 1976a).

There has been so much written about the Messinian Salinity Crisis in the last several years, so that awareness of the controversy seems to be less a problem than fatigue from discussion of it. I will not begin here with a description of the controversy. I will attempt, however, to reduce some of its complexity to attempt to show how the crisis relates to the events that preceded it so far as the Atlantic is concerned.

Ruggieri (1967) suggested that communication between Tethys or the Mediterranean (the names depend on ones belief in the degree of separateness) and the Atlantic ceased at the end of the Miocene and the basins either dried up or were converted to large inland sea-lakes. In either case the marine fauna of Tethys was probably destroyed. Soon afterwards, as thousands if not millions of cubic kilometers of evaporite deposits were being formed, the sea suddenly, perhaps catastrophically, invaded, even with faunal elements of the psychrosphere. This point of view has been called by Maria Cita (1973), W. F. B. Ryan (1973) and Ken Hsü (1973, in Ryan & Hsü) the 'dessication hypothesis'. The magnitude of importance of this event, if indeed it happened, cannot be overestimated for the study of biologic processes as well as for Neogene stratigraphy and for understanding geologic catastrophism. The simple exercise of trying to demonstrate paleontologically that it did or did not happen is of great value in refinement of procedures and principles. Beyond the study of the phylogeny of *Costa* (Benson, 1977), I have attempted in at least two other places (Benson, n.d.a., 1976d) to attack this problem. One of these study areas has been related to trying to demonstrate the absence of continuous marine sequences through the Messinian.

The other point of view from the dessication theory is that Tethys experienced a crisis to be sure, but that the connection with the Atlantic was not broken. The Tethyan marine faunas, although under severe stress, continued to exist in the Mediterranean region through the Late Miocene (the earliest Messinian) into the Pliocene. Some of the advocates of this position argue that the basins were deepened after the crisis, perhaps during the Middle Pliocene; and that the Mediterranean (or Tethys) before this time was relatively shallow, perhaps in part land. In Tunisia at Raf Raf, in Spain, in the Vera Basin section, at Eraclea Minoa in the upper Agrenozzola Formation, all of these reputedly marine sections have yielded *Cyprideis* faunas containing ostracodes that could not conceivably have lived in marine waters.

It is difficult to decide whether it requires more imagination to conceive of the present Mediterranean as an almost monotonous shallow seascape, or as a series of Dead Sea-like depressions of gigantic size. Fortunately this is not the question of greatest importance in my analyses, even though it may be of greatest historical interest. In my view, from above and below the evaporites in terms of the marine fossil evidence, the major point is proof or disproof of something happening at all. I keep reminding those who argue that the resoluton of connected paleontological evidence needs much improvement, not just the collection of faunas for there are many of these available, but the analogues by which ecologic succession can be shown, or a means of demonstrating either continuity or discontinuity in genetic linkage within the Mediterranean. The argument that the same species exists before or after the crisis requires considerable amplification. The presence of a planktonic fossil even within the record of the crisis itself must be shown not to be allochthonous.

I have written abundantly about the ostracodes and their part of history of this event, as have several other colleagues. My position has changed as new evidence was found. I believe that the solution to the problem lies in the realization of a mechanism common to several lines of evidence and at the same time is part of a large historical theory. It is my opinion that this mechanism is within the history of water-mass structure and especially the stratification that must have occurred when World Ocean influences were as strong or stronger than local influences.

The Mediterranean today, and Tethys before it, had unique structural properties related to its latitudinal position and orientation. When it had a high capacity for production of warm dense water-masses to flow into the Atlantic, it dominated its thermal character and consequently depressed the psychrospheric-thermospheric barrier. When restriction occurred at the Iberian Portal, that energy which had formerly been absorbed into simple Tethyan water-mass formation was concentrated into evaporite formation. This provided an energy deficit in the World Ocean, and the results can be seen in many forms. The relationship of Tethys with the Atlantic is like Nelson's crossing of the 'T' at Trafalgar (a naval maneuver of some fame). The key to understanding the crisis begins and ends at this juncture. The answer to the magnitude of the salinity crisis question does not lie very far away (geographically and as a model) from Trafalgar in southern Spain.

Novelty and history

The explanation of elephant bones in a fisherman's net in the North Sea requires some adjustment in either a fisherman's concept of history (unless he has had a geological education) or of elephants. *Bradleya dictyon* in Middle Miocene deposits in southern Spain (Benson, 1976a, b) and in DSDP Site 372 east of Manorca (Benson, n.d.c., f) requires either a belief that the Western Mediterranean was joined to the Atlantic at oceanic depths 12 million years ago, or that *Bradleya dictyon* was able to live in shallow water then when it can only live on the ocean floor today. Which system of explanation requires the least distortion of historical theory? Is it easier to imagine that elephants could swim in the ocean, or that at one time the North Sea was dry?

The power of novelty as evidence in paleontology is not as an addition to the weight of other evidence, but a refutation to a theory which seems easier to reject than does the fact of the evidence. I say 'fact', because this is the rank which such evidence must seem to warrant, even though by its very presence as a novelty it begs the question of its validity. The strongest evidence the ostracode student has in contributing to the solution of a historical problem is novelty. If he can be as easily convinced that ostracodes can fly and elephants can swim in the ocean as he may be that continents are permanent features on the face of the earth, he will have a difficult time reconstructing geologic history.

Conclusion?

This report (perhaps better described as an essay) was written as the completion of eight (or nine, if you count introductory defenses) reports on (1) the discoveries of deep-sea ostracodes in the Miocenes of Spain and in the floor of the Western Mediterranean (DSDP, Leg 42A; Benson, n.d.c.,1976a, Berggren, Benson et al., 1976); (2) a monograph on *Oblitacythereis*, its evolution, extinction and historical significance in the devolution of Tethys (Benson, n.d.b); (3) a study of the Cenozoic ostracodes of the Sao Paulo Plateau and the Rio Grande Rise (DSDP, Leg 39, Benson, n.d.a); (4) the bringing together of the works of nine paleontologists (including my own) on the 'Biodynamical Effects of the Messinian Salinity Crisis' (Benson, 1976b); (5) a new theoretical synthesis on the origin and development of the psychrosphere (Benson, n.d.b), and the oral presentation of some of the results of these studies here at Saalfelden. If it were possible to draw conclusions to all of these and other prior studies, I would say that the more one works with ostracodes the more one appreciates the importance of procedures and ideas, and that 'facts' about ostracodes are not in short supply. I am not convinced about the usefulness of some of these facts, however. As we seem to progress in our studies, the ostracodes always seem to remain shrouded in about the same degree of mystery as before, only the nature of the mystery changes. For my colleagues who search for facts about ostracodes, this experience must be frustrating.

I recall the remarks of Professor Sylvester-Bradley at the meeting in Pau in 1970 (remarks on p. 614, Oertli, 1971) (I now write this report in Pau, 1976) in which he sounded a general note of caution about the danger of conjecture in the form of assertions from correlations of morphologic trends relative to environmental gradients (in this case developing a hypothesis in such a way that it could be disproved). I would take this opportunity to give a rejoinder to this argument to suggest that until a causal mechanism can be demonstrated, the co-occurrence of data, no matter how consistent, is limited as a source of historical explanation. Perhaps this is justification for the invention of several of the morphologic and biogeographic ideas in the present summary. Yet they are testable or at least they are intended to be testable. I suggest that those who seek this test be as equally concerned with the plausability of the mechanisms as with the consistancy of the data. No generalization has ever been discovered from distributional data alone.

References

Benson, R. H. 1975. Morphologic stability in Ostracoda. Bull. Amer. Paleontology. 65, 282:13–46.

Benson, R. H. n.d.a. The evolution and extinctions of Oblitacythereis; a new costine ostracode (Trachyleberididae) from the Neogene of the Mediterranean and the South Atlantic. Smithsonian Contributions to Paleobiology.

Benson, R. H., n.d.b. In search of lost oceans: a paradox on discovery, Symposium on Paleobiogeography (Boucot & Gray, édit; 1976), Contributions to Biology, Oregon State University (Corvalis, Oregon).

Benson, R. H. n.d.c. The paleoecology of the ostracodes of DSDP Leg 42A, Initial Reports of the Deep-Sea Drilling Project, v. 42, 4 figures, 2 plates.

Benson, R. H. n.d.d. The Cenozoic ostracode faunas of the Sao Paulo Plateau and the Rio Grande Rise (DSDP Leg 39, Site 356 and 357). Initial Report of the Deep Sea Drilling Project, v. 39.

Benson, R. H. 1976a. Miocene Deep-Sea Ostracodes of the Iberian Portal and the Balearic Basin. Marine Micropaleontology 1: 249–262.

Benson, R. H. 1976b. Testing the Messinian Salinity Crisis Biodynamically: An Introduction. Palaeogeography, Palaeoclimatology, Palaeoecology 20: 1–9.

Benson, R. H. 1976c. Changes in the ostracodes of the Mediterranean with the Messinian Salinity Crisis. Palaeogepgraphy, Palaeoclimatology, Palaeoecology 20: 147–170.

Benson, R. H. 1977. The evolution of the ostracode Costa analyzed by 'Theta-Rho Difference'. In: G. Hartmann (ed.), Proceedings of the International Symposium on Evolution of Post-Paleozoic Ostradoca, Hamburg, 1974. Mitt. Hamburg. Zool. Mus. Inst.: 127–139.

Berggren, W. A., Benson, R. H. et al. 1976. Ostracodes of the El Cuervo Section (Middle Miocene, S. W. Andalusia, Spain), Marine Micropaleontology 1: 195–247.

Cita, M. B. 1973. Mediterranean evaporite: paleontological arguments for a deep-basin desiccation model. In: C. W. Drooger (ed.), Messinian Events in the Mediterranean, Utrecht, 1973, North-Holland, Amsterdam, pp. 206–228.

Neale, J. W. 1969. The taxonomy, morphology and ecology of Recent Ostracoda. Oliver & Boyd (Edinburgh), 553 pp.

Oertli, H. J. (ed.). 1971. Colloque sur la Paléoécologie des Ostracodes. Bulletin du Centre de Recherches Pau-SNPA 5:953.

Ruggieri, G. 1967. The Miocene and later evolution of the Mediterranean Sea. In: Aspects of Tethyan Biogeography. Systematics Assoc. (London), 7, pp. 283-290.

Ryan, W. B. F. 1973. Geodynamic implications of the Messinian Salinity Crisis. In: Drooger, C. W. (ed.), Messinian Events in the Mediterranean, Kon. Nederland. Akad. v. Wetenschappen, North-Holland Publ. Co (Amsterdam), pp. 26-43.

Ryan, W. B. F., Hsü, K. J., et al. 1973. Initial Reports of the Deep Sea Drilling Project, v. 13, pt. 2. Washington (U.S. Government Printing Office), pp. 517-1447.

Author's address:
Smithsonian Institution, Washington, D.C., USA

Sixth Intern. Ostracod Symposium, Saalfelden

TERTIARY OSTRACOD PROVINCES IN WESTERN EUROPE

M. C. KEEN

Abstract

Three provinces can be recognised in western Europe for Eocene ostracods: South Aquitaine, North Aquitaine, and the Anglo-Gallic Province. These were related to climate, but their individuality was emphasised by the local geography. At the present day this whole area lies within a single Celtic Province, but this may be a temporary affair following the last glaciation. Recent climatic changes have also aided the spread of ostracods to both sides of the Atlantic; no amphiatlantic species were present in the Eocene, they first appeared in the Miocene.

The study of Tertiary ostracods indicates the presence of several biogeographical regions such as the Gulf Coast of N. America, N.W. Europe, and north and west Africa. These are probably equivalent to the faunal provinces of the present day, but their exact status is not clear because there are too many gaps in our knowledge, both stratigraphical and geographical. However, anyone who studies more than one of these areas is immediately aware of the individuality of the faunas. More bewildering is the presence of very small provinces within an area such as western Europe. There are differences between ostracod faunas of the same age between England, Belgium, and the Paris Basin, but the differences within this region seem small when they are compared with the faunas of Aquitaine and of Germany. Even more surprising, the Eocene ostracods of northern and southern Aquitaine appear to be very different. These are intuitive impressions, so it was decided to attempt a quantitative comparison to find out how significant the differences really are. Many problems present themselves in this type of study. How far can faunal lists be trusted? Many species of genera such as *Bairdia, Cytherella, Cytherura, Cytheridea, Krithe, Leguminocythereis, Paijenborchella,* and *Schizocythere* have become umbrella species, recorded from many localities but often proving to be misidentifications; the reverse probably happens as well, so hiding similarities between faunas. Is the 'Middle' Eocene of one worker the same as that of another, especially with stratigraphical boundary faunas? How accurate are correlations? Because of these problems this paper is mostly restricted to areas whose faunas are known to the author. The method adopted was to compare the faunas of England, Belgium, Paris Basin, N. Aquitaine and S. Aquitaine and move rarely N. Italy (Ascoli 1969) and the area around Nantes (Campbon and Saffre Basins, Blondeau 1971) by calculating Jaccard's Coefficient of Correlation. The works of Apostolescu, Deltel, Ducasse, Haskins, and Keij were extensively drawn upon, together with personal studies. The figures used are given in Table 1. Because the number of species present and in common between areas is not accurately known and will undoubtedly change with further work, too much rel-

Table 1. Value of Jaccard's coefficient of correlation and no. of species considered

	S.A.	I	NA	C	PB	B	E
S. Aquitaine	54	–	.34	–	.08	.13	.03
	91	.26	.17	–	.04	–	.03
	126	–	.15	.12	.07	.08	.07
	65	–	.16	–	.05	.05	.05
Italy		–	–	–	–	–	–
		48	.06	–	.07	–	.04
		–	–	–	–	–	–
		–	–	–	–	–	–
N. Aquitaine			68	–	.08	.08	.03
			156	–	.05	–	.07
			152	.43	.28	.11	.11
			113	–	.23	.14	.08
Campbon				–	–	–	–
				–	–	–	–
				127	.29	.26	.28
				–	–	–	–
Paris Basin				Oligocene	50	.20	.26
				Upper Eocene	35	–	.26
				Middle Eocene	100	.45	.36
				Lower Eocene	44	.30	.28
Belgium						40	.26
				No. of Species		–	–
				Present		93	.40
						50	.44
England							13
							70
							40
							50

iance should not be given to the exact values. However, the presence of some 150 species with only 20 in common with a neighbouring area can still give significant results whether the exact number of species is 143 or 152. The results are presented in Fig. 1. The value of Jaccard's Coefficient indicates general relationships and differences and allows comparison of faunal distributions of different ages.

Palaeogene and Holocene Ostracod provinces in W. Europe

During the Eocene there appear to have been three distinct provinces: the Anglo-Gallic, North Aquitaine, and South Aquitaine. In Oligocene times the two Aquitainian provinces merged into a single entity. At the present day a single province stretches from Norway to northern Spain. These are all based on similar types of data, i.e. taking the whole recorded fauna regardless of environmental

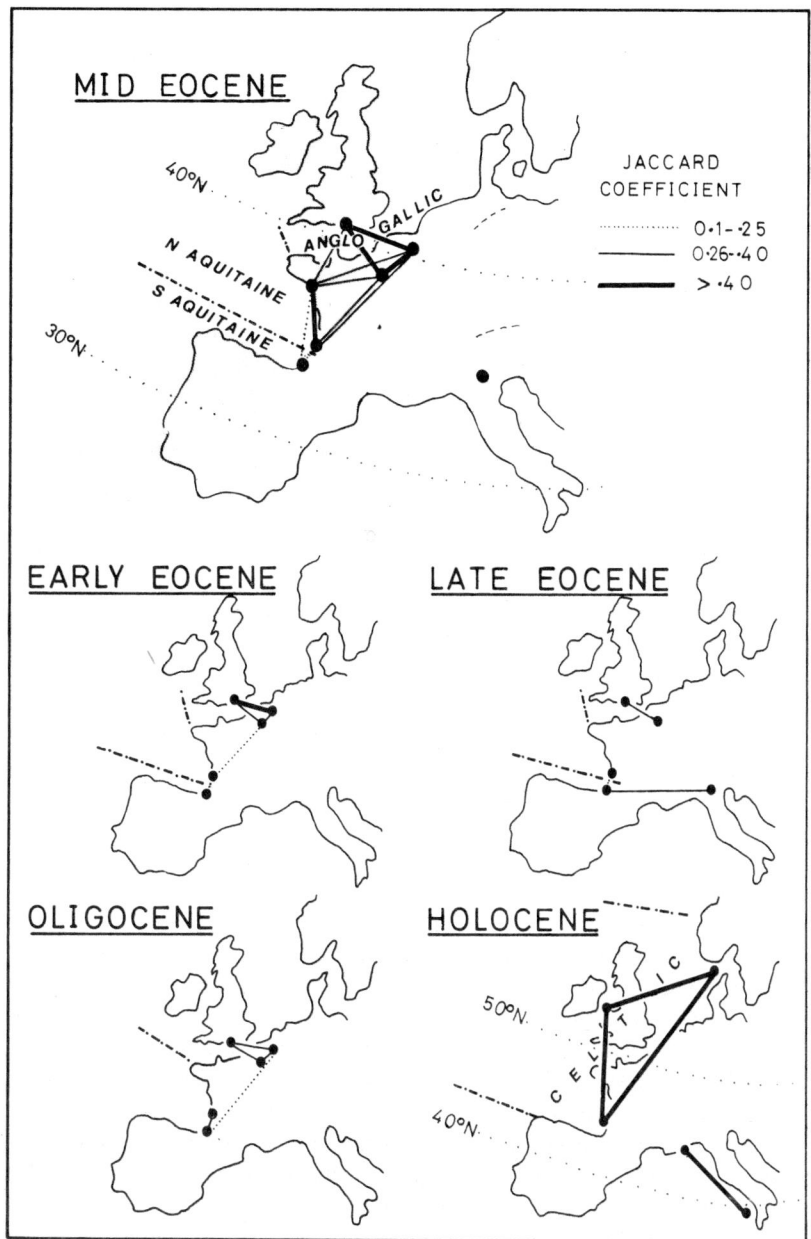

Fig. 1. Values of Jaccard's Coefficient between the various areas of deposition.

conditions within the marine realm. Thus faunas from sands, muds, limestones, deep (up to 200 m) and shallow water, areas of bare bottoms and sea grass meadows have all been included. The differences therefore indicate provincial differences rather than those of facies. If the study of the Palaeogene faunas had utilised genera rather than species, few differences would have been observed. Aquitaine had many species of *'Bradleya', Cytheretta,* and *Leguminocythereis* but they were different from many species of these genera living in the Anglo-Gallic Province.

The presence of three provinces in the Eocene and only one at the present day is puzzling. The Holocene data is taken from the works of Brady, Elofson, Neale, Muller, Norman, Robertson, Rome, Sars, Wagner, and Yassini. The values of J are higher for the recent than for the Palaeogene; this is probably because fossil deposits in different regions are rarely truly contemporaneous to within a few thousand years. The Lower Eocene data used for the Anglo-Gallic area almost certainly include some older faunas in England and Belgium than are present in the Paris Basin.

At the present day many provinces are arranged latitudinally along the N–S coast of the continents, and are related to climate and sea temperature, both of which are the result of interacting variables. Provinces can be recognised because species have a limited latitudinal range, with each province having a percentage of indigenous species. Today many European ostracod species range over 25–30° of latitude, with smaller ranges in the tropics (10–15°?). During the Eocene the range was 10° or less in mid latitudes with greater ranges seen in the tropics, e.g. 20° for species of the north Africa–west Africa province. This distribution is odd. The Eocene world was warmer than that of today with much broader temperate and subtropical belts. According to the model presented by Valentine (1967) this should mean broader provinces, not narrower.

Climatic influences can be seen in the Eocene distribution of larger Foraminifera such as *Alveolina, Nummulites* and *Orbitolites.* These warm water genera were abundant in the ancient Tethyan province of the Mediterranean, but only occur sporadically in the Anglo-Gallic Province which must have lain at the northern limits of their temperature tolerance. Southern Aquitaine has a much stronger 'Mediterranean' affinity in this respect than does northern Aquitaine. This is also seen in other invertebrate groups such as the echinoderms. The palaeogeography for Mid Eocene times is shown in Fig. 1, indicating a connection with Tethys through southern Aquitaine, which is controversial, and also another through northern Spain which is more generally accepted. Thus it is possible that a warm water current flowing through this region from Tethys met a colder southern flowing current of the north Atlantic system, so producing a temperature barrier for some benthonic organisms between north and south Aquitaine. Anyhow, whatever the cause there must have been some barrier. This would have been emphasised by the geography of the time because the N. Aquitaine Basin was a distinct embayment separated from Southern Aquitaine.

Proceeding northwards, the boundary between the North Aquitaine and Anglo-Gallic Provinces is more marked than would be expected by simple and gradual cooling of the water. The geography of the time is once again relevant, and in a reconstruction given by Curry (1966) the boundary for the Mid Eocene is shown as a narrow strait dotted with islands. The strait may have been completely closed at times, allowing the indigenous fauna of the Anglo-Gallic Province to firmly establish itself and resist competition from southern invaders. There is also evidence of more variable conditions in the Anglo-Gallic Province, with hyposaline waters at times in Hampshire and hypersaline in the Paris Basin. Once again, this would favour the local and better adapted species rather than immigrants. Thus the two provincial boundaries can be explained as latitudinal temperature-controlled changes emphasised by geographical factors. Northern Aquitaine can be regarded as transitional between the warm South Aquitaine-Tethyan province and the colder Anglo-Gallic.

In the Oligocene there was no geographical barrier between northern and southern Aquitaine, so the temparature barrier would have occurred further northwards. This would have eliminated the intermediate N. Aquitaine Province.

The distribution of a few important Middle Eocene species are shown in Table 2. Considering the faunas as a whole, a few generalisations are possible. *Bairdia* is much more abundant in Aquitaine, and *Triebelina* is restricted to Aquitaine; *Pokornyella*, represented by 13 species in North Aquitaine, is much commoner in Aquitaine. *Cytheridea, Cyamocytheridea,* and *Clithrocytheridea* are rare in south Aquitaine; *Pterygocythereis* is much more characteristic of the Anglo-Gallic Province, while *Eocytheropteron* is absent from Aquitaine.

Table 2. Distribution of important species within the Middle Eocene

Species	S. Aquitaine	N. Aquitaine	Campbon	Anglo-Gallic
Aquitaniella transversa Deltel	×			
Bradleya ordinata Deltel	×			
Cytherella consueta Deltel	×			
Echinocythereis multicostata Deltel	×			
Idiocythere aquitanica Deltel	×			
Kangarina tridens Deltel	×			
Krithe luyensis Deltel	×			
Monoceratina striata Deltel	×			
Pontocyprella aturica Deltel	×			
Trachyleberis parmula Deltel	×			
Bairdia crebra Deltel	×	×	×	
Bairdia cymbula Deltel	×	×	×	
Bairdia succincta Deltel	×	×	×	
Bairdia tenuis Deltel	×	×	×	
Triebelina punctata Deltel	×	×	×	
Uroleberis striatopunctata Ducasse	×	×	×	
Clithrocytheridea faboides (Bosquet)	×	×	×	×
Cytherelloidea dameriacensis Apostolescu	×	×	×	×
Cytheretta eocaenica Keij	×	×	×	×
Hermanites paijenborchiana Keij	×	×	×	×

Tabel 2 (cont.)

Species	S. Aquitaine	N. Aquitaine	Campbon	Anglo-Gallic
Krithe rutoti Keij	×	×	×	×
Leguminocythereis angulatopora (Reuss)	×	×	×	×
Leguminocythereis pertusa (Roemer)	×	×	×	×
Leguminocythereis striatopunctata (Roemer)	×	×	×	×
Oertliella aculeata (Bosquet)	×	×	×	×
Pokornyella ventricosa (Bosquet)	×	×	×	×
Quadracythere lamarckiana (Bosquet)	×	×	×	×
Schizocythere appendiculata Triebel	×	×	×	×
Schizocythere tessellata (Bosquet)	×	×	×	×
Schuleridea perforata (Roemer)	×	×	×	×
Trachyleberis horrescens (Bosquet)	×	×	×	×
Pokornyella talencis Ducassse		×		
Bradleya oertlii Ducasse		×	×	
Cytheretta sculpta Ducasse		×	×	
Cytheretta vulgaris Ducasse		×	×	
Echinocythereis septentrionalis Ducasse		×	×	
Leguminocythereis inflata Ducasse		×	×	
Leguminocythereis magna Ducasse		×	×	
Novocypris eocaenica Ducasse		×	×	
Pokornyella longicostis Ducasse		×	×	
Pterygocythereis aquitanica Ducasse		×	×	
Cytheridea atlantica Blondeau			×	
Bairdoppilata gliberti Keij		×	×	×
Cyamocytheridea hebertiana (Bosquet)		×	×	×
Cytherella pustulosa Keij		×	×	×
Monsmirabilia oblonga Apostolescu		×	×	×
Quadracythere orbignyana (Bosquet)		×	×	×
Cytheretta costellata (Roemer)			×	×
Pterygocythereis cornuta (Roemer)			×	×
Bradleya kaasschieteri Keij				×
Cytheretta crassivenia Apostolescu				×
Cytheretta forticosta Keen				×
Cytheridea rigida rigida Haskins				×
Echinothereis scabropapulosa (Jones)				×
Eocytheropteron sherborni Bowen				×
Monsmirabilia subovata Apostolescu				×
Paracypris contracta (Jones)				×
Pterygocythereis tuberosa Keij				×
Sphenocytheridea gracilis Keij				×

If we consider the situation at the present day, the presence of a single Celtic Province is not as straightforward as it seems. Neale & Howe (1975) have divided it into two subprovinces, a northern Brittanic and a southern Gascoynian. Peypouquet (1973) records several 'Mediterranean' ostracods from the area of Cap Breton in southern Aquitaine, which have arrived there fairly recently in geological time. This marks the northern limit of their range and is related to 'mediterranean water masses' in the area. This is presumably a readjustment taking place since the last glaciation when northern faunas migrated southwards. So perhaps the single Celtic Province is an anomaly caused by the last glaciation and given time the two subprovinces would become more clearly differentiated. The boundary between them may have lain further north than Britanny in the relatively recent geological past because the Mio-Pliocene ostracod faunas of St. Erth in Cornwall, and of Britanny contain many 'Mediterranean' species (Whatley, personnal communication).

Tertiary and Holocene North Atlantic faunas

The discussion so far has been concerned with latitudinal ranges. Longitudinal ranges are generally much greater. For example, many western European Tertiary species are present in the eastern Balkans, giving a range of some 1500 miles. In North America many species have a comparable range, from Texas to Maryland, so that the whole area of the Gulf Coast and Atlantic coastal plain may have belonged to a single province. This province only stretched through some 10° of latitude however.

At the present day many ostracods are present on both sides of the Atlantic. The distribution of these amphiatlantic species have been brilliantly shown in the charts produced by Hazel (1970) and in various works of John Neale. Neale & Howe (1975) have defined an Arctic Province with a circumpolar fauna present at lower latitudes in the west than in the east. Southwards it is replaced by the Celtic Province of Europe and the Nova Scotian Province of Canada; the latter has about one third amphiatlantic species (Hazel, 1975). Because of different hydrographic conditions the temporate provinces are wider in Europe than in North America. Succeeding provinces have very few amphiatlantic species, only 4% in the Virginian Province of North America. Thus, going southwards fewer species are present on both sides of the Atlantic. These distribution patterns are probably the result of the climatic fluctuations of the past few million years.

No amphiatlantic species are known for the Palaeogene, although genera and species groups present on both sides suggest some sort of communication. They first appear in the Miocene with several species occurring throughout the Carribean and southern Europe. This is not what would be expected considering the widening of the Atlantic since the late Cretaceous. However, climatic factors are just as important as geographic proximity, and when these are considered the results are not so surprising. The Eocene ostracods of North America occur within the Eocene latitudes of 15–25°N, while those of Europe occurred between 35–45°N. So to find equivalent faunas Europe should be compared with the region around Labrador and Newfoundland, areas lacking in described Tertiary ostracod faunas. Even then, by comparison with present day amphiatlantic distributions, these waters may have been too warm and too far south, so perhaps it was only in the Arctic regions that ostracods ranged from the old world to the new. This may indeed have been the migration route for the genera found on both sides of the Atlantic during the Palaeogene.

Conclusions

Great care must be exercised when using Holocene distribution patterns to interpret those of Tertiary ostracods. In a period of changing conditions, as today, the intermingling of a retreating cold water fauna and an advancing warm water faunas may lead to a temporary state of very broad latitudinal provinces until the status quo is restored. Comparison of faunas of different continents in order to judge their proximity to one another in the light of plate tectonics should be used with caution unless the full climatic conditions of the time are understood.

Résumé

On a décrit trois provinces d'ostracodes dans l'éocène de l'Europe d'ouest: Aquitaine méridionale, Aquitaine Septentrionale, et la Province Anglo-Gallique. La variation climatologique est proposée comme agent responsable de la répartition d'ostracodes, modifée par les facteurs géographiques. A l'époque actuelle on définit une seule province, la Province Celtique, de Norvège à l'Aquitaine, peut-être à cause de la dernière glaciation. Au cours du Paléogène on ne trouve aucun des espèces d'ostracodes en tous les deux côtes de l'Atlantique, mais à l'époque actuelle plusieurs éspèces sont communes aux côtés occidentales et orientales. Cette repartition est attribué à les changements climatiques du Pleistocène.

References

Ascoli, F. 1969. First data on the ostracod biostratigraphy of the Possagno and Brendola sections (Paleogene, N.E. Italy). In: Colloque sur l'éocene, Paris 1968, 3, Mem. Bur. Rech. geol. min. 69: 51-71.

Blondeau, M-A. 1971. Contribution à l'étude des Ostracodes éocènes des Bassins de Campbon et de Saffre (Loire-Atlantique). Thèse, Université de Nantes, 162 pp., 17 pls.

Curry, D. 1966. Problems of correlation in the Anglo-Paris-Belgian Basin. Proc. Geol. Ass. 76: 437-467.

Hazel, J. E. 1970. Atlantic continental shelf and slope of the United States – Ostracode zoogeography in the southern Nova Scotian and northern Virginian faunal provinces. Prop. pap. U.S. Geol. Surv. 529E, 21 pp.

Hazel, J. E. 1975. Ostracode biofacies in the Cape Hatteras, North Carolina, Area. Bull. Am. Paleont. 65: 463-487.

Neale, J. W. & Howe, H. V. 1975. The marine Ostracoda of Nova Zemlya and other high latitude faunas. Bull. Am. Paleont. 65: 381-417.

Peypouquet, J-P. 1973. Sur la presence d'éspèces mediterranées au niveau des étages circalittoral et epibathyal de la zone de Cap-Breton. Bull. Inst. Geol Bassin Aquitaine 13: 143-146.

Valentine, J. W. 1967. The influence of climatic fluctuations on species diversity within the Tethyan provincial system. In: Aspects of Tethyan Biogeography, Pub. Syst. Assoc. 7, pp. 153-166.

Author's address:

Dept. of Geology, University of Glasgow, Glasgow G12 8QQ, U.K.

Discussion

Mme. Krstić: Do the French ostracodologists have any comments to make on the straits through the South of France shown in the palaeogeographical reconstruction? I ask this because we Yugoslavians had no time to discuss the palaeogeography of Yugoslavia shown by Dr. Benson.

Dr. Keen: As no French ostracodologists have replied, perhaps I might add a few comments. The straits indicated are probably widely accepted for the Palaeogeocene and Early Eocene, but not for the Middle Eocene and later, when the most likely migration route was through northern Spain. The uplift of the Pyrenees in the Mid to Late Eocene presumably contributed to this change.

Dr. Neale: I was interested in Dr. Benson's firm statement that there was no gyroidal circulation ('Gulf Stream') in the North Atlantic in the Eocene. Since there appears to be sharp differences of opinion on this, I would be grateful if he would outline the evidence on which he based this statement.

Dr. Keen: I feel Dr. Benson should reply to this!

Dr. Whatley: You should, I think, be very wary of comparing Eocene Provinces with Recent ones. Most of the latter, based as they are on the distribution of dead, rather than living individuals, are misleading. This is particularly so in the western Atlantic where faunas, the product of glacial maxima, are used in such areas as the Bay of Biscay where they are inextricably mixed with faunas actually living there today.

Probably in the late Tertiary, land barriers were quite as important as ocean currents in controlling ostracod distributions. For example, in the late Pliocene, when Britain was a peninsula and the English Channel did not exist, the faunas of St. Erth in Cornwall and Redon in Britanny were Mediterranean/Lausitanian in type: highly diverse warm water faunas. The Coralline Crag of East Anglia, of approximately the same age, was, however, a low diversity cold water fauna even though present at almost the same latitude. Could you not, therefore, place more emphasis on land barriers in some of your reconstructions?

Dr. Keen: I agree with you entirely, and have used land barriers together with climatic changes and oceanic circulation in the reconstruction. As regards your first point, my main conclusion, as stated in the text, is the very one you make: namely, that caution should be used when taking Recent faunal provinces as models for Tertiary or earlier ones.

Dr. Colin: A similar situation to the one described for the Palaeogene can be seen in the French Cretaceous. Particularly in the Cenomanian, most of the species present in the Aquitanian Basin are absent in the central and northern parts of the Paris Basin, but are present in the southern Paris Basin (Poitou, and Touraine) as well as in Provence.

(No reply necessary)

Sixth Intern. Ostracod Symposium, Saalfelden

APPROACHES TO THE DIVERSITY OF ASSEMBLAGES OF OSTRACODA

ROGER L. KAESLER & PATRICK S. MULVANY

Abstract

Brillouin's equation from information theory is well suited for studying diversity and community structure of recent and fossil ostracodes. His equation gives the actual diversity per individual in a collection and not an estimate of the parametric diversity of a statistical universe. Because of the difficulties of defining the extent of the statistical universe from which samples of ostracodes are drawn, study of collections for their own sake is to be preferred. Diversity may also be partitioned hierarchically among taxonomic categories. Thus, species diversity may be viewed as the sum of diversities of superfamilies, families within superfamilies, genera within families, and species within genera. Hierarchical diversities of ostracodes from fifty-six samples from Todos Santos Bay, Baja California, showed that diversities at all taxonomic categories followed the same general pattern of distribution in the bay. This result suggests that ecological and paleoecological studies of categories higher than the species level provide valid information, rather than being too generalized as one might expect. Because geologic range usually increases among higher taxonomic categories, use of higher taxa in studies of community structure will facilitate comparisons made through geologic time.

Introduction

Data on the distribution and abundance of foraminifera and ostracodes of Todos Santos Bay in Baja California collected by Walton (1955) and Benson (1959) have been used by a number of other investigators as well: Kaesler (1966) for cluster analysis and quantitative biofacies analysis of both foraminifera and ostracodes, Niitsuma (1968) in a quantitative analysis of communities of benthic foraminifera, and Rowell (1972) for demonstrating the usefulness of relative entropy mapping. In our research we have once again used data from Benson's (1959) now classic study of the Ostracoda to investigate the usefulness of the higher taxa for understanding diversity and 'community structure' of assemblages of Ostracoda.

The problem is this: many ostracode species have quite short geological ranges. In general, genera have long ranges; families have still longer ones; and superfamilies extend back into the Paleozoic. Thus, one usually cannot compare two assemblages of species from quite different geologic times because many of the species in one assemblage may not have been extant at the time the other assemblage was formed. Similarly, studies of diversity of diachronous assemblages are not strictly comparable because the species contributing to the diversity may not have been coeval. It seems worthwhile, therefore, to determine how much information about the diversity of an assemblage is contributed by the various taxonomic categories – species, genus, family, and superfamily.

We show that the diversities at all the taxonomic categories are highly correlated. Thus it is reasonable to make comparisons between assemblages of different age on the basis of their diversities at higher taxonomic categories.

Materials and methods

Throughout our study we have used Brillouin's (1962) equation for species diversity adapted from information theory, first by Margalef (1958) (Table 1). In a series of papers and textbooks, E. C. Pielou has reviewed the use of indices of diversity from information theory (Pielou, 1966a,b,c, 1967, 1969, 1974, 1975). The following discussion is based on her analysis. She has stressed that Shannon's equation 2) is intended for analysis of a statistical universe large enough to be

Table 1. Equations for diversity from information theory. In the first three equations, s refers to the number of species, N refers to the number of individuals, N_i refers to the number of individuals in the ith species, and p_i refers to the proportion the ith species makes up of the statistical universe, not of the sample. In the fourth equation, $\alpha, \beta, \gamma,$ and δ are weighting coefficients; $O, F, G,$ and S refer, respectively, to order, family, genus, and species; $o, f,$ and g refer to the number of orders, families within orders, and genera within families, respectively; H is Brillouin's diversity; and the other symbols are the same as for equations 1), 2), and 3)

Brillouin (1962) $H = \frac{1}{N} \log_e \frac{N!}{N_1! N_2! \ldots N_s!}$ 1)

Shannon (1949) $H' = \sum_{i=1}^{s} p_i \log_e p_i$ 2)

Approximate $H'' = -\sum_{i=1}^{s} \frac{N_i}{N} \log_e \frac{N_i}{N}$ 3)

Hierarchical (modified from Pielou, 1967)

$H = \alpha H_0 + \beta \sum_{i=1}^{o} \frac{N_i}{N} H_{F,i} + \gamma \sum_{i=1}^{o} \sum_{j=1}^{f_i} \frac{N_{ij}}{N} H_{G,ij} + \delta \sum_{i=1}^{o} \sum_{j=1}^{f_i} \sum_{k=1}^{g_{ij}} \frac{N_{ijk}}{N} H_{S,ijk}$ 4)

considered effectively infinite in which the number of species s and their proportions p in the statistical universe, not in a sample from it, are both known. The approximate equation 3) has been used for two purposes, and it is not particularly well suited for either purpose. First, it has been used to estimate H', Shannon's diversity. H'' 3) is, in fact, a maximum likelihood estimator of H', but it is also a biased estimator. In order to correct for the bias, one must substract a correction term, $(s-1)/2N$. In practice the value of this term cannot be determined because s remains an unknown in virtually all real ecological or paleoecological problems. Second, H'' has been used as a substitute for H, a practice that began in precomputer days when factorials were difficult to determine. It is true that equation 3) for H'' can be derived from equation 1) for H if all the N_i's are very large so that the approximation $\log_e n! = n(\log_e n - 1)$ holds. Pielou (1969, p. 232) has pointed out that H'' '... is not a good approximation to H in practice, for unless *all* the N_j are very large (which seldom happens) the approximation to $\log_c n!$ used in deriving it is not sufficiently close.' H'' for a collection is always larger than H, and the magnitude of the bias decreases as the number of individuals in each species increases. For most collections of ostracodes, in which many species are represented by only a few individuals, the error introduced by using H'' rather than H is likely to be large enough to be important.

Shannon's (1949) equation 2) was originally intended to give the information content per symbol of a code; Brillouin's (1962) equation 1) gave the information content per symbol of a message. In carrying these equations into ecology, Pielou (1969) emphasized the difficulty of defining the extent of the statistical universe being sampled and of knowing s, the number of species. When we are dealing with a fully censused collection of ostracodes and are unable to define the limits of the parent statistical universe from which the ostracodes were collected, it is appropriate to regard the collection as a message from the ecosystem, and Brillouin's equation 1) is the appropriate one to use for computing species diversity.

Pielou (1967) has pointed out that diversity may be partitioned hierarchically, enabling one, for example, to determine the diversity of orders, families within orders, genera within families, and species within genera. The diversities at the various taxonomic levels are additive and sum to H, the species diversity. Clearly one gains a great deal more insight into the structure of a community if he knows which levels in the taxonomic hierarchy contribute most to the total diversity (see, e.g., Lloyd et al., 1968).

Pielou (1967) has proposed a hypothetical case in which two collections had the same number of species. One collection had one or a few genera, and the other collection had many genera. The second collection might be considered to be more diverse than the first, especially if the congeneric species in the first collection were thought to be ecologically very closely similar. The diversity may be modified to include such taxonomic and ecological considerations by the use of weighting coefficients. In the above example, the weighting coefficient applied to the generic diversity would have been larger than the coefficient for diversity of species within genera. The values to be assigned to weighting coefficients are arbitrary and must be determined by the investigator. In the absence of specific criteria for assigning weights, a value of one for equal weighting should be assigned, a procedure we followed in our research.

A matrix of Spearman's rank correlation coefficients (Siegel, 1956) was computed between the diversities of four taxonomic levels (superfamily, family, genus; and species), depth of water, and number of individuals in the samples (Table 2).

Results and discussion

Figure 1 shows the species diversity in Todos Santos Bay and the locations of the stations at which ostracodes were found. The pattern of diversity follows very closely the bathymetry of the bay (Benson, 1959, Figure 2, p. 9). The nearshore areas tended to have higher diversity than the middle part of the bay, which had

Table 2. Spearman's rank correlation coefficients between diversity at four taxonomic levels, depth of water, and number of individuals, *** indicates statistically significant correlation at p <0.001

	S	G	F	SF	D	N
Species	–					
Genus	0.987***	–				
Family	0.970***	0.987***	–			
Superfamily	0.461***	0.485***	0.503***	–		
Depth	–0.047	–0.060	–0.084	0.140	–	
Number	0.780***	0.793***	0.793***	0.343***	0.016	–

a coarse-silt to very-fine-sand substrate that was not suitable for the development of a diverse ostracode fauna. Nearshore areas designated by Benson as 'phytal zones' generally had lower diversity than other nearshore areas, but phytal zones away from the main shore generally had higher diversity than surrounding areas, for example around station D and near the southeastern edge of Islas de Todos Santos. Diversity was low in the deep canyon between Islas de Todos Santos and the peninsula, Punta Banda.

The generic diversities (Figure 2) followed the species diversities very closely and differed from them mainly in being slightly lower in value. In the southeastern corner of the bay around the islands, diversities at the two levels were virtually identical. Familial diversities (Figure 3) were also closely similar to species and generic diversity, with high diversities in the same areas. In general, the diversities decreased with increasing level in the taxonomic hierarchy except in the southeastern corner of the bay where diversities at all three levels were about the same.

Superfamilial diversities (Figure 4) differed greatly in magnitude from diversities at the other taxonomic levels, but the general pattern was the same. Areas of low species diversity generally had superfamilial diversities of zero. Areas of high species diversity had superfamilial diversities as high as 0.4 or 0.8. The primary exception to this pattern was the low nearshore superfamilial diversity, suggesting that not all superfamilies are adapted to the rigors of the nearshore, rocky tide-pool environment, as we know from other studies.

Table 2 shows Spearman's rank correlation coefficients between diversity at the four taxonomic levels, depth of water, and the number of individuals in the samples. In this table, correlation coefficients with an absolute value less than 0.25 are not statistically significant at the 0.05 level of probability. Note the very high correlation between diversities at the species, generic, and familial levels, all greater than 0.97. This result reinforces the interpretation based on Figures 1, 2, and 3 that the diversities at the three levels are very closely similar. The correlation of superfamilial diversities with diversities at the other three levels is much lower but still very highly significant statistically ($P < 0.001$). The correlations of diversities at all levels with depth of water were statistically nonsignificant. Diver-

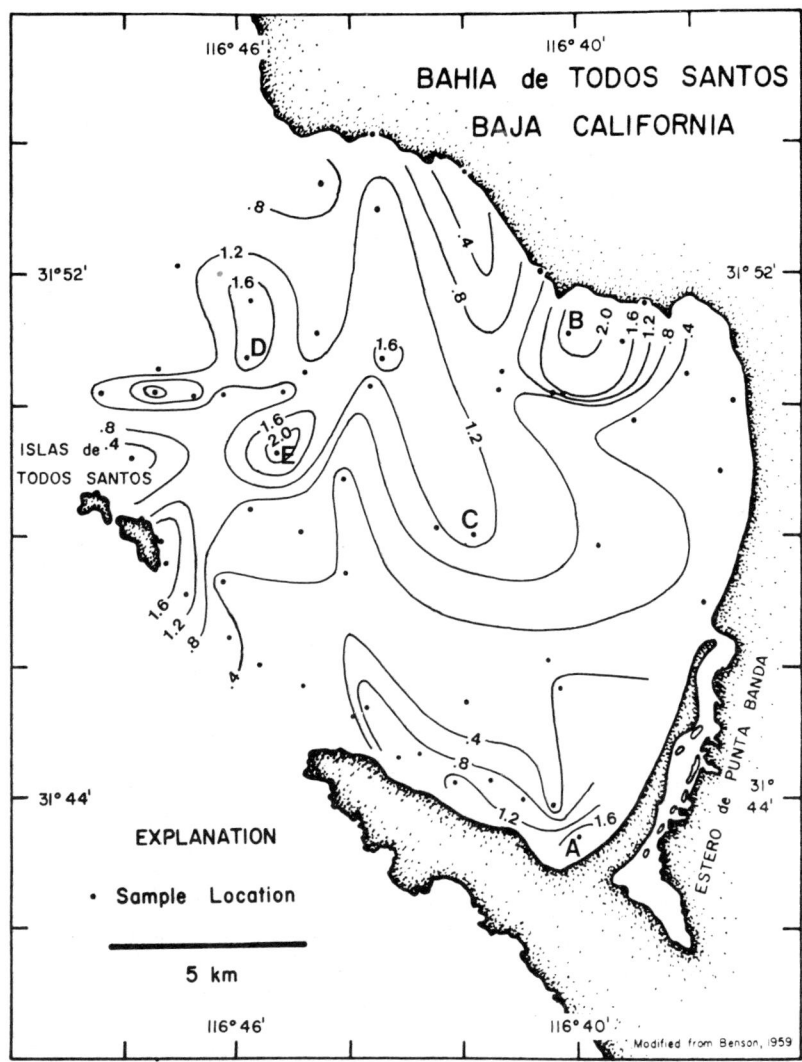

Fig. 1. Todos Santos Bay showing stations at which ostracodes were found and species diversity (H) contoured at an interval of 0.4. Letters A to E refer to samples shown in Figure 6; sample numbers are as follows: A, Benson's station 34; B, 108; C, 68; D, 81; E, 49.

sity neither increases nor decreases with depth but, instead, is affected by local environmental factors such as substrate and presence of attached plants. Diversities at all levels were strongly correlated with the number of individuals in the sample. This characteristic of Brillouin's index is judged to be advantageous since the number of individuals found in samples of constant size is important ecological information, but it may cause some problems with paleontological material and necessitate working with samples comprising a constant number of individ-

37

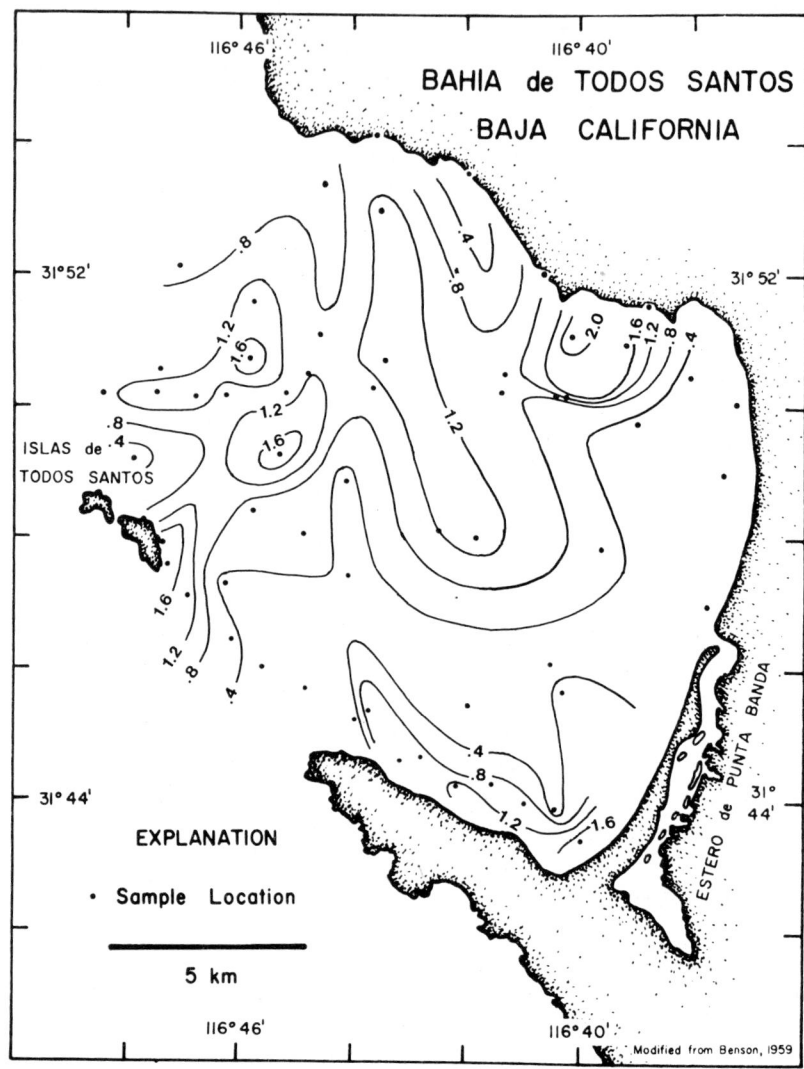

Fig. 2. Todos Santos Bay showing generic diversity contoured at an interval of 0.4.

uals. Note that number of individuals in the samples was not correlated with depth of water.

Figure 5 shows the increase in diversity with level in the taxonomic hierarchy for five selected samples, A to E. For each sample, the point with the lowest diversity and lowest number of taxa represents the superfamilial diversity, and the right-most point represents species diversity. The diversities were plotted against number of taxa rather than against ranked taxonomic category in order to show that the effect of increased number of taxa in any category generally increases the diversity. Except for the deepest sample (E), the increase in diversity from

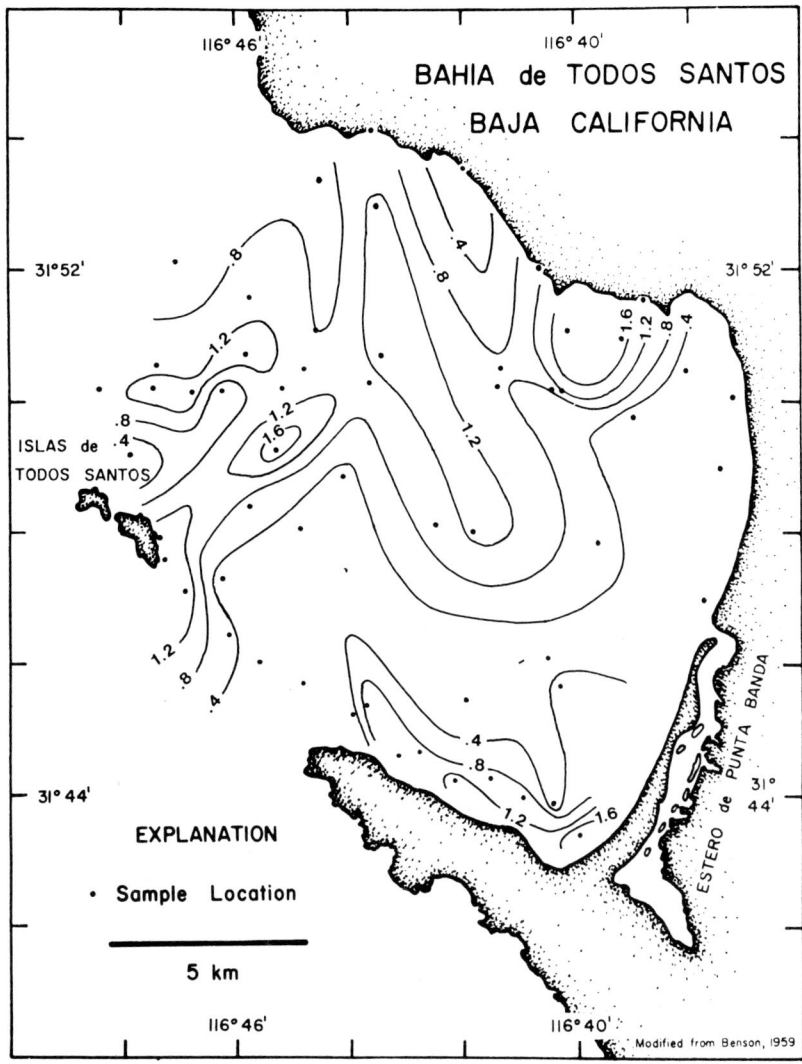

Fig. 3. Todos Santos Bay showing familial diversity contoured at an interval of 0.4.

the generic to the species level was negligible, and in all cases the increase from the familial to the generic level was quite small. Thus, familial diversity tells us almost as much about species diversity as species diversity itself.

These results have implications for those interested in the application of ostracode ecology and paleoecology. First, when the primary interest is in diversity of the assemblages being studied, it may be advantageous to work at the familial level or, perhaps, at the generic level. Note that genera need only be discriminated, not identified, in order to compute generic diversity. The cost of evaluating a great many ostracodes from sediment samples, well cuttings, and cores could

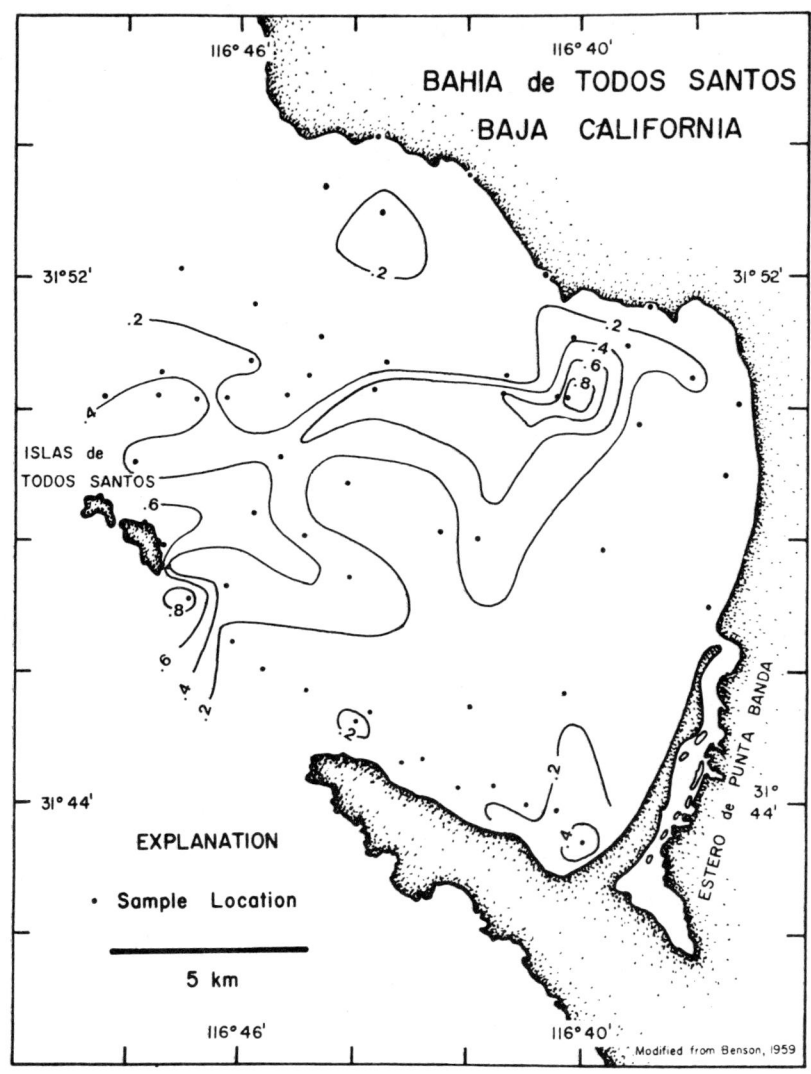

Fig. 4. Todos Santos Bay showing superfamilial diversity contoured at an interval of 0.2.

be greatly reduced; and it seems likely that work at a higher taxonomic level would provide almost as much information about the diversity.

Second, species diversities from different times in the geological past are difficult to compare because the species contributing to the diversities may have lived at different times. However, because diversities at higher taxonomic levels are highly correlated with species diversity and because higher taxa usually have longer ranges than species, diversities of the higher taxa are more readily comparable over long intervals of geologic time. One might compare Recent species

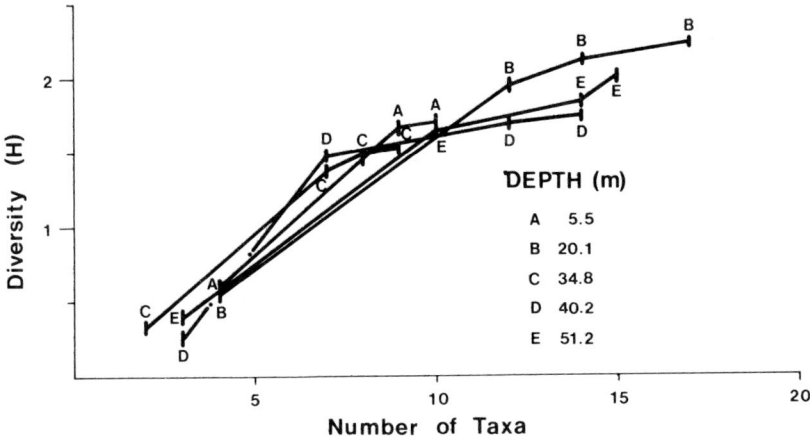

Fig. 5. Plot of diversity (H) against number of taxa in each of the taxonomic categories studied. The five samples, A to E, are identified in Figure 1. For each sample, the taxonomic level increases from left to right: superfamily, family, genus, species. Sample A had the same number of genera as species so that the diversity and number of taxa for the generic level and specific level were coincident.

diversities with species diversities from the Neogene, generic diversities with those from the Paleogene, familial diversities with Cretaceous ones, and Recent superfamilial diversities with ones from as far back as Jurassic, Triassic, and, with caution, the Late Paleozoic.

Conclusions

1. Brillouin's index of diversity is to be preferred for computing species diversity of collections of Ostracoda.
2. Partitioning species diversity into its additive taxonomic components provides insight into the taxonomic structure of the assemblage or ostracode community.
3. Familial, generic, and species diversities of ostracodes from Todos Santos Bay were very highly correlated, suggesting that study of diversity at taxonomic levels higher than the species would have provided almost the same information in substantially less time and at substantially lower cost.
4. If conclusion 3 above holds true for assemblages of Ostracoda in general, then those interested primarily in diversity may find similar advantages in the use of higher taxa. Moreover, diversities of higher taxonomic categories may be comparable over longer intervals of geologic time than species diversities.
5. In Todos Santos Bay, diversities at the four taxonomic levels studied were not correlated with depth, but all were very strongly correlated with number of individuals in the sample. Moreover, diversities at all levels were generally higher in those samples with more taxa in each hierarchical category.

Acknowledgments

Our work was supported in part by a Title II grant from the Office of Water Research and Technology to The University of Kansas (grant no. 14-31-0001-5200) and by two grants from the Union Carbide Corporation to The University of Kansas. All computation was done at The University of Kansas Computation Center.

References

Benson, R. H. 1959. Ecology of Recent ostracodes of the Todos Santos Bay region, Baja California, Mexico. Univ. Kansas Paleontological Contr., Arthropoda, Art. 1:1-80.
Brillouin, L. 1962. Science and information theory: Academic Press, New York, 2nd ed., 347 pp.
Kaesler, R. L. 1966. Quantitative re-evaluation of ecology and distribution of Recent Foraminifera and Ostracoda of Todos Santos Bay, Baja California, Mexico. Univ. Kansas Paleontological Contr., Paper 10:1-50.
Lloyd, Monte, Inger, R. F. & King, F. W. 1968. On the diversity of reptile and amphibian species in a Bornean rain forest. American Naturalist 102:497-515.
Margalef, D. R. 1958. Information theory in ecology. General Systems 3:36-71.
Niitsuma, Nobuaki. 1968. Analysis of the benthonic foraminiferal community (in Japanese). Fossils 16:25-33.
Pielou, E. C. 1966a. Shannon's formula as a measure of specific diversity: its use and misuse. American Naturalist 100:463-465.
Pielou, E. C. 1966b. Species-diversity and pattern-diversity in the study of ecological succession. Jour. Theoretical Biology 10:370-383.
Pielou, E. C. 1966c. The measurement of diversity in different types of biological collections. Jour. Theoretical Biology 13:131-144.
Pielou, E. C. 1967. The use of information theory in the study of the diversity of biological populations. Fifth Berkeley Symp. Math. Statistics and Probability, Proc. 4:163-177.
Pielou, E. C. 1969. An introduction to mathematical ecology. Wiley-Interscience, New York, 286 pp.
Pielou, E. C. 1974. Population and community ecology: principles and methods. Gordon and Breach, New York, 424 pp.
Pielou, E. C. 1975. Ecological diversity. Wiley-Interscience, New York, 165 pp.
Shannon, C. E. & Weaver, W. 1949. The mathematical theory of communications. Univ. Illinois Press, Urbana, 125 pp.
Siegel, Sidney. 1956. Nonparametric statistics for the behavioral sciences. McGraw-Hill, New York, 312 pp.
Walton, W. R. 1955. The ecology of living benthonic Foraminifera, Todos Santos Bay, Baja California. Jour. Paleontology 29:952-1018.

Authors' address:

Dept of Geology and Museum of Invertebrate Paleontology, The University of Kansas, Lawrence, Kansas 66045, USA.

Discussion

Thomas I. Kilenyi: As higher taxonomic groups are not based on adaptive characters, is there any justification in looking at the diversity indices of orders, superfamilies, etc?

Roger L. Kaesler: Yes, study of diversity at higher levels in the taxonomic hierarchy can be justified. Perhaps the best justification is an empirical one, the very high intracorrelations between the diversities computed at the various levels in the hierarchy. Given these high correlations, one could gain almost as much information about the structure of the ostracode portion of the community by studying diversities at the generic or familial levels as by determining species diversity. Moreover, in spite of the arbitrariness of the higher taxa, we hope that congeneric species are more closely related to each other than species in different genera. Phylogenetic relationship implies some similarity in niche. If one is interested primarily in still higher taxa, such as the family or superfamily, he can apply weighting factors to diminish the impact of congeneres on the total species diversity.

Thomas I. Kilenyi: What would you say is the main use of the diversity index?

Roger L. Kaesler: The diversity index we have used, Brillouin's index, gives species diversity per individual of a fully censused collection of organisms. The statistics diversity and evenness can be used to express the structure of a community of organisms or of a portion of such a community. In effect, they give one means of describing the shape of the histogram that shows how the abundances of different species are distributed.

Sixth Intern. Ostracod Symposium, Saalfelden

SOME SPECULATION ON THE SIGNIFICANCE OF CARAPACE LENGTH IN PLANKTONIC HALOCYPRID OSTRACODS

M.V. ANGEL

Introduction

Size frequency distribution in a population of organisms is a result of genotypic and phenotypic variability. The degree of plasticity of the genotype will regulate how much the organism can respond to environmental pressures. Planktonic halocyprids are generally thought not to have post-maturation moults, yet Poulsen (1973) shows that there are considerable geographical variations in size in many halocyprid species. Poulsen in synonymising *C. bispinosa* Claus and *C. secernenda* Varva argues that post-maturation moulting occurs. Logically following this argument many other sibling species should be synonymised, e.g. *C. spinirostris* with *C. porrecta* and the three species of the '*procera*' complex. However, adults and juveniles of *C. bispinosa* and *C. secernenda* are easily separated in live animals by the greater pigmentation of *C. secernenda*, and also by the greater development of the shoulder vaults in preserved material. Poulsen's synonymisation is not accepted here nor is the argument for post-maturation moulting considered proven.

Adult size must be determined genotypically by the potential for growth at each junenile instar, modified phenotypically by environmental factors such as food supply and temperature. It is argued in this paper that there is circumstantial evidence that competition for food is an important environmental factor in regulating juvenile growth rates.

Fowler (1909) in postulating Brook's Law suggested that growth increments of juvenile instars were a fixed percentage of the carapace length within a species. He also suggested that the percentage increase in length varies within a species geographically and seasonally, that is to say that there is phenotypic regulation of growth. Kesling (1952, 1953) verified for freshwater ostracods that linear dimensions increase by a factor approximating to $\sqrt[3]{2}$ i.e. 1.26, a result of the doubling of body volume at each moult. Rudjakov (1962) found in three halocyprid species *Conchoecia alata minor*, *Conchoecia elegans* and *Halocypria globosa*, the factor of linear increase was 1.49–1.64 for juvenile instars, 1.42–1.48 for the female maturation moult and 1.25–1.29 for the male moult. Hillman (1969) gave evidence for regularities in growth of the Antarctic species *C. serrulata*.

Material and methods

The data on the juvenile instar sizes are all drawn from material collected during Cruise 61 in 1974 by RRS Discovery within a one degree square centred around 44° N 13° W in the North-east Atlantic just south of the Bay of Biscay. The sam-

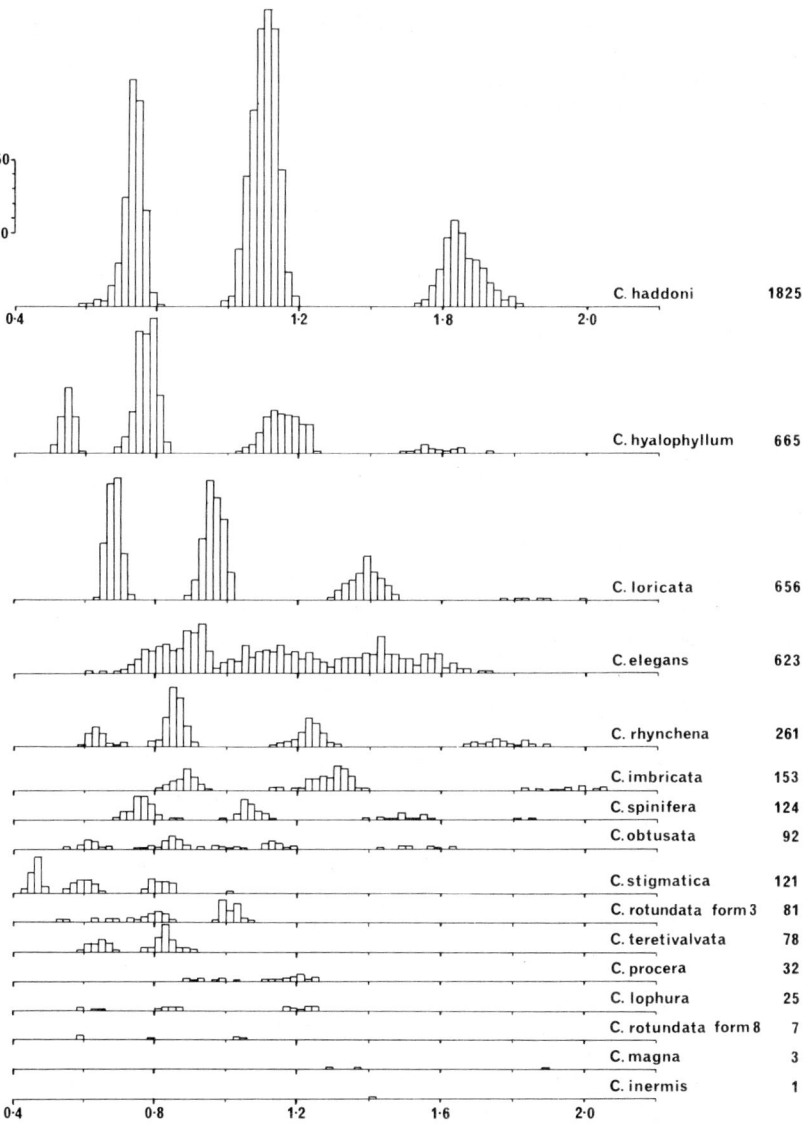

Fig. 1. The length – size distributions of the halocyprid ostracods in a 1/8 subsample of an RMT 1 catch, Discovery station 8507 haul 20, depth 300–400 m, 1021–1221 h, 4 April 1974, starting position 44°7.7′N 13°18.6′W. The total numbers of each species are given. The taxonomic problem of the *C. rotundata* complex is discussed in Angel & Fasham (1975).

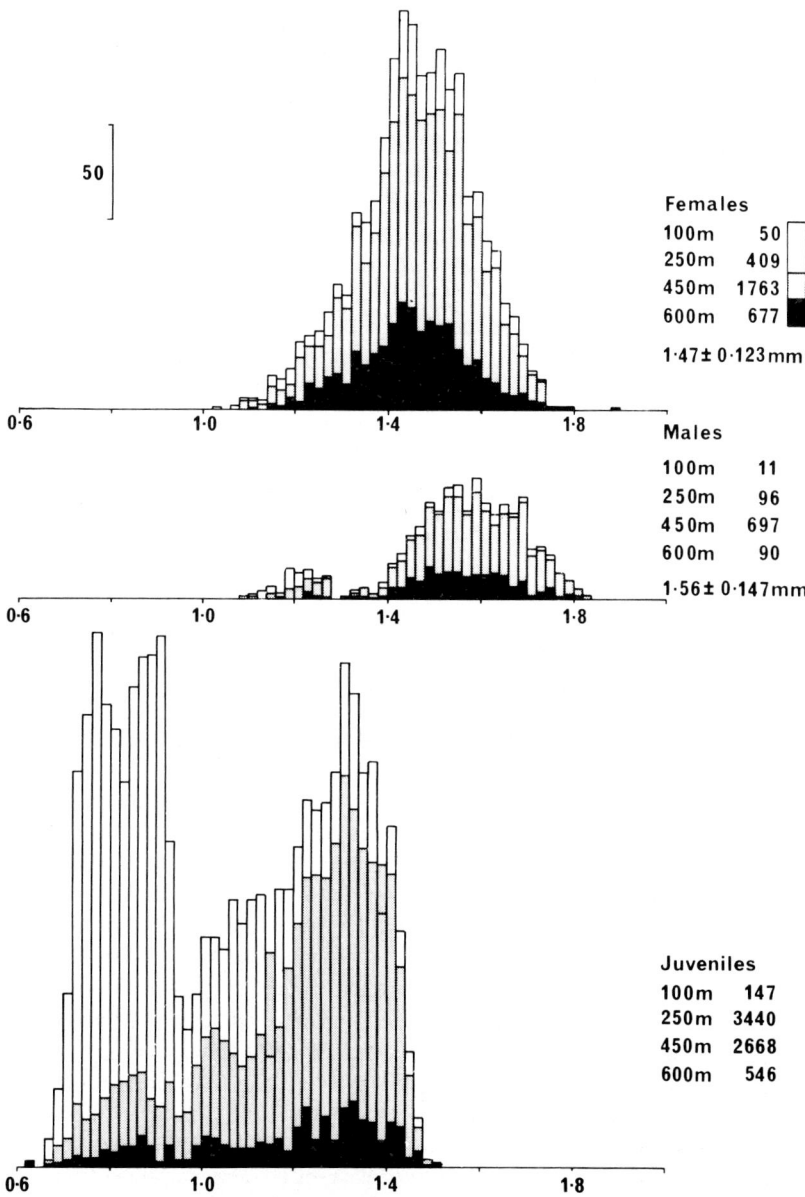

Fig. 2. The length-size distributions of all *C. elegans* Sars specimens taken in four series of tows taken over 48 hours at four depths 100 m (25 tows), 250 m (23 tows), 450 m (25 tows) and 600 m (23 tows) taken within a 5 km square in the vicinity of 44° N, 13° W between 6–19 April 1974. The total numbers taken in each series are shown together with the mean sizes of the aldults. There was no significant change in the adult mean lengths with depth, but the smaller juvenile stages did occur shallower than the later stages, as indicated by the different shading.

ples were collected using an RMT 1+8 net (Baker et al., 1973), towed at 2 knots for two hours. Only material from the fine mesh net (pore size 0.330 mm) has been analysed. Most of the hauls were taken from depth strata within the 300–600 m range. Samples were either totally sorted for halocyprids, or if initial inspection suggested a sample contained over 6000 specimens, one eighth sub-samples were taken. All the halocyprids were identified to species and measured with a Wild M5 microscope to the nearest 0.02 mm. Animals were measured by lying them on their backs and taking the longest measurement including any spines.

Results

Figure 1 shows the size distribution for the 16 species in haul 8507/20 taken from 305–400 mm at 1021–1221 hours on 4 April. All the species show distinct instar groupings with the exception of *Conchoecia elegans* Sars. Figure 2 shows the length frequency distribution for *C. elegans* from pooled data from throughout the water column. Poulsen (1973) described the mean lengths of adult males of *C. elegans* in the Atlantic as being 1.86 mm north of 60°N, 1.67 mm at 60–40°N and 1.13 at 40°N–40°S. The mean lengths of females were 1.62, 1.50, and 1.12 mm, respectively. Angel & Fashan (1975) also identified three size groupings in the N.E. Atlantic with the smallest forms occurring south of 30°N, the middle form from 30–60°N, and the largest form at 53°N and northwards; the larger animals tended to occur deeper in the water column than the middle form. The data presented in Figure 2 are interpreted as being the result of hybridisation between the largest and smallest forms producing a broad spread of both adult and juvenile sizes.

In Table 1 the lengths of the instars together with their standard deviations are tabulated for the eleven most abundant mesopelagic species in the vicinity of 44°N 13°W. *C. imbricata* Claus was omitted from this list because of the uncertainties of consistent measurements if the carapace spineds were broken. The number of animals measured at each stage for each species are shown together with the growth factor for each instar. Direct comparisons with Fowler's (1909) results are difficult, partly because of taxonomic confusion (Skogsberg, 1920), and partly because of his method of measurement omitted the rostrum and other carapace features. For the two species in which Stage III juveniles were retained by the net, the growth factor for moult III–IV was lower than for moult IV–V. In six species the growth factor was higher for moult V–VI, than for moult IV–V, in four it was much the same or a little less; only *C. obtusata* show a marked deviation from this pattern. Hillman's (1969) data for *C. serrulata* Müller from the Antarctic showed a similar pattern — 1.30 for moult I–II, 1.29 for II–III, 1.31 for IV–V, 1.43 for V–VI, 1.43 from VI to adult. (The last two values assume that the length of Stage VI animals was 0.977 mm as indicated by Hillman's figure and not the 0.777 mm given in his table.) Hillman did not distinguish between the sexes in *C. serrulata,* but in most of the Atlantic species the growth factor for the female maturation moult is smaller than the V–VI moult; *C. haddoni* and *C. obtusata* are the exceptions. The growth factor for the male maturation moult is lower than the female maturation moult except in *C. stigmatica. C. stigmatica* is one of the exceptional halocyprids in which the males are larger than the females. This difference in the growth factor could be a consequence of a sexual dimorphism in allometry, as the males are usually broader than the females. Both *C. stigmatica* and *C. teretivalvata* are more spherical in body shape, and in both species the growth factors were close to the value of $\sqrt[3]{2}$ i.e. 1.26, throughout their development. Unexpectedly the only other species showing a similarly low value is the rather elongate *C. inermis*.

Table 1. Mean carapace length in millimeters of all instars retained in a net of mesh 0.33 mm, of eleven of the more abundant mesopelagic halocyprids from the region of 44° N 13° W. Below each value is given in italics the number of observations and the growth factor for the previous moult in bold print

INSTAR Species	III	IV	V	VI	♀	♂
C. spinifera Claus	–	0.78 ± 0.29 *301*	1.09 ± 0.032 *271* **1.39**	1.50 ± 0.042 *133* **1.38**	2.08 ± 0.083 *50* **1.38**	1.82 ± 0.044 *54* **1.20**
C. stigmatica Müller	–	0.48 ± 0.015 *358*	0.61 ± 0.029 *406* **1.28**	0.81 ± 0.052 *311* **1.33**	1.00 ± 0.34 *99* **1.22**	1.13 ± 0.131 *23* **1.39**
C. obtusata Sars		0.62 ± 0.032 *49*	0.84 ± 0.033 *42* **1.36**	1.00 ± 0.036 *12* **1.18**	1.57 ± 0.065 *15* **1.57**	1.15 ± 0.027 *19* **1.15**
C. haddoni Brady & Norman (Northern form)		0.73 ± 0.035 *2213*	1.10 ± 0.041 *2222* **1.49**	1.68 ± 0.058 *716* **1.52**	2.75 ± 0.084 *108* **1.63**	2.23 ± 0.059 *163* **1.33**
C. inermis Claus		0.92 ± 0.031 *26*	1.23 ± 0.040 *38* **1.33**	1.64 ± 0.029 *18* **1.32**	2.07 ± 0.068 *55* **1.26**	1.93 ± 0.046 *21* **1.18**
C. rhynchena Müller	0.66 ± 0.023 *874*	0.88 ± 0.027 *1162* **1.33**	1.24 ± 0.040 *1084* **1.41**	1.79 ± 0.054 *818* **1.44**	2.43 ± 0.086 *413* **1.35**	2.34 ± 0.051 *233* **1.30**
C. loricata Müller		0.71 ± 0.028 *811*	0.98 ± 0.031 *1093* **1.37**	1.30 ± 0.040 *762* **1.43**	1.93 ± 0.077 *201* **1.37**	1.78 ± 0.055 *193* **1.26**
C. hyalophyllum Claus		0.36 ± 0.019 *643*	0.79 ± 0.029 *1160* **1.41**	1.17 ± 0.047 *473* **1.47**	1.66 ± 0.074 *97* **1.41**	1.58 ± 0.045 *126* **1.34**
C. lophura Müller	0.64 ± 0.021 *39*	0.87 ± 0.025 *86* **1.36**	1.24 ± 0.034 *139* **1.42**	1.81 ± 0.051 *62* **1.45**	2.41 ± 0.059 *33* **1.33**	2.35 ± 0.050 *65* **1.30**
C. teretivalvata Iles				0.64 ± 0.025 *66*	−0.83 ± 0.033 *78* **1.28**	–
C. ametra Müller	0.82, 0.84 *2*	1.13 ± 0.054 *55* **(1.34)**	1.75 ± 0.072 *91* **1.54**	2.66 ± 0.108 *80* **1.52**	3.45 ± 0.136 *24* **1.48**	3.43 ± 0.080 *18* **1.28**

49

The number of animals counted and measured can be used with caution to estimate mortalities between instars. The carapace length of the smallest instar retained varies between species depending on the cross-sectional shape of each. The spherical species such as *C. stigmatica* are retained at smaller sizes than elongate species. The numbers caught of Stage IV animals for all species are lower than Stage V, which suggests that extrusion through the meshes has biased the results. Avoidance by the larger animals is not thought to be a major source of error in sampling halocyprids.

Discussion

MacArthur (1972, p. 23) points out that it has been shown empirically that species that differ only in size seem to require the larger to be twice as heavy as the smaller in order to co-exist (see also Hutchinson, 1959). He then shows mathematically that co-existence becomes rapidly more precarious as the distance their resource mean values approaches $\sqrt{2}$ times their standard deviation (p.44). The data presented in Table 1 shows that the instars of halocyprids approximately double their volumes, but the standard deviations are much smaller than would be expected if the instars of each species were competing alone for any resource.. It is suggested here, that assuming the carapace length is related to the size or the type of resource exploited, that competition between instars of different species is reducing the variability of lengths of animals of each instar. A shift in the community structure may alter the 'length size niche' available to any given instar, so causing a deviation from the normal pattern of growth factors. Such competition pressure may have produced the aberrant growth factor for moult V–VI in *C. obtusata*. If the growth pattern is distorted beyond the limits of the genotype to recover, there will be geographical variation in the adult sizes; these length variations will be independent of temperature. A lack of temperature dependence of adult size is shown by *C. hyalophyllum*. At 44°N 13°W specimens caught at 400-500 m, where the water temperatures were 10.8°C–11.2°C cooler than at the surface, were identical in size to those caught off Fuerteventura (28°N 14°W) at the same depth where the in situ water temperature was 11.25-12.2°C and at night the animals migrated up to around 200 m where the water temperature was 15.38°C. A different situation occurs in *C. haddoni* where there is a marked difference in size between the sensu strictu form which occurs to the north of 40°N and the small 'southern form' (Skogsberg, 1920) which occurs in the upwelling region off the N.W. African coast. Poulsen (1973) suggests that the 'southern form' is tropical form restricted to 20°N–20°S in the Atlantic. However, it appears to be geographically isolated and there is probably justification in considering it a distinct sub-species.

The sexual dimorphism in carapace length in species for which there are sufficient measurements from a single locality (see Table 2) is consistent with the hypothesis that each sex is exploiting an overlapping but distinctive food resource and that they are behaving like competing species. The values of de in MacArthur's notation (i.e. the mean size difference beteen males and females) is greater than double the average standard deviations of their size distributions.

Table 2. The sexual dimorphism in carapace length of various halocyprid species, showing the mean difference in size between the sexes and the degree of overlap of their size distributions

	♀ mm	♂ mm	$\frac{♂}{♀} \times 100$	♀ − ♂ mm	$\frac{\sigma_♀ + \sigma_♂}{2}$
1 C. porrecta	1.69 ± 0.032	1.42 ± 0.019	84	0.27	0.036
1 C. spinirostris	1.15 ± 0.033	1.01 ± 0.023	87	0.14	0.047
2 C. hyalophyllum					
(28° N)	1.65 ± 0.059	1.57 ± 0.071	95	0.08	0.092
(44° N)	1.66 ± 0.074	1.58 ± 0.045	95	0.08	0.084
2 C. magna	1.89 ± 0.070	1.73 ± 0.071	92	0.16	0.100
2 C. parthenoda	1.67 ± 0.033	1.52 ± 0.028	91	0.15	0.043
3 C. pseudoparthenoda	1.84 ± 0.038	1.64 ± 0.037			
4 C. bispinosa	1.94 ± 0.095	1.78 ± 0.043	92	0.16	0.098
4 C. haddoni					
(southern)	2.36 ± 0.090	1.83 ± 0.050	78	0.53	0.099
(northern)	2.75 ± 0.084	2.23 ± 0.060	81	0.52	0.102
4 C. secernenda	2.55 ± 0.075	2.29 ± 0.180	90	0.26	0.180
5 C. procera	1.18 ± 0.021	1.02 ± 0.019	86	0.16	0.028
5 C. microprocera	1.00 ± 0.013	0.86 ± 0.013	86	0.14	0.018
5 C. macroprocera	1.30 ± 0.021	1.15 ± 0.021	88	0.15	0.030
C. spinifera	2.08 ± 0.083	1.82 ± 0.044	88	0.26	0.090
C. stigmatica	0.99 ± 0.034	1.13 ± 0.031	112	−0.14	0.046
C. obtusata	1.57 ± 0.065	1.15 ± 0.027	73	0.42	0.065
C. inermis	2.07 ± 0.068	1.93 ± 0.046	93	0.14	0.069
C. rhynchena	2.43 ± 0.086	2.34 ± 0.051	96	0.09	0.097
C. loricata	1.93 ± 0.077	1.78 ± 0.055	92	0.15	0.093
C. lophura	2.41 ± 0.059	2.35 ± 0.050	98	0.06	0.077
C. ametra	3.95 ± 0.136	3.42 ± 0.080	87	0.53	0.153

1. After Angel 1969a. 2. After Angel 1969b. 3. After Angel 1970a. 4. After Angel 1970b.
5. After Angel 1971.

C. haddoni (both forms) and *C. obtusata* are notable for the marked disparity in the sizes of the sexes.

If competition does regulate the size and the range of variation in each instar, it can only affect species occupying the same depth strata at the same time which feed synchronously. Competition may operate during juvenile instars as effectively as on adults. At mesopelagic depths off N.W. Spain the halocyprids would seem to have been exploiting 3 or 4 food resources. Resource partitioning can occur either through increased vertical structuring of the community or by increased cyclic feeding or vertical migrations. Angel & Fasham (1975) have shown how vertical stratification accompanied the increase of species richness at progressively lower latitudes in the N.E. Atlantic. They also point out that vertical migration is more extensive at middle latitudes. Thus the mechanism by which resource partitioning is achieved varies geographically, but it is not apparent why one or other mechanism is more effective in an area.

At high latitudes where environmental stress rather than competition is likely to be the dominant factor regulating the number of species present, a much greater variation in instar size would be expected. This would result from the absence of competitive restraints. The next obvious step, however, is to investigate the diets of the various species and how it changes during development.

The hypothesis also has another interesting consequence for two sibling species becoming sympatric. If the two species exploit similar resources and have similar feeding cycles, then their juvenile instars must have relatively little overlap in their size ranges. There is good evidence of inter-digitating instar sizes in *C. spinirostris* and *C. porrecta* in the Adriatic (Gooday & Angel in press). Non-overlapping juvenile sizes occur in the later instars of the two sibling species related to *C. edentata* Müller in the N.E. Atlantic although the earlier instars (i.e. Stage IV) are indistinguishable (Gooday, 1976). The juveniles of *C. incisa* Müller and *C. gaussi* Müller overlap in size but are separated vertically.

Summary

1. Study of the length frequency distributions of mesopelagic halocyprid ostracods from the Northeast Atlantic in the vicinity of 44°N 13°W confirms that in the majority of species juvenile instars occur in discrete, well separated size groups.
2. It is suggested that the spread in size range of each instar may be restrained by interspecific competition.
3. *Conchoecia elegans* Sars is the one exceptional species, a possible result of a large high latitude population hybridising with a small low latitude population.
4. It is shown that sexual dimorphism in carapace length within a species fits the theory that competition for a single food resource prevents all but a minimum overlap between two forms. It is very tentatively suggested that the function of the sexual dimorphism in size is the more efficient utilization of food resources.

References

Angel, M. V. 1969a. The ostracod Conchoecia porrecta Claus redescribed and compared with C. spinrostris Claus. Crustaceana 17: 35–44.

Angel, M. V. 1969b. The redescription of three halocyprid ostracods, Conchoecia hyalophyllum Claus, C. magma Claus and C. parthnoda Müller from the North Atlantic. Crustaceana 17: 45–63.

Angel, M. V. 1970. The redescription of Conchoecia bispinosa Claus, C. haddoni Brady and Norman and C. secernenda Vavra from the North Atlantic. Crustaceana 18: 147–166.

Angel, M. V. 1971a. Conchoecia pseudopathenoda (nov. sp.) a new halocyprid ostracod for the tropical North Atlantic. Bull. Br. Mus. (Nat. Hist.) Zool. 21 (8): 289–296.

Angel, M. V. 1971b. Conchoecia from the North Atlantic. The 'procera' group. Bull. Br. Mus. (Nat. Hist.) Zool. 21 (7): pp. 259–283

Angel, M. V. & Fasham, M. J. R. 1975 Analysis of the vertical and geographic distribution of the abundant species of planktonic ostracods in the North-east Atlantic. J. Mar. Biol. Assn U.K. 55: 709–737.

Baker, A. de C., Clarke, M. R. & Harris, M. J. 1973. The N.I.O. combination net (RMT 1+8) and further developments of rectangular midwater trawls. J. Mar. Biol. Assn U.K. 33: 167–184.

Fowler, G. H. 1909. Biscayan plankton Pt. XII. The Ostracoda. Trans. Linn. Soc. Lond. (Ser. 2 Zool.) 10: 219–358.
Gooday, A. J. 1976. The taxonomy of Conchoecia (Ostracoda, Halocyprididae) of the gaussi group and edentata group from the North-east Atlantic with a note on their ecology. Bull. Br. Mus. (Nat. Hist.) Zool. 30: 57–100.
Gooday, A. J. & Angel, M. V. (in press). Distribution of planktonic Ostracoda (Halocyprididae) in the North Adriatic with the description of a new subspecies, Conchoecia porrecta adriatica. Crustaceana.
Hillman, N. S. 1969. Ontogenic studies of Antarctic pelagic ostracods. Antarct. J. U.S. 4: 189–190.
Hutchinson, G. H. 1959. Homage to Santa Rosalie, or why are there so many kinds of animals. Am. Nat. 93: 143–159.
Kesling, R. V. 1952. Doubling in size of ostracod carapaces in each moult stage. J. Paleont. 26:
Kesling, R. V. 1953. A slide rule for the determination of instars in ostracod species. Contr. Mus. Pal. Univ. Mich., No. 11.
MacArthur, R. H. 1974. Geographical Ecology. Harper & Row, New York, 268 pp.
Poulsen, E. M. 1973. Ostracoda-Myodocopa. Part IIIb. Halocypriformes–Halocypridae Conchoecinae. Dana Report No. 84, 224 pp.
Rudjakov, Yu. A. 1962. Some growth regularities in pelagic ostracods of the family Halocypridae. Trud. Inst. Oceanol. 58: 167–171 (in Russian).
Skogsberg, T. 1920. Studies on marine ostracods. 1. Cypridinids, halocyprids and polycopids. Zoolog. bidrag fr. Uppsala, Suppl. 1: 1–784.

Authors address:
Institute of Oceanographic Sciences, Wormley, Godalming, Surrey, GU8 5UB, England

Discussion

Maddocks: After competing species drop out, remaining instars show increased range of size. Is this a disproportionate increase in range? Have you tried using the coefficient of variation or any other statistical measures? Secondly, the size distribution of *C. elegans* shows a solid aray of points without discrete clusters for instars. When I see this in *Macrocypris*, it is a sign that I have confused two species or that males and females differ in size but not shape. If you plot specimens of just one sex, do they show this continious distribution?

Angel: In answer to your first question, I have not yet got sufficient data to make it worthwhile analysing the spread in size range when competing species drop out. One problem is that it is still not known which species are in direct competition.

Your second question I can answer a little more satisfactorily. Specimens of *C. elegans* from 60° N 20° W show discrete separation of instars and the adult males average around 1.8 mm in carapace length. These are indistinguishable by any taxonomic character other than size and extremely subtle variation in shape from *C. elegans* from 30° N 23° W which also shows discrete separation of instars and the adult males average around 1.2 mm in capace length. At 44° N 13° W and at 40° N 20° W adult males average around 1.5–1.6 mm with a normal distribution of size from 1.1–1.9 mm; the adult females show the same sort of spread in size range. If there were two species involved the distribution would be either bimodal or highly skewed. An alternative interpretation would be that it is a multi-species swarm, but without supporting evidence from elsewhere I consider the interpretation that this is a hybrid swarm to be the most simple one. There is, as I have shown, a sexual dimorphism in carapace length and shape. It is no greater in *C. elegans* than in most other species. The dimorphism is only expressed from the final larval instar for which size distributions are bimodal or skewed in a few other species. So I do not think that the rather featureless size distributions of the larval instars of *C. elegans* can be a consequense of sexual dimorphism.

Sohn: If the vertical scale represents number of individuals, the number of each instar of each instar should be greater than the number of adults. Your diagrams show the reverse, can you explain?

Angel: The abundance of instar observed at any one time is a function of many parameters. First there is a sampling bias, the smaller an animal is the more likely it is to be extruded through the meshes of the net; the largere an animnal is the more effective it may be in avoiding capture by the net because it can swim faster. Duration of an instar is important since if an instar has a very short duration, it will not appear to be very abundant. Similarly if adults are extremely long-lived compared with the juvenile stages, they will appear disproportionately abundant, thus recruitment and loss (either by mortality or growth) of a particular stageis important. If breeding is seasonal rather than continuous, juvenile stages or adults may be entirely absent at certain times of year. Finally if the species undergo ontogenic migrations, samples may include only adults or only juveniles depending on the depth of sampling. For individual species I can offer guesses as to which of these factors is the more important in determining ratios between different instars and adults, but these would be guesses.

Kaesler: Do you have evidence of character displacement among sympatric species? If, so it might be strong evidence of selection to avoid competition for resources.

Angel: Perhaps the best example is in *Conchoecia oblonga* in which there are two morphological forms with a single character difference in the position of the right asymmetric carapace gland. In form A the gland is in the normal position, whereas in form B it is anteriorly displaced.

Juveniles as well as adults are easily separated by this one character. On the eastern side of the Atlantic at latitudes of around 30° N the form A is predominant. Moving across to Bermuda there is a clinal change in the structure of the whole ostracod community which is accompanied by a switch to dominance of the form B. There are also minor shifts in the adult carapace lengths of several species which may be a response to shifts in competition patterns.

In regions where current systems result in the mixing of two communities within the same water column, i.e. along the edges of the main oceanic gyral systems, species from different communities with apparently similar requirement tend to segregate vertically. Although this is not character displacement in the classical sense, I think it does indificate that it will be found in halocyprids when looked for in the right place.

Sixth Intern. Ostracod Symposium, Saalfelden

THE VARIABILITY OF THE SIEVE-PORES IN RECENT AND FOSSIL SPECIES OF CYPRIDEIS TOROSA (JONES, 1850) AS AN INDICATOR FOR SALINITY AND PALAEOSALINITY

AMNON ROSENFELD & BERND VESPER

Abstract

Observations were made of recent species of *Cyprideis torosa* from different saline environments in Northern Germany and Israel. This species possesses sieve-pores of different shapes: 'round', 'oblong', and 'irregular'. An inverse relationship between the percentage of 'round' pores and salinity appears to exist. 'Irregular' pores predominate in hypersaline waters. A correlation graph (Fig. 3) between pore shape and salinity is presented.

Applying this relationship to Pleistocene occurrences of *C. torosa* from Northern Germany (Holsteinian), Denmark (Eemian) and Israel (Lisan Formation), different palaeosaline environments of deposition were established. Other palaeoecological studies are in agreement with the results obtained in this investigation.

The method offers the possibility of palaeosalinity estimations for the Pleistocene. Further studies on the other salinity tolerant ostracod species in other stratigraphical ranges may supply a simple method for palaeosalinity determinations.

Introduction

Cyprideis torosa is a cosmopolitan holeuryhaline species (for a list of occurrences, see Vesper, 1972a) which tolerates a wide range of water salt concentrations (occurring in fresh, brackish, hypersaline water). The relationship between the morphology of the carapace and the water salt concentration has been investigated for this species on numerous occasions (Vesper 1972 a-b). See Vesper (1972b), for an account of the phenomenon of node formation.
From investigations of populations of the species from different localities, with water of different salt concentrations, it became apparent that variations occur in the shape of the opening of the sieve-pore. Thus this study was begun to establish whether there is a causal relationship between the variability of the shape of the sieve-pore of the ostracode valve and the salinity.

The soft body of the ostracode is completely enclosed in a shell, the valves of which are penetrated on the exterior surface by numerous pore-canals (the number of which may vary). Canals which run vertically through the valve towards the surface are referred to as superficial pore canals ('normal pores'). These can be divided into two categories on the basis of the opening: 'single pores' and 'sieve-pores'.

Whereas Müller (1894) described the sieve pore as a central canal, carrying a bristle and surrounded by 'dead end' canals ('blind canals'), other authors (Sandberg & Hay 1967, Sandberg & Plusquellec 1969, Omatsola 1970, Puri 1974) have established that the 'blind canals' of Müller similarly open to the exterior.

Puri & Dickau (1969) set up a classification of the 'normal pores' according to types. Based on this, the type of sieve pore which occurs in *Cyprideis torosa* can be described in the following way: a sieve plate with a well defined central or subcentral pore, within which is a bristle.

Material

The recent and fossil material used in this investigation originated from Northern Germany, Denmark (Fig. 1) and Israel (Fig. 2).

Sampling localities

Recent material:

I. Northern Germany: Schleswig-Holstein* (Fig. 1)
A. Schlei: a relatively narrow (width: 500–800 m), shallow (average depth: 4–5 m) water body, extends approximately 40 km inland from the Ostsee (Baltic Sea). Salt concentrations decrease continuously inland. Average values range from 0.7 to 15‰.
1. Kleine Breite: values for the salt concentration range from 0.7 to 1.8‰, depth 80 cm (samples taken in June, 1969).
2. Missunde: values for the salt concentration range from 6 to 8‰, depth 30 cm (samples taken in June, 1970).
3. Gut Olpenitz: values for the salt concentration range from 13 to 15‰, depth 30–50 cm (samples taken in June, 1970).
B. Lakes located along the western shore of the Ostsee (Baltic Sea)
4. Kleiner Binnensee (Hohwachter Bucht): a lake located along the shore with salt concentrations ranging from 0.2 to 2.8‰, depth 30 cm (samples taken in June, 1969).
5. Sehlendorfer Binnensee (inland lake): a shallow standing water body connected by a canal with the Ostsee (Baltic Sea); fluctuations in salt concentration (ranging from 0.3–16‰) depend on the season and possibilities of connection with the open part of the Baltic Sea (Ostsee); concentrations usually range from 8 to 16‰, depth 60 cm (samples taken in August, 1969).
C. Regions containing standing water bodies located near the shore of the Baltic Sea (Ostsee).
6. Bottsand (Kieler Förde): an area of dunes and fields, located along the coast, with shallow standing water bodies (maximum depth 1 m), average salt concentrations ranging from 15 to 17‰ (samples taken in October, 1969).
D. Regions along the North Sea coast.
7. Neufelder Koog: a ditch situated behind the dike; maximum depth 75 cm, salt concentration ranging from 0.4–4.2‰ (samples taken in June, 1969).

II. Israel (Fig. 2)
8. Lake Kinnereth (also known as the Sea of Galilea or Lake Tiberias): area 170 km², maximum depth 42 m; a part of a rift system, bordered in the east and west

---

* Detailed description of sampling sites given by Vesper (1972a).

Fig. 1. Map showing localities in Northern Germany and Southern Denmark.

by steep mountains. The lake is situated approximately 212 m below sea level. Salt concentrations range from 0.3 to 6‰, (see also Por 1968, Bentor 1961, Schattner 1973); samples collected in March 1969 at Oholo- (Bet-Yerah), from a depth of 1.5 m (see Lerner-Seggev 1968).

9. Sabkhat el Bardawil (also known as Bardawil Lagoon): area of approximately 600 km^2, separated from the Mediterranean by a long sand barrier, with a maximum depth of approximately 2.5 m; salt concentrations lie between 50 and 80‰ (Levy 1971, Por 1972, Levy 1974, Ehrlich 1975) (Values almost twice those of normal seawater); this is due to the intensive evaporation and solution of bottom sediments containing gypsum, under the arid subtropical conditions (Levy 1971). The sample collected in 1969 from the inner lagoon corresponds with sample number 11 in Ehrlich (1975, text Fig. 1).

Almost all ostracods originated from muddy sand sediments.

Sampling localities in the Pleistocene (see Fig. 1 and 2)

I. Northern Germany – Denmark

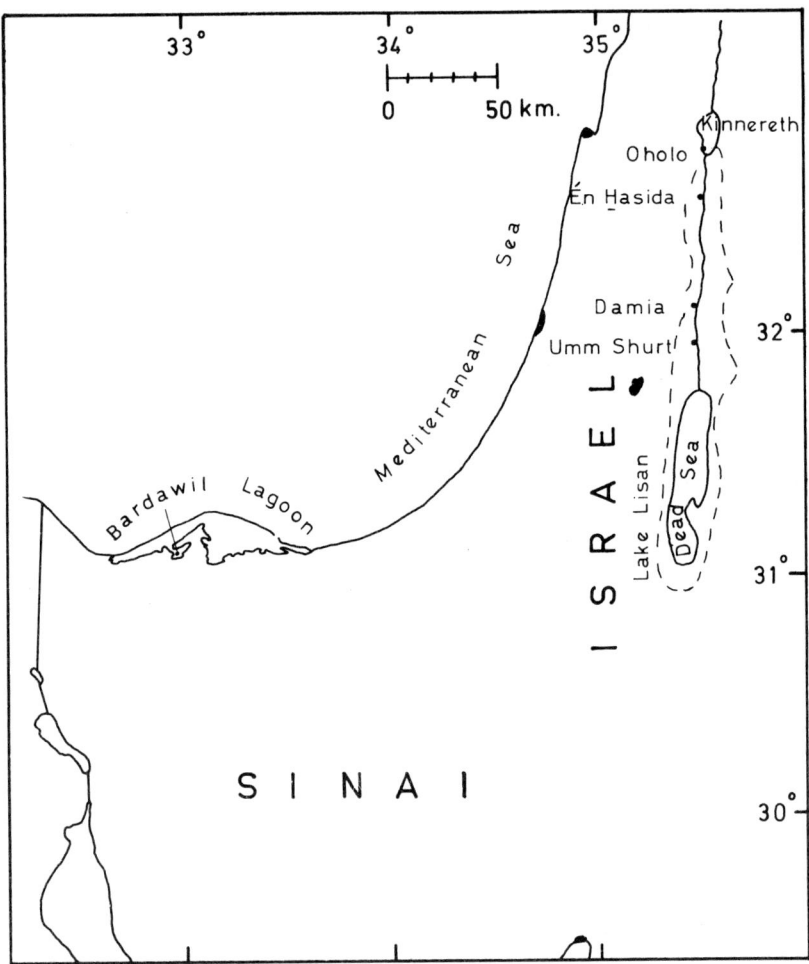

Fig. 2. Map showing localities in Israel: Broken outline: Lake Lisan as the fossil predecessor of the Dead Sea.

10. Clay pit Muldsberg (Schleswig-Holstein): Holstein-Interglacial (approximately 160,000–200,000 BP) the sample containing fossils (June 1975) corresponds approximately with sample D-9 from Woszidlo (1962, Table 2).

11. Broager (Denmark): steep cliff 4.5 km E.S.E. of Broager, near Steensigmoos (compare with Gottsche 1904, Lafrenz 1963). Eem-Interglacial (approximately 70,000–120,000? BP), the sample originates from the higher part of the 'tapes Sandes' (compare with Gottsche 1904, p. 181, 'Series e').

II. Israel
Laminated Member (Warwit) (Begin, Ehrlich & Nathan, 1974, Figs. 3, 6, 7) of the Lisan Formation (approximately 18,000–60,000 BP, after Neev & Emery 1968).

12. En Hasida sample: Geological survey of Israel (JB 2210 = T-4593). 'Diatomite Facies' (Begin, Ehrlich & Nathan 1974).
13. Near Damia, sample number JB-1704 = T-3888. 'Aragonite Facies', (Begin, Ehrlich & Nathan 1974); clayey sandstone.
14. Near Umm Shurt; sample number JB-1653=T-3838. 'Aragonite Facies' bordering the 'Gypsum Facies' (Begin, Ehrlich & Nathan, 1974). Clayey sandstone.
Remarks: The three samples 12-14 may be correlated with one another and accord to the same horizon. They differ somewhat, however, in the facies (refer to details of the respective facies under 12-14).

Methods of investigation

In each sample, 10-40 random specimens of the largest adult valves (both male and female) were investigated. To determine the form of the sieve pore, photographs of all but the most outlying pores, were taken using the R.E.M. (Plate 1).

After embedding in a glycerin-alcohol mixture, all pores on a valve (except the most outlying pores) were investigated using a light microscope (transmitted light) and diagrammed (drawing miror, magnification ×1000). The lengths and breadths of the pores were measured. On each valve, three different sieve pore shapes were recognized:
1. Sieve pore 'round' (= round, or almost round in form with a L/B relationship of $< 1,5$)
2. Sieve pore 'oblong' (= oblong, with a L/B relationship of $> 1,5$)
3. Sieve pore 'irregular' (= often almost triangular, y-shaped, or heart shaped)

For each locality, the percentage distribution of sieve pore types on the valves was determined; diagrammatic representations of such are presented in Fig. 3. Values of the maximum salinity recorded at each recent locality are also represented. Values for the palaeosalinity of a fossil sample can be estimated by comparing the percentage composition of sieve pore types recorded, with those determined in recent samples of known salinity (Fig. 4).

Summary of the results of investigations on recent material
(Fig. 3, Fig. 1-2)

90% of the sieve pores on the valves of the specimens from lake Kinnereth (salt concentration ranging from 0.3 to 0.6‰, freshwater) are round sieve pores, 5% 'oblong', and 5% are 'irregular'.

In Neufelder Koog (0.4 to 4.2‰ salt concentration), 55% of the sieve pores are of the round type, 32% oblong, 13% irregularly shaped.
In the Kleinen Breite (0.7 to 1.8‰ salt concentration), 65% of the sieve pores on the valve are round, 25% are oblong, 10% are irregularly shaped.
In the Kleinen Binnensee (0.2 to 2.8‰ salt concentration), a similar distribution was observed: 63% of the sieve pores are round, approximately 18.5% oblong, 18.5% irregularly shaped.
In Missunde (63 to 8‰ salt concentration), only approximately 50% of the

Plate 1. Variation of the sieve pores of *Cyprideis torosa* (line drawn near 1, indicating scale = 10 μ applies also to 3-9; scale line of 10 μ under 2 applies to 2 only.
1-3 – Sieve pores 'round'. 1) ♀, L. Neufelder Koog; 2) ♀, R. Lake Kinnereth (Oholo); 3) ♂, L. Gut Olpenitz.
4-6 – Sieve pores 'oblong' 4) ♂ R. Kleiner Binnensee; 5) ♀, R. Bottsand; 6) ♂, R. Bottsand.
7-9 – Sieve pores 'irregular'. 7) ♂, L. Bardawil; 8) ♂, L. Bardawil; 9) ♂, L. Kleiner Binnensee.
L = left valve; R = right valve.

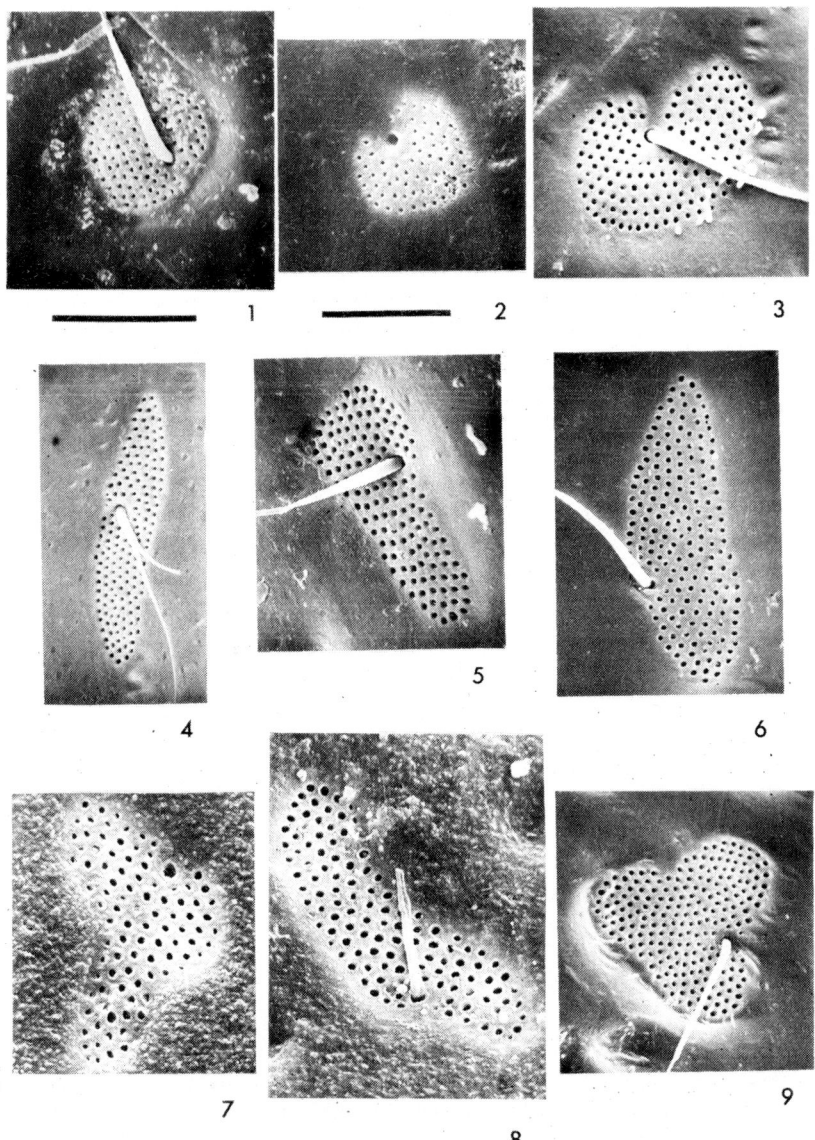

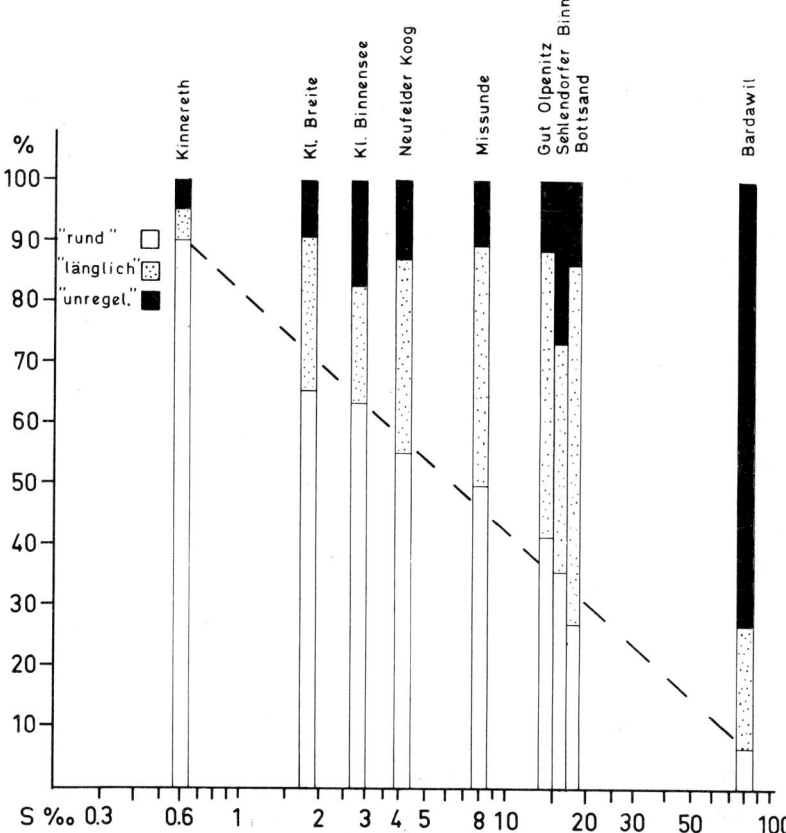

Fig. 3. Distribution of sievepore-types in % in recent populations of *Cyprideis torosa* at sites with different salinities (‰ in log.). White='round', dotted='oblong', black='irregular'.

sieve pores on a carapace are round, almost 40% are oblong (elongated), 10% are irregularly formed.

The valves of specimens from Gut Olpenitz (13 to 15‰ salt concentration) have only 41% of the pores rounded, 47% of the pores oblong, and only approximately 10% irregularly shaped.

In Sehlendorfer Binnensee (8 to 16‰ salt concentration), the sieve pores are distributed almost evenly between the three types.

Valves of the ostracodes from Bottand (approximately 15 to 17‰ salt concentration) possess 25% round sieve pores, approximately 60% oblong and 15% irregularly shaped.

In Sabkhat el Bardawil (inner lagoon), (50 to 80‰ salt concentration), the pore shape frequency distribution is approximately: 6% of the round form, 20% oblong, and 74% of irregular shape.

A comparison of all recent material investigated established the following points:

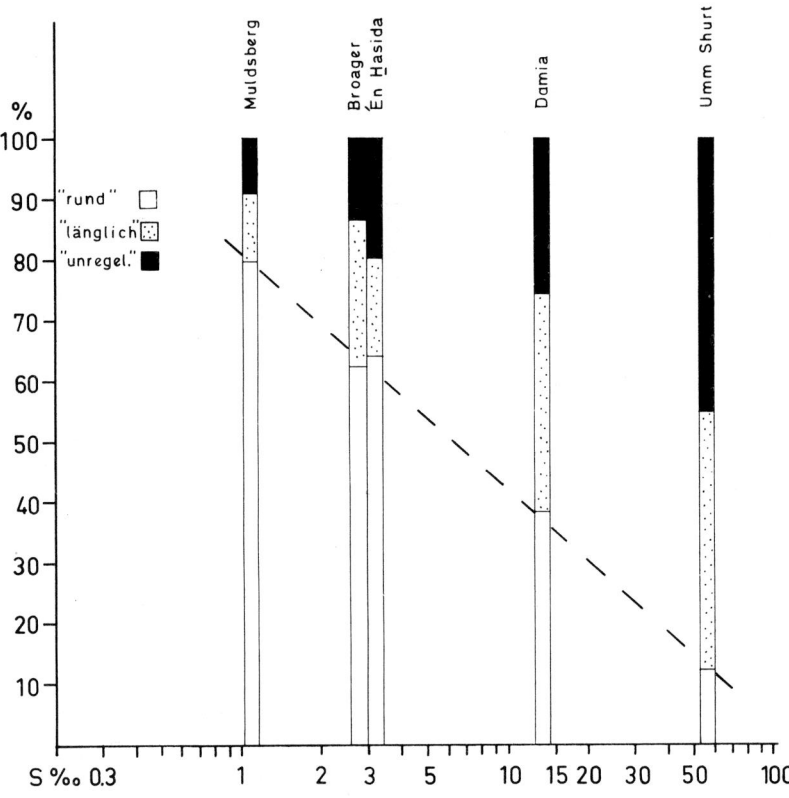

Fig. 4. Distribution of sievepore-types in % in fossil populations of *Cyprideis torosa* at different sampling localities. The palaeosalinities are derived from a comparison of the component percentages/salinities in Fig. 3. (Legend see Fig. 3).

1. An apparent dependency of the form of the sieve pore on the salt concentration exists.
2. At low salt concentration, round sieve pores are encountered at high frequencies, which decrease with increasing salinity; a simultaneous increase in the proportion of oblong pores takes place.
3. At hypersalinities (Sabkhat el Baradwil) the percentage of round sieve pores is relatively small; irregularly shaped sieve pores predominate.

Results on fossil material (Fig. 4, Fig. 1–2)

Approximately 80% of the sieve pores of specimens of *Cyprideis torosa* from the clay mine at Muldsberg are round, only approximately 10% are oblong and 10% are irregular in shape.

In the samples from Broager, more than 60% of the sieve pores are of the round type, approximately 25% are oblong and approximately 15% are irregular in shape.

The percentage distribution of the individual sieve pore types in samples from En Hasida reflects that recorded for the Broager.

In samples from Damia, approximately 40% of all sieve pores belong to the round category, 36% to the oblong, 24% to the irregular.

From the locality Umm Shurt only 12% of the sieve pores are round, approximately 44% are oblong, and 44% are irregular in shape.

As results are available for frequency distributions of sieve pore shapes of both recent (at known salinities) and fossil material, a comparison between the two should be made. Accordingly, the frequency distribution of pore shapes in fossil samples from En Hasida approximates that recorded in recent material from Kleine Breite, respectively Kleiner Binnensee, and Neufelder Koog, with a salinity value of approximately 3‰, slightly brackish water is thus indicated.

Using the same comparative criterion, an estimate value for the palaeosalinity of samples from Damia would be approximately 14 to 15‰, indicative of brackish water of medium salinity.

No recent material is available for a similar comparison of frequencies recorded for Umm Shurt. However, it can be stated on the basis of the proportion of round sieve pores that the palaeosalinity of this locality must have been in the hypersaline range (approximately 50–60‰, interpolated according to Bardawil values). The palaeosalinity estimated on the basis of sieve pore types corresponds with the results of Begin, Ehrlich & Nathan (1974, pp. 21–26).

The investigations on sediments, the occurrence of diatoms and the presence of freshwater gastropods in Lake Lisan, led Begin, Ehrlich & Nathan (1974) to assume that a gradient in salt concentration existed, increasing from the north towards the south; they designated the northern part of the lake as limnic. A consideration of the distribution of the sieve pore types shows that a decrease in numbers of round sieve pores takes place from the north to the south (En Hasida, Damia, Umm Shurt), with a simultaneous increase in the proportions of the 'oblong' and 'irregular' shaped sieve pores. This change is in accordance with the increase in the estimated salinities from north to south: En Hasida = fresh to slightly brackish water; Damia = mesohaline water; Umm Shurt = hypersaline water.

Salinity values estimated for Muldsberg material compare well with those for recent material from Lake Kinnereth. Accordingly, the Muldsberg Material (upper Holstein Interglacial) can be interpreted as originating in a freshwater environment. An estimate value for the palaeosalinity would be between 0.5‰ and 1‰.

Investigations of the Microfauna (Foraminifera and Ostracoda) from a number of ditches from the Holstein Interglacial in Northern Germany were carried out by Lange (1962). Samples from Wacken (approximately 2.5 km from Muldsberg) indicated an increasingly brackish environment (Lange 1962, pp. 229–230).

From these investigations of the upper part of the Holstein Interglacial, Woszidlo (1962) concluded that the distribution of the Foraminifera and the Ostracoda within the profile indicated that the sea became brackish (1962, p. 89). In connection with these investigations, Brockmann (1936) reports on his studies of diatoms in other localities that: 'all three kinds of deposits, freshwater, brackish-water and marine, occur'.

According to Grahle (1936, pp. 99–100), the presence of freshwater molluscs and numerous Chara-Oogonia similarly indicates a freshwater environment. His report on these investigations is concluded with the following words: 'In a small, separate basin, freshwater sediment deposits were covered with water' (seawater s. Ref.) (Grahle 1936, pp. 105).

Buch (1955) concluded from his investigations of the Foraminifera of the Holstein Interglacial from Denmark that the sea in question had become brackish during its history. All of the results quoted are in agreement with our prediction of slightly brackish to fresh water conditions in the upper part of the Holstein Interglacial in Muldsberg.

A comparison of material from Broager with that from the Kleinen Binnensee (Fig. 3) produces estimates of 2 to 3‰ for palaeosalinity values. The results obtained from other localities support our views. The Diatom contents at the sandy Eem deposits near Lübeck (close to the Kleiner Binnensee) (Heck & Brockmann 1950), indicate freshwater conditions. Dittmer (1951) described the Eem of the Treenetal and noted a number of freshwater molluscs accompanied by marine species. König (1953) investigated the diatoms from the Treene area, and also found both freshwater and brackishwater species.

In investigations on diatoms and pollen in the Northsea canal area, Von der Brelie (1951) established that '... species favouring high salt concentrations decrease rapidly in numbers upwards the section, whereas brackish water species increase to a corresponding degree'.

König (1954) found both salt and freshwater diatoms in the Eem Bohrung Klixbüller Koog, and the author concluded that 'freshwater was located nearby'.

Investigations on Foraminiferas out of the Eem ('Tapes Sand') from Steensigmoos (= Broager), Lafrenz (1963) demonstrated that the same fauna occurred there, as was recorded in the Bohrung Amersfoort I (= locus typicus of the Eem) by Van Voorthuysen' (1958). Van Voorthuysen (1958) estimated values for the salinity of less than 5‰ for the upper part of the Eem.

All these studies confirm the palaeosalinity value 2 to 3‰ for the upper Eem from Broager established by the present investigation.

Zusammenfassung

Es wurde an rezentem Material von *Cyprideis torosa* (Jones, 1850), aus Norddeutschland und Israel eine Beziehung zwischen der Siebporen-Form und dem Faktor Salzgehalt erkannt. Drei Siebporen-Typen wurden beobachtet: 'runde', 'längliche' und 'unregelmäßige'. Bei niedrigem Salzgehalt erscheint der prozentuale Anteil an 'runden' Siebporen hoch, er verringert sich mit zunehmender Salinität. Bei entsprechend höherer Salinität steigt der prozentuale Anteil an 'länglichen' und 'unregelmäßigen' Siebporen-Typen an. In hypersalinen Gewässern dominiert die 'unregelmäßige' Siebporen-Form.

Die an rezentem Material gewonnenen Ergebnisse lassen sich auf pleistozänes Material von *Cyprideis torosa* aus Norddeutschland (Holstein-Interglazial), Dänemark (Eem-Interglazial) und Israel (Lisan-Formation) übertragen. Damit wird es für die genannten Fundschichten möglich, die Paläosalinität abzuschätzen.

Auf andere Weise gewonnene Vorstellungen (in der Literatur) über Paläosalinitäten stimmen mit unseren Ergebnissen gut überein.

Die Methode bietet eine Möglichkeit für Paläosalinitäts-Bestimmungen im Pleistozän. Es ist wahrscheinlich, daß sie auf ältere Formationen und auf andere Ostracoden-Gruppen übertragen werden kann.

References

Begin, Z. B., Ehrlich, A. & Nathan, Y. 1974. Lake Lisan — the Pleistocene precursor of the Dead Sea. Geol. Surv. Israel Bull. 63: 1–30.

Bentor, L. K. 1961. Some geochemical aspects of the Dead Sea and the question of its age. Geochim. Cosmochim. Acta 25:239–260.

Brelie, G. von der. 1951. Die junginterglazialen Ablagerungen im Gebiet des Nord-Ostsee-Kanals. Schr. naturw. Ver. Schlesw.-Holst. 25: 100–107.

Brockmann, Chr. von. 1936. Die Diatomeen im marinen Altinterglazial von Westholstein und Hamburg. In: Grahle, 1936, pp. 97–99.

Buch, A. 1955. De marine interglaciale lag ved Inder Bergum — Foraminifernfauna og stratigrafi. Medd. Dansk. Geol. Foren. 12 (6): 593–652.

Dittmer, E. 1951. Das Eem des Treenetals. Schr. naturw. Ver. Schl.-Holst. 25: 91–99.

Dittmer, E. 1954. Eine eemzeitliche Austernbank bei Leck und die Entwicklung des Ober-Eems. Meyniana, Veröff. geol. Inst. Kiel 2: 70–79.

Ehrlich, A. 1975. Diatoms from the surface sediments of the Bardawil Lagoon, Northern Sinai. 3rd Symp. on Recent and Fossil Marine Diatoms, Kiel. R. Simonsen (ed.), pp. 253–282.

Gottsche, C. 1904. Über den Tapes-Sand von Steensigmoos. Z. Dt. geol. Ges. 56 Mber.: 181–184.

Grahle, H. O. 1936. Die Ablagerungen der Holstein-See (marines Interglazial I), ihre Verbreitung, Fossilführung und Schichtenfolge in Schleswig-Holstein. Abh.Preuss. Geol. Landesanst. N.F. 172: 110 pp.

Heck, H. L. & Brockmann, C. 1950. Eem-Ablagerungen bei Lübeck. Schr. naturw. Ver. Schl.-Holst. 24 (2): 80–86.

König, D. 1953. Diatomeen aus dem Eem des Treenetals. Schr. naturw. Ver. Schl.-Holst. 26 (2) 124–132.

König, D. 1954. Diatomeen in der Eembohrung Klixbüller Koog. In: Dittmer (ed.), Kiel, pp. 79–80.

Lafrenz, von H. R. 1963. Foraminiferen aus dem marinen Riss-Würm-Interglazial (Eem) in Schleswig-Holstein. Meyniana 13: 10–46.

Lange, W. 1962. Die Mikrofauna einiger Störmeer-Absätze (I. Interglazial) Schleswig-Holsteins. N.Jb. Geol. Paläont. Abh. 115: 222–292.

Lerner-Seggev, R. 1968. The Fauna of Ostracoda in Lake Tiberias. Israel J. Zool. 17: 117–143.

Levy, Y. 1971. Anomalies of Ca^{2+} and So_4^{2-} in the Bardawil Lagoon, Northern Sinai. Limnol. Ocean. 16 (6): 983–987.

Levy, Y. 1974. Sedimentary reflection of depositional environment in the Bardawil Lagoon, Northern Sinai. J. Sedim. Petrol. 14: 219–227.

Müller, G. W. 1894. Die Ostracoden des Golfes von Neapel und der angrenzenden Meeresteile. Fauna und Flora Golf Neapel, Monogr. 21, 404 pp.

Neev, D. & Emery, K.O. 1967. The Dead Sea, depositional processes and environments of evaporites. Geol. Survey Israel Bull. 41: 147 pp.

Omatsola, M. E. 1970. On structure and morphologic variation of normal pore system in recent Cytherid Ostracoda (Crustacea). Acta Zool. 51: 115–124.

Por, F. D. 1968. The invertebrate zoobenthos of Lake Tiberias qualitative aspects. Israel J. Zool. 17: 51–79.

Por, F. D. 1972. Hydrobiological notes on the high-salinity waters of the Sinai Peninsula. Marine Biol. 14: 111–119.

Puri, H. S. 1974. Normal pores and the phylogeny of Ostracoda. Geosc. and Man 6: 137–151.
Puri, H. S. & Dickau, B. E. 1969. Use of normal pores in taxonomy of Ostracoda. Gulf Coast Assoc. Geol. Soc., Trans. 19: 353–367.
Sandberg, P. A. & Hay, W. W. 1967. Study of microfossils by means of the scanning electron microscope. J. Paleontol. 41: 999–1001.
Sandberg, P. A. & Plusquellec, P. K. 1967. Structure and polymorphism of normal pores in Cytheracean Ostracoda (Crustacea). J. Paleontol. 43: 517–521.
Schattner, J. 1973. Physiography in Israel geography. Israel Pocket Library, pp. 24–93. Keter Publ. House, Ltd., Jerusalem. Comp. from Enc. Judaica.
Sohn, I. G. 1965. Late Quaternary Ostracodes from the Southern part of the Dead Sea, Israel. Ocean Sci. & Ocean Engng 1: 82–94.
Voorthuysen, J. H. van. 1958. Foraminiferen aus dem Eemien (Riss-Würm-Interglazial) in der Bohrung Amersfoort I (Locus-typicus). Meded. geol. Sticht. N.S. 11: 27–39.
Vesper, B. 1972a. Zur Morphologie und Ökologie von Cyprideis torosa (Jones 1850) (Crustacea, Ostracoda, Cytheridae) unter besonderer Berücksichtigung seiner Biometrie. Mitt. Hamburg Zool. Mus. Inst. 68: 21–77.
Vesper, B. 1972b. Zum Problem der Buckelbildung bei Cyprideis torosa (Jones, 1850) (Crustacea, Ostracoda, Cytheridae). Mitt. Hamburg Zool. Mus. Inst. 68: 79–94.
Vesper, B. 1975. To the problem of noding on Cyprideis torosa (Jones, 1850). In: Swain, F. M. (ed.), Biology and Paleobiology of Ostracoda, A Symposium — Univ. of Delaware. Bull. Amer. Paleontol. 65: 205–216.
Woszidlo, H. 1962. Foraminiferen und Ostrakoden aus dem marinen Elster-Saale-Interglazial in Schleswig-Holstein. Meyniana, Veröff. geol. Inst. Univ. Kiel 12: 65–96.

Authors' addresses:
A. Rosenfeld, Geologisch-Paläontol. Institut und Museum der Universität Kiel, Olshausenstrasse 40/60, 23 Kiel, B.R.D.
Present address: Geological Survey, Paleontological Div., Jerusalem, Israel.
B. Vesper, Zoologisches Institut und Zoologisches Museum Martin-Luther-King-Platz 3, 2 Hamburg 13, B.R.D.

Discussion

Carbonnel: Did the authors try an application to some Messinian species belonging to the genus *Cyprideis*?
Rosenfeld & Vesper: No, but we would like to if material would be available.
Van Harten: In the diagrams you showed us, no confidence intervals were given in connection with the percentages. Since confidence increases with an increasing number of observations, I should like to know the number of observations on which the diagrams are based.
Rosenfeld & Vesper: Every recent and fossil sample is based on 10–40 valves, from each valve 30–50 pores were counted, drawn and mesured. The percentages in the diagrams represent the mean values. The standard deviation is about 5%.
Sohn: Did you use males only, females only or both sexes? Only right or left valves, or both?
Rosenfeld & Vesper: We used both males and females, right and left valves. We did not find any essential difference in the sieve-pores between the sexes or in respect to the valves from the same sample.
Oertli: How many persons counted the different types of pores? I imagine that it is not always evident whether a pore is still round or already elongate/irregular.

Rosenfeld & Vesper: All the pores were drawn by both of us. The relation between length and height was measured in order to avoid difficulties in classifying the different groups: 'round' with l/h — less than 1.5, 'elongate' with l/h — more than 1.5 and 'irregular'. We admit that in some cases it was not easy to differentiate.

Oertli: Counting on photographs, wasn't there the danger of misidentifying pores near the borders, i.e., all those that were seen vertically? In this case all the pictures were not taken at exactly the same angle; also, errors may arise when dealing with more or less inflated populations.

Rosenfeld & Vesper: We did not count from photographs, we used a light microscope and we did not count pores on the margins or on the inflated zones of the valves. We tried to minimize the error and checked only pores perpendicular to the objective, e.g., from the central area of the valves.

Sixth Intern. Ostracod Symposium, Saalfelden

PRECOCIOUS SEXUAL DIMORPHISM IN FOSSIL AND RECENT OSTRACODA

R.C. WHATLEY & J.M. STEPHENS

Abstract

Despite the existence in the literature of various statements which categorically deny the existence of precocious sexual dimorphism in Ostracoda, it is herein demonstrated, from the works of a number of authors, that pre-adult sexual dimorphs are commonly encountered in the ontogeny of certain myodocopids and palaeocopids. This phenomenon is also demonstrated in six species of Middle Jurassic podocopids. The relationship between precocious sexual dimorphism and seasonal size variation is discussed.

Introduction

Precocious sexual dimorphism in the Ostracoda is a subject which has not had the benefit of serious study and has, rather in the same way as the similar problem of post-maturation moulting, been tacitly assumed to exist or not exist according to the vagaries of interest of the authors concerned. The problem is also aggravated by the fact that some authors refer to soft part anatomy and others to carapace morphology when speaking of sexual precosity.

What seems to be a generally accepted view is that precocious sexual dimorphism is not a feature in ontogeny of Ostracoda, especially the Podocopida, since one would not expect animals, which only achieve full maturity in terms of their reproductive organs at the maturation moult, to exhibit such dimorphism.

This viewpoint is given by a number of authors and proba bly most explicitly by Van Morkhoven (1962, vol. I, p. 94):

'Since the latter (sexual organs) do not develop fully before the ostracod reaches maturity, it may be easily understood that sexual dimorphism of the calcareous parts does not become clearly apparent until in the last (eight) larval stage, when the sexual organs are beginning to form, and that it reaches maximum development only in the adult stage, when the external sexual organs of the male become fully chitinized and the ovaries of the female begin to fill with ripe eggs.'

Other authors have, however, clearly demonstrated the existence of sexual dimorphism in the ontogeny of various species of Ostracoda, although they have usually referred it to some other phenomenon. Kornicker (1969, 1970) has, for example, referred to what he clearly understands to be precocious sexual dimorphism, as 'intraspecific variation' in various Recent myodocopid species. He cites the genus *Spinacopia* Kornicker 1969 as being one in which 'intraspecific variation' appears commonplace and (1969, p. 19) recognizes a female IVth instar of *S. variabilis* Kornicker as such by its possession of a carapace similar in shape to that of the adult female of the species 'but not globose in posterodorsal region.'

He also recognizes (p. 20) a male instar IV of the same species which although similar to the adult female in its carapace morphology, possesses a 'very weakly developed' copulatory organ. He is unable to determine the sex of an instart II of the same species (p. 19). In the same paper Kornicker (pp. 31–33) refers to male and female instars III and IV of *Spinacopia sandersdi* Kornicker and in his discussion of the ontogeny of the species, clearly indicates by such phrases as 'juveniles of both sexes' that he can demonstrate precocious sexual dimorphism in the species.

Kornicker in 1970 describes a 'juvenile male' (p. 24) of *Spinacopia antarctica* Kornicker based on carapace mophology and appendages and erects *S. octo* on the basis of juvenile males (p. 24). Again, (p. 32) Kornicker describes a 'juvenile female' instar II of *S. torus* Kornicker and erects the species *Euphilomedes rhabdion* on the basis of a female which 'may not be mature'. In the same paper (pp. 36–40), Kornicker describes two nomen nudum species of *Synasterope* Poulsen on 'juvenile' males and females.

In subsequent papers (1970a, 1974a, 1974b, 1975, 1975a, 1976) Kornicker refers to 'juvenile' or 'pre-adult' etc. males and females in various species of the following myodocopid genera, *Paravargula* Poulsen; *Cylindroleberis* Brady; *Parasterope* Poulsen; *Synasterope* Poulsen; *Prionotoleberis* Kornicker; *Polyleberis* Kornicker; *Cycloleberis* Skogsberg; *Skogsbergia* Poulsen; *Pseudophilomedes* Müller; *Sarsiella* Norman; *Rutiderma* Brady and Norman; *Bathyconchoecia* Deevey; *Bathyvargula* Poulsen; *Cypridinodes* Brady; *Doloria* Skogsberg; *Hadracypridina* Poulsen; *Vargula* Skogsberg; *Philomedes* Lilljeborg; *Euphilomedes* Poulsen; *Schleroconcha* Skogsberg; *Anathron* Kornicker; *Igene* Kornicker; *Archasterope* Poulsen; *Skogsbergiella* Kornicker; *Empoulsenia* Kornicker; *Bathleberis* Kornicker; *Asteropteron* Skogsberg; *Spinacopia* Kornicker; *Cymbicopia* Kornicker; *Scleraner* Kornicker; *Paradoloria* Poulsen; *Diasterope* Pouilsen. Precocious sexual dimorphism in the 32 genera mentioned above, is recognized by Kornicker on the basis of both hard and or soft part morphology. Since a number of other authors, notably Poulsen (1962, 1965, 1969), have recognized what they refer to as 'immature males and females' etc., it seems clearly established that precocious sexual dimorphism is a common ontogenetic feature in both major groups of the Myodocopina. In some members of this suborder, the fact that proto-males and proto-females are recognizably distinct morphologically, quite early on in ontogeny, may be due to the sexes being ecologically disjunct. Some species of *Philomedes*, for example, have benthic females and pelagic males and presumably the 'proto-females' and 'proto-males' undergo their development in these differing environments and it is, therefore, reasonable to suspect that their morphologies, both in terms of appendages and carapace, would reflect their differing ecological requirements.

It would seem, therefore, that the general view outlined by Van Morkhoven above is, at least in respect of the Myodocopina, erroneous.

Precocious sexual dimorphism has also been reported in the Paleocopida by a number of authors, notably Martinsson (1956), Jaanusson (1956), and Guber (1971). Martinsson (pp. 10–12) discusses 'dimorphism in preadult instars', both cruminal, velar, and in terms of size, in a number of palaeocopid genera such as *Beyrichia* M'Coy; *Clavoflabella* Martinsson; *Euprimites* Hessland; and *Oepikella*

Thorslund. He concludes that this dimorphism is due to sexual precosity in juveniles rather than, as had been suggested by Spjednaes (1951), that penultimate instars with brood pouches (in *Beyrichia*), represented adults living in different and perhaps warmer environment. Martinsson suggests (p. 12) that in the case of some *Clavoflabella* species that ' — the mere existence of dimorphic characters in this subadult instar suggests that the subadult dolonate specimens were fertile'.

This represents a new departure since in none of the many cases cited above of precocious sexual dimorphism in the Myodocopina, did any of the authors suggest or imply that the instars were sexually mature or fertile.

Jaanusson ('1956) reviews the situation concerning pre-adult sexual dimorphism in palaeocopids in some detail.

Guber (1971) recognizes various dimorphic structures of the adults as 'incipient dimorphic structures' as early as the A-3 instar, in various species of *Tetradella* Ulrich, and in his text-figures 4 and 5, of instar diagrams of two species of the genus, clearly demonstrates precocious sexual dimorphism. This is recognized in terms of various 'ornamental' features and also size relationships. In this Table 1 (page 13) he demonstrates that the earliest 3 or 4 instars are sexually undifferentiated (or undifferentiable); that the early dimorphs A-3 or A-2 contain a large % of proto-males (technomorphs) which decline in importance relative to the heteromorphs (proto-females) towards the adult stage. It is thought valuable to reproduce this diagram since in so many respects it is similar to evidence which the authors will present below in respect of certain Jurassic podocopids.

Table 1. Sex ratios (percent technomorphs to percent heteromorphs) for the instars of *Tetradella scotti* Guber and *T. quadrilirata* (Hall and Whitfield). After Guber 1971

Instar	*T. scotti*	*T. quadilirata*
Adult	10/90	9/91
Adult-1	19/81	14/86
Adult-2	53/47	40/60
Adult-3	76/234	100/0
Adult-4	100/0	100/0
Adult-5	100/0	100/0
Adult-6	100/0	100/0

N.B. Guber (p. 6) defines technomorph as 'inferrred males and unsexable juveniles'.

From all the above it would seem that precocious sexual dimorphism is a well-known and documented feature in the ontogeny of not a few palaeocopid Ostracoda. Apart from the reference by Martinsson (1956) cited above to possible fertility of juveniles, there does not seem to be any other assertion on implication that this incipient dimorphism was accompanied by precocious sexual maturity.

Apart from the instances mentioned above for the Palaeocopida and the Myodocopina, the authors have not been able to encounter any reference in the literature to pre-adult dimorphism in other groups of Ostracoda. We both suspect this to occur, however, quite widely in both living and fossil podocopids.

Precocious sexual dimorphism in podocopid Ostracoda

One of the authors (J.M.S.), in a study of Bathonian Ostracoda from Oxfordshire, England, has encountered a number of podocopid species which exhibit considerable variation in size and shape, and in some cases ornamention, not only in the adult stage, but also in several of the preceeding moult stages. The following species exhibit this phenomenon:

Glyptocythere quembeliana (Jones) 1884
G. binodosa Stepephens n. sp.
G. oscillum (Jones & Sherborn) 1888
Lophocythere (L.) *scabra scabra* (Triebel) 1951
L. (L.) fulgurata (Jones & Sherborn) 1888
Glabellacythere dolabra (Jones & Sherborn) 1888

The last species belongs to the Cytherideidae and the remainder to the Progonocytheridae. In the case of both *G. guembeliana* and *G. binodosa* all specimens measured were from the same sample. Because of relatively sparse material, however, all the other species are represented on the accompanying diagrams by specimens from more than one sample. These samples were, however, all collected from sediments of Bathonian age from the Alpha Cement Works Quarry at Shipton-upon-Cherwell, Oxfordshire.

Specimens were recognized as being juvenile, not just on the basis of size but also in their possession of juvenile features in respect of such characters as the hinge, inner lamella, degree of calcification, etc.

Glyptocythere guembeliana (Jones) 1884

In the ontogeny of this species it is possible to distinguish the following forms: adult males and females, 'protomales' and 'protofemales' and sexually undifferentiated instars. The morphological differences between the three forms may be summarized as follows:

i) 'Protomales' are longer than 'protofemales' and of equal or slightly greater height; this size relationship persists into the adult stage, as is shown in Table 2.
ii) The greatest width of the carapace is more posterior in males and 'protomales' than in members of the female lineage.
iii) The postero-ventral border in the RV of the male lineage is more angular than in the female one where it is broadly rounded.
iv) All valves possess a mid-dorsal swelling which projects above the hinge margin. This, however, is more pronounced in the female lineage because the hinge margin is shorter than in males and 'protomales'.-
v) Members of the female lineage develop a ventro-lateral blade-like structure with a crenulate margin on the right valve. This structure never appears in the male lineage, although it is occasionally found on the right valves of the younger sexually undifferentiated instars.
vi) Sexually undifferentiated instars are much more acuminate posteriorly than members of either the male or the female lineages. They also lack the mid-dorsal swelling, and are much less tumid.

Table 2. Ratio of mean length: height dimensions for *G. guembeliana*

Growth stage	Sex	Male and 'Protomale'	Female and 'Protofemale'	Sexually Undifferentiated instars
Adult	LV	1.909 : 1	1.699 : 1	
	RV	2.102 : 1	1.883 : 1	
A–1	LV	1.892 : 1	1.602 : 1	
	RV	2.079 : 1	1.853 : 1	
A–2	LV	1.758 : 1	1.587 : 1	
	RV	2.050 : 1	1.823 : 1	
A–3	LV	1.823 : 1	1.605 : 1	1.716 : 1
	RV	2.092 : 1	1.838 : 1	–
A–4	LV	1.835 : 1	1.568 : 1	1.683 : 1
	RV	1.205 : 1	1.789 : 1	–
A–5	LV	1.721 : 1	1.541 : 1	1.671 : 1
	RV	–	–	1.796 : 1
	LV			1.623 : 1
	RV			1.711 : 1
A–7	LV			1.684 : 1
	RV			1.759 : 1
A–8	LV			1.640 : 1
	RV			–

Text-figures 1 and 2 are instar diagrams of *G. guembeliana* based on left and right valves respectively. The divisions between growth stages are admittedly arbitrarily drawn, but a clear distinction can be seen between the male and female lineages from the A–5 stage onwards.

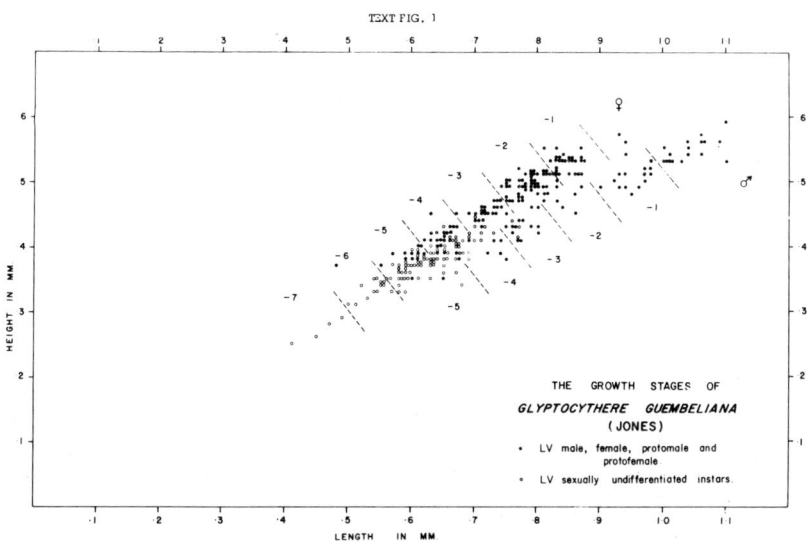

TEXT FIG. 1

THE GROWTH STAGES OF
GLYPTOCYTHERE GUEMBELIANA
(JONES)

• LV male, female, protomale and protofemale
∘ LV sexually undifferentiated instars

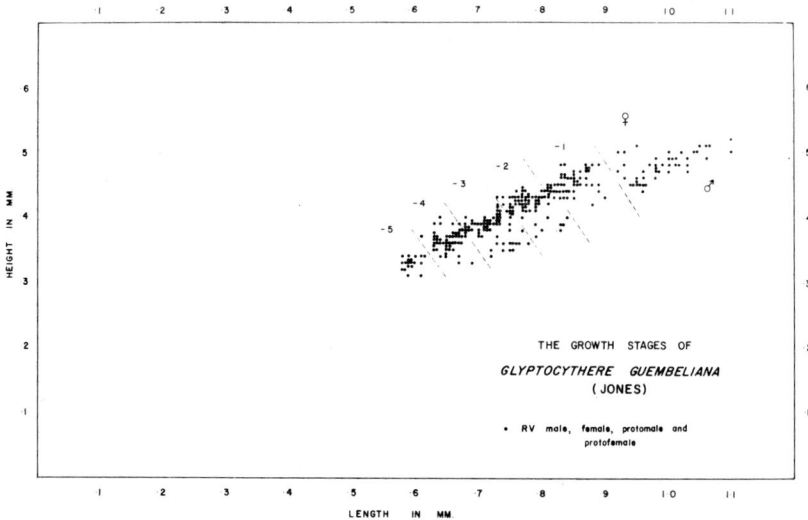

Table 3. The percentage of each form present at successive ontogenetic stages of *G. guembeliana*.

Sex	Instar	Adult	A–1	A–2	A–3	A–4	A–5	A–6	A–7
Male lineage		82.5	20.3	18.3	20.6	10.5	–	–	–
Female lineage		17.5	79.7	86.7	71.6	80.0	17.3	–	–
Unsexed instars					7.8	9.5	82.7	100.0	100.0

A notable feature of the polymorphism of this species is the fact that the disappearance of undifferentiated instars is not sudden and neither is the appearance of pre-adult dimorphs. At the A–5 stage 'protofemales' and undifferentiated instars coexist and in the A–3 and A–4 stages, all 3 varieties occur together. The undifferentiated forms, however, constitute a sharply declining percentage in each of these three stages and do not appear again after the A-3 stage whilst the pre-adult dimorphs become increasingly dominant, particularly the 'protofemales'. At the A–5 stage, 17.3% of instars are 'female' whilst the dominant 82.7% are undifferentiated. The female lineage has become dominant at the A–4 stage and remains so up to the time of the maturation moult when there is a sudden and inexplicable reversal which, in the adult stages leaves females outnumbered by males (♀ 17.5%; ♂ 82.5%).

Bate (1967, p. 50), in redescribing specimens of *G. guembelana* from the Jones collection, found a male dimorph of small size occuring with larger male dimorphs. He inferred that the size range of the adult males was very variable. The present authors consider that Bate in fact was dealing with a population containing adults and pre-adult dimorphs.

Table 4. The range and mean dimensions of the growth stages of *Glyptocythere guembeliana*. (in mm)

Instar		Sex	Male lineage length height	Female lineage length height	Sexually undifferentiated instars length height
Adult	LV	range mean	1.00–1.10 : 0.53–0.59 1.05 : 0.55 (14)	0.93–0.94 : 0.54–0.57 0.94 : 0.55 (3)	
	RV	range mean	0.98–1.10 : 0.47–0.52 1.03 : 0.49 (22)	0.87–0.95 : 0.45–0.51 0.904 : 0.48 (9)	
A–1	LV	range mean	0.90–1.01 : 0.48–0.53 0.961 : 0.508 (17)	0.81–0.87 : 0.51–0.55 0.841 : 0.525 (29)	
	RV	range mean	0.88–0.98 : 0.42–0.48 0.948 : 0.456 (22)	0.81–0.88 : 0.43–0.47 0.838 : 0.451 (46)	
A–2	LV	range mean	0.80–0.87 : 0.46–0.49 0.835 : 0.475 (6)	0.75–0.81 : 0.47–0.52 0.784 : 0.494 (43)	
	RV	range mean	0.79–0.84 : 0.38–0.43 0.818 : 0.399 (9)	0.74–0.81 : 0.41–0.44 0.771 : 0.423 (48)	
A–3	LV	range mean	0.77–0.80 : 0.41–0.46 0.791 : 0.434 (10)	0.71–0.77 : 0.41–0.47 0.727 : 0.453 (31)	0.68–0.77 : 0.41–0.45 0.731 : 0.426 (8)
	RV	range mean	0.74–0.79 : 0.35–0.38 0.753 : 0.360 (11)	0.67–0.75 : 0.37–0.42 0.724 : 0.394 (2)	–
A–4	LV	range mean	0.59–0.76 : 0.35–0.43 0.691 : 0.382 (11)	0.62–0.69 : 0.39–0.43 0.649 : 0.414 (20)	0.65–0.70 : 0.39–0.43 0.672 : 0.407 (9)
	RV	range mean	0.67–0.72 : 0.33–0.34 0.687 : 0.335 (4)	0.63–0.71 : 0.35–0.39 0.662 : 0.37 (56)	–
A–5	LV	range mean	–	0.55–0.61 : 0.37–0.40 0.595 : 0.386 (12)	0.58–0.69 : 0.35–0.40 0.640 : 0.383 (55)
	RV	range mean	–	–	0.57–0.65 : 0.30–0.35 0.598 : 0.333 (26)
A–6	LV	range mean			0.52–0.58 : 0.31–0.35 0.555 : 0.342 (19)
	RV	range mean			0.50–0.52 : 0.29–0.31 0.51 : 0.298 (5)

Instar		Sex	Male lineage length	height	Female lineage length	height	Sexually undifferentiated instars length	height
A-7	LV	range mean					0.45–0.51 : 0.26–0.31 0.48 : 0.285	(4)
	RV	range mean					0.47–0.48 : 0.27 0.475 : 0.27	(2)
A-8	LV	range mean					0.41 : 0.25	(1)
	RV	range mean						

Glyptocythere binodosa Stephens n. sp.*

This species exhibits in its ontogeny the same type of polymorphism as shown by the previous species. The major morphological differences can be summarized as follows:

i) Males are longer and slightly less high than females and this relationship is also seen in pre-adult dimorphs. This is shown in Table 5 below:

Table 5. Ratio of mean lenght: height dimensions for *G. binodosa*

Instar	Sex	Male and 'Protomale'	Female and 'Protofemale'	Undifferentiated instars
Adult	LV	1.829 : 1	1.603 : 1	
	RV	2.059 : 1	1.813 : 1	
A-1	LV	1.797 : 1	1.586 : 1	
	RV	2.074 : 1	1.799 : 1	
A-2	LV	1.796 : 1	1.548 : 1	
	RV	2.136 : 1	1.977 : 1	
A-3	LV	1.798 : 1	1.553 : 1	1.646 : 1
	RV	2.035 : 1	1.823 : 1	1.803 : 1
A-4	LV		1.602 : 1	1.678 : 1
	RV		1.804 : 1	1.751 : 1
A-5	LV			1.634 : 1
	RV			1.726 : 1
A-6	LV			1.642 : 1
	RV			1.757 : 1
A-7	LV			1.605 : 1
	RV			1.537 : 1

ii) The greatest width is more posterior in the male than in the female lineage.

* *Glyptocythere binodosa* has subsequently been published as *Glyptocythere penni*. Reference: Bate & Mayes (1977). On *Glyptocythere penni* new species. Stereo Atlas of Ostracod Shells. Vol. 4 (No. 1).

iii) The postero-ventral border of the RV is more sharply angular in the male than in the female lineage.

iv) A mid-dorsal swelling is present in all dimorphic valves but is most pronounced, probably because they are shorter, in the female lineage.

v) Whilst members of the male and female lineages are sub-rectangular and sub-quadrate in outline respectively, the sexually undifferentiated instars exhibit a characteristically immature outline, being broadly rounded anteriorly and strongly acuminate posteriorly.

vi) The dorsal margin is straight in sexually undifferentiated instars and lacks the mid-dorsal swelling mentioned in (iv) above.

vii) The ventro-lateral rib of the male and female adults and instars is very thick. In contrast, in the unsexed instars it is represented by two crenulate blade-like structures.

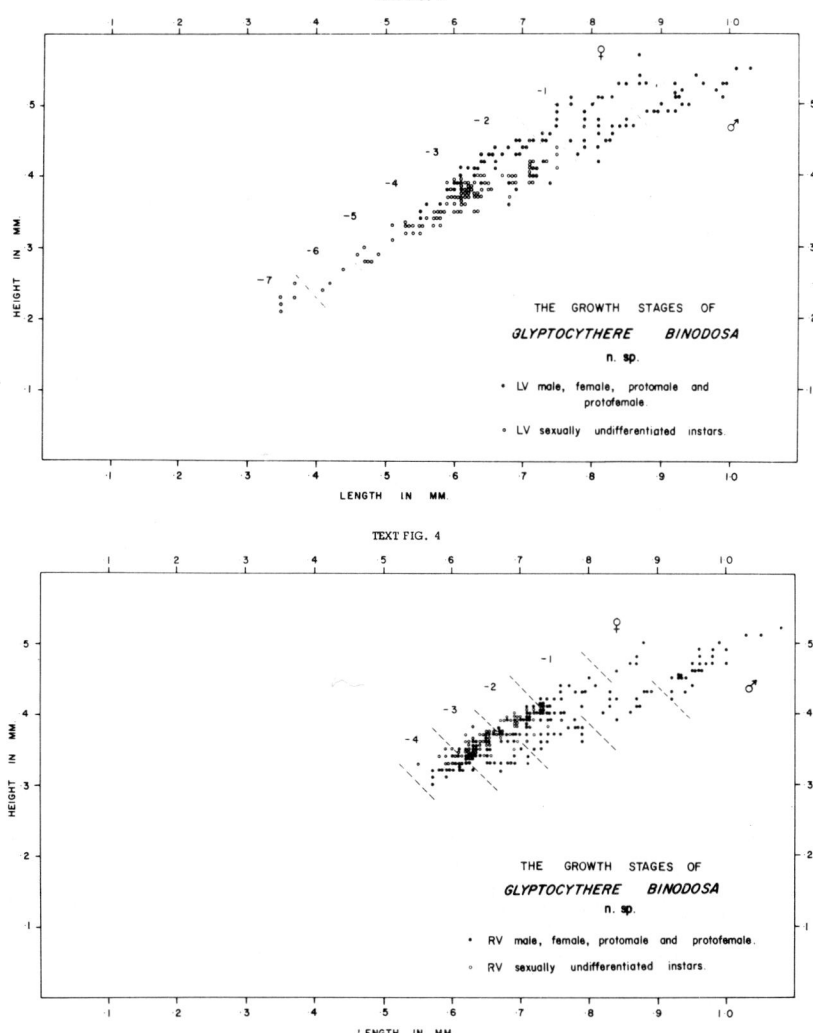

Text-Figs. 3 and 4 are instar diagrams of *G. binodosa*. Although the dimensions of the sexually undifferentiated forms may be the same as contemporary pre-adult dimorphs, they are readily distinguishable on the basis of the differing morphological characteristics outlined above, and, as in the case of *G. guembeliana*, distinct male and female and 'lines' can be seen from the A-3 instar stage onwards. The 'proto-female' again appears first, in the A-4 stage and the 'proto-male' in the subsequent A-3 stage. The sexually undifferentiated instars persist through both these stages and are also represented in the ante-penultimate populations but not in the penultimate stage. This is summarized in Table 6 below:

Table 6. The percentage of each form present at successive ontogenetic stages of *G. binodosa*

Sex	Instar Adult	A-1	A-2	A-3	A-4	A-5	A-6	A-7
Male lineage	76.7	35.7	20.6	9.9				
Female lineage	23.3	64.3	47.0	29.6	41			
Sexually undifferentiated instars	–	–	32.4	60.5	59	100	100	100

As in *G. guembeliana*, the sexually undifferentiated forms in the ontogeny of *G. binodosa* coexist for three stages with pre-adult 'females' and for two stages with pre-adult 'males'. The 'proto-female' appears one stage earlier than the 'proto-male' and is the dominant juvenile dimorph. At the maturation moult, however, as in the previously discussed species, there is a reversal of sex ratio from A-1 (64.3% ♀; 35.7% ♂) to the adult stage with 23.3% ♀ to 76.7% ♂.

Table 7. The range and mean dimensions of the growth stages of *Glyptocythere binodosa* (in mm)

Instar		Sex	Male lineage length : heigth	Female lineage length : height	Sexually undifferentiated instars length : height
Adult	LV	range mean	0.88–1.03 : 0.49–0.55 0.944 : 0.516 (20)	0.80–0.88 : 0.51–0.57 0.843 : 0.526 (10)	
	RV	range mean	0.92–1.08 : 0.44–0.52 0.968 : 0.47 (26)	0.84–0.88 : 0.46–0.50 0.865 : 0.477 (4)	
A-1	LV	range mean	0.81–0.89 : 0.45–0.49 0.841 : 0.468 (8)	0.71–0.81 : 0.45–0.51 0.75 : 0.473 (19)	
	RV	range mean	0.83–0.94 : 0.40–0.43 0.873 : 0.421 (12)	0.71–0.81 : 0.45 0.763 : 0.424 (17)	
A-2	LV	range mean	0.73–0.81 : 0.42–0.45 0.783 : 0.436 (7)	0.63–0.69 : 0.40–0.45 0.667 : 0.431 (16)	
	RV	range mean	0.72–0.84 : 0.35–0.41 0.769 : 0.36 (20)	0.67–0.76 : 0.36–0.41 0.712 : 0.389 (41)	

Instar		Sex	Male lineage length : heigth	Female lineage length : height	Sexually undifferentiated instars length : height
A-3	LV	range	0.68–0.71 : 0.36–0.40	0.51–0.66 : 0.37–0.41	0.59–0.75 : 0.37–044
		mean	0.694 : 0.386	0.612 : 0.394	0.637 : 0.387
			(5)	(20)	(52)
	RV	range	0.65–0.71 : 0.32–0.36	0.62–0.68 : 0.33–0.37	0.62–0.72 : 0.33–0.40
		mean	0.688 : 0.338	0.638 : 0.35	0.66 : 0.366
			(12)	(31)	(52)
A-4	LV	range		0.55–0.58 : 0.34–0.36	0.53–0.63 : 0.32–0.36
		mean		0.564 : 0.352	0.574 : 0.342
				(5)	(25)
	RV	range		0.57–0.63 : 0.30–0.35	0.55–0.61 : 0.32–0.35
		mean		0.588 : 0.326	0.592 : 0.338
				(20)	(11)
A-5	LV	range			0.51–0.55 : 0.31–0.33
		mean			0.523 : 0.32
					(3)
	RV	range			0.53–0.54 : 0.31
		mean			0.535 : 0.31
					(2)
A-6	LV	range			0.44–0.49 : 0.27–0.30
		mean			0.468 : 0.285
					(6)
	RV	range			0.45–0.51 : 0.25–0.29
		mean			0.478 : 0.272
					(10)
A-7	LV	range			0.35–0.42 : 0.21–0.25
		mean			0.374 : 0.233
					(7)
	RV	range			0.39–043 : 0.22–0.25
		mean			0.417 : 0.24
					(6)

Glyptocythere oscillum (Jones and Sherborn) 1888

Perhaps because it is a much smaller species, the morphological differences between the various polymorphs are less striking than in the preceeding two species. The main difference between male lineage and female lineage individuals is in terms of length and height. The former are considerably longer and less high than the latter.

Sexually undifferentiated instars differ from pre-adult dimorphs in being much more strongly acuminate postero-ventrally. Also the dimorphs have a markedly convex postero-ventral margin, particularly in the 'female' line. 'Proto-females' appear in the A-2 stage, where they coexist with undifferentiated instars. The latter are not represented in the penultimate instar populations which are made up entirely of pre-adult dimorphs, in the ratio of 16% ♂ to 84% ♀. Although the reversal of sex ratio at the maturation moult, exhibited by the preceeding two species of *Glyptocythere* does not take place in *G. oscillum* a substantial change in that ratio takes place, however, with, in the adult a ratio of 47% ♂ to 53% ♀.

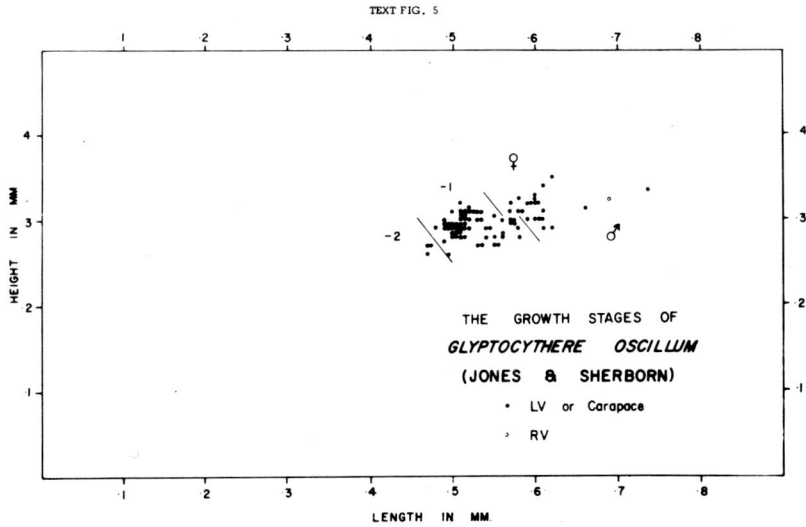

Table 8. The percentage of each form present at successive ontogenetic stages of *G. oscillum*.

Sex	Instar Adult	A-1	A-2	A-3
Male lineage	47	16	–	–
Female lineage	53	84	15	–
Sexually undifferentiated instars	–	–	85	100

It should be emphasized that, because of its relative rarity in the samples studied, the data for this species are much less acceptable than for the other two members of the genus. However, in Text-fig. 5, it can be seen that the male and female lineages are already distinct in the penultimate instar.

Lophocythere (L.) scabra scabra Triebel 1951

This is a large species (frequently over 1 mm in length in the adult) with a relatively large size difference between adult dimorphs, in which the male is considerably longer and less high than the female. The same degree of size difference occurs between contemporary pre-adult dimorphs. Sexually undifferentiated instars can be distinguished from the latter by their characteristically immature-outline with its narrow, triangular tapering posterior and sloped postero-ventral margin.

Table 9. Ratio of mean length: height dimensions for *L. (L.) scabra scabra*

			Mean of 3 samples
Adult	♂	LV	–
		RV	2.061:1 (1)
	♀	LV	1.733:1 (3)
		RV	–
A-1	♂	LV	2.073:1 (16)
		RV	2.304:1 (4)
	♀	LV	1.807:1 (12)
		RV	2.033:1 (8)
A-2	♂	LV	2.032:1 (16)
		RV	2.334:1 (10)
	♀	LV	1.763:1 (49)
		RV	2.010:1 (11)
A-3	♂	LV	2.033:1 (2)
		RV	1.944:1 (2)
	♀	LV	1.704:1 (3)
		RV	2.031:1 (29)
A-4		LV	2.032:1 (2)
		RV	1.472:1 (1)

Text-figures 6 and 7 are instar diagrams of *L. (L.) scabra scabra*. Two distinct lineages, leading to adult males and females respectively can be seen from the A-3 stage onwards. All earlier instars are sexually undifferentiated, but these latter forms do not persist to the A-3 stage where, throughout the remainder of the ontogeny of the species, all specimens are dimorphs.

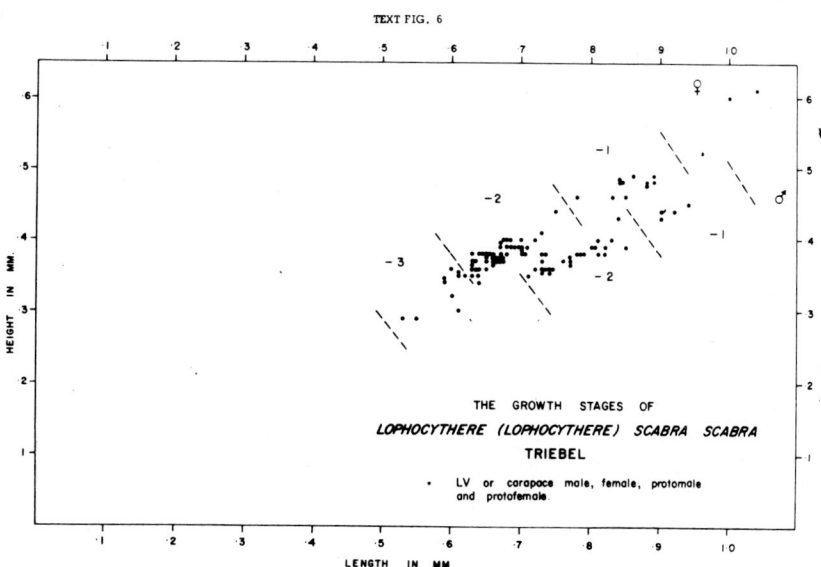

TEXT FIG. 6

THE GROWTH STAGES OF
LOPHOCYTHERE (LOPHOCYTHERE) SCABRA SCABRA
TRIEBEL

• LV or carapace male, female, protomale and protofemale.

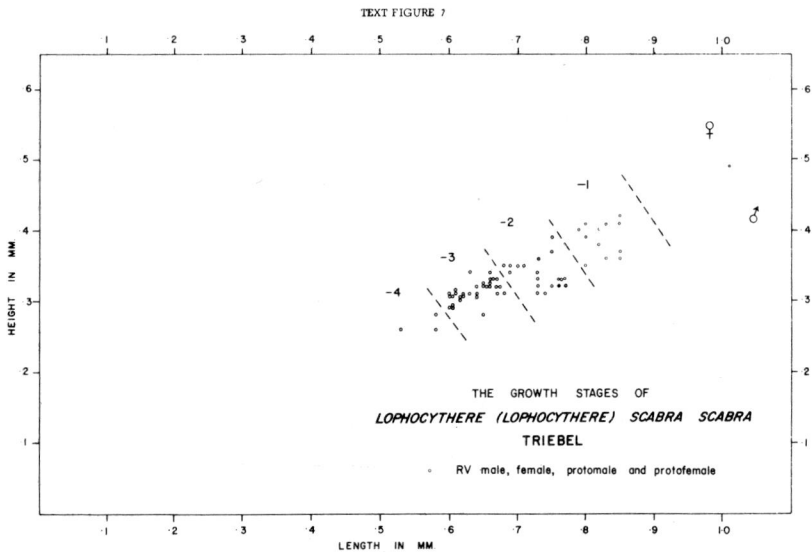

Table 10. The percentage of each form present at successive ontogenenetic stages of *L. (L.) scabra scabra*

Sex	Instar	Adult	A-1	A-2	A-3	A-4
Male lineage		25	49.5	29.6	8.1	–
Female lineage		75	50.5	70.4	91.9	–
Sexually undifferentiated instars		–	–	–	–	100

At the A-3 stage the 'proto-male' is very dominant, making up over 90% of the total instar population. After a decline in the dominance of the female line, a 50:50 parity is achieved at the A-1 stage. In the adult stage the females have a 3:1 dominance.

Table 11. The range and mean dimension of the growth stages of *Lophocythere (L.) scabra scabra* (in mm)

| | | | Mean of 3 samples | | |
			length	height	
♂	LV	range	–	–	
		mean	–	–	
Adult ♀	RV	range	1.01	0.49	
		mean	1.01	0.49	(1)
♀♀	LV	range	0.96-1.04	0.52-0.61	
		mean	1.002	0.578	(3)
	RV	range	–	–	
		mean	–	–	

82

				Mean of 3 samples		
			length	height	
A-1	♂	LV	range	0.78-0.94	0.38-0.45	
			mean	0.843	0.407	(16)
		RV	range	0.80-0.85	0.355-0.37	
			mean	0.832	0.361	(4)
	♀	LV	range	0.78-0.89	0.44-0.495	
			mean	0.841	0.474	(12)
		RV	range	0.79-0.85	0.38-0.42	
			mean	0.82	0.403	(8)
A-2	♂	LV	range	0.71-0.78	0.345-0.38	
			mean	0.742	0.365	(16)
		RV	range	0.725-0.77	0.305-0.335	
			mean	0.75	0.321	(10)
	♀	LV	range	0.61-0.73	0.34-0.405	
			mean	0.662	0.375	(49)
		RV	range	0.66-0.75	0.32-0.39	
			mean	0.702	0.349	(12)
A-3	♂	LV	range	0.605-0.61	0.305-0.32	
			mean	0.607	0.312	(2)
		RV	range	0.585-0.645	0.26-0.345	
			mean	0.615	0.302	(2)
	♀	LV	range	0.59-0.60	0.34-0.36	
			mean	0.593	0.348	(3)
		RV	range	0.585-0.68	0.28-0.34	
			mean	0.631	0.31	(29)
A-4		LV	range	0.535-0.55	0.285-0.29	
			mean	0.534	0.312	(2)
		RV	range	0.53	0.36	
			mean	0.53	0.36	(1)

Lophocythere (L.) fulgurata (Jones and Sherborn) 1888 also seems to exhibit precocious sexual dimorphism in its ontogeny. Unfortunately, this species is too rare in the material studied for the authors to demonstrate this adequately and therefore no diagrams are included.

Glabellacythere dolabra (Jones and Sherborn) 1888

This species is markedly dimorphic in the adult stage, the males being appreciably (as much as 0.15–0.20 mm) longer than females although the two sexes are of approximately the same height. The female is also more angular postero-ventrally and higher at the anterior cardinal angle than the male. These differences also serve to distinguish the pre-adult 'females' and 'males'.

Table 12. Range and mean length: height dimensions and ratio of length to height in late ontogenic stages of *Glabellacythere dolabra* (in mm)

			Range		Mean		Ratio
			Length	Height	Length	Height	L:H
Adult	♂	LV (6)	0.81-0.845	0.36-0.41	0.83	0.392	2.117:1
		RV (2)	0.78-0.82	0.365-0.38	0.80	0.372	2.150:1
	♀	LV (6)	0.61-0.67	0.34-0.39	0.638	0.366	1.743:1
		RV (5)	0.59-0.64	0.34-0.36	0.618	0.35	1.765:1
A-1	'♂'	LV (5)	0.69-0.745	0.32-0.345	0.713	0.336	2.122:1
		RV (4)	0.67-0.73	0.32-0.355	0.697	0.339	2.056:1
	'♀'	LV (5)	0.54-0.60	0.28-0.32	0.58	0.304	1.907:1
		RV (8)	0.50-0.555	0.28-0.32	0.527	0.301	1.750:1
A-2	'♂'	LV (4)	0.57-0.59	0.26-0.27	0.582	0.276	2.108:1
		RV (3)	0.53-0.57	0.26-0.30	0.547	0.277	1.974:1
	'♀'	LV (6)	0.46-0.51	0.24-0.27	0.488	0.258	1.891:1
		RV (1)	0.47	0.24	0.47	0.24	1.958:1

Table 13. The percentage of each form present at successive ontogenic stages of *G. dolabra*

sex	Instar	Adult	A-1	A-2	A-3
Male lineage		42 (8)	41 (9)	50 (7)	–
Female lineage		58 (11)	59 (13)	50 (7)	–
Sexually undifferentiated instars		–	–	–	100 (3)

Both pre-adult dimorphs appear together at the A-2 stage in equal proportions and these proportions change but little into the adult stage. Undifferentiated instars, of which only 3 were found, make up the entire population of the A-3 stage, and do not occur in subsequent stages.

Although the number of specimens available for study was unsatisfactorily few, Text-figs. 8 and 9 clearly show distinct pre-adult male and female lineages.

Discussion

Any interpretative studies on ostracod larval stages are complicated by such factors as seasonal effect on growth, and the possible existence of post-maturation moulting. Most of the instar diagrams produced for the species analyzed above do not consist of discrete clusters of points on the graph, which would be the theoretical ideal. Rather, they occur as a more general spread of points, with slight clustering about certain points. This phenomenon has been considered by Elof-

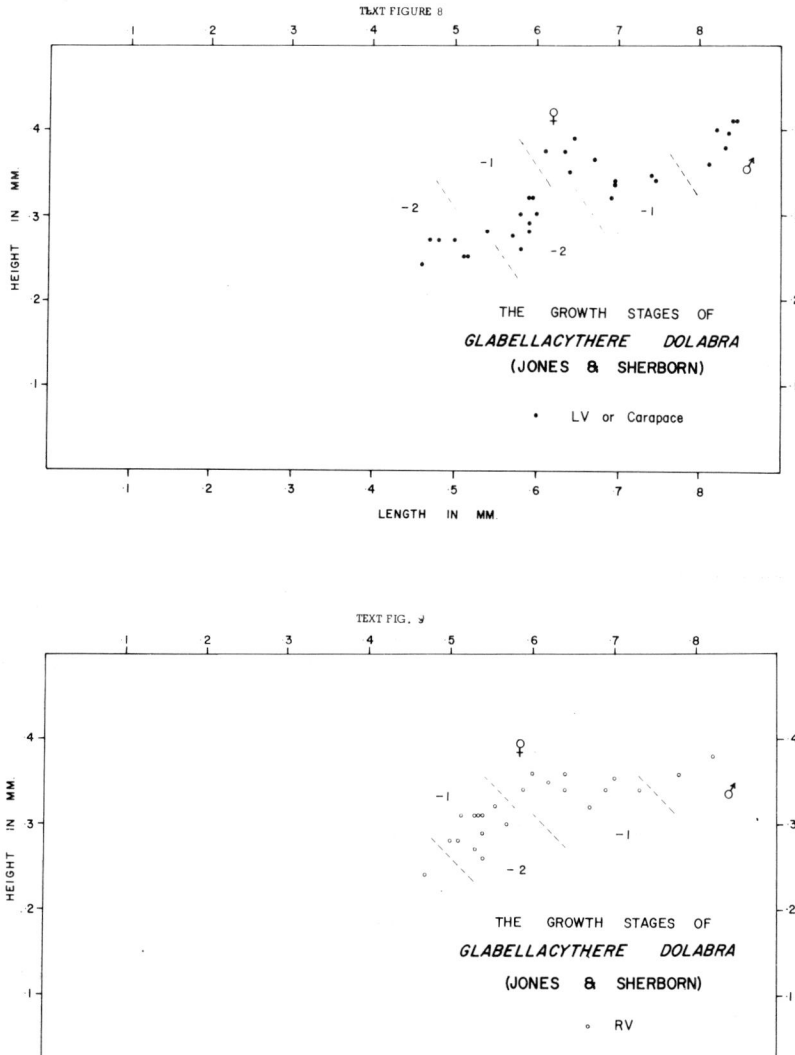

son (1941) and Szczechura (1971), and other authors to be caused by a seasonal variation in size. Elofson (op. cit.) found that the older larval stages of *Semicytherura nigrescens* (Baird), and also other ostracods, are almost invariably longer in winter than in summer, but he did not know whether this was also true of their respective adults, since individuals that had reached adulthood during different seasons were found together throughout the year.

Wall (1969, MS) found that on the Welsh Coast, a number of species exhibited a seasonal variation in size of instars, those of the spring being much smaller than those of the autumn. The species *Leptocythere pellucida* (Baird) is used to illustrate this:

Table 14. The range and mean dimensions of instars of *Leptocythere pellucia* from a single station at two seasons. After Wall, MS

		22.4.67		8.10.67	
		Length	Height	Length	Height
A-1	range	0.475-0.510	0.245-0.280	0.500-0.540	0.265-0.285
	mean	0.486	0.268 (10)	0.520	0.273 (10)
A-2	range	0.350-0.390	0.175-0.210	0.365-0.410	0.185-0.230
	mean	0.370	0.191 (10)	0.387	0.197 (10)
A-3	range	0.300-0.320	0.165-0.175	0.340-0.355	0.165-0.180
	mean	0.310	0.170 (7)	0.351	0.172 (12)
A-4	range	0.270-0.285	0.155-0.170	0.280-0.310	0.160-0.170
	mean	0.281	0.160 (8)	0.296	0.162 (7)
A-5	range	0.240-0.260	0.140	0.245-0.270	0.150
	mean	0.247	0.140 (6)	0.253	0.150 (4)

One of the present authors, RCW, has also observed seasonal size variation in adults of *Loxoconcha rhomboidea* (Fischer). Szczechura (1971), in laboratory studies involving the freshwater cyprid *Cyprinotus (H.) incongruens* (Ramdohr), found an extensive range in size and shape of dead specimens obtained from the bottom of her 'flower pot'. In comparing these valves with the carapaces of live individuals caught over a period of 3 years, it became clear that the sediment sample contained both summer and winter forms which differed from each other in size and shape. When both summer and winter forms were plotted on a graph, the result was a continuous size distribution chart, very similar to those figured for *Glyptocythere quembeliana* and *G. binodosa* above. Szczechura considers that a graph with well separated clusters of instar stages would be produced by a species with low tolerance of variation of environmental conditions; i.e. species living in very stable ecological conditions or species living only during one particular season of the year. The continuous size distribution diagram would be produced only by species with high tolerance of seasonal or other environmental change.

Since the Bathonian samples were collected from over some 8 cm. of strata, they obviously represent many seasons and this may account for the continuous nature of the diagrams. In such diagrams it is often very difficult to separate one growth stage from another and many of the lines drawn by the present authors are necessarily both arbitrary and tentative. It is not certain whether seasonal size variation in instars would be confused with precocious sexual dimorphism in ontogeny. If the sole criterion were of relative dimensions then this could possibly happen. When, however, the pre-adult dimorphs are also recognized on the basis of morphological characters possessed by the adult dimorphs, as in the case of the *Glyptocythere* species described above, then there is little chance of such confusion.

Plate 1. *Glyptocythere bindosa* Stephens n.sp. showing male and female dimorphs from A-4 to adult.

Post-maturation moulting can occur in Ostracoda and RCW has observed this phenomenon, albeit very rarely indeed, in a number of fossil and Recent species. For example, two adults in a population of several hundred of *Lophocythere (N.) multicostata* Oertli from the Oxfordian of southern England were found which were approximately twice the size of 'normal' adults. The present state of our knowledge on this subject is very unsatisfactory and despite the deliberations of such authors as Elofson (1941), Müller (1926), Hoff (1943), and others, the statement by Kornicker (1975a, p. 688) that '– no evidence is on hand to show con-

87

Plate 2. *Glyptocythere guembeliana* (Jones) showing male and female dimorphs from A-5 to adult.

clusively that postadult molting occurs in the Ostracoda. Sufficient work has been done, however, to conclude that if post-adult molting does occur, it is rare.' – is a very fair summary. There is no evidence, however, to suggest that this has occurred in the Jurassic species discussed above. In every case, the forms considered as penultimate growth stages of each of the species, exhibit juvenile characteristics in hingement, inner lamella, etc., and those forms therefore considered as adults, cannot be postmaturation moults.

Conclusions

The occurrence of pre-adult sexual dimorphs in myodocopid and palaeocopid Ostracoda seems, on the evidence cited in the introduction to this paper, quite incontrovertible. Indeed, in the Myodocopina, it would appear to be a common phenomenon in ontogeny. Little or no evidence is available to suggest that the precosity exhibited by instars in both these groups was anything other than morphological and, apart from the above cited suggestion by Martinsson (1956), no other authors have thought that the pre-adult dimorphs could be fertile. It should be clearly stated that the term precocious sexual dimorphism as used here, does not in any way imply precocious sexual maturity.

The authors feel that the evidence they have presented in respect of 6 Jurassic podocopids could probably be replicated in other prodocopids and intend to examine other species, both fossil and Recent, in an attempt to ascertain whether or not pre-adult sexual dimorphs can be recognized in other members of the Podocopina.

Many other Jurassic podocopids, examined by JMS from the Bathonian and RCW from the Callovian and Oxfordian do not, however, exhibit this phenomenon. Why, therefore, if the phenomenon is indeed due to precocious sexual dimorphism, should it manifest itself in the 6 species and not in other contemporary species? One reason which could be put forward to explain precocious sexual dimorphism in myodocopids, is that, in many cases, male and female lineages are ecologically disjunct during their development. This is much more readily acceptable in the more highly mobile myodocopids than the non-swimming Cytheracea, and there is little evidence from modern cytheraceans to support it. All instars are, of course, potentially male or female if they belong to syngamic species, and are destined to be one or other sex from the embryo. It is not difficult to understand that 'proto-males' and 'proto-females' should acquire incipient sexually dimorphic characters in ontogeny. It is perhaps more difficult to understand why, in other species with strongly dimorphic adults, there is no reflection of this dimorphism in the juveniles of the species.

The authors can equally find no satisfactory explanation for the abnormal sex ratios encountered in *G. guembeliana* and *G. binodosa,* and do not understand the significance of a reversal of sex ratio from $♀ > ♂$ in juveniles to $♂ > ♀$ in adults. One can invoke migration out of the sample area by females at about the time of the last moult but this type of suggestion creates as many problems as it resolves, i.e. where does copulation take place and how do the instars return to the original environment, etc.

Most syngamic species of cytheracean Ostracoda exhibit a dominance of females over males in the adult stage, as indeed do many other animals. A surplus of females seems necessary to the survival of many species. Some ostracod species, however, have more males than females and this has been remarked upon by Van Morkhoven (1962, vol. 1, p. 94):

'In Recent assemblages the males in the majority of cases are found to be greatly outnumbered by the females. The exact ratio differs from species to species, and would seem also to be considerably influenced by seasonal factors. The males of certain forms are extremely rare; occasionally, however, there are about equal numbers of the two sexes, or very rarely a preponderance of males, but in the large majority of cases, the females are about three to ten times as abundant as the males.'

The authors can only suggest that the *Glyptocythere* species mentioned above belong to that small group of Ostracoda in which the males preponderate. We remained unable, however, to understand the apparent reversal of sex ratio from juvenile to adult in these species.

Acknowledgements

The authors wish to thank Dr. D. R. Wall for permission to abstract data (Table 14) from his Ph.D. thesis. They are indebted to Mrs. Joanne Maltman, Mr. Arnold Thawley and Mr. Howard Williams who were responsible for typing, draughting and photography respectively.

References

Bate, R. H. 1967. The Bathonian Upper Estuarine Series of Eastern England. Part 1: Ostracoda. Bull. Br. Mus. nat. Hist., Geology 14(2): 5-66.
Elofson, O. 1941. Zur Kenntnis der marinen Ostracoden Schwedens. Uppsala Univ. Zool. Bidr. 19: 215-534.
Guber, A. L. 1971. Problems of sexual dimorphism, population structure and taxonomy of the Ordovician genus Tetradella Ostracoda. Jour. Paleontol. 45(1): 6-22.
Hoff, C. C. 1943. Two new ostracods of the genus Entocythere and records of previously described species. Jour. Washington Acad. Sci. 33(9): 276-286.
Jaanusson, V. 1956. Middle Ordovician Ostracodes in centtrtral and southern Sweden. Bull. Geol. Inst. Uppsala 37: 173-441.
Kornicker, L. S. 1969. Morphology, Ontogeny and Intraspecific Variation of Spiracopia a new genus of Myodocopid Ostracod (Sarsiellidae). Smithsonian Contr. Zool. 8: 1-50.
Kornicker, L. S. 1970. Ostracoda (Myodocopina) from the Peru-Chile Trench and the Antarctic Ocean. Smithsonian Contr. Zool. 32: 1-42.
Kornicker, L. S. 1970a. Myodocopid Ostracoda (Cypridinacea) from the Philippine Islands. Smithsonian Contr. Zool. 39: 1-32.
Kornicker, L. S. & Caraion, F. E. 1974. West African Myodocopid Ostracoda (Cylindroleberidae). Smithsonian Contr. Zool. 179: 1-78.
Kornicker, L. S. 1974a. Ostracoda (Myodocopina) of Cape Cod Bay, Massachusetts. Smithsonian Contr. Zool. 173: 1-20.
Kornicker, L. S. 1974b. Revision of the Cypridinacea of the Gulf of Naples (Ostracoda). Smithsonian Contr. Zool. 178: 1-74.
Kornicker, L. S. 1975. Ivory Coast Ostracoda (suborder Myodocopina). Smithsonian Contr. Zool. 197: 46.
Kornicker, L. S. & Angel, M. V. 1975. Morphology and Ontogeny of Bathyconchoecia septemspinosa Angel, 1970 (Ostracoda: Halocyprididae). Smithsonian Contr. Zool. 195: 1-21.
Kornicker, L. S. 1975a. Antarctic Ostracoda (Myodocopina). Parts 1 & 2. Smithsonian Contr. Zool. 163: 1-720.
Martinsson, A. 1956. Ontogeny and development of dimorphism in some Silurian ostracods. Bull. Geol. Inst. Uppsala 37: 2-42.
Van Morkhoven, F. P. C. M. 1962. Post-Palaeozoic Ostracoda. Their morphology, taxonomy and economic use. Vol. 1. Elsevier: VII+204 pp.

Müller, G. W. 1926. Ostracoda-Muschelkrebse. In: Ordnung 2, Handbuch der Zoologie, pp. 399-434.
Poulsen, E. M. 1962. Ostracoda-Myodocopa, 1: Cypridiformes-Cypridinidae. In: Dana Report 57, 1-414, Copenhagen, Carlsberg Foundation.
Poulsen, E. M. 1965. Ostracoda-Myodocopa, 2: Cypridiniformes-Rutidermatidae, Sarsiellidae and Asteropidae. In: Dana Report 65, 1-484. Copenhagen, Carlsberg Foundation.
Poulsen, E. M. 1969. Ostracoda-Myodocopa, 3a: Halocypriformes-Thaumatocypridae and Halocypridae. In: Dana Report 75, 1-100. Copenhagen, Carlsberg Foundation.
Spjeldnaes, N. 1951. Ontogeny of Beyrichia jonesi Boll. Jour. Paleontol. 25: 745-755.
Szczechura, J. 1971. Seasonal changes in a reared fresh-water species, Cyprinotus (Heterocypris) incongruens, and their importance in the interpretation of variability in fossil Ostracodes. Paléoécologie Ostracodes Pau 1970; Bull. Centre. Rech. Pau – SNPA, 5 suppl.: 191-205.
Wall, D. R. 1969. The taxonomy and ecology of Recent and Quaternary Ostracoda from the Southern Irish Sea. Ph. D. Thesis, University of Wales, Unpublished.

Additional reference

Kornicker, L. S. 1976. Myodocopid Ostracoda from South Africa. Smithsonian Contr. Zool. 214: 1-39.

Authors' address:

Dept of Geology, UCW, Aberystwyth, Wales, U.K.

Sixth Intern. Ostracod Symposium, Saalfelden

SOME HEMICYTHERINAE FROM THE TERTIARY OF PATAGONIA (ARGENTINA), THEIR MORPHOLOGICAL RELATIONSHIP AND STRATIGRAPHICAL DISTRIBUTION

VICENTE HUGO VALICENTI

Abstract

Twenty-one species of Hemicytherinae, belonging to ten genera, from the Eocene-Oligocene of Patagonia, Argentina, are discussed. These species have distinct stratigraphical distributions in both the San Julian (Eocene) and the Monte Leon (Oligocene) Formations. The stratigraphical ranges of the genera *Australicythere, Coquimba, Nanocoquimba, Cornucoquimba,* and *Ambostracon* are determined. First records of these genera from the Eocene and of the species *Bradleya normani* from the Oligocene are made.

Patagonacythere and Ambostracon are shown to be congeneric. *Ambrostracon* is divided into two subgenera, its geographical and stratigraphical distributions are also studied. Reticular silhouettes following Liebau's (1969) method of mapping the fossae and pore conuli are given for some fossil and living species of *Australicythere* and *Ambostracon*.

Introduction

Due to the splitting of taxa, the actual state of the taxonomy of the hemicytherids is rather confused, and the nomenclature in many cases no longer reflects the true relationships. This splitting results mainly because of:
1. differences in the importance that authors place on the various morphologic features used in taxonomy;
2. the rigidity of many of the schemes used in taxonomy which may be valid in some cases but not in others.

Some of the genera within the Hemicytherinae have been based upon monotypic examples with resultant rigid generic diagnosis, and the interspecific variation within some genera seems to have been frequently overlooked. It is obvious that the type species of the following genera: *Australicythere, Ambostracon, Radimella, Orionina, Caudites, Hermanites, Quadracythere Hornibrookella, Coquimba,* and *Cornucoquimba* are easy to distinguish. However, many other species could be placed within a number of genera with equal justification.

Material and age

The material used in the present study came from the type locality of the San Julian and Monte Leon Formations, in Santa Cruz Province, Argentina. The San Julian Formation is almost exclusively restricted to the coastal region of the Santa Cruz Province; its thickness is about 70 meters in the type locality (Bajo de San Julian). The Monte Leon Formation is restricted to the coastal region of Santa Cruz Province, and its thickness is about 100 meters in the type locality (Monte Entrance).

For both formations Bertels (1970) used planctonic foraminifera to establish an age correlative with that of the European Chattian stage (Upper Oligocene). Nevertheless, Camacho (1974), using other asssociations of invertebrates, concluded that while the Monte Leon Formation is definitely Oligocene, the San Julian Formation could be Lower or Middle Eocene.

The Hemicytherinae

The genus Australicythere Benson (1964) for which three living species have been recorded at the present, is represented in the Eocene and Oligocene of Patagonia by five new species, here referred to as species 1 to 5. The generic assignment of these species was not easy. It was necessary to amend the original diagnosis of Benson (1964) (this will be dealt within a later paper) to avoid creating more taxa with the consequence of further complicating the already chaotic state of the Hemicytherinae, presently comprised of more than 50 genera.

If the *Australicythere* species are arranged in the lienar morphological series: *sp 1 – sp. 2 – sp. 5 – sp. 3 – sp. 4 – polylyca,* they exhibit a progressive reduction in the strength of the longitudinal components and an increase in the vertical components of the ornament (see Text-fig. 1). All these species have: a conspicuous subcentral tubercle, a posterior vertical rib, a posterodorsal loop of variable development, sieve-type normal pore canals, anterior and posterior marginal denticles, eye tubercle, the same type of holanphidont hinge with the posteromedian element being crenulate, numerous medially expanded marginal pore canals and the same type of muscle scar patterns with three frontal scars and six adductor muscle scars. Nevertheless, it is possible to observe that the third scar from the top is, in some individuals, extremely elongated and undivided.

All these features are present in *Urocythereis californicum* Hazel (1962), from the Pliocene-Recent of California, and this species, which is morphologically similar to *Australicythere sp. 1* should be referred to as the genus *Australicythere*. *Cythere devexa* Müller (1908), which was placed within *Patagonacythere* by Benson (1964), is more similar to *Australicythere polylyca* (Müller 1908), than to *Patagonacythere tricostata* Hartmann (1962). This species could complete the morphological series *polylyca-Cythereis megalodiscus* Skogsberg (1928) – *devexa*. Nevertheless, it lacks the second order reticulation, and the posterovertical ridge is not very conspicuous.

The morphological similarities between the different species of *Australicythere* can be more easily understood by mapping the fossae and the pore conuli, following Liebau's (1969) scheme (see Text-fig. 1). The TZ9 pore conuli is always prominent. In *sp. 4,* the mesh N_2 is missing; in *sp. 1,* the meshers L_4 and L_5 are reduced and in *californicum* the meshes $L_2, L_3, L_4,$ and L_5 are fused to the meshes $K_5, K_6, K_7,$ and K_8 respectively, and the rib originating between the meshes A_1-B_2 and B_1-C_2 is slightly raised to form a short ocular ridge.

The genus *Ambostracon* Hazel (1962), is present in the Argentinian Eocene – Oligocene with three new species *Ambostracon (Ambostracon) sp. 1, Ambostracon (Patagonacythere) sp. 2* and *Ambostracon (Patagonacythere) sp. 3.*

From a comparison of the two genotypes, *Patagonacythere tricostata* Hartmann

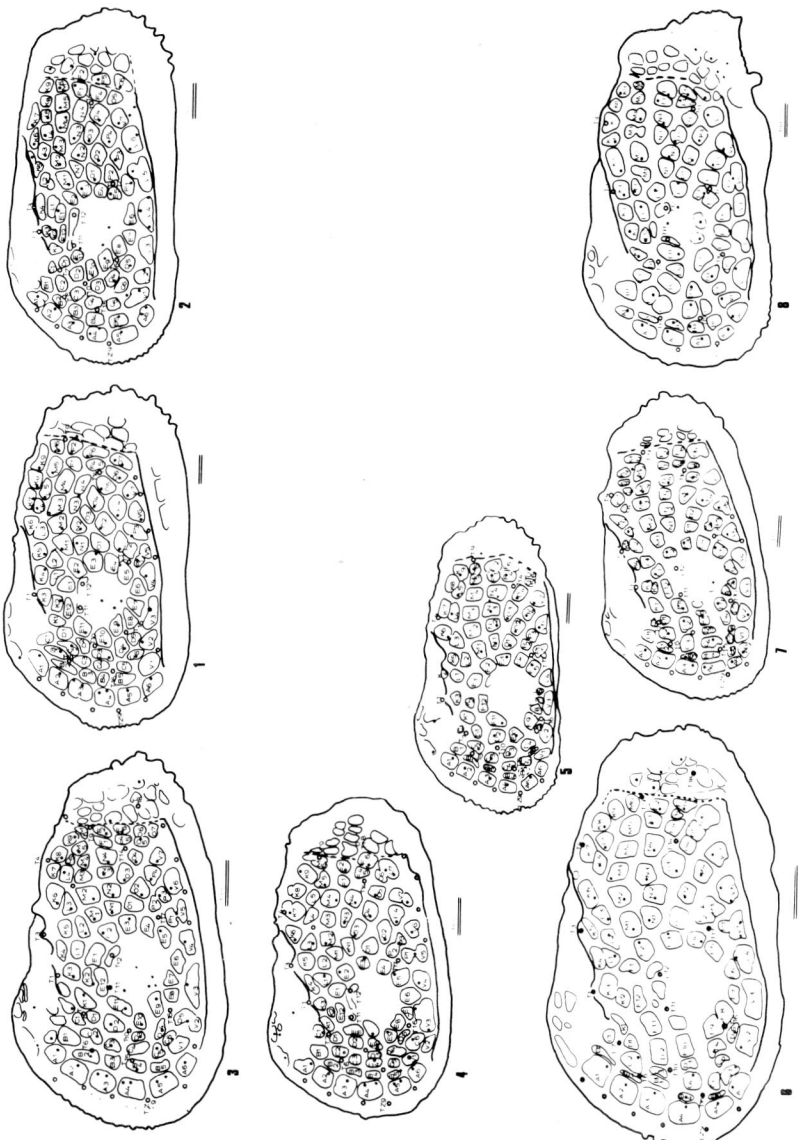

Text-fig. 1. Reticular silhouettes of: 1) *Australicythere polylyca*; 2) *A. devexa*; 3) *A.* sp. *3*; 4) *A.* sp. *4*; 5) *A.* sp. *5*; 6) *A.* sp. *2*; 7) *A.* sp. *1*; 8) *A. californicum*. The fossae and the pore conuli are coded according to Liebau's (1969) scheme. Graphic scale: 0.1 mm.

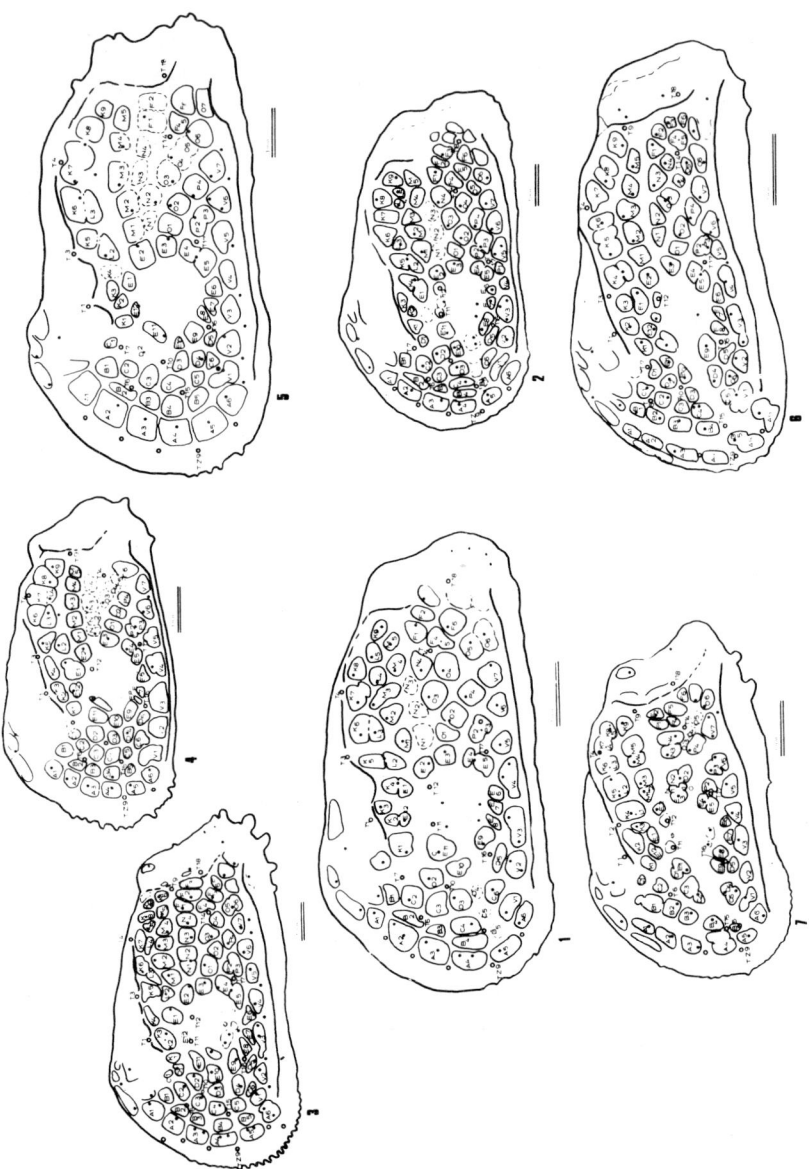

Text–fig. 2. Reticular silhouettes of: *Ambostracon (Patagonacythere) tricostata*; 2) *A. (P.)* sp. (undescribed Recent Argentina); 3) *A. (P.) wyvillethomsoni*; 4) *A. (P.)* sp. *2*; 5) *A. (P.)* sp. *3*; 6) *A. (Ambostracon)* sp. *1*; 7) *A. (A.) glaucum*. The fossae and the pore conuli are coded according to Liebau's (1969) scheme. Graphic scale: 0.1 mm.

(1962) and *Ambostracon constatum* Hazel (1962), it is possible to establish that both species belong to the same genus. Both have the same distribution of the principal ridges, which is their main characteristic. In *tricostata*, the surface between the main ridges is reticutate, in costatum, this surface seems to be celate; *tricostata* is subtrapezoidal and extremely elongated, and *costatum* is more subquadrate in lateral view; *costatum* has an anterior marginal rib as well as a thick but interrupted submarginal one (ocular ridge), *tricostata* has the marginal rib, but the submarginal rib is only weakly developed. In both species, this rib extends from the eye tubercle to the anterior end of the antero-ventral ridge.

The muscle scar patterns are similar in both species, they have three frontal scars in a 'U'-shaped formation, and the adductor muscle scars are arranged as follows: the uppermost is subrounded, the dorsomedian divided, the ventro median very elongate with an indistinct division posterior to its mid-length; the lowermost is subelongate. In other species of this group, the ventro-median adductor scar is divided.

Both species possess the same type of normal pore canals, which, when seen internally, have almost exactly identical distributions in the area around the muscle scars.

The single genus henceforth referred to as *Abostracon* can, nevertheless, be divided into two subgenera on the basis of the degree of development of the anterior submarginal ridge (ocular ridge). In *Ambostracon (Patagonacythere)*, the ridge between the fossae group A and B (in Liebau's code) is weakly developed. In *Ambostracon (Ambostracon)* this rib became an ocular ridge which varies in strength within the different species.

At the present time *Ambostracon* possesses at least twenty-nine species. The known stratigraphic range is from the Eocene to the Recent, and apparently both subgenera originated in Southern South America. The actual known geographical distribution (see Table 1) is the Atlantic side of South America and the Pacific side of North and Central America. However, two species of *Ambostracon* have been found in the Pliocene of Cornwall and Brittany (R. C. Whatley, Pers. communication, 1976). The genus has been recorded in South Africa, Australia and Japan and other, probably related, species from the Miocene of Turkey.

Ambostracon is a genus with strong sexual dimorphism and highly variable ovnamentation, whose species can be divided into at least four morphological groups:

The first group includes all those species morphologically similar to *Cythereis glaucum* Skogsberg (1928), i.e. those species in which the ocular ridge is more conspicuous. The longitudinal ridge is always present, The L fossae of the reticulation are joined to the nearest K fossae, which are enlarged. In some species, the meshes of the anterior half of the carapace tend to form groups. Therefore, the recognition of the A, B, C, D, and E meshes is rendered rather difficult but not impossible. Some of these fossae are obscured at the anterior part of the longitudinal ridge. The following species are included in this group:
Ambostracon costatum Hazel (1962)
Cythereis glaucum Skogsberg (1928)
Ambostracon (Ambostracon) sp. 1

Table 1. Stratigraphical and geographical distribution of *Australicythere* and *Ambostracon*.

	SOUTH AMERICA	ANCTARTICA	NORTH AND CENTRAL AMERICA (PACIFIC)	EUROPE	SOUTH AFRICA	AUSTRALIA	JAPAN
RECENT	5 6	7 8 9					
PLEISTOCENE		F G H			27	28	29
PLIOCENE	1 2 3 4		10 11 12 13 14 15 16 17 18 19 20 21 22 23 24	25 26			
MIOCENE	B C D J E						
OLIGOCENE	A						
EOCENE							

Ambostracon (*Ambostracon*)	*Ambostracon* (*Patagonacythere*)	*Australicythere*

1 *Ambostracon (Ambostracon) sp 1*
2 *Ambostracon (Patagonacythere) sp 2*
3 *Ambostracon (Patagonacythere) sp 3*
4 *Ambostracon (Ambostracon) sp 1* (Rossi de Garcia, 1970)
5 *Ambostracon (Patagonacythere) tricostata* Hartmann, 1962
6 *Ambostracon (Patagonacythere) sp* (undescribed Rec Argentina)
7 *Ambostracon (Patagonacythere) wyvillethomsoni* (Bady, 1880)
8 *Ambostracon (Patagonacythere) longiductus* (Skogsberg, 1928)
9 *Ambostracon (Patagonacythere) longiductus antarticum* (Benson), 1964)
10 *Ambostracon (Ambostracon) costatum* Hazel, 1962
11 *Ambostracon (Ambostracon) glaucum* (Skogsberg, 1928)
12 *Ambostracon (Ambostracon) diegoensis* (LeRoy, 1943)
13 *Ambostracon (Ambostracon) sp E* Valentine, 1976
14 *Ambostracon (Ambostracon) sp G* Valentine, 1976
15 *Ambostracon (Ambostracon) sp J* Valentine, 1976
16 *Ambostracon (Ambostracon) sp K* Valentine, 1976
17 *Ambostracon (Ambostracon) sp L* Valentine, 1976
18 *Ambostracon (Ambostracon) sp M* Valentine, 1976
19 *Ambostracon (Ambostracon) sp O* Valentine, 1976
20 *Ambostracon (Ambostracon) sp D* Valentine, 1976
21 *Ambostracon (Abostracon) vermilloensis* Swain, 1967
22 *Ambostracon (Ambostracon) hulingsi* McKenzie and Swain, 1967
23 *Ambostracon (Ambostracon) sp* Swain and Gilby, 1974
24 *Ambostracon (Patagonacythere) sp A* Valentine, 1976
25 *Ambostracon (Ambostracon) sp* (undescribed Pliocene Brittany and Cornwall)
Ambostracon (Ambostracon) sp (undescribed Pliocene Brittany and Cornwall)
27 *Ambostracon (Ambostracon) flabellicostata* (Bady, 1880)
28 *Ambostracon (Patagonacythere) pumila* (Brady, 1880)
29 *Ambostracon (Patagonacythere) japonicus* (Ishizaki, 1971)

A *Australicythere sp 1*
B *Australicythere sp 2*
C *Australicythere sp 4*
D *Australicythere sp 5*
E *Australicythere sp (Bradleya aff praecrasa* Rossi de Garcia, 1970)
F *Australicythere californicum* (Hazel, 1962)
G *Australicythere polylyca* (Muller, 1908)
H *Australicythere devexa* (Muller, 1908)
I *Australicythere megalodiscus* (Skogsberg, 1928)
J *Australicythere sp 3*

Costa? sp. 1 Rossi de Garcia (1970)
Cythereis flabellicostata Brady (1880)
Ambostracon vermilloensis Swain (1967)
Ambostracon hulingsi McKenzie and Swain (1967)
Cythere diegoensis LeRoy (1943)
Ambostracon sp. Swain and Gilby (1974)
Ambostracon species D, E, G, J, K, L, M, O, Valentine (1976)

The second group, the *Ambostracon (Patagonacythere) tricostata* group, is strongly reticulate, with the development of a conspicuous longitudinal ridge running from the M_5–F_1 fossae to the anteroventral area. This ridge crosses the subcentral tubercle. The N_1, N_2 and N_3 fossae are usually obscured (celate) by this ridge, but they may be observed in transmitted light. The mesh L_1 is missing; L_2 is usually joined to K_5; L_4 and L_5 are reduced in size. The following species are included in this group:

Table 2. Stratigraphical distribution of the *Hemicytherinae* in the San Julian and Monte Leon formations.

EOCENE	OLIGOCENE	
San Julian Fm.	Monte Leon Fm.	
		Cornucoquimba sp. 1
		Hermanites cf. dohmi
		Nanocoquimba sp. 1
		Australicythere sp. 1
		Urocythereis sp. 1
		Ambostracon (Ambostracon) sp. 1
		Aurila sp. 1
		Australicythere sp. 2
		Coquimba ? sp. 2
		Urocythereis sp. 2
		Ambostracon (Patagonacythere) sp. 2
		Coquimba sp. 1
		Aurila sp. 3
		Australicythere sp. 3
		Australicythere sp. 4
		Australicythere sp. 5
		Cornucoquimba sp. 2
		Ambostracon (Patagonacythere) sp. 3
		Aurila sp. 2
		Caudites sp.
		Bradleya normani

Patagonacythere tricostata Hartmann (1962)
Cythere pumila Brady (1880)
Caudites japonicus Ishizaki (1971)
Ambostracon sp. *A* Valentine (1976)
Ambostracon (Patagonacythere) sp. (undescribed Recent Argentina)

The third group, the *Ambostracon (Patagonacythere) wyvillethomsoni* group is comprised of those species in which the longitudinal and vertical components of the reticulation are developed more or less to the same degree. The fossae N_1, N_2 and N_3 are conspicuous. This group includes:

Cythereis wyvillethomsoni Brady (1880)
Cythereis (Cythereis) longiductus Skogsberg (1928)
Patagonacythere longiductus antarticus Benson (1964)

The fourth group includes those species which are morphologically transitional between the *tricostata* group and the *wyvillethomsoni* group. The surface is highly

celate, and thus the M and N fossae are obscured. The anterior marginal rib is wide and bears a second order reticulation. This group includes:
Ambostracon (Patagonacythere) sp. *2*
Ambostracon (Patagonacythere) sp. *3*

Although *Ambostracon (Patagonacythere) tricostata* and *Australicythere polylyca* are quite distinct morphological entities, intermediate species may be found for which the generic assignment becomes rather difficult. For example, *Ambstracon (Ambostracon)* sp. *D* Valentine (1976) is morphologically similar to *Australicythere californicum* (Hazel, 1962). In the morphological series *tricostata-wyvillethomsoni-longiductus longiductus-longiductus antarticus,* the N meshes become more conspicuous fron *tricostata* to *antarticus,* the L meshes become more differentiated and the intensity of the reticulation diminishes. In this way, the morphological distance between *tricostata* and *longiductus antarticus* is greater, and the latter is more similar to *Australicythere devexa* than to *Ambostracon (Patagonacythere) tricostata.*

The coquimbids (*Coquiminae* Ohmert, 1968 – although the present author does not accept their discrete subfamilial status), are presented by all three known genera with five new species. *Coquimba* Ohmert (1968), has two species: *Coquimba* sp. *1* and *Coquimba* sp. *2*. The former species is very similar to *coquimba piscicula* Ohmert (1968), from the Pliocene of Chile, and it seems to be its direct ancestor. Although *Coquimba* sp. *2* is related to the Chilean Pliocene species *Coquimba bicostata* Ohmert (1968), it is not a typical representative of the genus, since in some respects it resembles *Ambostracon (Patagonacythere).*

Nanocoquimba Ohmert (1978), has only one species *Nanocoquimba* sp. *1,* which is thought to be the direct ancestor of the Chilean species *Nanocoquimba apiata* Ohmert (1968).

Cornucoquimba Ohmert (1968), is represented by two new species: *Cornucoquimba* sp. *1* and *Cornucoquimba* sp. *2.* the latter resembles *cornucoquimba angulosa* Ohmert (1968), from the Pliocene of Chile, and could be considered its direct ancestor.

In his original description of this group of species, Ohmert assumed that they probably originated in what he called 'the faunal revolution at the Miocene-Pliocene boundary', but from the present record, it can be seen that they were already well established in the Eocene and Oligocene.

The genus *Urocythereis* Ruggieri (1950) is represented in both formations by two new species: *Urocythereis sp. 1* and *Urocythereis* sp. *2*. The former is one of the most abundant and typical species and seems to be confined to the Eocene-Oligocene of the east of Santa Cruz Province.

The genus *Bradleya* Hornibrook (1952), is represented by only one species *Bradleya normani* (Brady, 1880), and it only occurs in the upper part of the Monte Leon Formation. The lenghts and heights of the few specimens found fall outside the ranges given by Benson (1972) for Pliocene and living material. A possible explanation for their smaller size could be the fact that the water bodies in which deposition of the Monte Leon Formation took place were shallower than those occuring in younger areas in which the species lives. The present record is the first for this species from the Oligocene.

The genus *Aurila Pokorny (1955), is represented by three species Aurila* sp. *1,*

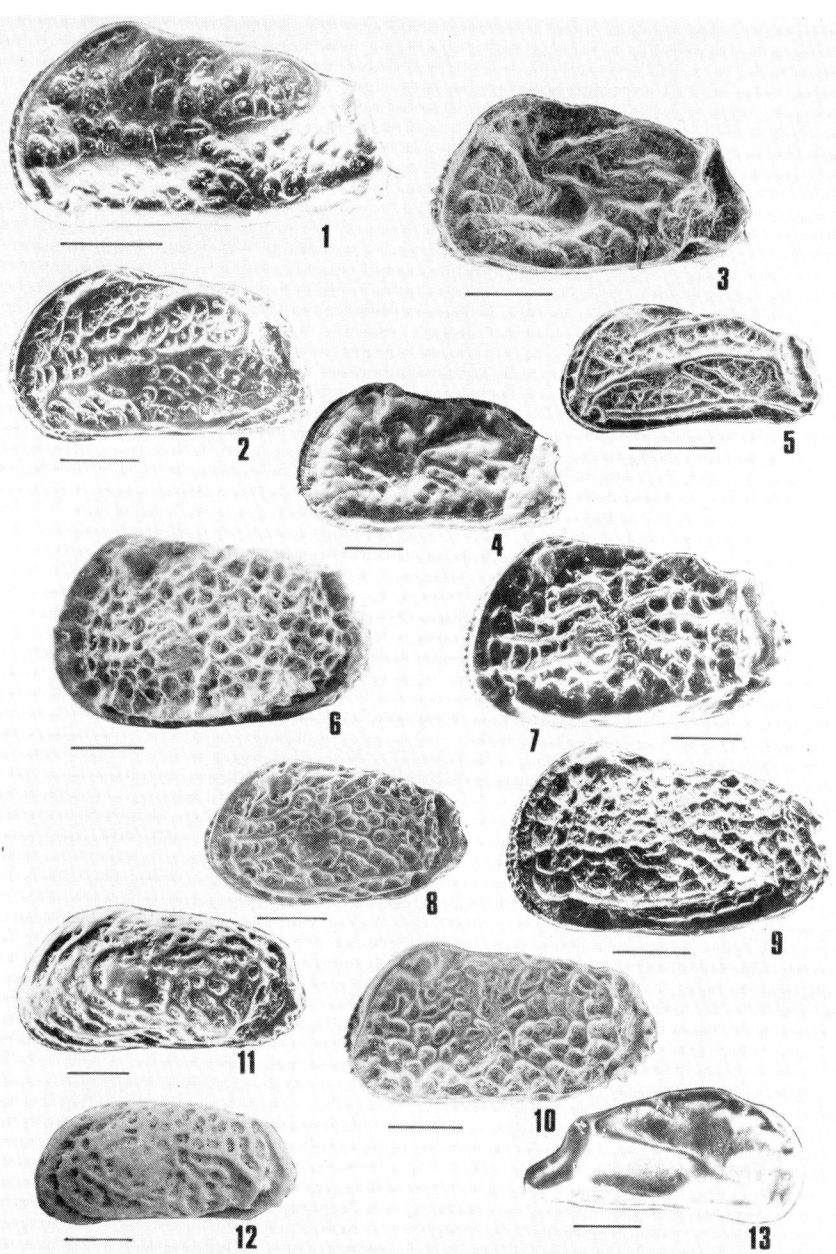

Plate 1. (Graphic scale: 0.2 mm)
1) *Ambostracon (Patagonacythere) tricostata*, left valve, female. Recent. Puerto Deseado. 2) *Ambostracon (Patagonacythere)* sp. (not yet described) left valve, female. Recent. Porvenir. 3) *Ambostracon (Patagonacythere)* sp. *2*, left valve, female. Momte Leon Fm. 4) *Ambostracon (Patagonacythere)* sp. *3*, carapace, female. Monte Leon Fm. 5) *Ambostracon (Ambostracon)* sp. *1*, left valve, male. Monte Leon Fm. 6) *Australicythere* sp *4*, carapace, female. Monte Leon Fm. 7) *Australicythere* sp *1*, carapace, female. San Julian Fm. 8) *Australicythere* sp. *2*, left valve, female. Monte Leon Fm. 9) *Australicythere* sp. *3*, left valve, male. Monte Leon Fm. 10) *Australicythere* sp. *5*, carapace, female. Monte Leon Fm. 11) *Urocythereis* spl *2*, carapace, female. Monte Leon Fm. 12) *Urocythereis* sp. *1*, left valve, male. San Julian Fm. 13) *Caudites* sp., right valve, female. Monte Leon Fm.

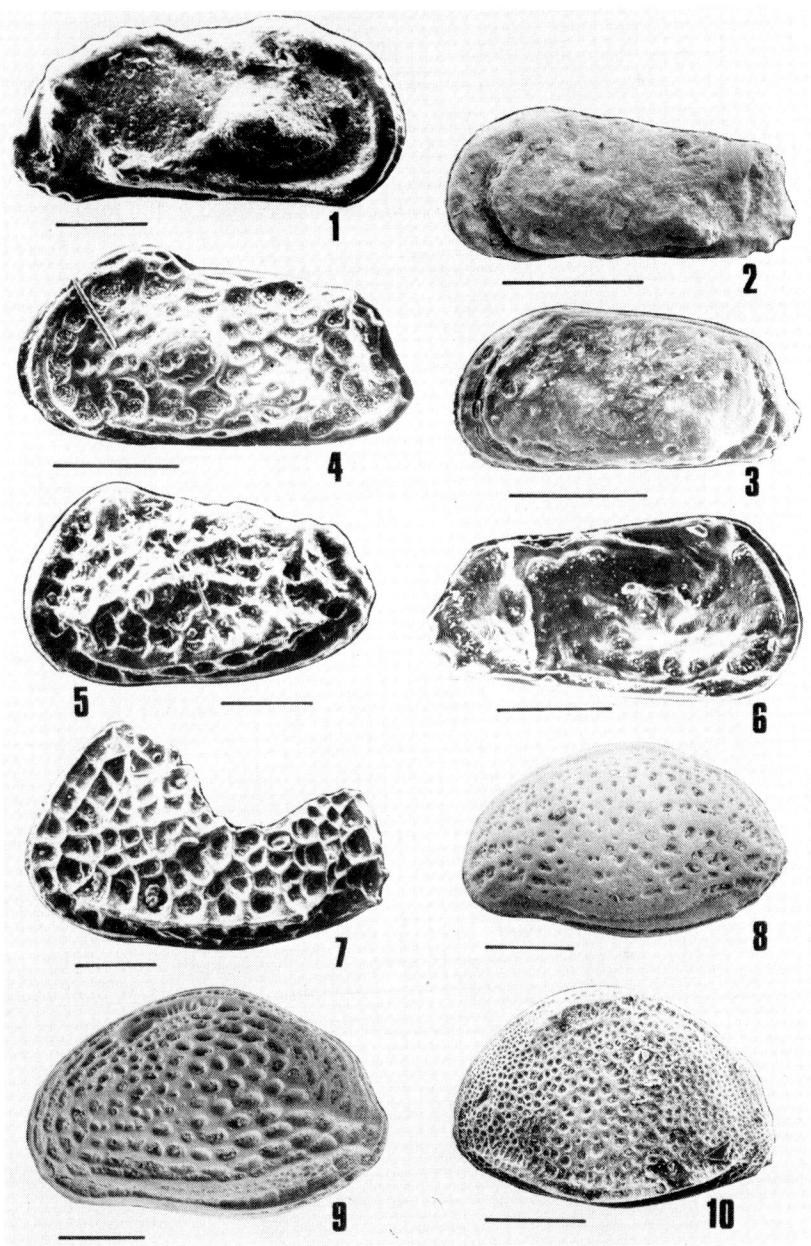

Plate 2. (Graphic scale: 0.2 mm)
1) *Cornucoquimba* sp. *1*, right valve, male. San Julian Fm. 2) *Coquimba* sp. *1*, left valve, male. Monte Leon Fm. 3) *Nanocoquimba* sp. *1, carapace, female*. San Julian Fm. 4) *Cornucoquimba* sp. *2*, left valve, female. Monte Leon Fm. 5) *Coquimba* sp. *2*, carapace, male. San Julian Fm. 6) *Hermanites* cf. *dohmi*, right valve, male. San Julian Fm. 7) *Bradleya normani*, left valve. Monte Leon Fm. 8) *Aurila* sp. *1*, carapace, male. Monte Leon Fm. 9) *Aurila* sp. *2*, carapace, female. Monte Leon Fm. 10) *Aurila* sp. *3*, carapace, female, Monte Leon Fm.

Aurila sp *2*, and *Aurila* sp. *3*. The two former species are morphologically similar to the genotype *Aurila convexa* (Baird 1850). All are abundant in the Monte Leon Formation, but *Aurila* sp. *1* has also been found in the San Julian formation, albeit rarely. The latter is a typical primitive *Aurila* and in some respects resembles *Pokornyella* Oertli (1956). It possesses comparatively few marginal pore canals: about 40 anteriorly and 30 posteriorly, whilst *Aurila* sp. *2* has an average of 80 anteriorly and 75 posteriorly. *Aurila* sp. *1* may be considered ancestral to *Aurila* sp. *2*.

Only one species of the *Hermanites* Puri (1955) was found: in the samples *Hermanites* cf. *dohmi* (Howe and Chambers, 1935). The species was very rare. It has been recorded in the Jackson Eocene of Louisiana, Mississippi and Alabama, and in the Uppermost Alazan formation (Oligocene) of Mexico. In Argentina, it only occurs in the San Julian Formation. This is the first record of it in the southern Hemisphere.

Caudites Coryell and fields (1937) is represented by only one species, *caudites* sp. Rossi de Garcia (1970), which is very rare. It only occurs in the upper part of the Monte Leon Formation and in the Entre Rios formation (Miocene) further north. This appears to be the oldest record of *Caudites* in Southern America.

Acknowledgements

I would like to thank: Dr. Robin Whatley, UCW for supervision of my work and for critical reading of the manuscript, for the loan of Recent hemicytherid species from Argentinian waters and for ideas on the *Ambostracon-Australicythere* relationship, particularly with reference to Skogsberg's species; Dr. J. E. Hazel, USGS, who kindly sent me S. E. M. micrographs of the holotype of *Ambostracon costatum*; Dr. P. C. Valentine, USGS, who sent me S.E.M. micrographs of many species of *Ambostracon* from a paper (in press), from which I took many examples for the discussion on the *Patagonacythere-Ambostracon* problem; Dr. P. C. Sylvester-Bradley who encouraged me to use Liebau's system; Lic. A. Moguilevsky for her helpful encouragement. I would also like to thank Messrs H. Williams and A. Pugh of the UCW Geology Department for photographic assistance.

References

Becker, D. 1964. Micropaleontologia del Superpatagoniense de las localidades, Las Cueva y Monte Entrance (Provincia de Santa Cruz). Ameghiniana III (10): 319-451.

Benson, R. H. 1964. Recent Cytheracean Ostracodes from McMurdo Sound and the Ross Sea, Antarctica. Univ. of Kansas, Palaeont. Contrib. Anthropoda, Art. 6, pp. 1–36.

Benson, R. H. 1972. The Bradleya problem, with descriptions of two new psychopheric Ostracode genera, Angrenocythere and Poseidonamicus (Ostracoda). Smithsonian Contrib. Paleobiol. 12, 138 pp.

Benson, R. H. & Maddocks, R. S. 1964. Recent Ostracodes of Knysna Estuary, Cape Province, Union of South Africa. Univ. of Kansas Paleont. Contrib., Anthropoda, Art 5, pp. 1–39.

Bertels, A. 1970. Sobre el 'Piso Patagoniano' y la representacion de la epoca del Oligoceno

en Patagonia austral, Republica Argentina. Rev Asoc. Geol. Argentina. Rev. Asoc. Geol. Argentina XXV (4): 495–501.
Bertels, A. 1975. Ostracode ecology during the Upper Cretaceous and Cenozoic in Argentna. Bull. Amer. Paleont. 65 (282) 317-349.
Bold, W. A. van den. 1957. Oligo-Miocene Ostracoda from Southern Trinidad. Micropal. 3 (3): 231–244.
Bold, W. A. van den. 1958. Ostracoda of the Basso Formation of Trinidad. Micropal. 4 (4): 291–418.
Bold, W. A. van den. 1963. Upper Miocene and Pliocene Ostracode of Trinidad. Micropal. 9 (4): 361–424.
Bold, W. A. van den. 1966a. Miocene and Pliocene Ostracoda from Northeastern Venezuela. Verh. Kon. Nederl. Akad. Wetenschp. Serie 1, 23 (2): 43 pp.
Bold, W. A. van den. 1966b. Upper Miocene Ostracoda from the Tubara Formation (Northern Colombia). Micrpal. 12 (3): 360–363.
Brady, G. S. 1880. Report on the Ostracoda dredged by HMS Challenger during the Years 1873-1876. Report on the Scientific Results of the Voyage of HMS Challenger. Zoology I: 1–184.
Camacho, H. 1956. La transgresion patagoniense de la costa atlantica entre Comodoro Rivadavia y el curso inferior del Rio Chubut. Rev. Asoc. Geol. Argentina XI (1): 23–45.
Camacho, H. 1967. Las transgresiones del Cretacico Superior y del Terciario de la Argentina. Rev. Asoc. Geol. Argentina. XXII (4): 253–280.
Camacho, H. 1974. Bioestratigrafia de las Formaciones Marinas del Eoceno y Oligoceno de la Patagonia. Anal. Acad. Ci. Ex. Fis. Nat. Buenos Aires 26: 39–57.
Hartmann, G. 1964. Neontological and Paleontological classification of Ostracoda. Pubbl. Staz. Zool. Napoli 33, suppl.: 550–587.
Hartmann-Schröder, G. & Hartmann, G. 1962. Zur Kenntnis des Eulitorals der Chilenischen Pazifikküste und der Argentinischen Küste südpatagonies unter besonderer Berücksichtigung der Plychaeten und Ostracoden. Teil III. Ostracoden des Eulitorals. Mitt. Hamburg. Zool. Mus. Inst., Ergänz. zu Bd 60, pp. 169–270.
Hartmann, G. & Puri, H.S. 1974. Summary of Neontological and Neontological classification of Ostracoda. Mitt. Hamburg. Zool. Mus. Inst. 70: 7–73.
Hazel, J. E. 1962. Two new Hemicytherid Ostracods from the Lower Pleistocene of California. J. Paleont. 36 (4): 822–826.
Hazel, J. E. 1967. Classification and distribution of the Recent Hemicytheridae and Trachyleberididae (Ostracoda) off Northeastern North America. U.S. Geol. Surv., Prof. Paper 564, 49 pp.
Ishizaki, K. 1971. Ostracodes from Aomori Bay, Aomori Prefecture, northeast Honshu, Japan. Tohuku Univ. sci. Repts., Sendai, Japan, Ser. 2 (Geol.), vol. 43, No.1.
Liebau, Al 1969. Homologisierende Korrelationen von Trachyleberididen–Ornamenten (Ostracoda, Cytheracea). Neues Jahrbuch für Geologie und Paleont. Vol. 7, pp. 390–402.
Liebau, A. Comment on suprageneric taxa of the Trachyleberididae (Ostracoda, Cytheracea). Paläont. Abh. 148 (3): 353–379.
McKenzie, K. 1967. Recent Ostracoda from Port Phillip Bay, Victoria. Proc. Roy. Soc. Victoria 80, Part 1: 61–108.
Moore, R. C. (ed.). 1961. Crustacea, Ostracoda. Treatise on invertbrate Paleontology. Q. Arthropoda 3: 1–442.
Neale, J. W. 1967. An Ostracod Fauna from Halley Bay, Coats Land, British Antarctic Territory. Brit. Antarctic Survey Scient. Reports No. 48, pp. 1–50.
Ohmert, W. 1968. Die Coquimbinae, eine neue Unterfamilie der Hemicytheridae (Ostracoda) aus dem Pliozän von Chile. Mitt. Bayer. Staatssamml. Paläont. Hist. Geol. 8: 127–165.
Omatsola, M. E. 1972. Recent and Subrecent Trachyleberididae and Hemicytheridae (Ostr., Crust.) from the Western Niger Delta, Nigeria. Bull. Geol. Inst. Univ. Uppsala N.S. 3 (4): 37–120.
Pokorny, v. 1964. The taxonomic delimitation of the subfamilies Trachyleberidinae and Hemicytherinae (Ostracoda, Crustacea). Acta Universit. Carol., Geologica 3: 275–284.

Rossi de Garcia, E. 1966. Contribucion al conomiento de los ostracodos de la Argentina. I. Formacion Entre Rios, Victoria, Provincia de Entre Rios. Rev. Asoc. Geol. Argentina XXI (3): 194–208.

Rossi de Garcia, E. 1967. Contribucion al conomiento de los ostracodos Cenozoicos de la Argentina. Parte II. Ostracodos del Cordon Litoral Loma del Tajamar. Rev. Asoc. Geol. Argentina. XXII (3): 203–208.

Rossi de Garcia, E. 1969. Algunos ostracodos del Entrerriense de Parana, Provincia de Entre Rios, Republica Argentina. Rev. Asoc. Geol. Argentina XXIV (3): 276–280.

Rossi de Garcia, E. 1970. Ostracodes du Miocène de la République Argentine ('Entrerriense' de la Peninsule Valdez). IV Colloque Africaine de Micropaleontologie, pp. 391–417.

Skogsberg, T. 1928. Studies on marine ostracods. Part II. External morphology of the genus Cythereis, with dercriptions of twenty one new species. Calif. Acad. Sci., Occ. Papers, S. Francisco, No. 15, pp. 1–155.

Swain, F. M. 1967. Ostracods from the Gulf of California. Geol. Soc. Amer., Mem. 101: 1–139.

Swain, F. M. 1969. Taxonomy and ecology of near-shore Ostracoda from the Pacific coast of North and Central America. Tax., Morph., and ecol. Rec. Ostr., Hull, pp. 423–474.

Sylvester-Bradley, P. C. & Benson, R.H. 1971. Terminology for surface features in ornate Ostracodes. Lethaia 4 (3): 249–266.

Valentine, P. C. 1976. Zoogeography of Holocene Ostracoda off Western North America and Paleoclimatic implications. United States Geol. Surv., Prof. Paper No. 916 (in press).

Van Morkhoven, F. P. C. M. 1962. Post Palaeozoic Ostracoda, Vol. 1. General. Amsterdam, Elsevier Publ., 204 pp.

Van Morkhoven, F. P. C. M. 1963. Post Palaeozoic Ostracoda, Vol. 2. Their Morphology, Taxonomy and Economic Use. Amsterdam, Elsevier Publ., 478 pp.

Author's address:

Department of Geology, University College of Wales, Aberystwyth, Dyfed, Great Britain

Sixth Intern. Ostracod Symposium, Saalfelden

CARAPACE ORNAMENTATION OF THE OSTRACODA CYTHERACEA: PRINCIPLES OF EVOLUTION AND FUNCTIONAL SIGNIFICANCE*

ALEXANDER LIEBAU**

Abstract

The components of the fine sculpture of the Ostracoda Cytheracea are classified according to their variability and evolvability. Three evolutionary levels are observed with respect to the differentiation of the ornament genetics. The indicative first-order reticulation is most primitive in the Limnocytheridae, transitional in the Bythocytheridae, and very advanced in the Trachyleberididae and related Cytherecea.

Introduction

This paper presents a summary of results of comparative studies on the carapace sculpture of Trachyleberididae and of other Cytheracea. A detailed illustrated account of this subject is in preparation.

Introductory outlines of the problems of sculpture homologisations are found in Pokorný 1964 and 1969, Benson 1972, and Liebau 1969, 1971 and 1975a and b. A preliminary paper on the ornament classification, as defined here, was presented at the Delaware Symposium in 1972 (see Liebau 1975b). The descriptive terminology is mainly adapted from Sylvester-Bradley & Benson (1971).

Evolutionary levels of sculpture components

The so-called 'Williston's law', as interpreted by Remane et al. (1973: 100; 'Zahlenreduktionsgesetz'), describes a general trend in the evolution of multielemental patterns or sets of serial structures: in the primitive stage, the patterns or structures consist of numerous elements varying in number, then a fixation of the number is observed, and from that point onwards only a reduction seems to be possible. Ostracode characters, which could be studied in this respect, are: e.g. muscle scars, pore systems and mesh and spine patterns. Especially for reticulations, i.e. components of the fine sculpture, a more detailed sequence of evolutionary levels is outlined here:

'Proto' level

The elements (meshes, pore cones etc.) of the ornament component show intra-

---

* Konstruktionsmorphologie No. 68. No 67 = Schmalfuss, H. Morphologie und Funktion der tergalen Rippen bei Landisopoden (Oniscoidea, Isopoda, Crustacea). Zoomorphologie (in press).
** The studies have been supported by the Deutsche Forschungsgemeinschaft. Since 1974 they have been included in the research program (SFB 53) 'Paleoecology', which is established at the Institute for Geology and Paleontology at the University of Tübingen.

specific variation in number and arrangement. Genetic changes may affect the whole ornament component or inexactly defined parts of it, but not single elements. Therefore, single elements do not possess individual characters of e.g. intraspecific constancy and cannot be used individually for taxonomical purposes.
Examples: protoreticulations, protocostulations, protoconations and, as far as observed, all components consisting of extremely small elements (microreticulation, papillations, foveolation etc.).
Comment: Proto-ornamentation represents the primitive stage of ornament genetics. It is observed in practically all ornate Paleozoic ostracodes and in the conservative groups (e.g. Cytherellacea) among the post-Paleozoic ostracodes.

'Meso' level

The elements are constant in number and more or less in their arrangement, but are not 'individualized' with regard to the genetic background. Genetic changes, as in proto-ornaments, do not affect single elements.
Examples: mesoreticulation, mesoconation.
Comment: Meso-ornamentation forms an intermediate stage between the primitive and the highly advanced ornament classes. Reticulation patterns which seem to represent a transition between proto- and macroreticulation are observed in some Bythocytheridae. The anterior spine row of the marginal denticulation of most of the Trachyleberididae ('TX'-row in Liebau 1975a: Fig. 1) is, at present, the best example of a mesoconation.

'Macro' level

The elements are constant in number and arrangement. Genetic changes may affect single elements, the whole ornament component or inexactly defined parts of it.
Examples: macroreticulation, macrocostulation (obviously always derived from the macroreticulation), macroconation.
Comment: The genetic background of a *'Limburgina*-type' macroreticulation (typical for Trachyleberididae s.l.) is so differentiated that about 200 fossae (= meshes) per valve or approximately 400 muri (mesh walls) may be separately changed or reduced, and in this way each may represent taxonomical characters. Occurrence: Trachyleberididae s.l. and other advanced groups among the Cytheracea s.l. (see below).

Advanced Cytheracea

The most complex and advanced ostracode ornamentation studied, occurs in the Trachyleberididae s.l., i.e. a prominent group of the Cytheracea. Here the following sculpture components are distinguished:
fine sculpture (= ornament; cf. microsculpture sensu Bolz 1971):
    macroreticulation (excl. myogene elements)
    microreticulation

nannoreticulation = foveolation
macrocostulation
macroconation (excl. myogene elements)
mesoconation: marginal row
mesoconation: 'normal' mesocones
microconation
nannoconation (2 systems?)
myogene tubercles and/or fossae

coarse sculpture (approx. macrosculpture sensu Bolz 1971):
eye node
sulcation/lobation (incl. muscle node)
costation
'hollow tubercles' (here especially due to cold water influence)

The components of the fine sculpture, as far as studied in detail, are described here with special reference to their 'evolvability', i.e. how they can evolve. This 'evolvability' has been observed in phylogenetic lineages (main examples in the genera *Oertliella, Mosaeleberis, Pokornyella, Limburgina, Alteratrachyleberis, Cletocythereis* s.l.). It has also been deduced from examples of intraspecific variation or constancy, respectively, which yielded general information on the ornament genetics. A summary of the results is presented here. A more detailed account, in connection with phylomorphogenetic examples, will be published separately.

Macroreticulation

A component of the fine sculpture consisting of rounded or polygonal fassae (macrofossae) which show intraspecific constancy* in number and arrangement. Genetic changes may affect single fossae or muri.

The *Limburgina*-type macroreticulation is characterized by a mesh pattern which can be traced mesh by mesh through the major part or even all subfamilies of the Trachyleberididae. The standard pattern for these ornament homologizations is that of *Limburgina ornata* (see Liebau 1969, 1971).

'Spongy' macroreticulations with equal-sized fossae and high, distally broadened muri are often adapted in order to protect the sieve pores or to double the shell thickness especially in the turbulent zone of shallow seas.

Another way of strengthening the carapace is the development of a (macro-) costulation: certain chains of muri are thickened and raised, other muri are reduced. Except for the peripheral flange, all costulae follow the macroreticulation pattern, also when their apparent main function is to connect pore cones. Because the number of meshes and mesh rows is to restricted, there are also limits for the number of costulae: about 23 is the maximum between the dorsal and ventral margin at half length of a valve, and 9 or 10 for costulae parallel to the anterior margin in front of the antennal scar. This is also true for costulations in species

---

* Exceptions, as in similar cases: pathological examples, variation due to intraspecific genetic changes and 'false numerical variation' caused by phenotypic disappearance (= extremely weak expression) of parts of the pattern.

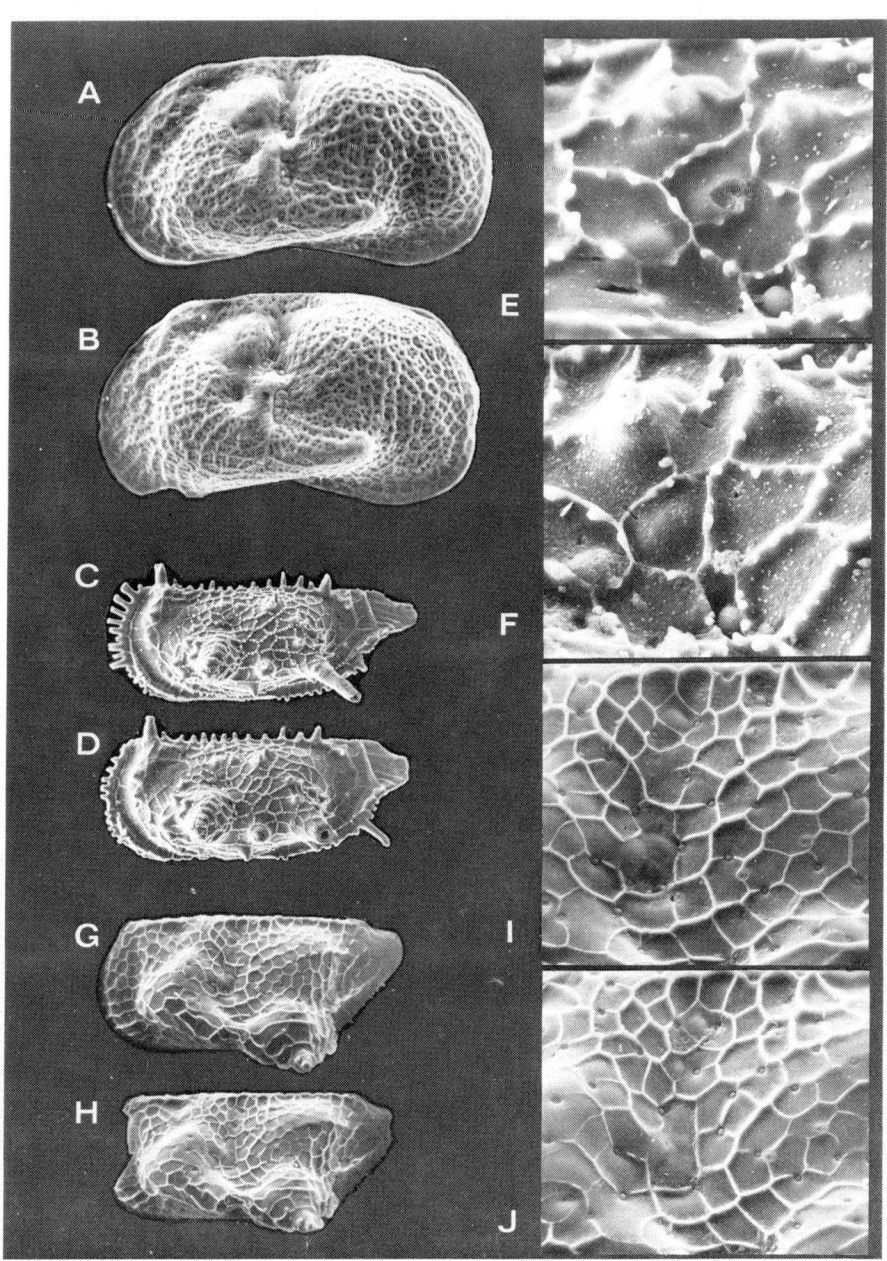

◀ Plate 1. Primitive reticulations of Limnocytheridae (A, B) and Bythocytheridae (C-J).
  A,B) *Limnocythere inopinata* (Baird, 1843) forma 'incisa' (= non-torose variant), left valves, Recent, Tegeler See, Berlin. Length of specimens 0.54 (A) and 0.60 mm (B).
    Protoreticulation, microreticulation.
  C,D) *Nemoceratina (Pariceratina)* sp., left valves, U.Aptian, Sarstedt near Hannover, N.Germany. Length of both specimens 0.79 mm.
  E,F) Enlarged protions (width: 0.103 mm) of figs. C and D.
    Microconate protoreticulation, locally transitional to a constant mesh arrangement. Two pore cones (macrocones?). Hemispherical tubercle in a sunken mesh: myocone belonging to a lower dorsal muscle scar.
  G,H) *Bythoceratina* sp., left valves, Recent, Runaway Bay, Jamaica. Length of specimens 0.53 (G) and 0.49 mm (H9).
  I,J) Enlarged portions (width: 0.195 mm) of figs. G and H.
    Transition proto-/macroreticulation, mesh pattern nearly constant. (The individual differences are rare, but significant of a certain tolerance in the connexion of meshes).

groups without typically reticulate members: in such cases the genetic background of the reticulation seems to be still present in spite of the phenotypic disappearance of the pattern.

Similarly all costae of the Trachyleberididae can be interpreted as raised rows of meshes. For instance, the median rib of *Protocythere* is derived from the M-meshes (of the *Limburgina*-type pattern, see Fig. 1 in Liebau 1969), while the similar ribs in *Cythereis, Mosaeleberis* and *Limburgina* are formed by the N-meshes. The occurrence of tricostate forms in many groups of the Trachyleberididae does not reflect the existence of an old basic costation pattern but is due to a functionally controlled tendency.

The macroreticulation with its 'genetically individualized' muri can easily be transformed, within very few evolutionary steps, into all kinds of costae or costulae. This 'box-of-bricks principle' includes the possibility of reverse processes: costae can be replaced by mesh rows again, every costulation pattern can be redeveloped to an undifferentiated reticulation. Excellent models for these ornament changes are found in many ontogenies.

Phenotypic disappearance by microreticulization: In a pre- or post-adaptive stage, the macroreticulation becomes reduced and may be replaced by an entirely smooth shell surface or by microreticulation. 'Microreticulization', i.e. the phenotypic replacement of macro- by microreticulation, as well as the reverse process, the 'rejuvenation' of the macroreticulation, belong to the most common ornament changes which are observed in Cytheracean lineages (for a good example see Oertli 1964). The occurrence of microreticulation is correlated with a certain stage of weak shell calcification. This explains the observation that many genetically controlled microreticulation changes are 'imitated' by phenocopies, i.e. phenotypically analogous developments caused by ecological influences only; both microreticulization and rejuvenation of the macroreticulation may represent both evolutionary and ecologically controlled processes. For the same reason, the ontogenetic development of the ornament is usually a good model for a 'rejuvenation' of the macroreticulation (the weakly calcified instars tend to develop a fine microreticulation).

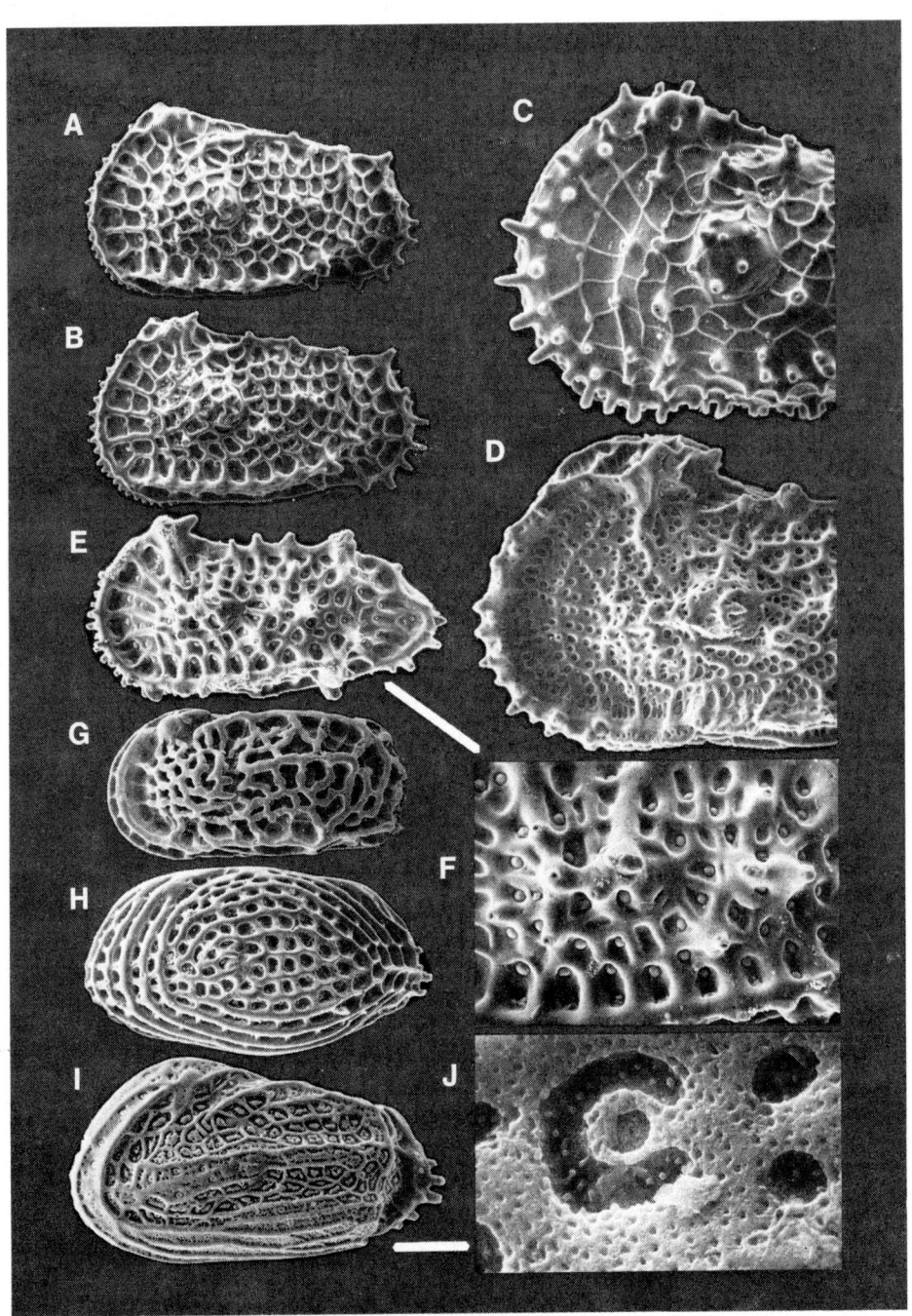

Plate 2. *Limburgina*-type macroreticulation (Trachyleberididae s.l.).

A,B) *Limburgina ornata* (Bosquet, 1847), left valves of larvae (last instar), U.Maastrichtian, Benzenrade near Maastricht, Netherlands. Length of specimens 0.72 (A) and 0.68 mm (B).
  Typical macroreticulation, not differentiated except for slightly broadened muri. The anterior row of marginal denticles consist of mesocones.

C) *Cythereis* (s.l.) *ciplyensis* (Marlière, 1958), anterior part of left larval valve (last instar), Danian of Curfs quarry, Berg near Maastricht, Netherlands. Height: 0.40 mm.
  Pre- (or post-) adaptive macroreticulation in a modified *Limburgina*-pattern. Anterior margin with 3 rows of pore cones, the first (= adsagittal) one composed by mesocones and connected with the true radial pore canals. The rest of the pore cones are macrocones. Minute tubercles placed upon mesh walls indicate intramural pores, which are homologous to the mesh-pores of *Oertliella* and *Mosaeleberis* (s. below).

D) *Rehacythereis* sp., anterior part of left valve, U.Aptian, Sarstedt near Hannover, N.Germany. Height: 0.43 mm.
  Macrofossae filled with microreticulation, parts of the pattern being entirely microreticulized. Macrocones inconspicuous and similar to the intramural pores.

E) *Oertliella* sp., left valve (♂), U.Santonian, Casas de Canelles, Lérida, Spain. Length of specimen 0.95 mm.
  Macroreticulation pattern close to that of *Limburgina*, differing by certain mesh fusions only.

F) Same species and locality as fig. E, central area of left valve (♀). Width of enlarged portion: 0.50 mm.
  Most of the macrofossae provided with large hemispherical mesh-pores (sieve-type). In certain macrofossae the pores are constantly missing. Pairs of mesh pores indicate collective (= fused) fossae. Large macrocones with distal pores.

G) *Hammatocythere* sp., left valve (♂), Biarritzian (late M.Eocene), Bois de Barbe near Blaye, W.France. Length of specimen 1.11 mm.
  Differentiated macroreticulation adapted for the protection of the carapace in the marine turbulence zone. Posterior part of the reticulation dominated by large collective fossae, some of them still provided with relictic muri of the elemental macrofossae.

H) *Alteratrachyleberis? striatopunctata* (Roemer, 1838), left valve (♀), Lutetian, Grignon, Paris Basin. Length of specimen 0.94 mm.
  Limburgina-type macroreticulation transformed into a costulation. (For the homologization of the mesh patterns see Liebau 1971: 82).

I) *Mosaeleberis?* sp., left valve (♀), early Cenomanian, Mülheim-Broich, N.Germany. Length of specimen 0.63 mm.
  Macroreticulation partly microreticulized and/or replaced by costulae. Two variants of the microreticulation occur, one of them resulting in cuneiform fillings of the macrofossae. Normal microfossae are found on the median costa and the ventral side. Note transitions between rib pattern, costulation and macroreticulation.

J) Detail from fig. I (mid-height of the valve below eye node). Width of enlarged portion 0.036 mm.
  Horse-shoe groove derived from microfossae surrounding mesh-pore. At the right a pair of normal microfossae. Nanno-ornamentation: foveolae (= nannofossae) covering raised part of ornament, papillae (= nannocones) at the sola of microfossae.

Plate 1 and 2: All measurements without spines. Except for plate 2, Figs. A-C, all specimens are adults.

The following rules refer to the relation between macro- and microreticulation:
1) There is an upper limit for the number (about 200) of macrofossae per valve. If additional meshes are developed these are microfossae, i.e. varying in number and arrangement.
2) There is an upper size limit for microfossae: whenever fossae within a microreticulation exceed a certain size (in relation to the valve surface) they develop the characters (especially constant positions) of the macroreticulation.
3) Fossae surrounded by microcones belong to the macroreticulation.

'Celation' and similar phenomena: In many species groups and even within populations, the macroreticulation is smoothed, i.e. replaced by an entirely smooth or just conate valve surface. This process is described under the terms of 'celation' in Sylvester-Bradley & Benson (1971) or the effect of a 'facteur lisse' by Carbonnel (1975). In both studies, no differentiation is made between smoothing as a result of extreme microreticulization and the direct transformation from a macroreticulate into a smooth valve surface, although these types of smoothing are usually at opposite ends of morphogenetic, phylomorphogenetic and ecology-controlled developments (the examples in the papers mentioned show both types of smoothing). This means that there are at least two types of smoothing of the reticulation, which are correlated with different stages of the shell calcification. A genetically fixed ornament smoothing is characteristic for many Brachycytherinae, but there are also examples for a phenotypic rejuvenation of the macroreticulation *(Opimocythere* gr. *elonga, Pterygocythereis* gr. *tuberculata)*.

Relation to pore systems: Many Hemicytherinae and some relatives of *Cythereis* show a direct correlation of the *Limburgina*-type macroreticulation with a mesh-pore system. In Liebau (1969, 1971a) it was supposed that this was a primitive character, indicating the origin of the macroreticulation, but now there is evidence that these mesh-pores were derived from intramural ones during the earliest Cretaceous.

Microreticulation

A component of the fine sculpture consisting of small rounded or polygonal meshes or pits (microfossae) which show intraspecic variation in number and arrangement. Microfossae are always smaller than neighbouring macrofossae and are not related to a pore system. Their number increases in cases of environmentally determined weak shell calcification.

Microreticulation is present in practically all ornate families of the Cythracea. It occurs in ecolabile and ecostable variants. Its appearance is presumably correlated with a certain kind of weak shell calcification. Nevertheless, 'microreticulate' does not always mean 'thin-shelled' and there are many examples of heavily calcified microreticulate carapaces. It is possible that the strength or structure of only one shell layer is responsible for the origin of the microreticulation and that other layers supply the shell thickness. In some cases, SEM photographs indicate a conspicuous electron absorption at the bottom of the microfosae apparently corresponding to a higher organic content of an interior shell layer (see also Herrig, 1965: 416 ff., who describes the shell structure of microreticulate Trachyleberididae).

'Microreticulization', i.e. the replacement of macro- by microreticulation, is described further on in this paper.

Nannoreticulation

A component of the fine sculpture consisting of minute pits or meshes (nannofossae) varying in number and arrangement. In contrast to the microreticulation nannofossae may also cover the mesh walls of other reticulation systems (muri of macro- and microfossae).

The occurrence of nannoreticulation seems to be correlated with a strong shell calcification (which does not always exclude the presence of an ecostable microreticulation). The term 'foveolation' in Sylvester-Bradley & Benson (1971) refers to the same ornament component, but is defined differently. Probably the same sculpture component is described as the faveolate type (favus = honey comb) of the 'mineralized minutisculpture' in Langer (1973:33).

Macrocostulation

A component of the fine sculpture consisting of macrocostulae, i.e., ledges with constant position (at least for their central or highest portions).

For the relations between macroreticulation and (macro-) costulation see above. Linear costulae are the normal ones. They are derived from mesh wall chains and only in a few cases do they represent 'costulized' ribs (costae). (Typical ribs (= costae) are distinctly broader than costulae; the term 'carina' should be reserved for elements transitional between costae and costulae.)
Cycloid costulae surround groups of macrofossae, which may be fused thus forming 'collective fossae'.

Macroconation

A component of the fine sculpture consisting of pore cones (or spines or tubercles similar to pore cones) which have constant positions and may be individually affected by genetic changes.

Myogene and ophthalmogene sculpture elements are not included in the definition of the macroconation. All macrocones studied have turned out to be pore cones. Among them are complex pore cones, containing two or more pore canals, the rest are simple ones (one pore canal only). Simple macrocones situated close together may fuse into a single complex macrocone. The reverse evolutionary process is also observed. Presence, shape and size of single macrocones may be characteristic for species and subgenera.

In the Trachyleberididae, two or three phylomorgenetic generations of macrocones are present. The older one, pre-Cretaceous in age, is observed in nearly all subfamilies and consists of T1-T18 plus T26 (see Liebau 1969: Fig. 1) and probably TZ9. A number of smaller macrocones which occur in many Cythereidinae n.s.* and Trachyleberidinae apparently were developed in the lowermost Cretaceous, somewhat preceding or simultaneous with the majority of the marginal macrocones (TY- and TZ-rows, perhaps except TZ9; see Liebau 1975a).

Together with the origin of the cythereidine intramural and mesh pores and with the raised quantity of true marginal pore canals in advanced Hemicytherinae (Hemicytherini, Aurilini), there are four to five systems of pores which show a sudden increase in number of elements in the evolution of the Trachyleberididae. These phases are followed, at least in some branches of the family, by gradual reductions in the number of pores, a trend which better corresponds with 'Williston's law' than the originating processes.

Cone pore structure: The pores belonging to the macrocones are normally described as simple ones, i.e. without a sieve-type distal opening. A small sieve plate with a loop-shaped subcentral main opening was observed in cone pores of *Protocythere (Saxocythere) tricostata* and a *Rehacythereis* sp. (both from the N. German Aptian) and in an Oertlielline instar from the Gosau Coniacian. As far as is known, cone pores form the bases of sensory bristles.

Mesoconation

A component of the fine sculpture consisting of pore cones (or tubercles or spines similar to pore cones) which have constant positions without being individually affected by genetic changes.

Mesocone characters are represented by the adsagittal row of anterior marginal denticles at least of the Cythereidinae, Trachyleberidinae and primitive Hemicytherinae (TX-row in Liebau 1975a: Fig. 1). A questionable case is the 'disjunctive spines' (Sylvester-Bradley & Benson 1971). Also the 'spongiogene' spinelets which protect the mesh openings in spongy reticulations are often constant in arrangement (one pair upon each mesh wall).

Spongiogene and disjunctive spines can also be considered as protrusions of the mural crests of the macroreticulation. They disappear, in contrast to the conjunctive spines, when the reticulation is smoothed or replaced by microreticulation.

Microconation

A component of the fine sculpture consisting of small spines of tubercles (microcones) varying in number and arrangement. As far as observed, microcones have no pore cone function, they disappear in cases of environmentally deter-

---

* Cythereidinae n.s.: Mauritsinini and Parvacythereidini are grouped together with the Cythereidini *(Cythereis, Rehacythereis)* in the subfamily Cythereidinae s.l. (fam. Trachyleberididae). The reasons are: there are all transitions in sculpture and muscle scar pattern between *Cythereis, Rehacythereis, Matronella*, several undescribed genera, *Parvacythereis, Curfsina* and *Mauritsina*. The V-shaped dorsal adductor scar observed in Mauritsinini is also present in *Rehacythereis* and in a Santonian species of *Cythereis*. Brachycytherinae s.s. (= Brachycytherini in Liebau 1975a) are regarded as a closely related, but well distinguishable subfamily.

mined weak shell calcification and normally do not occur together with microfossae on the same carapace.

Microcones may show uniserial, biserial and areal arrangement. The uniserial (or 'string-of-pearls') type is best observed in late larval stages of *Echinocythereis* species, the biserial one forms part in many spongy ornamentations (with transitions to the 'spongiogene mesoconation', see above), and the areal microcones may cover especially mesh sola like a meadow formed by minute spines or tubercles. The development of all these types depends on a stable shell calcification. As a rule, microcones are reduced in cold or brackish water and e.g. in populations living upon a substrate poor in oxygen. This means that microconation and microreticulation belong to opposite calcification stages. Nevertheless, there are some examples, all belonging to the Mauritsinini and Parvacythereidini, which show a direct transition from microconation into microreticulation on one and the same carapace (*Kikliocythere* gr. *labyrinthica, Paracaudites* spp., Mauritsininorum gen. spp.). This phenomenon is quite exceptional compared with some thousand examples from several families showing all the 'normal behavior' of the micro-ornamentation.

'Nannoconation'

There are one or two systems of spines or papillae still smaller than microcones. The rest of the Cytheracean families have not been studied in detail. Nevertheless, the presence of both macro- and microreticulation and macrocostulation and usually also traces of a macroconation seem to relate many other families with the Trachyleberididae s.l. As a whole, the following families apparently have approximately the same kind of ornament genetics:
Trachyleberididae s.l. (including also Brachycytherinae, Hemicytherinae, Cytherettinae, Palaeocytherideinae, Protocytherinae, Mauritsinini etc.),
Progonocytheridae s.l.,
Cytherideidae s.l.,
Cytheridae (incl. Schizocytherinae),
Cytheruridae (incl. Paracytherideinae)
Loxoconchidae,
Leptocytheridae.

Primitive Cytheracea

Limnocytheridae and Bythocytheridae, two families among the Cytheracea with well ornamented representatives, do not show a true macroreticulation. The mesh patterns observed in these families correspond to the meso- and proto-levels defined above, i.e. the meshes or muri are not 'genetically individualized' although otherwise quite similar to macrofossae in their general appearance.

Protoreticulation, -costulation and -conation correspond to the components of the macro-ornamentation in the size and usually also the function of their elements, but the protofossae, protocostulae and protocones vary in their number and have no definite positions:

Protoreticulation

A component of the fine sculpture consisting of rounded or polygonal fossae ('protofossae') which show intraspecific variation in number and arrangement.

Protofossae are comparable in size to macrofossae. They may surround groups of microfossae. Protoreticulations may derive from or be transformed into protocostulations.

Protocostulation

A component of the fine sculpture consisting of protocostulae, i.e., thin ledges verying in number and arrangement.

Good examples for protoreticulation and protocostulation are observed in Silurian *Clavofabella* species. The protoreticulation of a *Moorites* species from the Carboniferous is connected with a sieve pore system. In many cases, however, the distinction between proto- and microreticulation is more or less arbitrary; they are both 'varioreticulations'. A similar problem arises from the 'finger-print' type of costulations.

Protoconation

A component of the fine sculpture consisting of pore cones (or tubercles or spines similar to pore cones) varying in number and arrangement.

Typical protoconations occur in many Paleocopida, but also in some Platycopida. In some Beyrichiidae proto- and macrocones are combined (Liebau 1975b):.

The sculpture of *Limnocythere* consists of sulci, 'hollow tubercles' and two systems of fossae. The smaller fossae are grouped within the larger ones forming a 'second-order reticulation' overlain by the coarse 'first-order reticulation' (see Sylvester-Bradley & Benson 1971) as it is observed for the combination of micro- and macroreticulation in advanced Cytheracea. In *Limnocythere*, however, the large fossae vary in number, too; they belong to a protoreticulation. The same combination of micro- and protoreticulation was observed in a *Theriosynoecum* species. *Timiriasevia mackerrowi*, as figured in Clements (1974), is a good example for a protocostulation, showing the typical finger-print type of interindividual variation and being derived from a protoreticulation (see instar figured l.c.). This ornament variation (varying first-order fossae, finger-print variation of costulation) is impossible for the 'modern' Cytheracea. In contrast to advanced Cytheracea, Limnocytheridae have had obvious difficulties in developing constant costae or costulae; the ventral rib of several *Theriosynoecum* species is an ecostabilized connection between hollow tubercles, as it is observed in some Cyprideidini.*

Post-Triassic reticulate Bythocytheridae, as far as studied, possess subconstant mesh patterns. Often major mesh groups are found in approximately the same arrangement in a number of specimens, but some 1–5% of the meshes cannot be

---

* Cyprideidini = *Cyprideis* group.

identified in the interindividual comparisons. There is also a conspicuous tolerance in the connection of the muri. Groups of protocostulae were observed, but also typical macrocostulae, especially parallel to the valve margins and in prominent parts of the carapace. Microreticulation and -conation seem to occur alternatively as in advanced Cytheracea. As a whole there are primitive characters, but also transitions to the ornament genetics observed in advanced Cytheracea.

Conclusions

With regard to the differentiation of the genetic background, three evolutionary levels are observed for the fine sculpture of the Cytheracea.

The best indicator of the evolutionary development of the ornamentation is the coarse or first-order reticulation, which is represented by a primitive protoreticulation in the Limnocytheridae, by a transition from a proto- to a mesoreticulation in the Bythocytheridae studied, and by a macroreticulation in the Trachyleberididae and other advanced Cytheracea. These types of reticulation obviously form a phylomorphogenetic lineage which is representative for a fundamental evolutionary process: the evolution of a multielemental pattern from element variation towards element constancy. The striking similarities between e.g. the protoreticulation of *Limnocythere inopinata* and the macroreticulation of a *Cytheromorpha* species mask the huge differences with regard to ornament genetics and evolvabilities. The 'genetically individualized' elements of the macro-ornamentation permit evolutionary changes in about every scale. The main advantage of the macroreticulation is the possibility to construct constantly arranged costulae or costae – originating from mesh wall chains and entire mesh rows – in a few phylogenetic steps in every part of the carapace surface.

Because of the functional importance of reticulation, costulation and costation for the strengthening of the carapace, the genetically most advanced sculpture was apparently developed in the marine turbulent zone. More primitive types of ornament genetics were maintained in the quiet waters of the marine deep sublittoral and bathyal zones – where Bythocytheridae preferably are found – and of freshwater lakes, the biotope of the Limnocytheridae.

Zusammenfassung

Die Komponenten der Feinskulptur der Ostracoda Cytheracea werden nach ihrer Variabilität und Evolvibilität definiert und geordnet. Hinsichtlich der Differenzierung des genetischen Fundaments können drei Evolutionsniveaus unterschieden werden. Besonders kennzeichnend ist der Entwicklungsstand der Grobretikulation; diese ist primitiv bei den Limnocytheridae, vermittelnd bei den untersuchten Bythocytheridae und extrem abgeleitet bei den Trachyleberididae und ihren Verwandten.

References

Benson, R. H. 1972. The Bradleya problem, with descriptions of two new psychrospheric ostracode genera, Agrenocythere and Poseidonamicus (Ostracoda: Crustacea). Smithsonian Contrib. Paleobiol. 12: 1-138.

Carbonnel, G. 1975. Le facteur lisse chez certains ostracodes tertiaires: un index de paléotempérature. Bull. Amer. Paleont. 65 (282): 285-301.

Clements, G. 1974. On Timiriaesevia mackerrowi Bate. In: Sylvester-Bradley, P. C. & Siveter D. J., A Stereo-Atlas of Ostracod Shells, Vol. 2. Leicester (Dept. Geol. Univ.).

Herrig, E. 1965. Cythereis reticulata varia ssp. n., eine neue Ostracoden-Unterart aus der Rügener Schreibkreide (Unter-Maastricht). Ber. geol. Ges. DDR 10: 403-419.

Langer, W. 1973. Zur Ultrastruktur, Mikromorphologie und Taphonomie des Ostracoda-Carapax. Paleontogr. Abt. A. 144: 1-54.

Liebau, A. 1969. Homologiesirende Korrelationen von Trachyleberididen-Ornamenten (Ostracoda, Cytheracea). N. Jb. Geol. Palaeont. Mh. 1969: 390-402.

Liebau, A. 1971. Homologe Skulpturmuster bei Trachyleberididen und verwandten Ostrakoden. Thesis, T.U.B., Berlin, 118 pp.

Liebau, A. 1975a. Comment on suprageneric taxa of the Trachyleberididae s.n. (Ostracoda, Cytheracea). N.Jb. Geol. Paläont. Abh. 148: 353-379.

Liebau, A. 1975b. The left-right variation of the ostracode ornament. Bull. Amer. Paleont. 65 (282): 77-86.

Oertli, H. J. 1966. Etude des Ostracodes du Crétacé Supérieur du bassin cotier de Tarfaya. Notes Mém. Serv. Géol. No. 175, pp. 267-278.

Pokorný, V. 1964. New species of Pterygocythereis (Ostracoda, Crustacea) from the Upper Cretaceous of Bohemia. Acta Univ. Carolinae, Geologica 1966 (4): 305-320.

Pokorný, V. 1969. Radimella gen. n., a new genus of the Hemicytherinae (Ostracodae, Crust.). Acta Univ. Carolina, Geologica 1968 (4): 359-373.

Remane, A., Storch, V. & Welsch, U. 1973. Evolution. Tatsachen und Probleme der Abstammungslehre. München, Dtsch. Taschenbuch-Verl., 241 pp.

Sylvester-Bradley, P. C. & Benson, R. H. 1971. Terminology for surface features in ornate ostracodes. Lethaia 4: 249-286.

Author's address:

Geol.-Paläont. Institut, Sigwartstrasse 10, D-7400 Tübingen 1, B.R.D.

Sixth Intern. Ostracod Symposium, Saalfelden

L'ORNAMENTATION CHEZ LES LOXOCONCHIDAE: UN MOYEN DE PREDICTION (PALEO)ECOLOGIQUE

G. CARBONNEL & M. JACOBZONE

Abstract

The connections between the ornamentation (smooth, coarsely pitted, reticulate, ribbed and tuberculated) in some *Loxoconchidae* and ecological parameters are established, thanks to cluster analysis and correspondence analysis. These have shown the influence of salinity upon coarsely pitted species, the influence of temperature and depth upon reticulate/costulated ones, and a few coarsely pitted species. Paleoecological application is attempted for two species.

Avant-propos

Chargé récemment dans le cadre de la 'Commission d'étude des Ostracodes récents' de la famille des *Loxoconchidae,* nous avons entrepris en 1972 le recensement, sur cartes perforées, des nouvelles espèces actuelles et fossiles.

Un sondage a permis, la même année, à chaque auteur vivant de préciser luimême, les conditions écologiques ou paléoécologiques des espèces qu'il avait créées.

Cette étude est ainsi le prolongement de cette enquête. Elle ne porte que sur des espèces actuelles du genre *Loxoconcha* et accessoirement celles des genres *Hirschmannia* et *Elofsonia*

1. But de cette étude

Le point de départ résulte de la constation suivante: il semble exister une relation préférentielle (partielle ou totale?) entre l'ornementation des ostracodes actuels et certains paramètres écologiques. Déjà O. Elofson (1941) l'avait entrevue. Une réponse affirmative permettrait alors des applications paléoécologiques fructueuses.

On a mélé ici, aux données actuelles majoritaires, de rares espèces fossiles (voir 3) comme *L. atlantica* Hazel.

2. Variables analysées

Elles sont de 2 ordres. Les premières concernent certains paramètres écologiques, les secondes le type d'ornementation. On se rappellera que lors de l'enquête qui a précédé cette analyse, les auteurs vivants ont, eux-mêmes, (sauf exception) défini les caractéristiques écologiques et ornementales de leurs espèces.

2.1. *Variables écologiques*

2.1.1. Salinité

On a retenu 3 possibilités
- saumâtre pour une salinité < 18 ‰
- marin/saumâtre pour une salinité comprise entre 18‰ et 30‰
- marin pour une salinité > 30‰
- Aucune donnée ne manque pour les espèces étudiées.

2.1.2. Température

Chaque auteur pouvait choisir entre la température moyenne caractéristique ou les températures extrêmes. On note 23% de données absentes*.

2.1.3. Profondeur

Chaque auteur pouvait également indiquer la profondeur moyenne ou extrême. On note seulement 0,2% de données absentes*.

2.1.4. Nature du substrat

On a regroupé d'une part, les substrats sableux et à fraction grossière, d'autre part, les vases, boues, argiles et fraction fine. On note 1% de données absentes*.

2.1.5. Présence de végétation sous-marine

La présence d'algues ou de phanérogames dans le milieu n'implique pas nécessairement que l'ostracode vivait sur l'algue ou à ses dépens.

2.2. *Variables morphologiques*

Dans cette analyse, 5 types d'ornementation sont retenus:
- type lisse ou légèrement ponctué (15 espèces): représenté par *Lox. valerii* Caraion, 1964 (fig. 1.1).
  (= smooth/pitted)
- type fortement ponctué (6 espèces): représenté par *Lox. subalata* Brady, 1911 (fig. 1.2).
  (= coarsely pitted)
- type réticulé (19 espèces): représenté par *Lox. tosaensis* Ishizaki 1968 (fig. 1.3).
  (= reticulate)
- type costulé (9 espèces): représenté par *Lox. optima* Ishizaki, 1968 (fig. 1.4).
  (= ribbed)
- type tuberculé (1 espèce): représenté par *Lox. mediterranea*, Müller, 1894 (fig. 1.5).
  (= tuberculated)

Certains types peuvent coexister simultanément chez une espèce. C'est le cas de l'ornementation réticulée/costulée chez *Lox. hattorii* Ishizaki. Cette possibilité est prise en considération dans la suite de l'analyse. L'attribution de l'ornementation d'une espèce à l'un des types définis est souvent très difficile.

---

* Après compléments bibliographiques réalisés à partir du fichier de cartes perforées du Centre de Lyon.

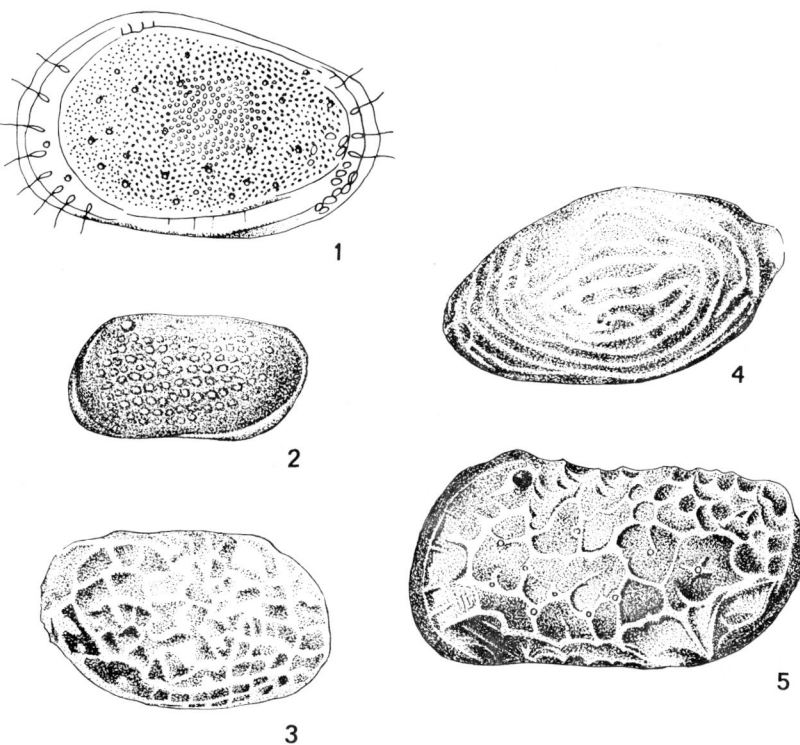

fig. 1. Différents types d'ornementation chez les *Loxoconchidae*.

3. Liste du matériel étudié et données écologiques et morphologiques

La totalité des résultats de ce sondage n'a pu faire l'objet de cette analyse en raison de la disparité des informations ainsi recueillies. Seules ont été retenues les espèces (en principe actuelles) dont la majorité des paramètres écologiques était connue.

On trouvera, ci-dessous, la liste des espèces (au nombre de 42). Deux n'appartiennent pas au genre *Loxoconcha*. Il s'agit d'*Elofsonia baltica* et *Hirschmannia tamarindus*.

1. A	Maddocks		11. granulata	Sars
2. alata	Brady		12. hatorii	Ishizaki
3. atlantica	Hazel		13. impressa	(Baird)
4. B	Maddocks		14. japonica	Ishizaki
5. B	Maddocks		15. kattoi	Ishizaki
6. Elofsonia baltica	(Hirschmann)		16. laeta	Ishizaki
7. bulgarica	Caraion		17. kitani	Ishizaki
8. elliptica	Brady		18. levis	Müller
9. fragilis	Sars		19. longipes	Sars
10. galilea	Lerner-Seggev		20. mediterranea	Müller

123

21. modesta	Ishizaki		31. sperata	Williams
22. multipunctata	Normann		32. stellifera	Müller
23. nana	Marinov		33. subrhomboïdea	Brady
24. OC	Maddocks		34. subulata	Brady
25. OE	Maddocks		35. Hirschmannia tamarindus	(Jones)
26. OF	Maddocks		36. tosaensis	Ishizaki
27. optima	Ishizaki		37. uranouchensis	Ishizaki
28. pulchra	Ishizaki		38. valerii	Caraion
29. pusilla	Brady et Robertson		39. versicolor	Müller
30. semistriata	Kingma		40. viva	Ishizaki
41. warneri	Williams		42. zamia	Ishizaki

Leur répartition géographique est choisie suffisament vaste pour recouvrir les différents océans. Ce choix, naturellement arbitraire, pourra faire l'objet de modifications ultérieures. C'est à partir de l'ensemble des données brutes du tableau 1 que le problème de la relation préférentielle entre un ou plusieurs types d'ornementation et certaines conditions écologiques devra être résolu.

4. Méthodes d'études: analyses multivariées

Elles sont uniquement du type multivarié à l'opposé des études réalisées par O. Elofson sur un sujet voisin. Elles nécessitent la codification des informations.

4.1. Codification des informations

On ne pouvait envisager de traiter ce problème en conservant les valeurs brutes, moyennes ou extrêmes fournies par les auteurs. La codification a pris en considération certains seuils retenus comme caractéristiques par quelques auteurs, en particulier O. Elofson (1941). La codification est binaire (01: présence, 00 = absence)*.
- seuils de salinité: ils correspondent aux seuils définis, lors du sondage (voir 2.1.1.).
- seuils de profondeur: 3 seuils ont été définis
P1 < 3 – 4 m: correspondant aux espèces franchement littorales.
P2 < 15 – 20 m: espèces sublittorales
P3 > 15 – 20 m: espèces infralittorales
- seuils de température: c'est la résistane à la température minimale qui conditionne, dans cette étude, la diversification de l'ostracofaune.

On peut ainsi, en suivant O. Elofson, séparer des espèces résistant à une température de $0°$ C = t1; $2°/3°$ C = t2; $4°$ C = t3. L'influence d'un seuil supérieur de température n'a pas été retenu. Le choix des seuils est provisoire et révisable. Il résulte de l'existence d'études autoécologiques des espèces, euvre loin d'être achevée.

---
* La matrice des données ainsi codées sera utilisée dans les calculs.

Variables écologiques	Variables morphologiques	Lisse/faiblement ponctué	Fortement ponctué	Réticulé	Costulé	Tuberculé	Total
Salinité	marin	3	4	15	7	1	30
	saumâtre	3	–	–	–	–	3
	mar/saum.	9	2	4	2	–	17
Profondeur	P1	5	–	2	1	1	9
	P2	4	2	11	4	–	21
	P3	6	3	6	4	–	19
Température	t1	2	1	–	–	–	3
	t2	5	–	–	–	–	5
	t3	4	2	17	7	–	30
	Total	15	6	19	9	1	

Tabl. 1 — **Ensemble des combinaisons variables écologiques (à l'exception de la nature du substrat et des algues) et variables morphologiques.**

4.2. Analyse de groupe – Résultats

C'est une procédure bien connue (Sneath & Sokal 1965, Kaesler 1966, Maddocks 1966, Hazel 1970, et Carbonnel 1973). Elle conduit à l'obtention d'un dendogramme. Dans ce problème seules nous préoccupent les relations existant entre l'ornementation et les variations écologiques des 42 espèces (analyse de mode Q).

Le dendogramme de la fig. 2 permet de séparer 3 groupes morphoécologiques pour des taux de liaison variant de 0,15 à 0,55.

Groupe morphoécologique 1: Ce sont des espèces lisses vivant dans des milieux de salinité variable (marin à saumâtre), pour une profondeur quelconque, sur un substrat fin (vases, boues, argiles).

Groupe morphoécologique 2: Ce sont des espèces réticulées, vivant à une profondeur limitée à 20 mètres dont la température minimale est supérieure à 4° et sur fond sableux (sables, fraction grossière). Les espèces costulées s'aggrègent à l'ensemble de ces 2 groupes.

Groupe morphoécologique 3: ce sont des espèces fortement ponctuées dont le seuil thermique inférieur peut s'abaisser jusqu'à 0°C. Le dendogramme ne permet pas de préciser les limites des autres paramètres écologiques.

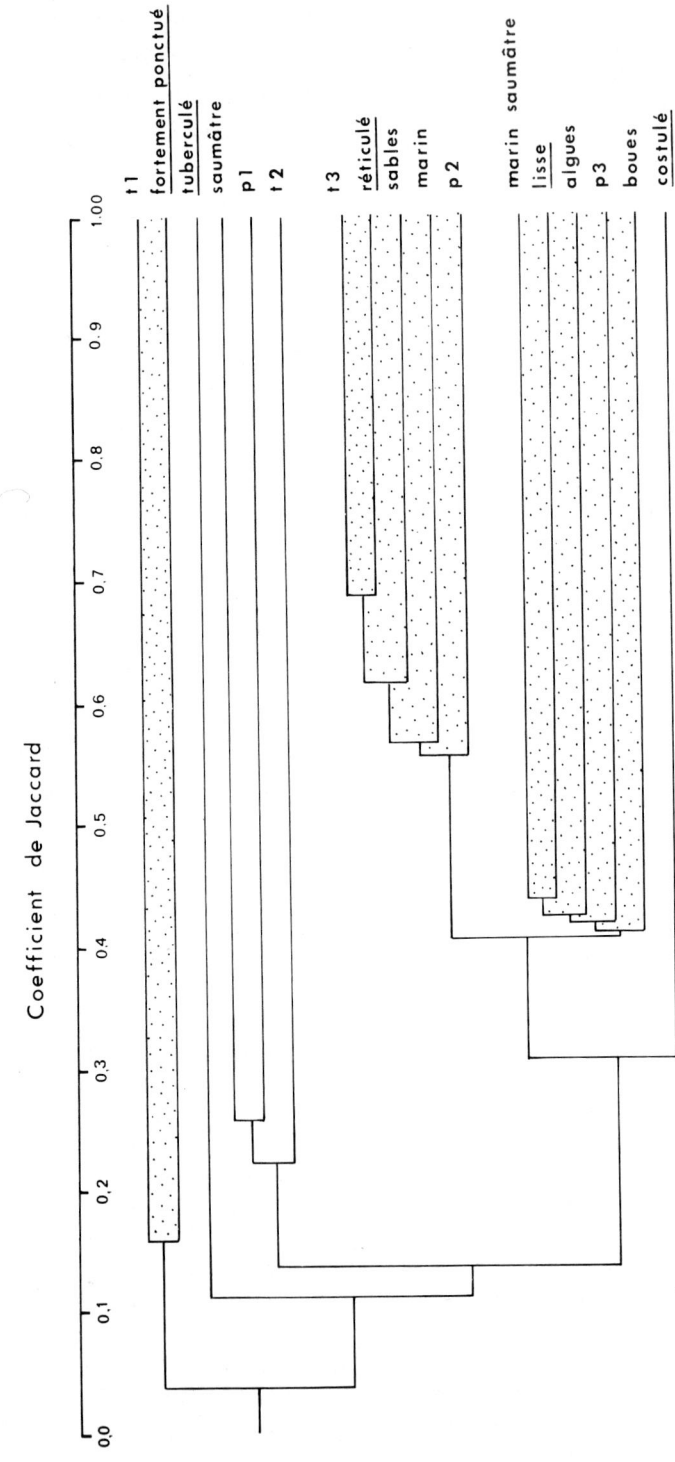

Fig. 2 — **Dendogramme (coefficient de Jaccard, W.P.G.M.) des groupes morpho-écologiques chez des espèces actuelles appartenant au *Loxoconchidae*.**

4.3 Analyse des correspondances de Benzécri (A.C.)

4.3.1 Elle repose sur la diagonalisation d'une matrice carrée symétrique. Cette méthode permet de découvrir la structure que recèle un tableau de données. En effet
- soit les ensembles finis I et J représentant les variables et les individus.
- soit k (i, j) des observations faites sur l'évènement (i, j), on établit à l'aide de lois statistiques un tableau de données. On détermine ensuite des facteurs ou couples de fonctions définies l'une sur I, l'autre sur J.

Ainsi conformément aux principes de la statistique, une analyse des correspondances comporte
1: un choix de données à savoir: tableaux de mesures, de contingence, de fréquence ou tableau de description logique... Ce dernier cas est utilisé dans ce travail. Il comporte également leur codage.
2: L'élaboration automatique, d'après un nuage de profils (= lois conditionnelles) munis de la distance du X^2 (= distance distributionnelle), des valeurs propres, des facteurs, des listes de contributions et des nuages de points.
3: L'interprétation qui n'est pas totalement objective. Elle est du ressort du spécialiste de la question analysée.
C'est une des analyses statistiques qui nous permet de pendre connaissance des valeurs sur les ensembles I et J des premiers facteurs (valeurs propres, taux d'inertie...) grâce à ces diagrammes plans.
L'interprétation facteurs (valeurs propres et taux d'inertie) est valable jusqu'à une valeur assez faible de l'ordre de 10^{-3}.

L'approfondissement des résultats obtenus devrait être réalisé à l'aide des listes de contributions, non retenus dans cette étude.

4.3.2. Extraction des facteurs
L'analyse des correspondances réalisées à partir des 14 variables, permet d'en isoler 4 (ou facteurs indépendants) qui, ensemble, expliquent 50% de la variabilité totale. A l'opposé les 10 autres facteurs expliquent seulement l'autre moitié de la variabilité.

Facteur 1. La représentation sur l'axe du 1er facteur des points variables est indiquée sur la figure 3. Elle oppose les variables morphologiques rét culée costulée au paramètre saumâtre. On peut donc en déduire que le espèces du genre *Loxoconcha* réticulée/costulée sont sténohalines, de faible pro deur et de seuil thermique minimum supérieur à 4° C.

L'analyse des correspondances précise la position de la variable costulée en l'associant à la variable réticulée. Ce n'était pas le cas avec l'analyse de groupe.

Facteurs 2 et 3. La représentation graphique sur l'axe du 2ème facteur des points variables est indiquée sur la figure 3. Elle oppose les variables fortement ponctuée et saumâtre au seuil thermique minimum de 2° C.

On observe également sur l'axe du 3ème facteur la même opposition entre la variable fortement ponctuée et le seuil thermique minimum de 2°C. De plus, ici,

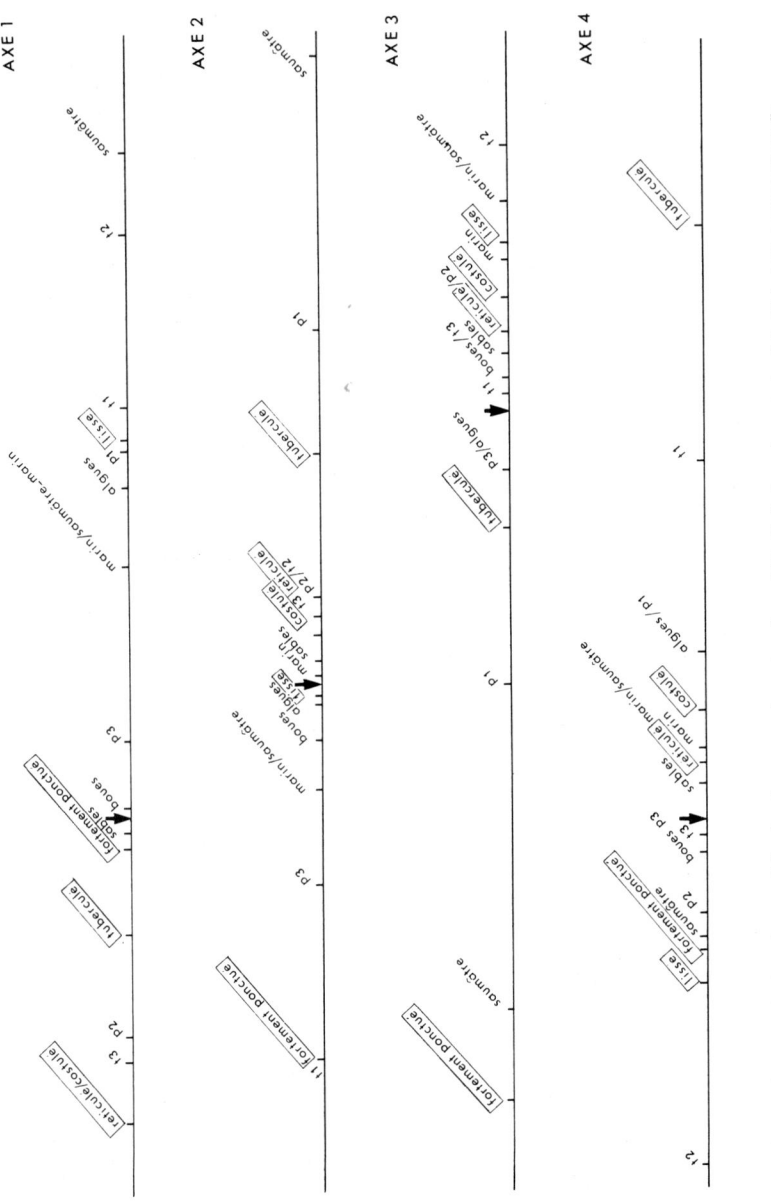

Fig. 3 – Position sur les 4 premiers axes de l'analyse des correspondances des points variables (écologiques et ornementaux). La flèche indique une position nulle.

la variable saumâtre est proche du facteur morphologique fortement ponctué. Il en était très éloigné sur l'axe 2. Cette opposition n'est qu'apparente.

Les espèces qui se situent autour de ces variations sont, en partie, différentes. Cela traduit une bivalence des espèces fortement ponctuées.

En conséquence, les espèces fortement ponctuées sont saumâtres ou sténohallines. Dans ce dernier cas, elles vivent à une profondeur supérieure à 20 m et peuvent supporter 0°C.

A l'inverse, dans le cas où elles sont saumâtres, le seuil thermique est sans influence sur leur répartition.

Facteur 4. La représentation graphique sur l'axe du 4ème facteur des points variables est indiquée sur la figure 3. Elle oppose les variables morphologiques tuberculée et lisse entre elles mais également celles-ci au seuil thermique minimum de 0°C et 2°C.

En raison du très petit nombre d'espèce tuberculée étudiée (1 espèce) ces indications sont à retenir avec réserve.

Plus intéressantes sont les informations concernant les espèces lisses. Celles-ci seraient plus euryécologiques que les autres.

Autres facteurs. Les autres variables, comme la nature du substrat où la présence de végétation sous-marine ne semblent pas jouer de rôle notable sur la distribution du type d'ornementation.
L'absence de relation privilégiée substrat/ornementation va à l'encontre des idées généralement admises.

L'analyse de groupe (dendogramme fig. 2) à l'inverse de l'analyse des correspondances assignait une place plus grande à la nature du substrat et à la végétation sous-marine.

4.4. *Résultats généraux: variables limitantes*

Le tableau 2 résume les compatibilités et les incompatibilités déduites de l'analyse des correspondances.

La salinité n'agit que sur les espèces à ornementation réticulée et fortement ponctuée.

Le seuil thermique minimum de 0°C conditionne le type fortement ponctué
de 2–3°C conditionne le type lisse
de 4°C conditionne le type réticulé/costulé.

La profondeur (dont la liaison avec la température est évidente) agit sur les espèces réticulées/costulées ($<$ 20 m) et sur certaines espèces fortement ponctuées ($>$ 20 m).

Les espèces lisses, quel que soit le genre, auraient, à la suite de cette analyse, la tolérance écologique la plus large. Les travaux de Theisen (1966, p. 234–239) sur *Lox. elliptica* le confirment.

5. Essai d'applications paléoécologiques

Nous prendrons deux exemples: *Lox. grignanensis* Carb. du Miocène inférieur de la Téthys et *Lox. hodonica* Pok. du Miocène supérieur de la Paratéthys.

5.1. *Cas de Lox. grignanensis Carb.*

Elle présente une ornementation réticulée/tuberculée. D'après le tableau 2, elle

---

En raison du nombre réduit de cas tuberculé, ce caractère est ignoré.

type ornementation / variable écologique	lisse faiblement ponctué	fortement ponctué	réticulé / costulé	tuberculé
saumâtre		(1) ● (2) +	●	
marin				
marin à saumâtre		●(2)		
0°		+(1)?		
2°	+	● (2)	●	● ?
4°			+	
0 à 3 - 4 m				
0 à 15 - 20 m			+	
> 20 m		+(1)		
Nbre de conditions morphologiques	14	6	28	1

Tableau 2 — Compatibilités + et incompatibilités ● revélées par l'analyse des correspondances.

serait marine, avec un seuil thermique minimum supérieur à 4°C et vivrait à une profondeur comprise entre 0 et 20 mètres.

La faune associée (cf. Carbonnel 1969, table 21) ne s'oppose pas à une telle interpretation. On émettra cependant des réserves sur la salinité du milieu déduite du type de l'ornementation par rapport à celle suggérée par la faune associée. Elle serait plus basse.

5.2. *Cas de Lox. hodonica Pok.*

Elle appartient également au type réticulé. Les conditions paléoécologiques seraient celles de *Lox. grignanensis*. Ici encore un désaccord se manifeste sur la salinité déduite de l'ornementation et de la faune associée, qui serait également plus basse. Ces désaccords impliquent deux hypothèses. Dans la première, les données brutes de base sont insuffisantes pour permettre l'établissement du tableau des incompatibilités. C'est en partie exact.

Dans la seconde hypothèse, la réponse des espèces à la salinité s'est modifiée au cours du Mio-Pliocène. On ne peut à priori la refuser.

6. Conclusions générales

– Ces méthodes ne sont pas, dans l'état actuel, suffisamment précises, particulièrement en ce qui concerne les types d'ornementation. Ces derniers devront être subdivisés. Egalement insuffisants sont les découpages écologiques retenus: par exemple seuils thermiques dont la valeur maximale n'a pas été prise en comsidération et nature du substrat.

A ces réserves près, cet essai démontre néanmoins la fiabilité des relations ornementation/milieu. Il faut poursuivre l'utilisation de cette méthode en affinant les paramètres écologiques retenus.

On pourra en outre l'étendre à d'autres genres tels *Semicytherura* ou *Cytherura*. Ceux-ci en effet présentent une ornementation très variée et sont capables de vivre dans des conditions écologiques larges.

Résumé

Les relations entre l'ornementation (lisse, fortement ponctuée, réticulée, costulée et tuberculée) de certains *Loxoconchidae* et les paramètres écologiques sont établies par l'analyse de groupe et l'analyse des correspondances. Elles ont montré: l'action de la salinité sur les espèces fortement ponctuées; l'action de la température (seuil thermique minimum) sur tous les types d'ornementation enfin l'action de la profondeur sur les espèces réticulées/costulées et certaines espèces fortement ponctuées. L'application paléoécologique de ces relations à deux espèces miocènes est tentée.

Remerciements

Ces recherches ont été réalisées grâce à l'aide du Swedish-Natural Science Research Council et du Centre National de la Recherche Scientifique, France. Le support informatique a été fourni par le Professeur R. Reyment Uppsala, Suède, et le Centre de Calcul de Lyon.

Bibliographie

Benzécri, J. P., et al. 1973. L'analyse des données. 1. La Taxinomie, 2. Correspondances. Dunod édit., Paris, 619 p.
Carbonnel, G. 1969. Les Ostracodes du Miocène rhodanien. Systématique. Biostratigraphie écologique, Paléobiologie. Thèse. Docum. Lab. Géol. Fac. Sci., Lyon, Fasc. 1-2, 469 p..
Carbonnel, G. 1973. L'analyse de groupe en paléoécologie en biostratigraphie. Application aux Ostracodes (Crustacea) miocènes. Arch. Sc. Genève 26 (1): 23–69.
Elofson, O. 1941. Zur Kenntniss der marinen Ostracoden Schwedens, mit Berücksichtigung des Skageraks. Uppsala Univ. Zool. Bidr. 19: 215–534.
Hazel, J. E. 1970. Binary coefficients and clustering in biostratigraphy. Geol. Soc. Amer. Bull. 81: 3237–3252.
Kaesler, R. L. 1966. Quantitative reevaluation of ecology and distribution of recent forami-

nifera and ostracoda of Todos Santos Bay, Baja California, Mexico. Univ. Kansas, Paleontol. Contrib., Topeka: 1–50.

Maddocks, R. F. 1966. Distribution patterns of living and subfossil pocopid ostracodes in the Nosy-Bé area, northern Madagascar. Univ. Kansas Paleontol. Contrib. Topeka, Paper No 12, 72 pp.

Sokal, R. R. & Sneath, P. H. A. 1963. Principles of numerical taxonomy. Freeman & Co., San Francisco, 359 p.

Theisen, B. F. 1966. The life history of seven species of Ostracodes from a Danish brackish-water locality. Medd. Danm. Fiskeri- og Havunders., Kobenhavn, N.S., 4 No. 8: 215–270.

Adresses des auteurs:

G. Carbonnel, Université Lyon I., Dépt Sc. Terre, et Laboratoire Associé au C.N.R.S., 15–43, Bd du 11 novembre, 69621 Villeurbanne, France

M. Jacobzone, Institut Sciences Financières, 15–43, Bd du 11 novembre, 69621 Villeurbanne, France

Discussion

Athersuch: Has the division into groups any taxonomic significance?
Carbonnel: None, because this study has been realised in one genus only: the *Loxoconcha*.

Sixth Intern. Ostracod Symposium, Saalfelden

ON THE DEVELOPMENT OF THE MUSCLE-SCAR PATTERNS IN TRIASSIC OSTRACODA

EDITH KRISTAN-TOLLMANN

Abstract

A short report on the development of the muscle-scar patterns of the Healdiidae and Cytherellidae from the alpine Trias is presented. Emphasis is placed on the evolutionary trend towards a progressive reduction in the scars proceeding from the outside to the inside of the field with simultaneous regular orientation. This occurs in both families at the same time according to the same principle with similar pattern. The closeness of the relationship between the Healdiidae and Cytherellidae is stressed and the Suborder Platycopina being untenable is discarded. The ontogenetic development of the scar structure of a number of representatives of the genus Healdiidae appears to be in contradiction with the phylogenetical trends towards scar reduction.

Introduction

Of the few characteristics offerred by fossil Ostracodes for determination of phylogenetic and taxonomic positions, the muscle-scar patterns are of the primary importance. Each scar on the ostracode valve represents an individual fibre of the muscle (fibre bundle).Thus, certain conclusions on the state of the muscle can be drawn from the number and arrangement of the scars. During the phylogenetic development of some ostracodes, the muscle-scar pattern has undergone considerable alterations, and one of the first recognized trends in ostracode evolution was towards a reduction in the number of fibres of the muscle.

The investigations on the muscle scars of triassic ostracodes, especially those of the Healdiidae and Cytherellidae, were carried out in order to establish, or at least to provide a basis for establishing, the taxonomic positions of the triassic genera also by means of the muscle scar patterns. For example, some of the questions which arose during the investigation were: do the paleozoic *Healdia* extend into the Upper Trias or into the Lias? ...is Ogmoconcha a younger synonym for *Hungarella?* ... are the triassic species member of *Ogmoconcha* or *Hungarella?* The Healdiidae are the largest and most abundant ostracode group in marl from the Upper Trias, they thus provide the best possibility to date for investigations of the muscle-scar patterns. For such an investigation a large quantity of material is necessary as the interior of the valves are usually corroded or are covered with a layer of calcite. In some cases, of 1000 specimens sampled, only 1–3 of the valves had distinct remains of scars. In addition to the specific respectively generic problems, the investigations also revealed important aspects on the phylogenetic position of the triassic Healdiidae and Cytherellidae. These results will be discussed briefly.

Hungarella and *Ogmoconcha* were established as distinct independent genera of the Healdiidae, using the Triebel type material, the author's triassic material and the photos of the type species from Mehes which were still available. These studies were carried out at the Forschungsinstitut and Museum Senckenberg, Frankfurt/Main with the financial support of a Humboldt Scholarship in 1971. This topic will be briefly dealt with in the reply to one of the discussion questions of Dr. Kilenyi. The first detailed investigations on the muscle scar patterns of the alpine triassic Healdiidae were carried out during a two month stay at the Geologic-Paleontologic Institut of the University Frankfurt/Main, also with a Humboldt Scholarship, using Scan drawings. Scan photos (magnification $\times 500$) of the muscle scar patterns of the interior of the valves were taken and drawings produced from the photos. A set of representative drawings is shown in Figs. 3 and 4. For a better view of the scars, the valves were inclined ventrally at an angle of 30 degrees and thus the muscle-scar patterns shown on the photographs or drawings are somewhat compressed vertically as compared with the originals.

The muscle scar patterns of Triassic Healdiidae

All the muscle scar patterns of the triassic Healdiidae, irrespective of the genera, are based on the folllowing principle: in the middle of the field are two vertical rows of large horizontally elongated to rounded scars, which are surrounded by a circle of smaller scars. This pattern differs markedly from those of the younger paleozoic Healdiidae, in which the scars are clumped irregularly... the largest scars are located in the centre, with a decrease in scar size occuring towards the periphery. In contrast, some Healdiidae genera from the Lias, situated at the end of a long phylogenetic row of large Healdiidae families, just prior to their disappearance, have developed another formation. Only the two rows of scars remain; the number in each is often reduced to five, the circle has disappeared completely (see Fig. 1, upper row).

Based on the investigations carried out to date on the muscle-scar patterns of the Healdiidae, principally from J. Gründel on liassic and from M.N. Gramm on paleozoic and triassic material, and on the present studies of the Healdiidae of the alpine Triassic, the results of which confirm respectively complement others available, the following tendencies in development during the evolution of the muscle scar patterns from the Paleozoic to the Lias can be recognized (Fig. 1):
a) reduction in the total sizes of the field of the muscle scar;
b) reduction in the number of scars towards the inner part;
c) transformation of the distribution of the scars, from clumped to regular rowed arrangements.

ad b) From the paleozoic to the liassic Healdiidae, continous progressive reduction in the scars occurs, proceeding from the outside towards the inside the muscle-scar pattern. Initially, the small, irregularly dispersed scars at the periphery (see Fig. 2, Fig. 1.2) are reduced to a single circle of individual densely crowded scars (see Fig. 2, Fig. 3.4, Fig. 3). In some cases, the scars may still be double (Fig. 3, Fig. 10, 12–14). The number of scars in the circle undergoes progressive reduction, the arrangement appears to be very flexible (Fig. 4, Fig. 2, Fig. 5) until finally no scars are present (Fig. 1, right diagram). Similarly, the large scars in the middle

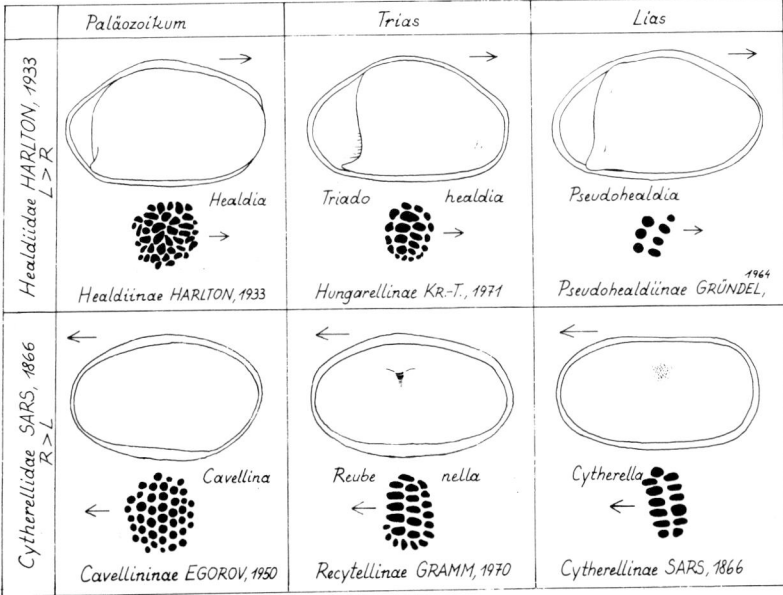

Fig. 1. The phylogenetic development of the muscle-scar patterns of the Healdiidae and Cytherellidae.

of the field undergo reduction in number from seven or more/row to three, respectively, two/row.

ad c) At the end of the Permian beginning of the Triassic the first fundamental change occurred in the muscle scar pattern from a clumped to an ordered arrangement of scars in two vertical rows, surrounded by a circle of such; at the end of the Triassic beginning of the Jura, the second change took place ... to a simple classical arrangement of scars in two exclusive rows. From the Triassic onwards, both rows are composed of horizontally elongated to rundish scars which are arranged more or less vertically with a slight concave anterior curvature.

The transformations in the muscle-scar patterns just described, neither proceeds simultaneously in all groups of genera, nor do they always occur exactly at the changeover between periods. Basically, the Healdiidae can be divided into three typical developmental groups:

a) Paleozoic groups with typical 'paleozoic' of healdioid muscle scar patterns (scars clumped or bunched).

b) Triassic groups with their main development in the Trias, however with extension into the Lias. Typical 'triassic' or hungarelloid muscle scar patterns (scars arranged into two interior rows, surrounded by an external circle).

c) Liassic groups with typical 'liassic' or pseudohealdioid muscle scar patterns (two rows of scars only).

For taxonomic purposes, the scars of the two vertical rows are especially significant, there is little variation and the number remains approximately constant

135

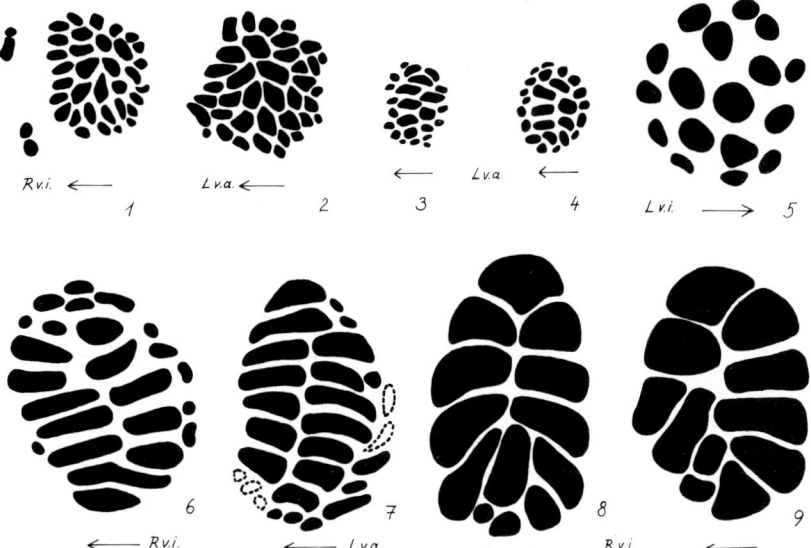

Fig. 2. Muscle-scar patterns of the Healdiidae and Cytherellidae. R v.i.= inside view of the right valve; L v.a.= outside view of the left valve. 1: *Healdia* sp., Paleozoic, America; from material of Triebel, Senckenberg-Museum, Frankfurt/Main. 2: *Healdia simplex?* Pennsylvanian, America; from material of Triebel, Senckenberg-Museum. 3: *Hungarella lunata* Kristan-Tollmann, Upper Trias, Cordevol, Cassian beds (x 84), Pedraces, Italy. 4: *Hungarella limbata* (Reuss), Cassian beds (x 23), Pralongia S, South Tyrol Dolomites, Italy. 5: *Ogmoconcha contractula* Triebel, Lias, Clay mine Osterfeld near Goslar, BRD. 6: *Signohealdia robusta* Kristan-Tollmann, very early larvae; Upper Trias, Sevat, Zlambach beds, Roßmoosalm, Salzkammergut, Austria. 7: *Leviella raibliana* (Gümbel), early larvae; Upper Trias, Karn, Raibl beds, Raibl, Yugoslavia. 8,9: *Leviella* sp., deep germanic Lias, material Tübingen, BRD.

within a genus. Thus, for example, of the upper triassic genera, Signohealdia has 6–7 scars/row (Fig. 3), *Hungarella* has 3 to 5 scars in the front row, 4 to 5 (usually 5) in the back row, *Ogmoconcha* 3 scars/row. The number of scars in the circle varies more markedly. Only a general statement can be made, namely that in the upper triassic and liassic Healdiidae, the circle is comprised of few, widely separated scars, whereas in the lower triassic genera it is made up of many, densely packed scars. Accordingly, the variation in size and arrangement of the individual muscle scars within the circle and also within the two rows, is of little importance, however, the total field, characteristic for each genus and formed in a similar fashion, and the number of scars/row within it, is.

Ontogeny

The ontogenetic development of the muscle scar patterns contrasts with its phylogenetic in which there is a gradual reduction in scar number with the apparently ideal final formation of one or two rows of few scars. A brief reference to such

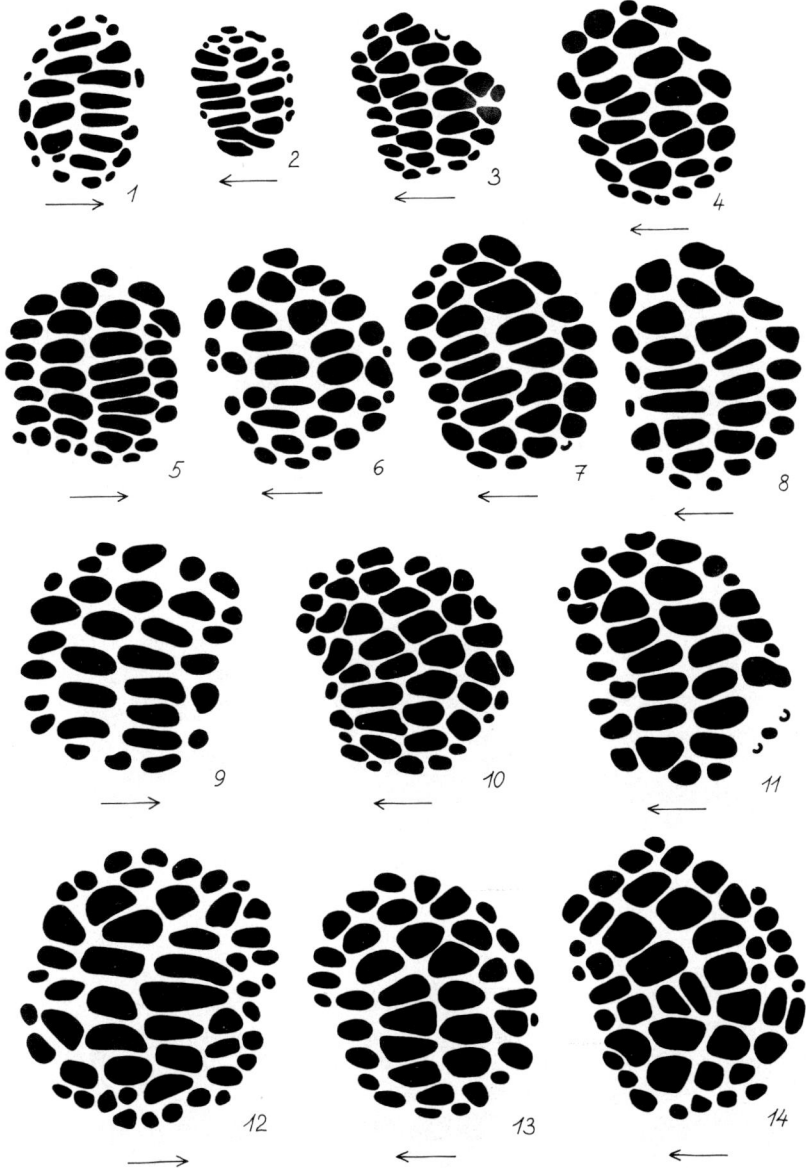

Fig. 3. Muscle-scar patterns of larvae of *Signohealdia robusta* Kristan-Tollmann; Inside view, drawn form Scan photos. Valves tilted ventrally at 30 degrees, such that the muscle field shown is somewhat shorter than the original. Magnification ×250. 2: From Roßmoosalm, all others Schneckenkogel. Upper Trias, Sevat, Zlambach marl; Salzkammergut, Austria.

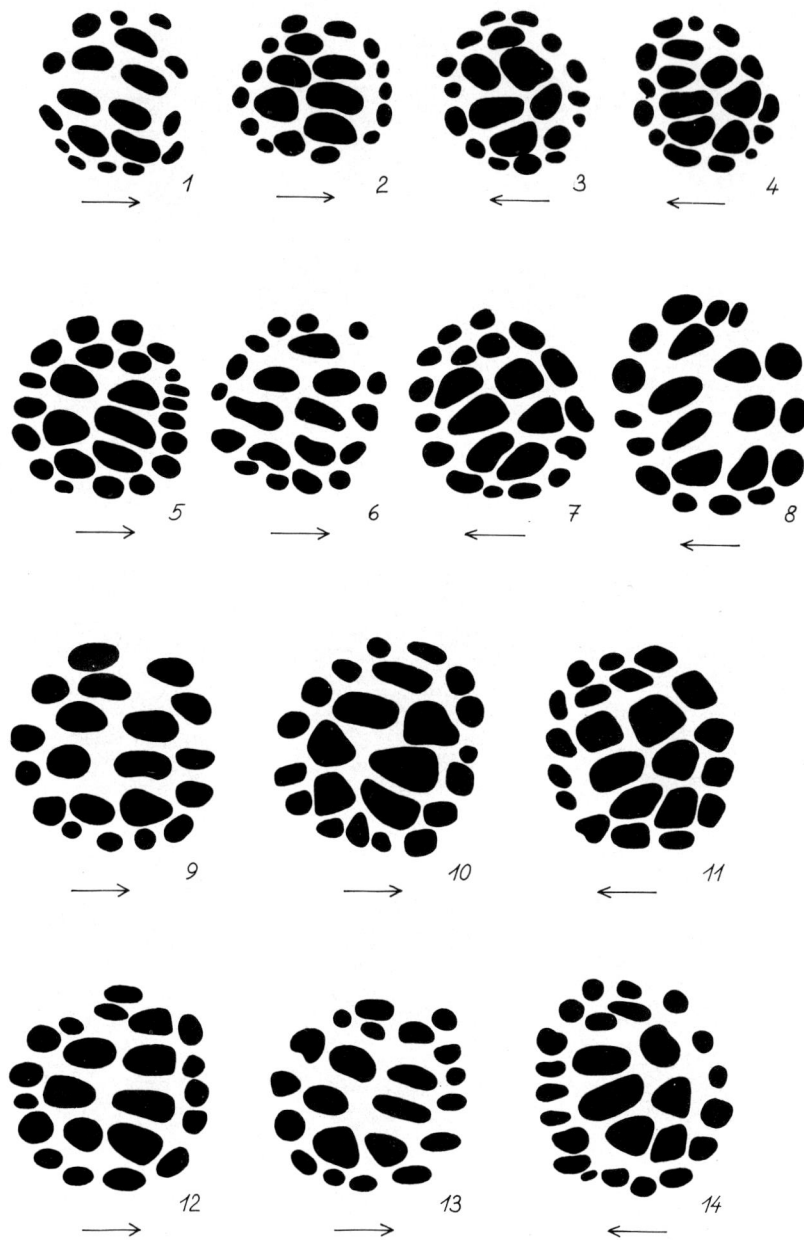

Fig. 4. Muscle-scar patterns of *Torohealdia opisthocostata* Kristan-Tollmann; Inside view, as in Fig. 3. Magnification ×250., 1–8, 10: Larvae, 9, 12–14: adult. Upper Trias, Sevat, Zlambach marl; Schneckenkogel, Salzkammergut, Austria.

was made in connection with Signohealdia (E. Kristan-Tollmann 1973a, p. 151), and the example shown in Fig. 3 provide an initial impression of such.

Although the earliest larvae of *Signohealdia robusta* (Fig. 3, Fig. 1.2) already possess the complete number of large scars/row, those of the circle are lacking to a large extent. In later larvae, the circle is gradually completed (Fig. 3, Fig. 3–8) and finally in the last larvae, the circle formed by densely packed to double scars (Fig. 9–14). Thus, during ontogenetic development, scar number increases markedly. The same phenomena was observed in *Ogmoconcha amalthei rotunda* Dreyer by J. Gründel 1972, p. 152 ff., who considers that apparently a proterogenetic developmental modus is present.

Characteristically, *O. amalthei rotunda,* whose generic allocation must still be clarified, is at least closely related to *Hermiella* (nom nud.). Due to lack of knowledge on the ontogenetic development of the muscle-scar of the latter species, the author has grouped it with *Signohealdia* and probably also with *Torohealdia* for purely morphological reasons, and would like to separate this group from the other Healdiidae which originate directly from *Healdia* (Fig. 5). Many questions still remain to be answered on the possibility of a polyphyletic origin of the Healdiidae. Referene to such will be made by the representation of two separate branches.

The evolution of the muscle-scar pattern of the Cytherellidae

Within the Cytherellidae, the muscle-scar patterns show the same tendencies in development as do those of the Healdiidae (Fig. 1,5). As already suspected by Gramm and Gründel, the younger paleozoic Cavellinids, the triassic, liassic and recent Cytherellidae represent one phylogenetic series. The initially irregularly clumped scar distribution transforms into one in which the scars are arranged in numerous vertical rows, which, throughout the Triassic, undergo reduction in number, beginning at the field edge, to four, then to three and finally to two rows of large scars with a half row or a half circle of small scars. In the deep Lias, the last small scars finally disappear, leaving two rows of large scars. This final stage of the typical Cytherellidae muscle scar pattern exists in recent forms.

Accordingly, the following facts may be stated:
a) The evolutionary trend towards reduction in the number of muscle fibres is the same in the Healdiidae and Cytherellidae.
b) In both families, the evolutionary development of the muscle field proceeds at a similar rate, according to a similar pattern, and achieves the same end formation.
In both cases, one can refer to a:
paleozoic muscle field   scars arranged in clumps
triassic muscle field   scars arranged in two rows, with relicts of the other surrounding scars
liassic muscle field   scars arranged in two rows.

In the alpine species investigated up till the present, the essentially triassic genera, *Reubenella* and *Leviella* (previously separated according to external morphological features) also differ in the formation of their muscle scar fields. The

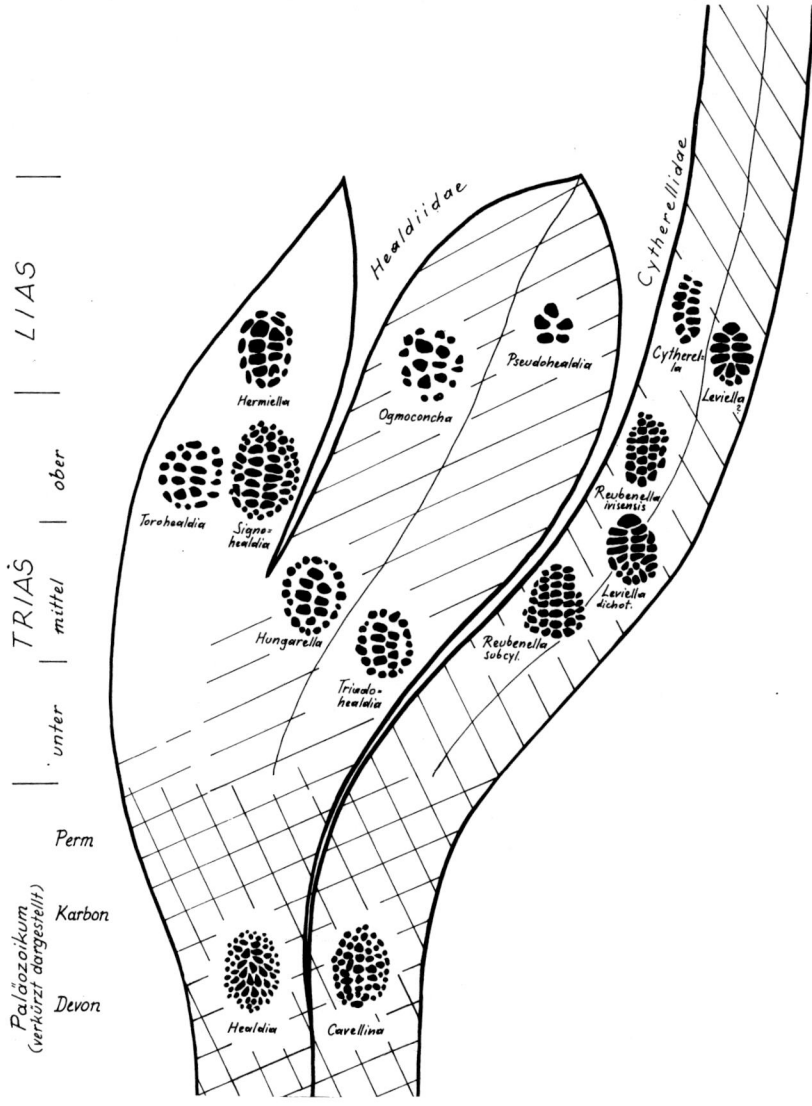

Fig. 5. Phylogenie of the Healdiidae and Cytherellidae (only european forms included for the Mesozoic).

Reubenella has several rows of rounded scars, *Leviella* has two rows of large, horizontally elongated scars to which many small scars are attached laterally, lower down underneath, in older triassic species (E. Kristan-Tollmann 1973 b, Fig.1). In a *Leviella* from the deeper germanic Trias (Fig. 2, Fig. 8,9), almost all small lower scars are reduced, however, the lowest largest scars are placed vertically or inclined at an angle indicating that the species can be considered a link to the more recent species with only two rows of horizontal scars.

Position of the Suborder Platycopina

H.W. Scott & P.C. Sylvester-Bradley maintain the Suborder Platycopina Sars, 1866, in the Treatise 1961, Q 380, and mention in the following discussion that these Platycopina are based exclusively on the single family Cytherellidae Sars, 1866. The Cytherellidae are also very similar to the Cavellinidae, the difference between the two consisting of the development of the muscle scar pattern: the latter have numerous small scars in a circular or elongated heap, arranged irregularly; the former, in contrast, have approximately ten or more single scars arranged in a double row. Moreover, the distribution of the Cavellinidae extends from the Silurian to the middle of the Permian whereas the Cytherellidae are known from the Jura to Recent. It thus appears that the two groups do not overlap in time. However, both authors emphasize that one of the trends in Ostracode evolution is towards reduction in the number of the muscle fibres, and that the Cytherellidae appear to stem directly from the Cavellinidae ... as indicated by the simple process of the reduction of the muscle scars.

Knowledge of the muscle scar pattern of the triassic Cytherellidae allows the postulate that the triassic Cytherellidae represent the connection between the young paleozoic Cavellinidae and the younger mesozoic Cytherellidae; the muscle scar patterns of the triassic Cytherellidae, in undergoing a reduction in row number from several to two, can thus be considered direct missing links. As the direct origin of the Cavellinidae-Cytherellidae is known, the Suborder Platycopina has become untenable and thus is discarded.

The question of the derivation of the groups

When the question as to the origin and relationships of the Healdiidae and Cytherellidae is raised, it can be stated that both families are closely related to one another ... also as shown by the muscle-scar patterns. Further, the few ontogenetic examples known at the present time indicate that at least a part of both families have the same genetic background and thus a similar root: the very early larvae of *Signohealdia robusta* (Fig. 2, Fig. 6) and *Leviella raibliana* (Fig. 2, Fig. 7) are very similar. Both have two rows of numerous small, horizontally elongated, large scars which are partly encircled by a few small scars. During ontogenetic development, the circle is first formed by numerous scars (Fig. 3). This ontogenetic development contrasts with the general phylogenetic development towards scar reduction in the Healdiidae and Cytherellidae, as described above. One is forced to make the hypothetical assumption that the ontogeny of the thus described species indicates a stem form possessing only two rows of large horizontally elongated scars, without small scars. For example *Darwinula* could come into question, not only for reasons of the muscle scar pattern, but also of the hingement, the shell morphology and shell relationship (species both with the left valve smaller than the right one and with the left larger than the right are of concern here). This is only a hypothesis, for which many supporting facts, especially on the ontogeny of the paleozoic and also on the mesozoic species, are lacking.

References

Dreyer, E.1967. Mikrofossilien des Rät und Lias von SW-Brandenburg. Jb. Geol. 1:491–531.
Gramm, M.N. 1969. Adduktoren-Abdrücke von Healdiidae (Ostracoda) aus der Mitteltrias von Süd-Primorje. Dokl. Ak. Nauk SSSR 186: 457–460 (in Russian).
Gramm, M.N. 1970. Die Adduktoren-Abdrücke der triadischen Cytherelliden, Ostracoda, von Primorje und einige Fragen der Theorie der Phylontogenie. Paleont. Shurnal SSSR 1970 (1): 88–103 (in Russian).
Gramm, M.N. 1970. Ostracoden aus der Familie Healdiidae aus der Trias Süd-Primorjes. In: Krasnow, Je.W. (ed.), Triassic Invertebrates and Plants of the East of the USSR, pp. 41–97. Wladiwostok, Daln. Geol.Inst. Akad. Nauk SSSR (in Russian).
Gründel, J. 1964. Zur Gattung Healdia (Ostracoda) und zu einigen verwandten Formen aus dem unteren Jura. Geologie 13: 456–477.
Gründel, J. 1968. Zur Gliederung der Familie Healdiidae (Ostracoda) und zu ihrer Stellung innerhalb der Ordnung Podocopida. Ber. dt. Ges. geol. Wiss. A. Geol. Paläont. 13: 225–232.
Gründel, J. 1972. Zur Variation der Schliessmuskelnarben bei Ostracoden (demonstriert an jurassischen Vertretern der Healdiidae). 1968 Proceed. IPU, XXIII Int. Geol. Congr., pp. 149–158.
Kristan-Tollmann, E. 1971. Zur phylogenetischen und stratigraphischen Stellung der triadischen Healdiiden (Ostracoda). Erdoel-Erdgas-Z.87: 428–438.
Kristan-Tollmann, E. 1973a. Zur phylogenetischen und stratigraphischen Stellung der triadischen Healdiiden (Ostracoda) II. Erdoel-Erdgas-Z. 89: 150–155.
Kristan-Tollmann, E. 1973b, Zur Ausbildung des Schliessmuskelfeldes bei triadischen Cytherellidae (Ostracoda). N.Jb. Geol.Paläont. Mh. 1973: 351–373.
Moore, R.C. (ed.). 1961. Treatise on Invertebrate Paleontology, Q, Ostracoda. Univ. Kansas Press.
Sohn, I.G. 1968. Triassic Ostracodes from Makhtesh Ramon, Israel. Bull. Geol. Surv. Israel 44: 77 pp.

Author's address:

Scheibenbergstrasse 53/6, 1180 Wien, Austria

Discussion

Kilenyi: On your phylogenetic diagram you showed *Hungarella* in the Trias being followed in the Lias by *Ogmoconcha*. Do you consider these two taxa separate genera, and if so why?
Kristan-Tollmann: According to my investigations, *Hungarella* and *Ogmochonca* are two distinct genera. They differ in both the exterior shell morphology and in the development of the muscle field.
1. In *Hungarella*, the morphology of the left valve differs considerably from that of the right (for example, *Hungarella limbata* (Reuss)), *Ogmoconcha*, however, exhibits valve symmetry if one neglects an occasional by occurring posteroventral thorn on the right valve.

2. In addition, although *Hungarella* occasionally has a ledge, lip or thorn on the anterior of the right shell and a thorn posteriorly, it never has the numerous supplementary little thorns present on both valves (as for example in *Ogmoconcha contractula* Triebel).

3. In the muscle field of *Hungarella*, both vertical rows have 8–10 scars; three scars rows are typical for *Ogmoconcha*. As all the investigations carried out up to the present have shown, in the triassic and jurassic Healdiidae, the number of the large scars of the rows in the muscle field remains markedly constant within a genus and has proved of considerable taxonomic value. Further, the number of scars encircling both scar rows in *Ogmoconcha* is much lower than in *Hungarella*.

III. ECOLOGY AND ZOOGEOGRAPHY

Sixth Intern. Ostracod Symposium, Saalfelden

ZOOGEOGRAPHY OF MACROCYPRIDIDAE (OSTRACODA)

ROSALIE F. MADDOCKS

Abstract

Distribution patterns of 53 living species of Macrocyprididae, belonging chiefly to *Macrocypris* and *Macrocyprina*, are reviewed here. The family is morphologically and ecologically conservative, and many species have broad geographic and depth ranges. Ontogenetic development, sexual dimorphism, increased size with depth, and poorly understood ecologic and geographic factors contribute to intraspecific variation. In many cases, soft parts are essential for consistent differentiation of males and females, adults and juveniles, and individual species. Narrowly defined local species may be combined into wide-ranging superspecies. Taxonomic decisions concerning Macrocyprididae must be based on large populations from numerous localities so that these geographic components of variation can be evaluated.

Introduction

The Family Macrocyprididae is the smallest and most homogeneous of the three families belonging to the Superfamily Cypridacea (Order Podocopida, Subclass Ostracoda; the other families are Cyprididae and Pontocyprididae). At least 53 are living today, and the true diversity may be as many as 100 species. The maps (Figs. 1, 2) show the known distribution of living Macrocyprididae. Apparently Macrocyprididae are absent from the Arctic Ocean, but the large gaps in the Indian and Pacific oceans merely reflect the lack of adequate collections and investigations in those regions. About 30% of all published faunal studies include at least one species of Macrocyprididae, but true incidence is probably slightly higher. Macrocyprididae do not occur in intertidal or brackish waters and are sparsely represented in shallow sediments of the continental shelves.

At least 65 fossil species have been proposed in *Macrocypris*, but most are misidentified, and for most of the remainder we do not know the hinge or muscle-scar pattern. Perhaps 15 species (2 Jurassic, 7 Cretaceous, 6 Tertiary) are authentic, but certainly many good species remain to be described, especially in deep-sea cores. Kozur (1971) traced the origin of Macrocyprididae from Devonian *Camdenidea* and *Acratina* through Carboniferous *Acratia* and Triassic *Praemacrocypris*.

Family characters

Macrocyprididae are the largest marine podocopids (to 3 mm length), sharing a smooth elongate carapace with distinctive outline. The right valve is larger than the left, overlapping it conspicuously in the ventral and anterodorsal regions. The

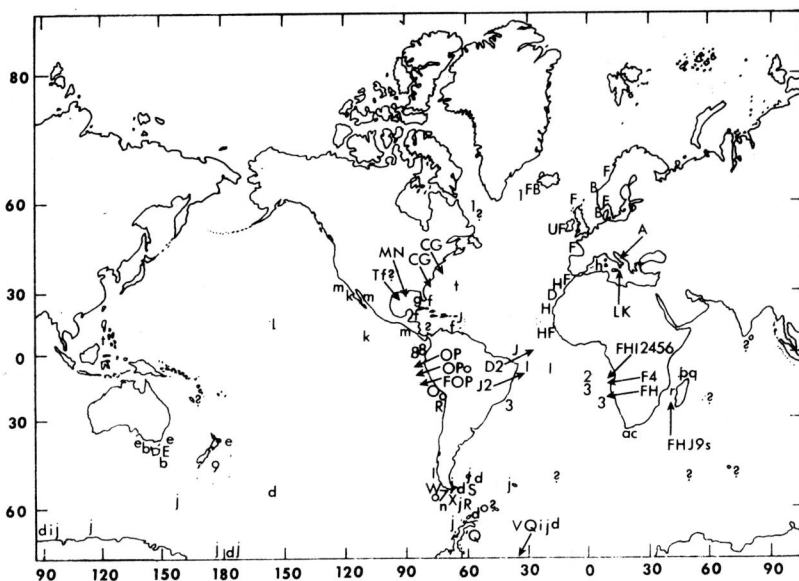

Fig. 1. Known geographic distribution of 53 new and previously described species of Macrocyprididae. The concentration of Macrocyprididae in Atlantic, Antarctic and coastal waters results from the great quantity of literature and collections for these regions. Macrocyprididae are presumably widespread also in the Indian Ocean, Pacific Ocean, Indo-Pacific islands, and east Asian shelf seas, though they may be truly absent from the Arctic Ocean. Total diversity of Macrocyprididae may be projected from this map at approximately 100 species. Explanation of symbols: capital letters and numerals designate species of *Macrocypris*, lower-case letters designate species of *Macrocyprina*, question marks designate misidentified, unillustrated species, chiefly *'decora'*, *'maculata'* and *'similis'*. Species of *Macrocypris*: A = *adriatica* Breman, 1975; B = *angusta* Sars, 1866; C = *bathyalensis* Hulings, 1967; D = *canariensis* Brady, 1880; E = *gracilis* Chapman, 1915; F = *minna* (Baird), 1850; G = *sapeloensis* Darby, 1965; H = *siliquosa* Brady, 1886; I = *similis* Brady, 1880; J = *tenuicauda* Brady, 1880; K = *gracilis* Seguenza of Puri, Bonaduce & Gervasio, 1969; L = *trigona* Seguenza of Puri, Bonaduce & Gervasio, 1969; M = *Macrocyprina* sp. of Howe & van den Bold, 1975; N = *Macrocyprissa* sp. of Howe & van den Bold, 1975; O-X, 1-9 = new species in collections now being studied. Species of *Macrocyprina*: a = *africana* (Müller), 1908; b = *decora* (Brady), 1866; c = *dispar* (Müller), 1908; d = *inaequalis* (Müller), 1908; e = *maculata* (Brady), 1866; f = *propinqua* Triebel, 1960; g = *schmitti* (Tressler), 1949; h = *succinea* (Müller), 1894; i = *tensa* (Müller), 1908; j = *turbida* (Müller), 1908; k = *vargata* Allison & Holden, 1971; l = *gracilis* (Brady) of Holden, 1967; m = *pacifica* (LeRoy) of Swain, 1967; n-t = new species in collections now being studied; ? = unverifiable, mostly *'decora'*, *'maculata'* and *'similis'*.

five-part hinge, entirely crenulate in most modern representatives, is unique, without homologue in other podocopids. The adductor muscle-scar pattern is also unique and highly conservative, consisting of an upper row of three scars and a lower group of about 9 scars. In some specimens it can be tentatively homologized with the paracyprid scar (Cyprididae) or with the bythocyprid scar (Bairdiacea) (see Fig. 3). There are always two small frontal scars and two mandibular scars, as well as several large dorsal body scars. Extremely broad anterior and pos-

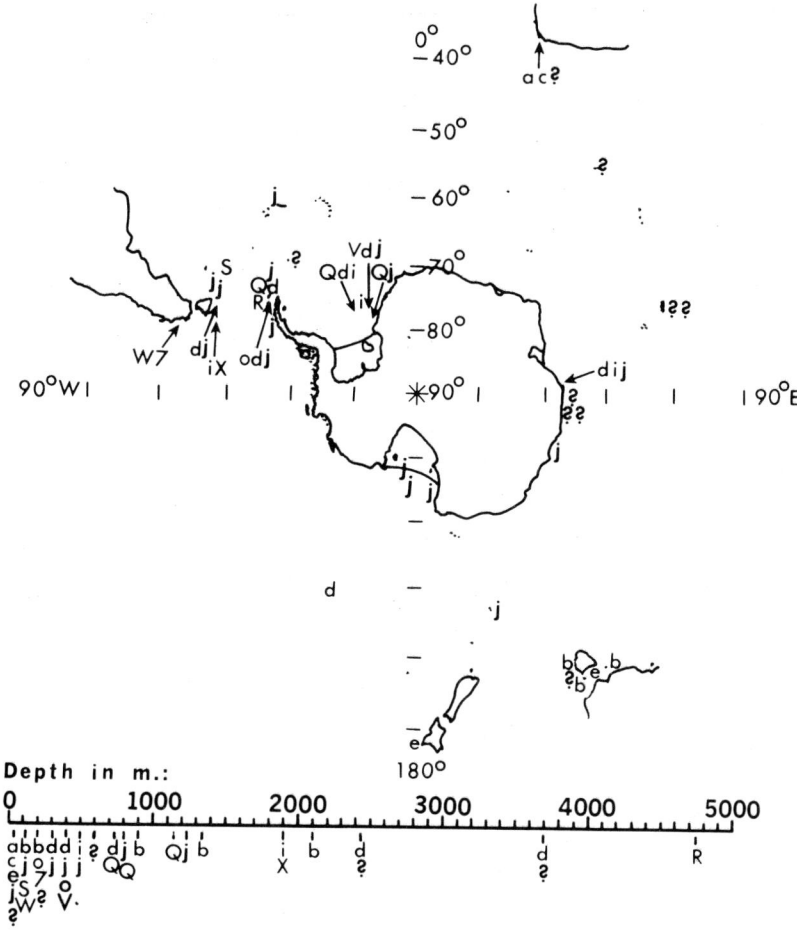

Fig. 2. Geographic depth distribution of 15 species of Antarctic and southern Macrocyprididae. For explanation of symbols, see Figure 1.

terior duplicatures are characteristic, with very deep vestibules. About half the species of *Macrocypris* have branching radial pore canals, which are useful for recognizing species but probably have little generic significance. The exterior has numerous funnel-pores, of varying sizes and shapes, with predominantly simple setae, but some species have both simple and tassel-setae (Fig. 3).

Macrocyprididae show sexual dimorphism in the carapace as well as in the soft parts. In some species the female is proportionately higher than the male. In other species the female is both higher and longer than the male but has the same shape. Thus a length/height scatter diagram may show either the usual podocopid branched pattern, with males diverging after about the 7th instar (ex., *M. siliquosa*, Fig. 4), or a nearly solid single line (ex., *M. bathyalensis*, fig. 5). In the second

149

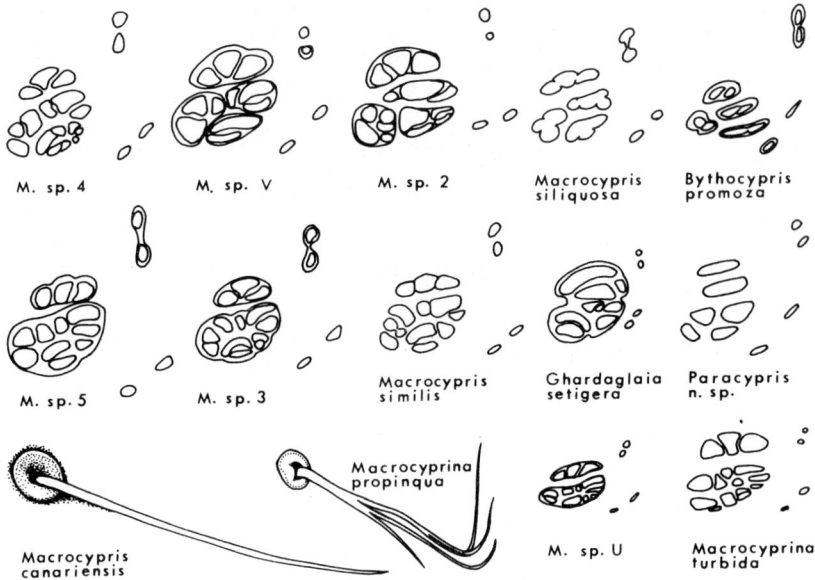

Fig. 3. Muscle-scar patterns (right valve exterior) of 9 species of Macrocyprididae, compared with bythocyprid and paracyprid scars. Two common types of funnel-pores with setae are also shown; most species have numerous simple setae plus just one or two setal tassels.

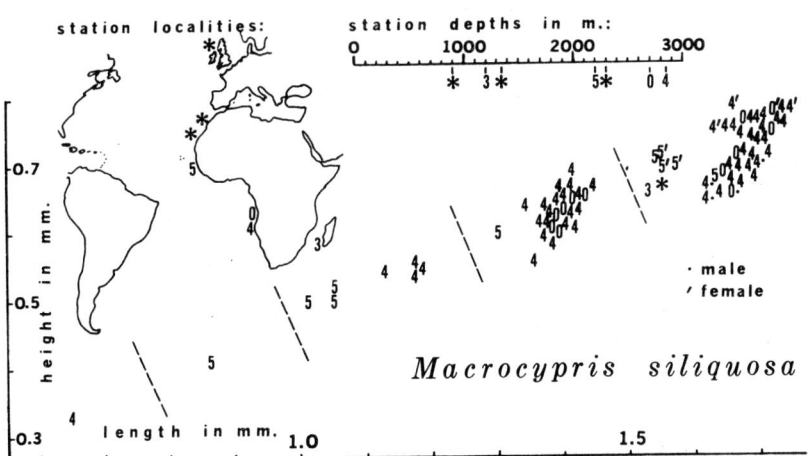

Fig. 4. *Macrocypris siliquosa* Brady, 1886. Geographic and depth distribution, with carapace measurements of 75 specimens. Many other measured specimens fall within the clusters delineated here and are not plotted for lack of space. Size increases with depth. Females are higher than males. Explanation of symbols: 0 = AI-42-200, 3 = IIOE 363K, 4 = AII-42-194, 5 = AII-31-145. Asterisks indicate previous reports.

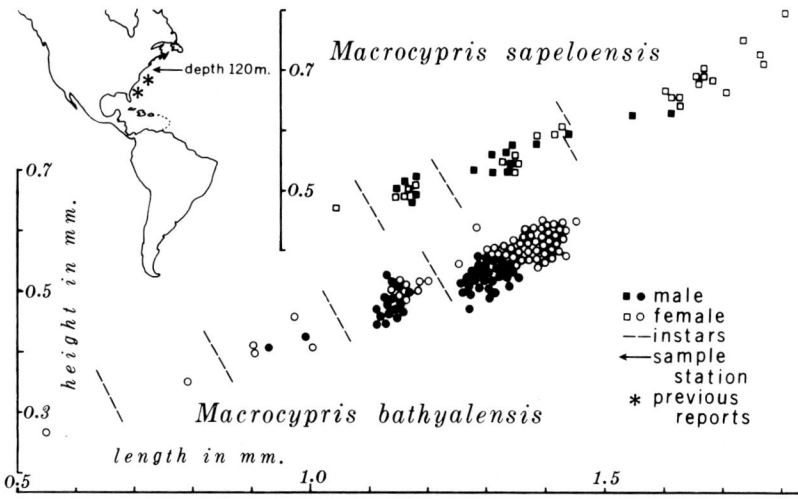

Fig. 5. *Macrocypris bathyalensis* Hulings, 1967, and *Macrocypris sapeloensis* Darby, 1965. These two species occur together at AII-40-172 and AII-40-173. Asterisks indicate previous reports. Note that carapace measurements for the two species share the same horizontal scales but different vertical scales. In reality, they plot along the same diagonal line, and it is not easy to distinguish specimens of the two species except by soft parts or radial pore canals. Females are higher and longer than males but have nearly the same proportions; the sexes and instars of the two species are difficult to discriminate from carapaces alone. Many additional adult specimens are not plotted here for lack of space; they fall into the clusters already delineated.

case, it may be difficult to tell males from adult or juvenile females unless soft parts are present.

Unlike most other Podocopida, including all other Cypridacea, juvenile Macrocyprididae have a calcified inner lamella. The width of the calcified inner lamella increases steadily with each instar; by the fifth (A-4) instar it is as broad as the adult of any other podocopid. The fused zone also appears early and widens, with short radial pore canals by the 7th (A-2) instar. In a species that will have branching radial pore canals as an adult, the late instars have straight canals. So far as the carapace is concerned, there is no way to distinguish such a juvenile from the adult of another species. It is essential, then, that taxonomic decisions concerning Macrocyprididae be made on large populations of specimens, including growth series, rather than on isolated specimens. It is likely that the types of some species are actually juveniles (ex., *M. similis,* Figs. 6, 7). Maturation of soft parts follows the normal sequence for Cypridacea, however, and if soft parts are preserved, recognizing juveniles is easy.

Macrocyprididae crawl on the sediment surface or attached organisms. All are sediment-eaters, with large balls of sediment in the gut like those of Bairdiacea. All are stenohaline, never found in brackish water or the intertidal zone. Most species are stenothermal. Appendage anatomy is highly conservative, indicating similar conservatism of life habits. Usually only one or two species are present

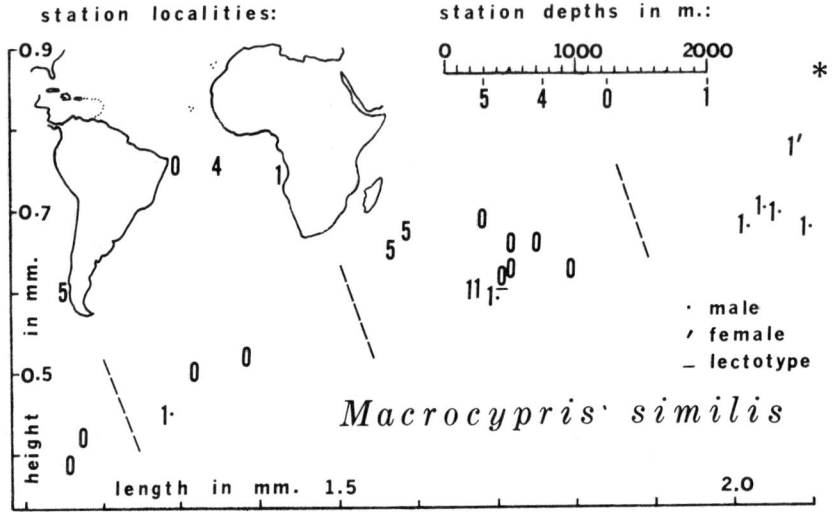

Fig. 6. *Macrocypris similis* Brady, 1880. Geographic and depth distribution, with carapace measurements of 21 specimens. There is no clear dependence of size on depth. Females are higher than males. Explanation of symbols: 0 = Challenger 120, from which H. S. Puri in 1967 selected the lectotype specimen, which appears to be a juvenile; 1 = AII-42-201; 4 = Challenger 344; 5 = Challenger 305; an asterisk represents the measurements and illustrated specimen of Brady (1880, pl. 2, fig. 2a), which is apparently no longer in the type material at either the British Museum or the Hancock Museum. Dubious reports of this species by several authors are not included here.

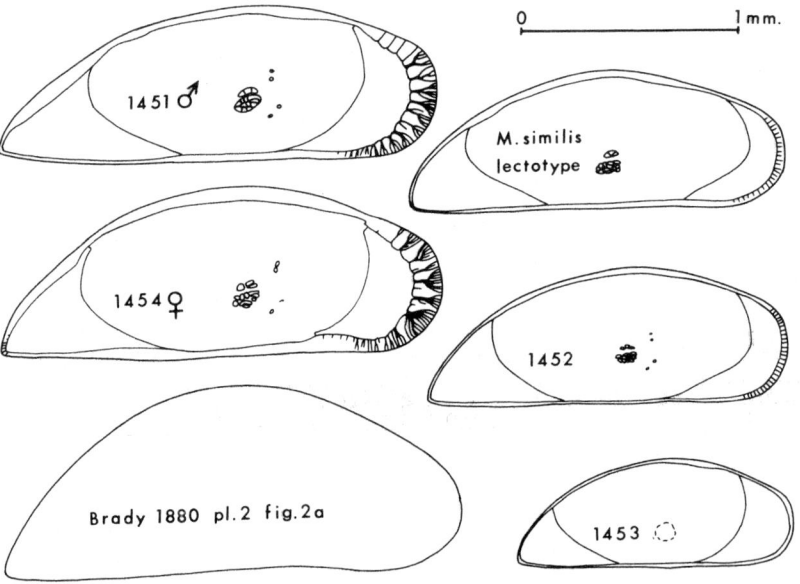

Fig. 7. *Macrocypris similis* Brady, 1880. All drawings to same scale. Numbers indicate specimens from AII-42-201.

in a local fauna, further suggesting competitive exclusion and rather broad ecologic niches. Two species at a single locality usually differ greatly in size, so that adults of the smaller species approximate (and perhaps compete with?) instars-VII or -VIII of the larger species (see Figs. 5, 8). Males are common, averaging 40% of adults. Most adult females have sperm in the seminal receptacles and 3 to 8 developing eggs in the ovaries. The female does not brood eggs in the domicilium. Juveniles average 20% of live collections, with late instars increasingly more numerous than early ones. Juveniles range from 50 to 90 % of dead carapaces. These figures may tell more about sampling methods than survivorship.

Genera

Macrocypris is distinguished by its large carapace with acutely pointed posterior end, tightly coiled vas deferens, and two-segmented maxillar palp. Soft-part characters have been described previously only for *M. minna* and *sapeloensis*. *Macrocypris* is especially characteristic of bathyal and abyssal depths, and certain species are characteristic elements of the psychrospheric fauna. A few species are found in outer sublittoral (50-200 m) habitats, chiefly in Antarctic and boreal waters.

Macrocyprina has a smaller carapace with bluntly rounded posterior end, a loosely looped vas deferens, and three-segmented maxillar palp. Soft-part characters have been fully or partly described previously for 8 species. It is typically a shelf and slope form, with rare records as deep as 3700 m. It is most abundant in two very distinct habitats, tropical reef-flats and Antarctic waters.

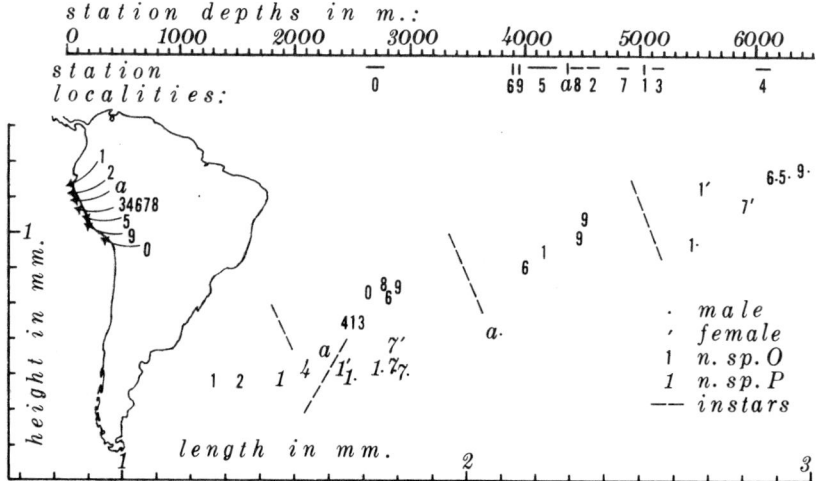

Fig. 8. Two new abyssal species of *Macrocypris* from the Peru-Chile Trench. Geographic and depth distributions, with carapace measurements for 19 specimens of species 0 (Roman type) and 10 specimens of species P (Italic type). Size decreases markedly with increased depth in both species. Females are only slightly higher than males. Explanation of symbols: 1 = AB-11-36; 2 = AB-11-59; 3 = AB-11-109; 4 = AB-11-113; 5 = AB-11-157; 6 = AB-11-169; 7 = AB-11-179; 8 = AB-11-196; 9 = Eltanin-3-48; 0 = Eltanin-3-50; a = AB-11-95.

Macrocypria has a doubly acutely terminated, elongate carapace and highly asymmetrical furca. The only known species is *M. angusta,* which is rare between 30 and 320 m. in the boreal east Atlantic. *Macrocyprissa* has a wide fused zone with branching radial pore canals and subcylindrical shape; its soft parts are unknown. *M. cylindracea* (Bornemann) is the Oligocene type species. Many new species have soft parts that fit well into *Macrocypris* or *Macrocyprina,* and it is expected that their description will confirm the validity of those genera. But *Macrocypria* and *Macrocyprissa* are not, as yet, consistently distinguishable from *Macrocypris*.

Species distribution patterns

As usual for smooth-shelled ostracodes, it is extremely difficult to recognize species accurately and consistently. Small differences in carapace curvature, though often highly reliable, are difficult to describe and illustrate. Species are best distinguished by comparing specimens, not illustrations, from large populations. The effects of ontogeny, sexual dimorphism, and phenotypic variation must first be recognized and subtracted before species can be defined. The existence of geographic character clines presents another strong warning — that species decisions should be based on many samples from numerous locations. If single specimens from different localities are compared, they always prove to have 'significant differences', but these differences have little meaning unless evaluated in zoogeographic context.

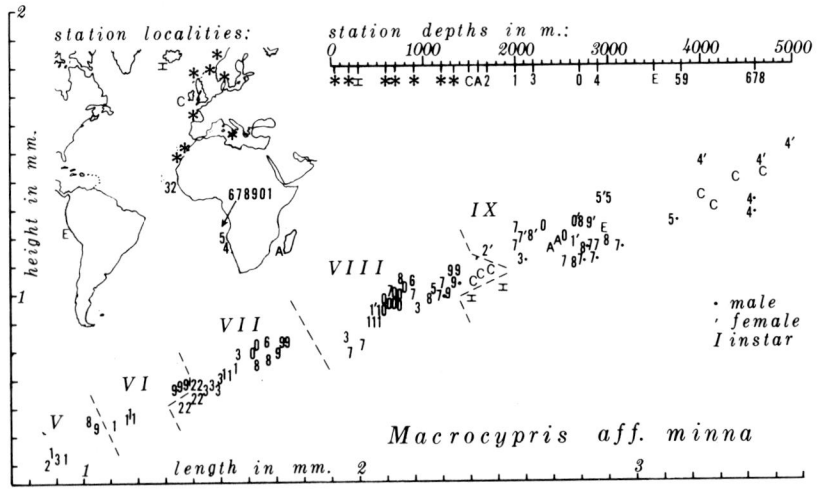

Fig. 9. *Macrocypris minna* (Baird, 1850) s. l. Geographic and depth distributions, with carapace measurements of 116 specimens. Females are higher than males and, in some cases, may be actually longer. The trend of increased size with depth is clearly expressed at some stations, violated at others, and these data probably include more than one species. Available soft parts do not yet permit these species to be discriminated. Explanation of symbols: 0 = AII-42-200; 1 = AII-42-201; 2 = AII-31-142; 3 = AII-31-145; 4 = AII-42-194; 5 = AII-42-195; 6 = AII-42-196; 7 = AII-42-197; 8 = AII-42-198; 9 = AII-42-199; A = IIOE 363K; C = Ch-106-313; I = Ingolf 85. Asterisks show previous reports of this species.

Soft parts appear to be even more conservative than the carapace. Most appendages change very little in structure, although the furca, fifth leg and seventh leg are useful. But, at present, only the male genitalia show enough change between species and consistency within species to be reliable for species identification. The corresponding females are not yet consistently distinguishable. This is a deplorable situation, and I hope that it will change with further work.

On the other hand, perhaps the male genitalia are more susceptible than other structures to geographic or ecophenotypic variation, and the small local popula-

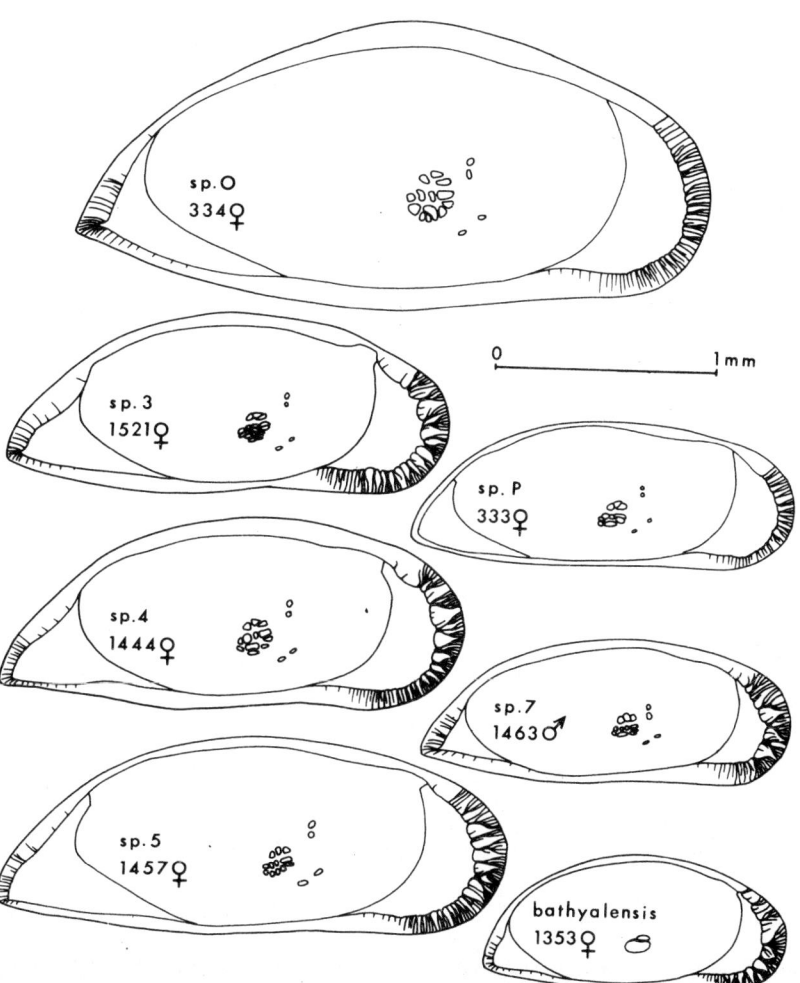

Fig. 10. Seven species of *Macrocypris* with oblong shape and branching radial pore canals, all to same scale. Within- and between-station intraspecific variation further complicates recognition of these species. Other morphs, not illustrated here, continue this variation in two directions, culminating in the quadrate *M. adriatica* and the siliquose *M. similis*. Numbers designate individual specimens in my records.

tions distinguished by these characters are demes or subspecies of a polytypic species. This would explain the allopatric distributions of most species in the present data. By this theory, true species of *Macrocypris* would have extremely broad geographic and depth ranges. This solution is very attractive, and it corresponds better to the concept applied to fossil species. Unfortunately, it cannot be proven. In fact, combining these species appears to obliterate rather than resolve the geographic character clines.

Such populations are here treated as species, grouped into superspecies (groups of closely related, largely allopatric species). *M. minna s. l.* (Fig. 9) may be one such superspecies. Another superspecies may include *M. adriatica, similis, bathyalensis*, new species O, P, and 1-9, all with oblong shape and branching radial pore canals (fig. 10).

The easiest character cline to illustrate here is also the most common one, increased size with depth. This is clearly shown in collections of *M. siliquosa* (Fig. 4), *M. tenuicauda* (Fig. 11), and some populations of *M. minna* s. l. (Fig. 9). On the other hand, this trend is exactly reversed in two new abyssal species in the Peru-Chile Trench (Fig. 8). Whatever its real cause, the phenomenon operates only within a single species. In combined size/depth data for many species, the trend is almost obliterated, showing only a weak tendency for the largest species to be lower bathyal. There is also a weak interaction with latitude, with the largest species concentrated in the tropical South Atlantic, smallest species in the boreal Atlantic and Antarctic.

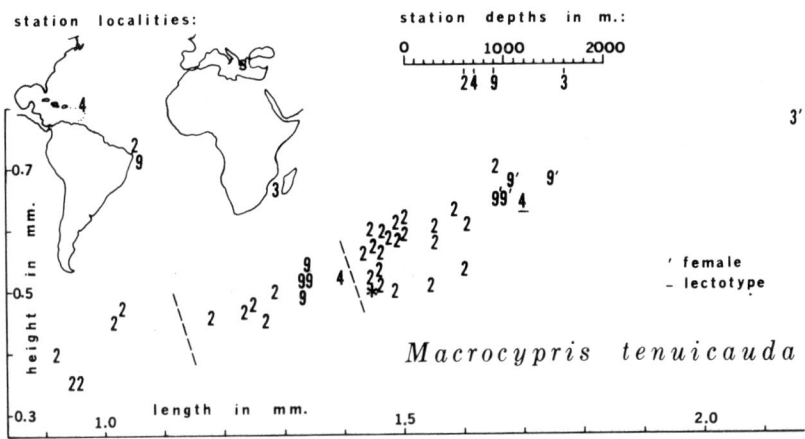

Fig. 11. *Macrocypris tenuicauda* Brady, 1880. Geographic and depth distribution with carapace measurements of 41 specimens. Size increases with depth. No males have yet been found. The measured Challenger specimens include mostly isolated valves; right valves are distinctly lower than left valves. Explanation of symbols: 2 = Challenger 122, from which H. S. Puri selected the lectotype in 1967; 3 = IIOE 363D; 4 = Challenger 24; 9 = AII-31-159; S = a Pleistocene report by Sissingh (1972); an asterisk represents the dimensions given by Brady.

Acknowledgments

This paper is a progress report on a much larger project, as yet unfinished, a taxonomic review of living Macrocyprididae. This project is based almost entirely on collections of others, and I am grateful to the persons and institutions who have donated or loaned specimens for study, most especially the British Museum (Natural History), the Hancock Museum at Newcastle-upon-Tyne, the Stanford University Museum of Paleontology, the Smithsonian Institution, the Zoological Museum in Copenhagen, Richard H. Benson, Thomas Bright, John Holden, Gary Kocurek. Louis S. Kornicker, James Teeter, and Torben Wolff. This summary is also based on extensive previous literature, which cannot be properly credited here in limited space. For the same reason, exact station locations are not listed here but will be supplied privately on request.

References

Kozur, Heinz. 1971. Die Bairdiacea der Trias. Teil III: Einige neue Arten triassischer Bairdiacea und Bemerkungen zur Herkunft der Macrocyprididae (Cypridacea). Geol. Paläont. Mitt. Innsbruck 1 (6): 1-18.

Author's address:

Geology Department, University of Houston, Houston, Texas 77004, USA

DIVERSITY OF BENTHIC MYODOCOPID OSTRACODES

LOUIS S. KORNICKER

Abstract

The number of myodocopid species per sample at shelf, bathyal, and abyssal depths is about the same, but the range is greater at shallower depths. The most commonly encountered number of species per sample is 3. Along the coasts of North America the number of species in embayments increases slightly from North to South along the Atlantic and Pacific Coasts and from West to East along the Gulf coast, with the greatest increase taking place in the tropical southern tips of Florida and Baja California. The number of species in embayments is much less than that at shelf depths. Diversity in southern oceans, when considered on a global basis, decreases with an increase in latitude markedly at shelf depths, less at bathyal depths and is about the same at abyssal depths.

Introduction

To date, somewhere between 400 and 500 species of benthic Myodocopina (Superfamily Cypridinacea) have been described. This paper examines and compares the number of species inhabiting various geographic areas. First to be considered is the number of species occupying a very small area, namely that area from which a single sample was collected. Second to be considered is the number of species occupying a specific geographic subdivision such as a bay or the continental shelf off a particular coast. Most data for the second study are from North America. And third to be considered is the number of species occupying specific latitudinal zones. Data for the third study are derived from the area between Antarctica and latitude 35°S. Changes in diversity with depth are considered in all phases of the study.

Number of species in the small area covered by a sampling device

Although this paper does not examine myodocopid abundance, it is necessary to consider the affect of abundance on diversity. Myodocopids are not evenly distributed. Some bottom areas have very few specimens or seem completely devoid of myodocopids, whereas in other areas they are abundant. Barren areas are found in almost all environments, but seem especially common in the deep sea, especially below 3000 m. Needless to say, the diversity of myodocopid species is zero in barren areas. The following analysis considers only those areas in which myodocopids are fairly abundant. It is based on samples containing at least 25 specimens, except for a few smaller samples included because of having a high diversity. It has been necessary because of sparsity of samples to combine data from samples with several hundred specimens with data from samples containing a much smaller number of specimens. However, some of the samples in each of the environments compared contained over a hundred specimens. I do not believe that the disparity in the number of species per sample has materially affected the results. It would seem that with myodocopids, a good indication of diversity may be realized with a sample containing only 25 specimens. In fact, the largest number of species in a single sample – 8 species – was encountered in a sample containing only 17 specimens, whereas, the smallest number – 1 species – was encountered in a sample containing over 1000 specimens.

Most of the samples used in the present analysis were obtained in trawls, but a few were collected in grab samplers taking a single bite of substrate. The latter sample represents a much smaller area of substrate than the former. But this does not seem to have materially affected the data, because samples collected in trawls and with a grab in a similar environment had about the same number of species, although generally more specimens were collected in the trawl.

The number of species per sample in collections from all depths and all oceans, but mainly from the Atlantic Ocean, is shown in Figure 1. From this it may be seen that the number of species per sample varies from 1 to 8, and that the most commonly encountered number of species per sample is 3.

The number of species per sample for samples from the continental shelf (0–200 m), bathyal depths (200–2000 m), and abyssal depths (2000–5000 m) are compared in Figure 2. The most commonly encountered number of species per sample is 3 at all depths. The data also suggest that the variability in the number of species per sample decreases with depth, but this could be caused by sample error due to the sparsity of samples.

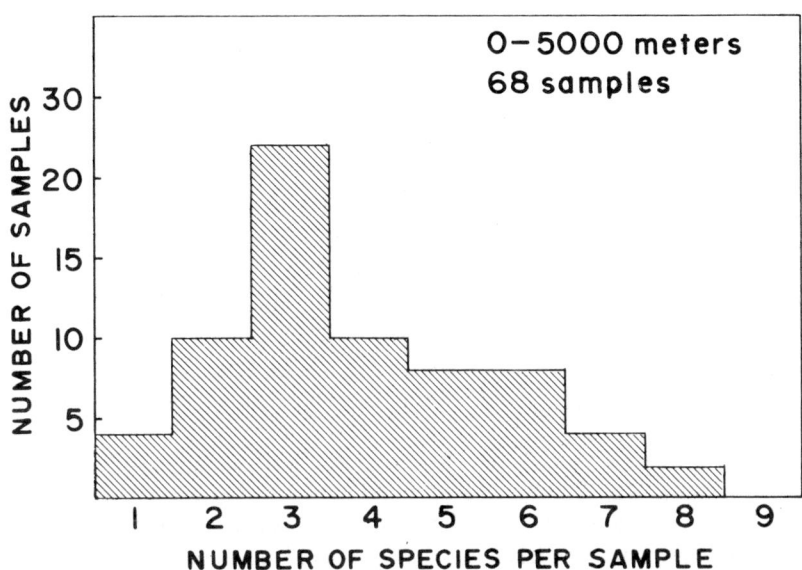

Fig. 1. Frequency distribution of species in samples from all depths and oceans. (Most samples with at least 25 specimens.)

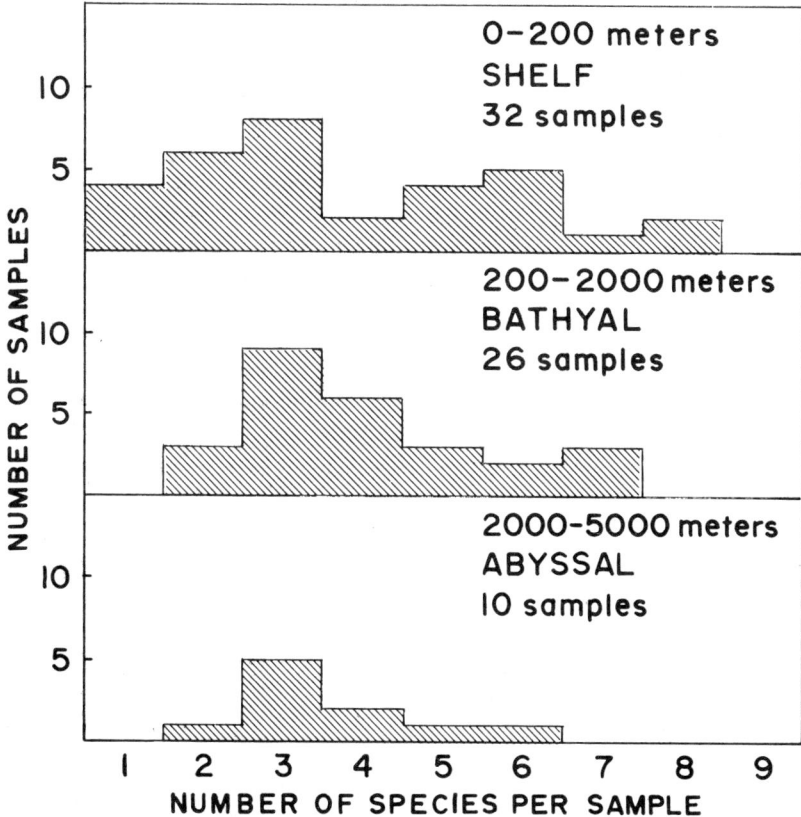

Fig. 2. Frequency distribution of species in samples from various depths zones. (Most samples with at least 25 specimens.)

Because some consider the abyssal zone to start at 3000 m rather than 2000 m, Figure 3 was constructed to show the distribution of the number of species per sample using 3000 m as the boundary between the bathyal and abyssal zones. The results are similar to those using 2000 m as the bathyal-abyssal boundary, but the apparent decrease in variability in the number of species per sample with increasing depth is more marked.

Sanders (1968), Sanders & Hessler (1969), Hessler (1974) and others (e.g. Buzas & Gibson 1969) have shown that a number of taxa increase in diversity in the deep sea. Benson (1975), on the other hand, found podocopid ostracodes to decrease in diversity in the deep sea. The present data indicate that the average number of myodocopid ostracodes per sample remains relatively constant with depth: shelf, 3.8 species per sample; bathyal depths (200-2000 m), 4 species per sample, and abyssal depths (2000-5000 m), 3.6 species per sample.

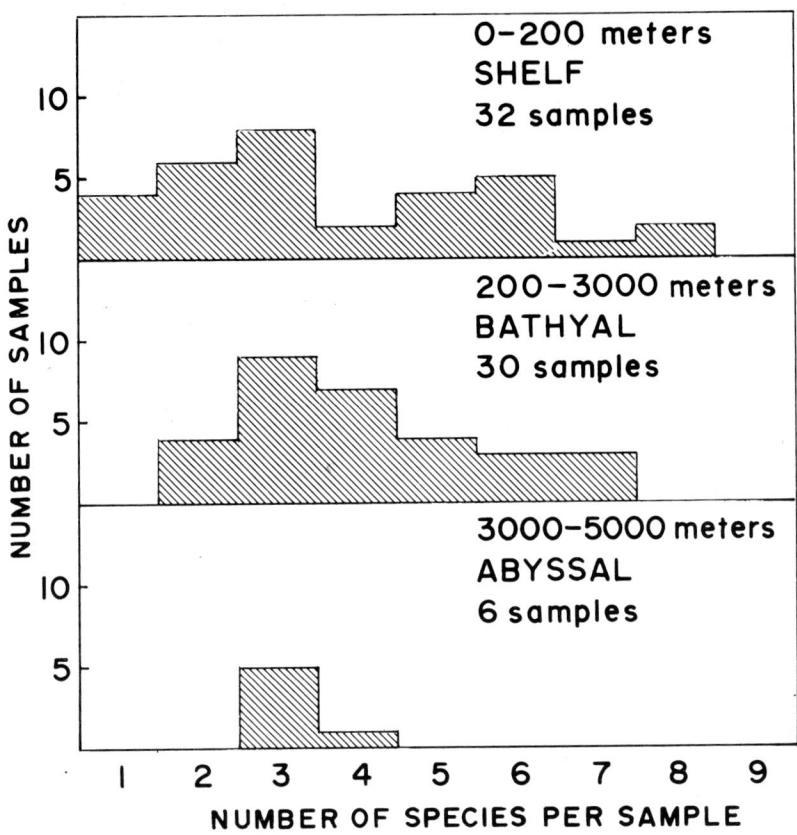

Fig. 3. Frequency distribution of species in samples from various depth zones. (Most samples with at least 25 specimens.)

The number of species per specific geographic subdivision

The North American Coast (Figure 4)

The coast of North America contains numerous estuaries, lagoons, bays open to the sea, and bays with restricted circulations caused by narrow connections with the sea. The continental shelves are broad along the Atlantic and Gulf Coasts and somewhat narrower along the Pacific Coast. The water temperature increases southward becoming tropical in southern Florida and Baja California. The tropical Bahama Banks lies about 50 miles east of the tip of Florida.

Atlantic Coast: The northernmost locality from which adequate data are available is Halifax Inlet, Nova Scotia, which contains 2 myodocopid species (Siddiqui & Grigg, 1975:369). these occurred in a littoral assemblage where salinities ranged form 28 parts per thousand to less than 1 part per thousand in the spring.

Cape Cod Bay, Massachusetts, also contains only 2 myodocopid species (Kornicker, 1974a). This large bay has a broad opening to the north and is about 56 m

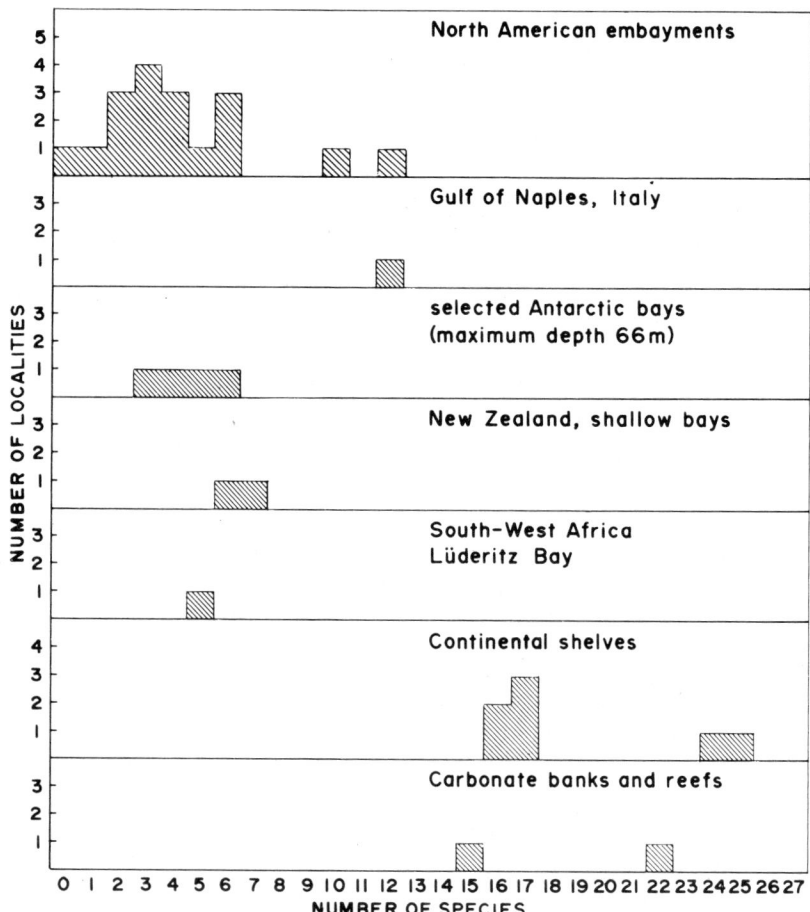

Fig. 4. The number of species in coastal embayments and on continental shelves of North America, and on the Bahama Banks. Numbers enclosed by squares indicate coastal embayments, numbers enclosed by circles indicate continental shelves and the Bahama Banks.

deep at its mouth. The salinities at which ostracodes were collected ranged from 31.2-32.6 parts per thousand.

Hadley Harbor, Massachusetts, is a small complex of tidal channels in the Elizabeth Islands, which lie just south of Cape Cod. It covers approximately one-third of a square kilometer of sea bottom. The salinities vary from just under 31 parts per thousand to just over 33 parts per thousand (Parker 1975). The complex contains 3 myodocopid species (Parker 1975 reported 4, but I consider 2 of them to be synonyms).

Martha's Vineyard is a small offshore island southwest of Hadley Harbor. It contains several small inlets and marine ponds partly separated from the sea by low bars and spits. Collections I have examined from the island indicate that the inlets and ponds support 4 myodocopid species, one of which may be considered extremely rare. The salinities are probably close to normal marine, but could be temporarily lowered during rainy periods.

Long Island Sound is a long narrow bay separating Connecticut to the north from Long Island to the south. Collections I have examined from the area (courtesy of M. Bowen, New York Ocean Science Laboratory) indicate that the Sound contains 3 myodocopid species. The salinities are probably normal marine.

Delaware Bay is a narrow bay forming the mouth of the Delaware River. According to Les Watling (personal communication, January, 1976), no myodocopid ostracodes were found in a collection of over 200 bottom samples from all depths in the bay. He believes that this could be due to turbid conditions in the bay which prevents growth of plants.

Rehobeth Bay, Delaware, is a small lagoonal like bay just south of the southern end of Delaware Bay. Collections I have examined from the bay (courtesy of Les Watling, University of Delaware Marine Laboratories) indicate that the bay contains 3 myodocopid species. Salinities in the bay are probably normal marine.

Chesapeake Bay, Maryland, is a large bay south of Delaware Bay. Wass (1965:28) reported 3 myodocopid species in the bay.

The Indian and Banana Rivers, Florida, are actually shallow lagoons extending about 253 km along the middle part of the eastern coast of Florida and are separated from the ocean by narrow barrier Islands. The average depth is 1.5 m. The salinity varies with rainfall, runoff, evaporation, and proximity of acces to the ocean. Collections I have examined from the area (courtesy of personnel of the Harbor Branch Foundation, Fort Pierce, Florida) contained 4 myodocopid species. The salinities at which they were collected varied from 18 to 42 parts per thousand.

Biscayne Bay is a broad open bay at the tropical tip of Florida. A small collection of ostracodes (collected by Dr. F. M. Bayer) contained 7 species of Myodocopa. Additional collections should reveal additional species.

The total number of species encountered in the Atlantic bay and estuaries between Halifax and the southern tip of Florida was 11. There is a general increase in the number of species from north to south with 2 species at the northernmost locality, Halifax, Nova Scotia, and 7 species at the southernmost locality, Biscayne Bay, Florida. This is attributed to the greater variability of temperature at higher latitudes.

The only data available for the number of species on the Atlantic continental shelf is that of Darby (1965) who reported 17 myodocopid species off the coast of Georgia. Darby's data suggest that the diversity of myodocopids is much greater on the continental shelf than in coastal embayments. A similar relationship for Podocopa was reported by Hazel (1975) who found diversity lowest in inshore shoal areas and highest on the outer shelf in the vicinity of North Carolina.

High diversity is encountered in the Bahamas, off Florida, where Kornicker (1958) reported 22 myodocopid species.

Gulf Coast

The types of embayments along the coast of the Gulf of Mexico are similar to those along the Atlantic coast. The low temperatures encountered along the northern part of the Atlantic coast are seldom encountered along the Gulf coast.

Placida Harbor, Florida, is a branch of Charlotte Harbor complex which is a drowned estuary partly separated from the Gulf by a series of barrier islands. It is located just below the middle of the west coast of Florida. Collections I have examined from the area (courtesy of R. Cressey, Smithsonian Institution) contained 11 myodocopid species. Salinities in the area sampled ranged from about 18 to 33.5 parts per thousand. Depths varied from about 1 to 3 meters.

Alligator Harbor, Florida, is a small shallow bay in the northern coast of Florida. Salinities are probably normal marine, becoming lower at times of heavy rainfall. Collections I have examined from the area (courtesy of Darrell K. Jones) contained 5 myodocopid species.

St. Andrew's Bay is a complex of bays just west of Alligator Harbor. Waller (masters thesis, 1961) reported 10 myodocopid species in the bay. The thesis contains no descriptions and I have not been able to locate the collections; therefore, additional collections are needed to verify the number of species inhabiting the bay.

Aransas Bay, Texas, is a small shallow bay in the coastal bend area of Texas. Kornicker and Wise (1962) reported 3 myodocopid species in the genus *Sarsiella* from the bay. I do not recall encountering any other myodocopids in the bay. Temperatures in the bay ranged from about 0° to 40°C and salinities from about 16 to 37 parts per thousand.

The total number of species encountered in the bays of the Gulf coast from Florida to Texas is 11. The diversity decreases from east to west from Florida to Texas, and this is also attributed to the greater variability in temperature in Texas than in Florida.

The only data available for the continental shelf of the Gulf is a collection of 47 samples from 50 to 60 m depth off the coast of Texas. I have examined the collection (courtesy of David Gettleson, Texas A&M University) and found 16 species. This data suggest that the diversity of myodocopids is much greater on the shelf than in adjacent coastal embayments.

Pacific Coast

Strait of Georgia, Departure Bay and Ganges Harbor, Vancouver Island, British Columbia, Canada: Lucas (1931) reported a single myodocopid species from Departure Bay at a depth of 20–30 m, and Smith (1952) reported another species from Ganges Harbor, at a depth of 5.5 to 7.3 m. It seems probable that both species occur together in the bays of Vancouver Island.

Puget Sound, Washington, is more-or-less a southern extension of the Strait of Georgia. Lie (1968) reported 3 myodocopid species from this area.

Tomales Bay, California, is a slitlike bay located just north of the middle of the western coast of California. Collections that I have examined from this bay contained 4 myodocopid species.

San Francisco Bay, California, is a large deep bay connected to the ocean by a very narrow opening that lies just south of Tomales Bay. Jones (1961) reported only 1 myodocopid species in the bay. The specimens were collected from water with a temperature range of 17.2° to 18.8°C, and a salinity range of 22 to 30 parts per thousand.

Half Moon Bay, California, is a small open bay just south of the entrance to San Francisco Bay. Collections that I have examined (courtesy of M. P. Wilderman, Marine Ecological Institute) contained 5 myodocopid species.

Monterey Bay, California, is just a large indentation in the coast at about the middle of the west Californian coast. Collections that I have examined (courtesy of P. H. Slattery, Moss Landing Marine Laboratories) from depths of 9 to 21 m contained 6 myodocopid species.

Bahía de San Quintín is a narrow bay along the western coast of Baja California separated from the Pacific Ocean by a very narrow mouth. Collections I examined from the bay (courtesy of Dr. J. Laurens Barnard) contained 6 myodocopid species. Maximum depth in the bay is 11 m, but most of the bay is less than 2 m deep. Salinities in the bay are normal marine except in the northern tip where it may become slightly hypersaline.

Scammon Lagoon is a bay similar to Bahía de San Quintín and is located on the west coast of Baja California about 325 km south of Bahía de San Quintín. McKenzie (1965) reported 6 myodocopid species from Scammon Lagoon.

In summary, the diversity of ostracodes along the Pacific coast increases from 2 or 3 species in British Columbia to the north to 6 species in Baja California to the south. This is attributed to the greater variability in temperature in the north. A similar north-to-south increase in mollusk diversity was reported by Fisher (1960).

The Gulf of California separates Baja California from the mainland of Mexico. Collections I examined from Bahía de los Angelos (courtesy of Dr. J. Laurens Barnard), a small open bay near the northern end of the east coast of Baja California, contained 12 myodocopid species. The Gulf of California contains water with normal marine salinities. Apparently, myodocopids are more diverse in bays of the Gulf than they are in bays along the Pacific coast of Baja California.

Baker (1974:131) reported 25 species on the continental shelf of Southern California. The total number of species recorded in embayments of the Pacific Coast from British Columbia to Baja California is about 12. This indicates that the diversity of myodocopids is much greater on the continental shelf than in the coastal embayments.

Bays (Figure 5)

As discussed above, bays along the North American coasts contain 0 to 12 myodocopid species. Bays in other areas of the world contain about the same number of myodocopid species: Gulf of Naples, Italy, 12 species (Müller 1894, Kornicker 1974b); Luderitz Bay, South-West Africa, 5 species (Klie 1940, Kornicker 1976); Lyttelton Harbor, New Zealand, 7 species (Kornicker 1975:30); Akaroa Harbor, New Zealand, 6 species (Kornicker 1975:30); Discovery Bay, Greenwich Island, South Shetland Islands, Antarctica, 6 species (Kornicker 1975:27); Port Lockry, Wiencke Island, Palmer Archipelago, Antarctica, 4 species (Kornicker 1975:27); Arthur Harbor, Anvers Island, Palmer Archipelago, Antarctica, 5 species (Kornicker 1975:27).

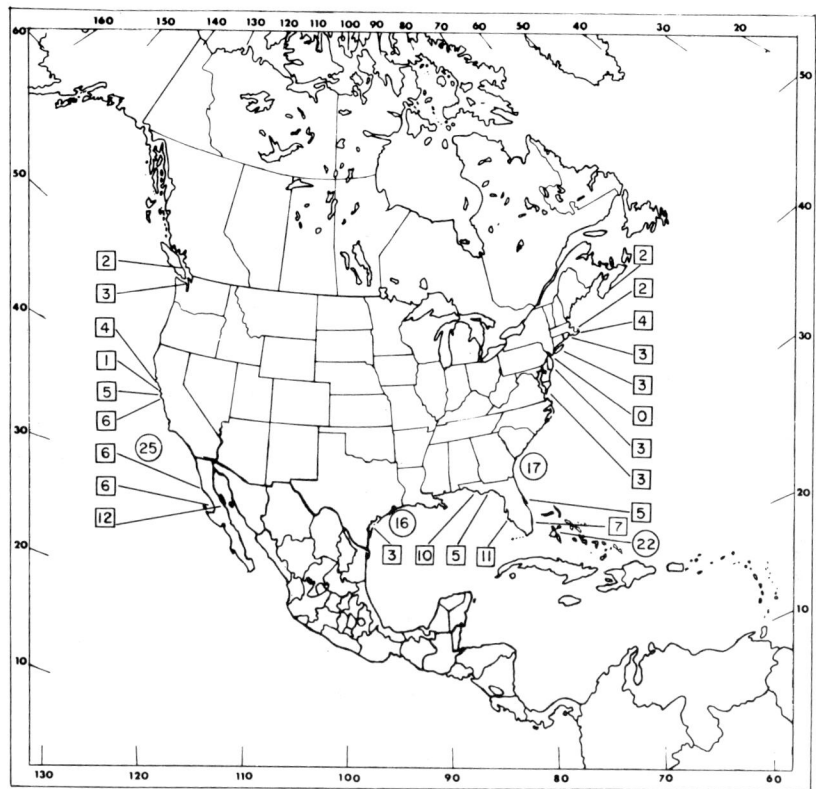

Fig. 5. Comparison of the number of species occupying various geographical areas.

Continental Shelves (Figure 5)

The continental shelves appear to contain a greater diversity of myodocopid species than present in embayments of the adjacent coasts. The largest number of species so far recorded occurs on the west African shelf off Mauritania, which contains 24 species (Kornicker & Caraion, 1974, + 2 papers in preparation). The continental shelf of the Gulf of Mexico also contains a high diversity of myodocopid species because 16 species were collected in a very small area off the Texas coast (herein). Baker (1974) reported 25 species off the coast of Southern California. Kornicker (1975: 26-30) recorded 16 species on the Argentina shelf, 17 species on the Chilean shelf, and 17 species on the shelf of the Antarctic continent. Darby (1965) reported 17 species on the shelf off Georgia.

The high diversity of the shelves, despite the low number of species collected per sample, suggests that the kinds of species change more rapidly from place-to-place on the shelf than in the embayments along the adjacent coast.

Carbonate banks and coral reefs (Figure 5)

These environments usually occur in either tropical or subtropical regions and are known to support a diverse fauna. Kornicker (1958) reported 22 myodocopid species in a relatively small area on the Great Bahama Bank. Collections of myodocopid ostracodes I examined (courtesy of B. Thomassin, Station Marine D'Endoume et Centre D'Océanographie, Marseille, France) from Tulear Reef, Madagascar, contained 15 species.

Diversity in southern oceans

In a recent paper (Kornicker 1975) I reported on myodocopid ostracodes between Antarctica and the latitude 35°S. The area was divided into 3 zones (Figure 6): 1, the Antarctic zone bounded by the Antarctic coast to the south and the Antarctic convergence to the north; 2, the Subantarctic zone bounded by the Antarctic convergence to the south and the subantarctic boundary to the north, the latter boundary was defined by Hedgpeth (1969) and is based on a study of the distribution of many taxa; and 3, the Subantarctic-to-35°S zone bounded by the subantarctic boundary to the south and the latitude 35°S to the north, the latter boundary is arbitrarily defined. Although the areas within each zone are not quite equal, they are used here as a basis for comparing diversity trends in the southern oceans.

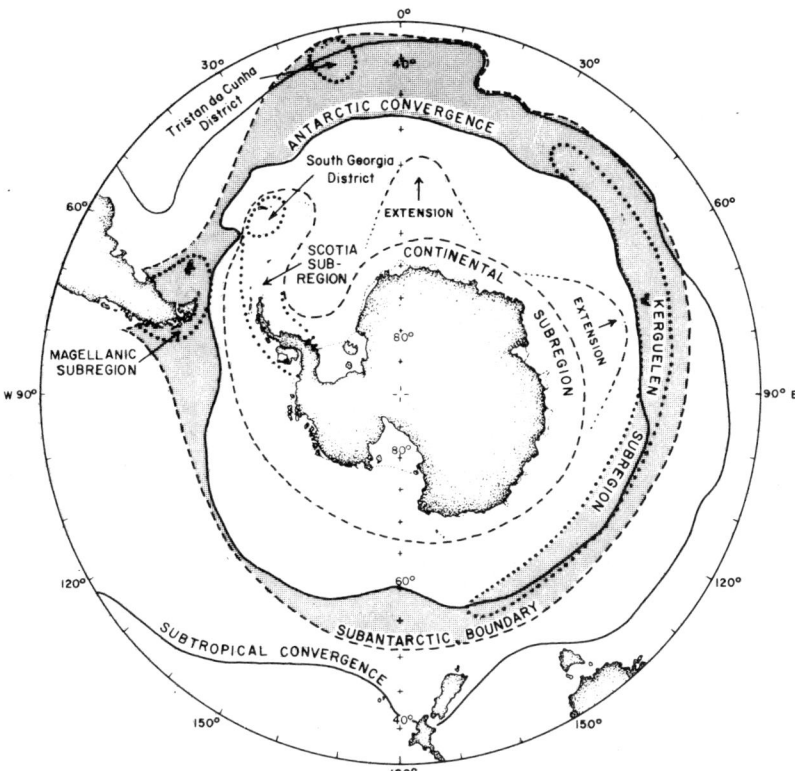

Fig. 6. Biogeographic zones: Antarctic, Subantarctic, Subantarctic-to-35°S. (From Kornicker 1975, Figure 6).

Only samples that contained myodocopids are considered in the present analysis. The samples were obtained with various types of trawls and grab samplers, but an insufficient number of samples were collected with any particular device to warrant resticting the analysis to samples collected with that device. The present analysis may be justified because each of the environments compared were sampled with the same diverse equipment, but the results must be questioned. The analysis should be considered a first approach to a complex problem.

The number of species at shelf, bathyal, and abyssal depths in each zone are compared in Table 1. In the present analysis the shelf is defined as delimited by depths of 0–200 m and includes bays. As a topographic unit, the outer edge of Antarctic shelf is at a depth of 400–500 m (Knox 1970:83), so that the term shelf, when referred to the Antarctic continent actually refers to the upper part of the topographic shelf, and the term bathyal includes the lower part of the topographic shelf. The number of shelf species increased from 15 species in the Antarctic zone to 20 in the Subantarctic zone, and to 55 in the Subantarctic-to-35°S zone (Table 1). This northerly increase took place even though the number of samples collected also decreased in a northerly direction. In the Antarctic zone the number of bathyal species is greater than the number of shelf species, even though fewer samples were collected in the bathyal zone. The remaining areas are difficult to compare because of the disparity in the number of samples collected in each area.

Table 1. Distribution and diversity of myodocopid ostracodes in the southern oceans

Region	Number of species	Number of samples	Diversity index*
Antarctic			
Shelf (0-200 m)	15	135	7.04
Bathyal (200-2000 m)	21	101	10.48
Abyssal (2000-6000 m)	13	24	9.42
Subantarctic			
Shelf	20	63	11.12
Bathyal	19	41	11.78
Abyssal	3	4	4.98
Subantarctic-to-35°S			
Shelf	55	90	28.14
Bathyal	20	25	14.31
Abyssal	8	8	8.86

* Diversity index = number of species/log number of samples.

An unsophisticated formula that has been advanced for the calculation of a diversity index is d = number of species/log of area (Whittaker 1975:95). Because the area sampled is a function of the number of samples collected, I have used, in an attempt to compare diversities using the data on hand, an untested formula, d = number species/log number of samples (Table 1).

The diversity at shelf depths is considerably higher in the Subantarctic-to-35°S zone than in the Subanatarctic and Antarctic zones (Figure 7). The difference in

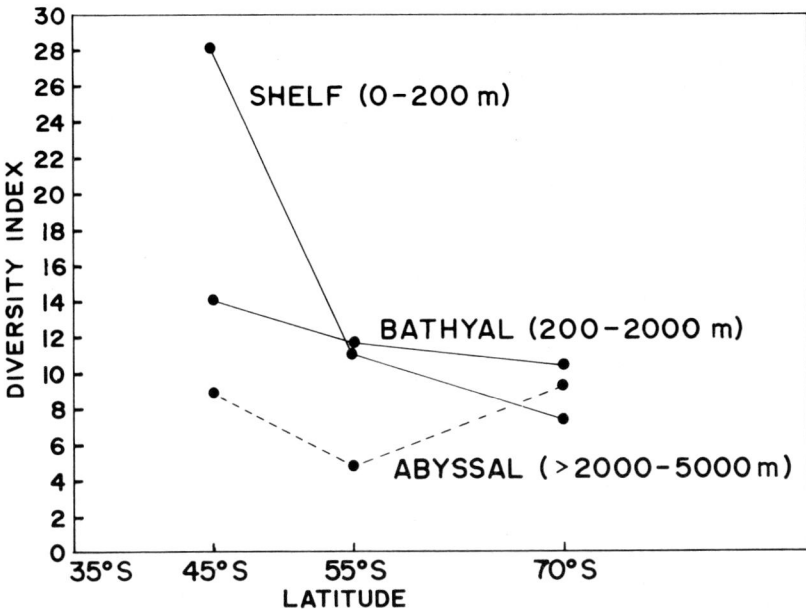

Fig. 7. Latitudinal distribution of diversity in the Antarctic, Subantarctic, and Subantarctic-to-35°S zones at shelf, bathyal, and abyssal depths.

diversity is less marked at bathyal depths, and is not apparent at abyssal depths. This is attributed to latitudinal differences in environment having a greater effect on shelf biota than on bathyal and abyssal biota.

The decrease in diversity with an increase in latitude at shelf and bathyal depths may be attributed to greater stability in environment in lower latitudes, but another factor probably plays a part. In the Antarctic zone the shelf and slope from continuous bands around the Antarctic continent. The lack of major topographic barriers permits easy dispersal of organisms. Thus, many of the animals inhabiting the shelf and slope are circumpolar. On the other hand, the seas adjacent to land in the Subantarctic and Subantarctic-to-35°S zones are separated by wide and deep water permitting allopatric speciation in the seas adjacent to the various islands and continents in those zones.

When diversity is considered as a function of depth in each of the 3 zones, it is seen that the effect of depth differs in ach zone (Figure 8). This is a consequence of the large differences in diversity at shelf depths compared to both bathyal and abyssal depths.

Summary and conclusions

The number of species per sample at shelf, bathyal, and abyssal depths are about the same, but the range is greater at shallower depths. Although the number of species per sample is not greatly different in coastal embayments and on shelves, the shelves contain a larger number of species than present in the embayments.

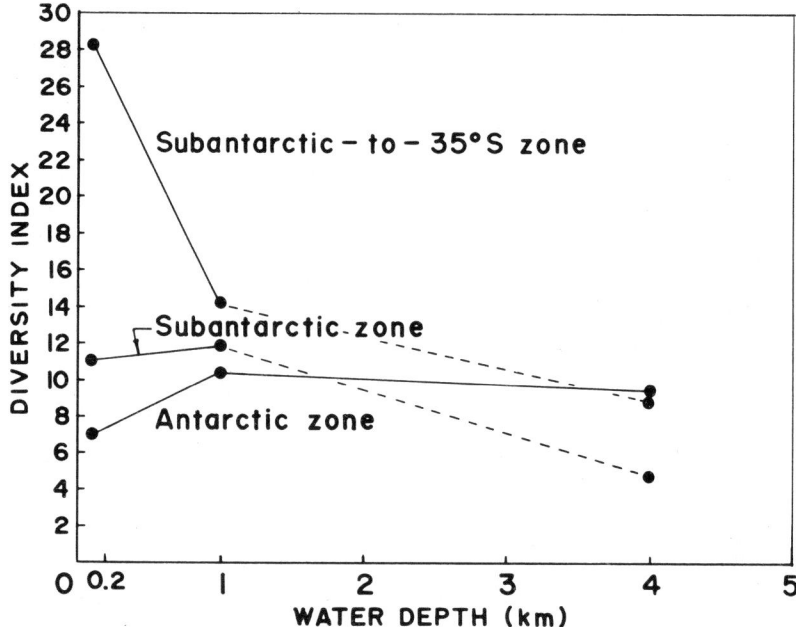

Fig. 8. Depth distribution of diversity in the Antarctic, Subantarctic, and Subantarctic-to-35°S latitudinal zones.

This suggests that the kinds of species change more rapidly from place-to-place on the shelves than in coastal embayments.

The diversity of myodocopid ostracodes in coastal embayments of North America increases in lower latitudes. This is attributed to greater stability of the environment in lower latitudes.

In southern oceans considered on a global basis, the diversity of myodocopid ostracodes increases greatly at lower latitudes at shelf depths, less at bathyal depths, and not at all at abyssal depths. This decrease in latitudinal affect with increase in depth is attributed to greater uniformity of environment at increasing depths. The decrease in diversity at shelf and slope depths with increase in latitude is attributed to both the greater stability in environment in lower latitudes and to the presence of widely separated land masses in the lower latitudes.

The relationship of depth and diversity varies in each zone because the latitudinal difference in environment in each zone has a much greater effect on shelf biota than on bathyal and abyssal biota.

Diversity in the Subantarctic-to-35°S zone decreases markedly with depth, whereas, in the Subantarctic and Antarctic zones diversity increases slightly at slope depths. This is a consequence of the large differences in diversity at shelf depths compared to bathyal and abyssal depths.

Acknowledgements

I wish to thank Drs. Howard L. Sanders, Robert R. Hessler, and Robert J. Menzies for many of the samples upon which this report is based. I am grateful to Drs. Thomas E. Bowman, Martin A. Buzas, and Richard M. Forester for criticizing the manuscript.

References

Baker, James H. 1974. Distribution, Ecology, and Life Histories of Selected Cypridinacea from the Southern California Mainland Shelf. The Texas Journal of Science 25:131 [abstract].

Benson, Richard H. 1975. Morphological Stability in Ostracoda. pp. 14-46, in: Biology and Paleobiology of Ostracoda, Bulletin American Paleontology, 65 (282).

Buzas, Martin A. & Thomas G. Gibson. 1969. Species Diversity: Benthic Foraminifera in Western North Atlantic. Science 163:72-75.

Darby, D. G. 1965. Ecology and Taxonomy of Ostracoda in the Vicinity of Sapelo Island, georgia. Report no. 2 in Four Reports of Ostracod Investigations, 77 pages. Ann Arbor, Mich. University of Michigan. [Offste report].

Fischer, Alfred G. 1960. Latitudinal Variations in Organic Diversity. Evolution, 14(1):64-81.

Hazel, Joseph E. 1975. Pattern of Marine Ostracode Diversity in the Cape Hatteras, North Carolina, Area. Journal of Paleontology, 49 (4):731-744.

Hedgpeth, Joel W. 1969. Distribution of Selected Groups of Marine Invertebrates in Waters South of 35 Degrees Latitude. Antarctic Map Folio Series, 11:1-9. American Geographical Society.

Hessler, Robert R. 1974. The structure of Deep Benthic Communities from Central Oceanic Waters. The Biology of the Oceanic Pacific: 79-93. Oregon State University Press.

Jones, M. E. 1961. A Quantitative Evaluation of the Benthic Fauna off Point Richmond, California. University of California Publications in Zoology 67 (3):219-320.

Klie, Walter. 1940. Beiträge zur Fauna Eulitorials von Deutsch-Südwest-Africas. Kieler Meesresforschungen 3:403-448.

Knox, G. A. 1970. Antarctic Marine Ecosystems. Antarctic Ecology 1:69-96.

Kornicker, Louis S. 1958. Ecology and Taxonomy of Recent Marine Ostracoda in the Bimini Area, Great Bahama Bank. Publications of the Institute of Marine Science (The University of Texas) 5:194-300.

Kornicker, Louis S. 1974a. Ostracoda (Myodocopina) of Cape Cod Bay, Massachusetts. Smithsonian Contributions to Zoology 173:1-20.

Kornicker, Louis S. 1974b. Revision of the Cypridinacea of the Gulf of Naples (Ostracoda). Smithsonian Contributions to Zoology 178:1-64.

Kornicker, Louis S. 1975. Antarctic Ostracoda (Myodocopina). Smithsonian Contributions to Zoology 163:1-720.

Kornicker, Louis S. 1976. Myodocopid Ostracoda from Southern Africa. Smithsonian Contributions to Zoology 214:1-39.

Kornicker, Louis S. & Francisca Elena Caraion. 1974. West African Myodocopid Ostracoda (Cylindroleberididae). Smithsonian Contributions to Zoology 179:1-78.

Kornicker, Louis S. & Charles D. Wise. 1962. Sarsiella (Ostracoda) in Texas Bays and Lagoons. Crustaceana 4 (1):59-74.

Lie, Ulf. 1968. A Quantitative Study of Benthic Infauna in Puget Sound, Washington, USA, in 1963-1964. Skriderektoratets Skrifter Serie Havundersokelser, 14 (5):556 pp.

Lucas, Verna Z. 1931. Some Ostracoda of the Vancouver Island Region. Contributions to Canadian Biology and Fisheries 6:399-410.

McKenzie, K. G. 1965. Myodocopid Ostracoda (Cypridinacea) from the Scammon Lagoon, Baja California, Mexico, and their Ecological Associations. Crustaceana 9 (1):57-70.

Müller, G. W. 1894. Die Ostracoden des Golfes von Neapel und der angrenzenden Meeres-Abschnitte. In Fauna und Flora des Golfes von Neapel 21:404.

Parker, Robert H. 1975. The Study of Benthic Communities. Elsevier Oceanography Series, 9:279 pp. Elsevier Scientific Publishing Company, Amsterdam – Oxford – New York.

Sanders, Howard L. 1968. Marine Benthic Diversity: A Comparative Study. The American Naturalist 102 (925):243-281.
Sanders, Howard L. & Robert R. Hessler. 1969. Ecology of the Deep-Sea Benthos. Science 163:1419-1424.
Siddiqui, Q. A. & U. M. Grigg. 1975. A preliminary Survey of the Ostracodes of Halifax Inlet. pp. 369-380 in Swain et al. (ed.), Biology and Paleobiology of Ostracoda, Bulletins of American Paleontology 65 (282).
Smith, Verna Z. 1951. Further Ostracoda of the Vancouver Island Region. Journal of the Fisheries Research Board of Canada 9 (1):16-41.
Waller, R. A. m.s. Ostracoda of St. Andrew Bay, Florida. Master's thesis submitted to Florida State University, 1961.
Wass, Marvin L. 1965. Check List of the Marine Invertebrates of Virginia. Special Scientific Report No. 24:1-58. Virginia Institute of Marine Science, Gloucester Point, Virginia.
Whittaker, R. H. 1975. Communities and Ecosystems, second edition, 387 pp. Macmillan Publishing Co., Inc. New York.

Author's address:

National Museum of Natural History, Smithsonian Institution, Washington D.C. 20560, USA

Sixth Intern. Ostracod Symposium, Saalfelden

DISTRIBUTION OF MARINE PODOCOPID OSTRACODA IN THE GULF OF MEXICO AND THE CARIBBEAN

W. A. VAN DEN BOLD

Abstract

Two aspects of ostracode-distribution are dealt with here: Geographic - and Depth distribution. The shallow-water fauna of the Gulf of Mexico is characterized by species of *Hulingsina*, *Actinocythereis* and *'Haplocytheridea'*, whereas the Caribbean fauna contains as typical representatives species of *Quadracythere* and tuberculate species of *Loxoconcha*. In between the two regions occurs a kind of transition zone with species of *'Haplocytheridea'*, reticulate *Cytheretta* and some Campylocytheridae, a fauna which appears typical of the shallow carbonate platforms of Yucatan, Cuba, Florida and the Bahamas. Within the Caribbean sea proper, only a few species are restricted in distribution, to a much lesser degree than in the Neogene, when e.g. the *Costa variabilocostata* group was characteristic for the South American subprovince: *Loxoconcha lapidiscola* Hartmann (Puerto Rico to Nicaragua) and *'Campylocythere' perieri* (Brady) (South Cuba to Venezuela) both follow the arc of the Lesser Antilles; in Central America *Reussicythere howei* n. name, occurs from Panama to Belize. The genus *Cativella*, which was widespread and generally common in the Caribbean during the Neogene is now abundant only in isolated areas (both ends of the Cayman Trench, Panama, Venezuela).

Although a rough bathymetric zonation has been established for some time, detailed results are still uncertain and show to some extend contradictory evidence in different areas. This is in part due to scarcity of deepwater material. Only in the Gulf of Mexico data are available that reach 4000m depth; off the Cuban north coast samples go down to 1700 m, off Jamaica to 340 m, on the Paria-Orinoco shelf down to about 200 m and off the islands north of Venezuela to 84 m; for other areas, this report is restricted to shallower water ostracodes.

Introduction

In the thirty-odd years that I have been studying the fossil ostracodes of the Caribbean, I have paid little attention to the Recent ones, except where their taxonomy and ecology could give clues to solving problems concerning the fossil ones. During that time I have collected ostracodes from over one hundred Holocene localities in the Caribbean, Bahamas and the Gulf of Mexico. Other samples were collected and/or donated by H.G. Kugler, P.J. Bermúdez, R.S. Steward, J. Selliers de Civrieux, H.M. Bolli, J.W. Teeter and others. Additional Recent samples, collected by students and colleagues of the late H.V. Howe, are now in the collections of Louisiana State University; C.H. Moore provided sample material from the reef-slope north of Jamaica. Recently, I had the opportunity to study material from different sources in the collections of F.P.C.M. van Morkhoven, mainly from the Gulf of Mexico, but also material collected by Wagenaar Hummelinck in the Caribbean, bringing the total number of samples to just over four hundred. Therefore, it seems, that the time has come to bring these together in a study on the distribution of Recent ostracodes of this region.

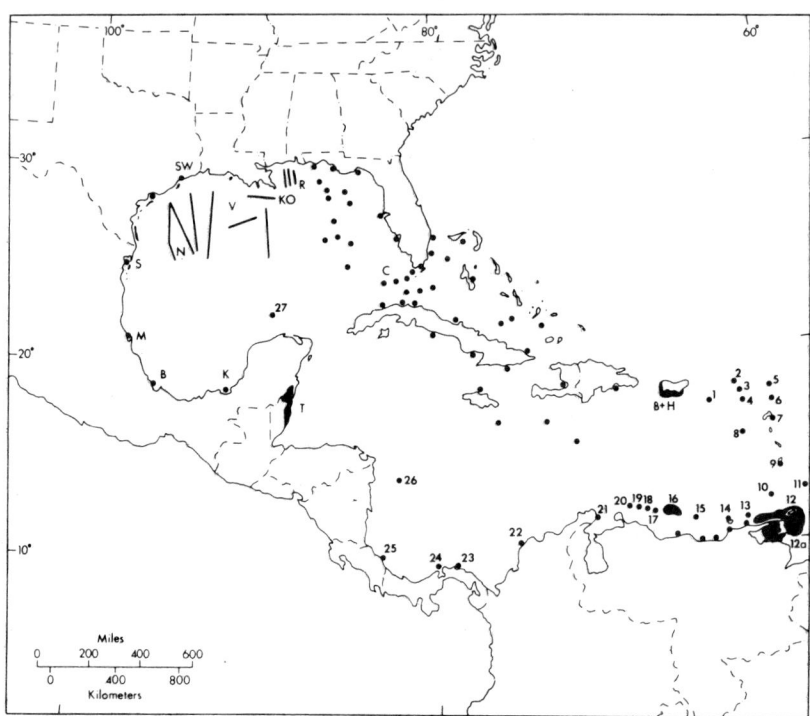

Fig. 1. Location map.
Authors of original studies of material are indicated by capital letters: B: G.S. Brady; B+H: J. Baker and N.C. Hulings; C: R.H. Benson and G.L. Coleman; K: P.R. Krutak; KO: M. Kontrovitz; M: G.A. Morales; N: R. Neal, H. Johns and R. Erickson; P: H.S. Puri; R: M.A. Rafle; S: P.A. Sandberg; SW: F.M. Swain and V: F.P.C.M. van Morkhoven.

A number of localities has been identified by number: 1: St. Croix; 2; St. Martin; 3: St. Barthélemy; 4: St. Eustachius; 5: Barbuda; 6: Antigua; 7: Guadeloupe; 8: Aves island; 9: St. Lucia; 10: Grenada; 11: Barbados; 12: Paria-Trinidad-Orinoco shelf; 12a: Gulf of Paria; 13: Los Frailes; 14: Margarita, Cubagua and Coche; 15: Tortuga; 16: Los Roques; 17: Las Aves; 18: Bonaire and Klein Bonaire; 19: Curaçao; 20: Aruba; 21: Tucacas bay; 22: Cartagena; 23: San Blas (Panama); 24: Colón harbour, Las Minas Bay (Panama); 25: Puerto Limón (Costa Rica); 26: Miskito Keys (Nicaragua); 27: Alacrán reef.

Even with this large amount of material, it is apparent, that there are still considerable gaps in our knowledge on the geographic distribution. Text-fig. 1 shows that there are a few areas in which concentrated collecting has been done, e.g. the Gulf of Mexico and the Paria-Trinidad-Orinoco shelf, and that the coverage of the Antilles and the north coast of Venezuela is quite adequate. In contrast, only one sample exists from the area between the Goajira peninsula and Panama, and none from the Caribbean coast between Costa Rica and British Honduras, so that the eastern Caribbean is poorly represented.

Zoogeographic (sub) provinces

The first attempt to establish Holocene ostracode-faunal provinces in this region was made by H.S. Puri (1967), who recognized: a Gulf of Mexico -, Bahamas -, South Florida – and a Venezuela (sub) province. As this was done within the scope of a general, world-wide review, no attempt was made to delineate these provinces and sub-provinces and their boundaries were not discussed. Their relative positions are even somewhat uncertain and the faunal evidence on which they were based was very incomplete and, to a certain extent, misleading. For instance, all species mentioned for the Venezuelan subprovince had already been reported from other parts of the Caribbean, and there is no clear differentiation between the South Florida and Bahamas subprovinces.

In the Neogene, the Caribbean faunal province can be divided into three subprovinces: Antillean, South American and Central American (van den Bold 1970). However, with the disappearance of some typical species in the late Neogene a more thorough mixing of the different faunal elements occurs from east to west across the Caribbean, so that a clear differentiation is not possible in the Holocene. Even the boundary with the Gulf of Mexico fauna becomes a more or less wide transition-zone. The fauna consists mainly of that of the shallow carbonate platforms between Florida, Cuba and Yucatán and includes that of the Bahamas, made up of several representatives of the Campylocytheridae, *Actinocythereis*, ridged and reticulate *Cytheretta* (or *Protocytheretta*) and '*Haplocytherideae*' ex gr. *setipunctata*. This last group will be described under a new name by J.E. Hazel.

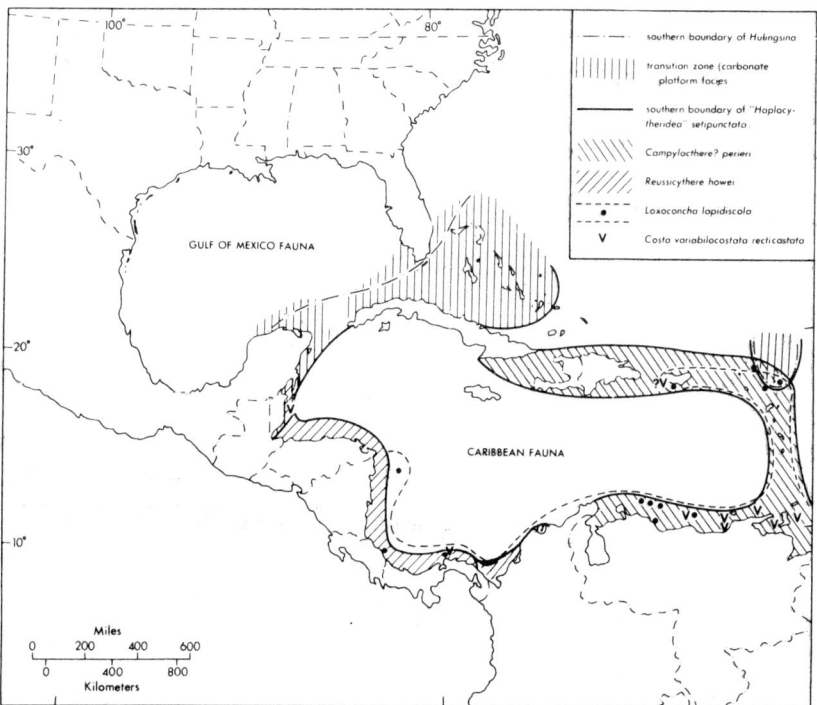

Fig. 2. Distribution of the ostracode biogeographic provinces in Recent sediments of the Gulf of Mexico and the Caribbean and of some ostracodes within these provinces.

In a recent paper (Bold 1976), I refer them questionably to *Hemicyprideis* to show their distinction from true *Haplocytheridae*.

The Gulf of Mexico exhibits marked differences in fauna between the predominantly carbonate or clastic areas, but always has, in addition to species from the transition zone *Hulingsina* and '*Haplocytheridae*' *bradyi*. The latter species has been found on one occasion in the Bahamas. A list of species for the Gulf of Mexico can be found in Hulings (1967). For later additions, see Howe & Van den Bold (1975), whose early Holocene or late Pleistocene fauna is almost identical with the Recent. More details will be available when the dissertation of Kontrovitz on the ostracodes of the Mississippi delta region appears. Van Morkhoven (1972) did not make specific identifications, but his table shows a complete record of the genera.

Typical representatives of the Caribbean fauna are tuberculate species of *Loxoconcha* (or *Loxocornuculum*), *Quadracythere* of the *producta* group and many other forms that constitute a rather uniform fauna that extends well into the transition-zone, into South Florida and even into the Gulf of Mexico. Examples of such forms are *Hermanites*, *Morkhovenia*, *Jugosocythereis*, *Caribella*, *Pellucistoma*, and in carbonate areas, ornate Bairdiidae. Some are also found occasionally in the Gulf of Mexico. The genus *Orionina* is largely confined to the Caribbean, but *C. bradyi* generally occurs more northerly, especially in the transition-zone,

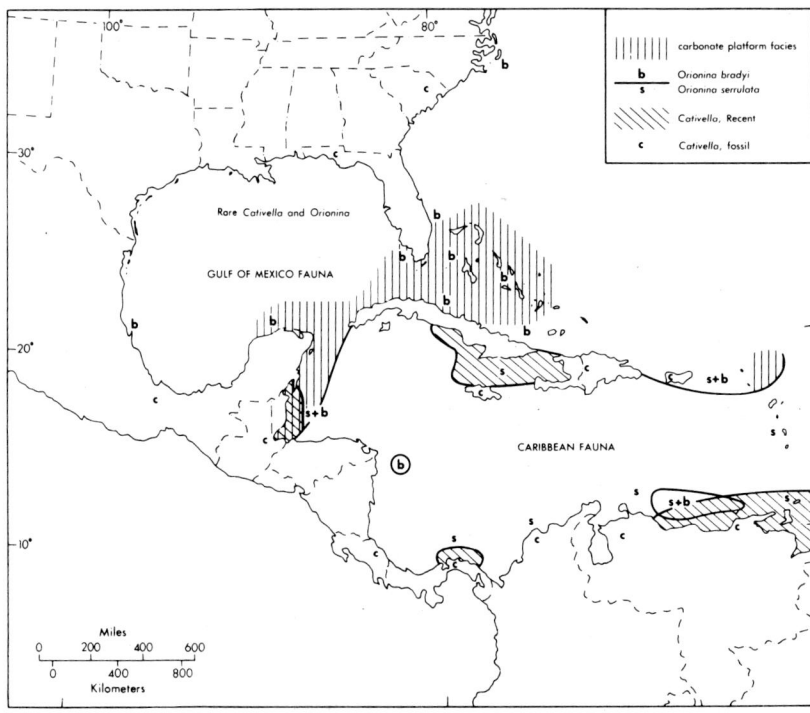

Fig. 3. Distribution of the genera Orionina and Cativella in the Gulf of Mexico and the Caribbean.

whereas *O. serrulata* appears typical for the Caribbean proper. (Text-fig. 3). A mixture of Caribbean and Gulf of Mexico elements is present in the area of Antigua (northern Lesser Antilles) (Text-fig. 2, 3). In this general area, both species of *Orionina* are found but other records have been made from the Venezuelan north coast, and *O. bradyi* has been found exclusively in a sample from Miskito Keys, off Nicaragua.

The genus *Cativella*, which extended as far north as the Carolinas in the late Tertiary, now appears to be restricted to the Caribbean (north of Venezuela, both ends of the Cayman trench) and is represented in the Gulf of Mexico only by a rather rare form which lives in slightly deeper water (Text-fig. 3).

The genus *Costa*, which was used to separate the different faunal sub-provinces in the Caribbean (Bold 1970) has now almost disappeared. In the Greater Antilles, only a deepwater species of the *C. maquayensis* group (*C. bellipulex*) remains (originally described from the Pleistocene of the Gulf of Mexico). Two subspecies of *Costa variabilocostata* (South American subprovince) occur in Recent sediments, but they are also found along the Central American coast as far north as British Honduras (Teeter 1975) and one species has been reported by Baker & Hulings (1966) from Puero Rico. I have not seen the species in the samples from that area, available to me, and consider this ocurrence as questionable (Text-fig. 2) because so far this species has not been found in samples from the Lesser Antilles.

Species that show some restriction in their Holocene Caribbean distribution are (Text-fig. 2): *Loxoconcha lapidiscola* Hartmann (Puerto Rico, Lesser Antilles, southern Caribbean, Central America to Miskito Keys); *Campylocythere? perieri* (Brady) (South Cuba, Hispanola, Puerto Rico, Lesser Antilles and southern Caribbean to Aruba); and *Reussicythere howei* (new name for *R. reussi* (Brady), see note at the end of this paper (Central America from Panama to British Honduras).

Bathymetry

Rothwell (1949) reported a more or less barren zone between 10 and 20 m depth in the Gulf of Mexico. This is confirmed by Van Morkhoven (1972). However, near the north coast of South America, there is no sign of a barren zone, nor is one reported from Puerto Rico (Baker & Hulings,1966). Perhaps this is a characteristic of the Gulf of Mexico.

Van Morkhoven (1972) was the first to make a well documented depth-distribution study of ostracodes in the western Gulf of Mexico, but only an abridged version was published. Van Morkhoven allowed me access to his types and to the original, more detailed, distribution chart and, moreover, loaned me his material and depth-disbribution chart of a similar study of the eastern Gulf. These are the only studies in the area which go down to a depth of more than 4000 m. A preliminary study of hydrographer bottom samples, down to 3400 m depth, was carried out by three graduate students of Louisiana State University (Richard Erickson, Hillary Johns & Ronalds Neal). Although the samples were taken from sites located to Van Morkhoven's more westerly region (see Text-fig. 1), results of their analyses differ considerably from those of van Morkhoven. In part, this must definitely be ascribed to downslope movement and redeposition of sediments. Another possibility is that Van Morkhoven's traverse, almost at the edge of the Mississippi River plume, carried a higher load of nutrients than the Hydrographer traverses, which may have affected the ostracodes distribution. These and other possibilities are now being investigated by Neal.

In very shallow water (less than 1 m), van Morkhoven reports *Neocytherideis, Neomonoceratina, Loxoconcha* and *Campylocythere* the most characteristic genera. Most of these are also present in the same depth region in the Caribbean, with the exception of *Neocytherideis* which is reported only below 40 m from the Mississippi-Alabama coast (Rafler, MS thesis, Washington University, information received through the courtesy of Dorothy Echols). *Campylocythere*, as stated above, appears to be more typical of the shallow carbonate environment.

In the enriched zone between 20 and 100 m, van Morkhoven finds *Cytherelloidea, Basslerites, Macrocyprina* (all of which occur from sealevel downwards in the Caribbean), *Cativella* (approx. below 10 m in the Caribbean) and *Paracytheridea* (below 60 m off South America).

Neocaudites and *Parakrithe* appear in depths between 100 and 200 m (these are already present at 10 m depth in the South Caribbean).

On the Paria-Trinidad-Orinoco shelf, the following general succession with depth is found (Table 1, dashed lines):

Very shallow (less than 10 m): *Tanella, Perissocytheridea, Callistocythere, Campylocythere?, Neocaudites, Hemicytherura, Neomonoceratina, Paracytheroma,* 10–40 m: *Morkhovenia, Pumilocytheridea, Orionina, Caudites,* further *Ruggieria?, Occultocythereis* and *Costa variabilocostata recticosata.*

Forms first present at 40 m: *Munseyella, Aurila, Quadracythere, Hermanites, Jugosocythereis, Puriana, Pterygocythereis, Ambocythere, Bythoceratina.* At 60 m: *Bythocythere, Pseudocythere, Paracytheridea* and *Costa variabilocostata laticostata.*

Below 60m: *Krithe. Costa variabilocostata recticostata* disappears.

Several discrepencies between this profile and that for the region off Los Roques, some 600 km further west, exist (Table 1 solid lines). On the other hand, however, it is fairly similar to the profile obtained for region of the North Coast of Jamaica, in a carbonate environment, but here it is obvious, that much displacement of faunas took place down the very steep slope towards the Cayman Trench (Table 2). Similar occurences of shallow-water ostracodes in deep water sediments are noticed in the Gulf of Mexico, where for instance, Erickson, Johns & Neal find specimens of *Limnocythere* at 40 m depth and of *Perissocytheridea* and *Cyprideis* even down 628 m. Off the south coast of Cuba, the fauna found in a core of Atlantis 3345 at 1278 m depth, has a relatively large number of shallow-water forms *(Quadracythere producta, Radimella wantlandi, Loxoconcha fischeri, Macrocyprina* sp.) together with deep-water forms such as *Bythocypris* and *Krithe* of the *trinidadensis* type. Similar examples are reported by Breman (1976) in the Adriatic. In relation to such, I would also like to point out the resemblance

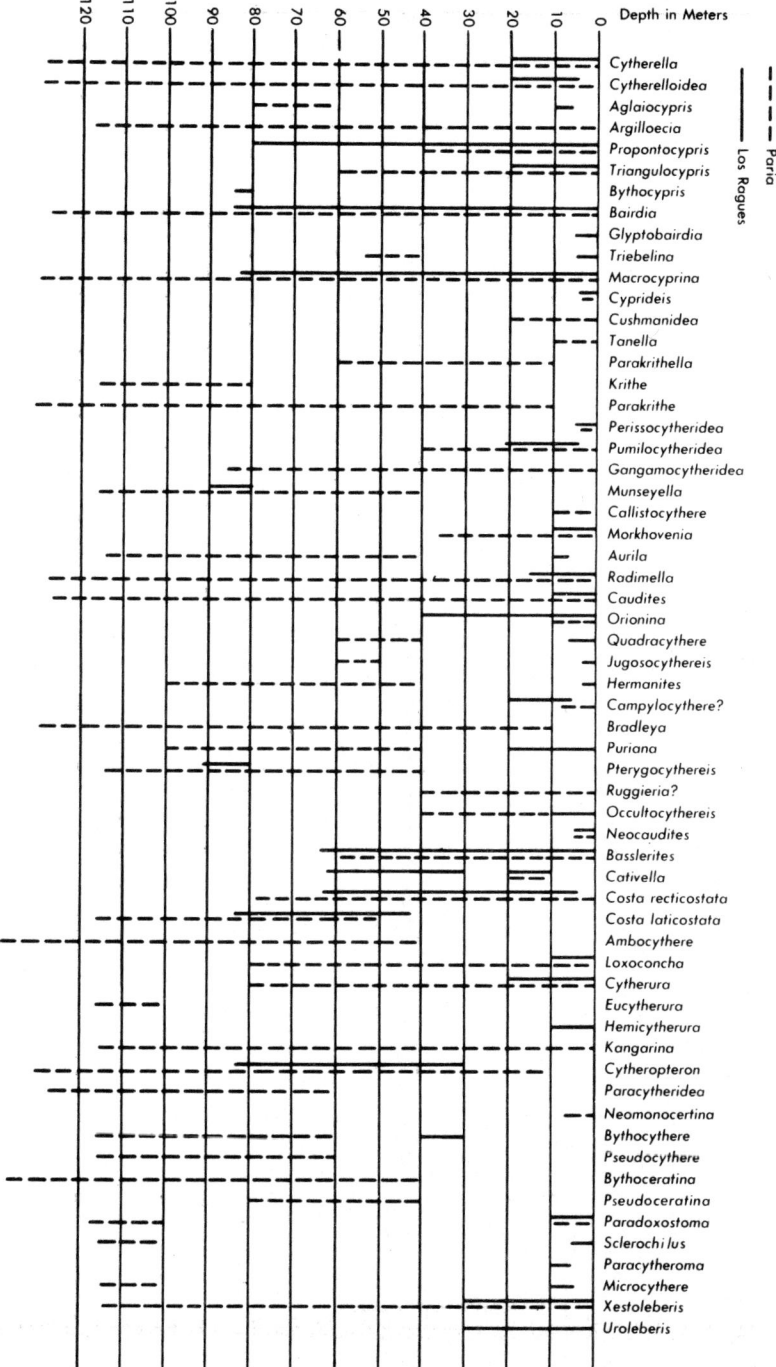

Table 1. Bathymetric distribution of ostracodes on the Paria-Trinidad-Orinoco shelf, and around Los Roques (southern Caribbean).

Table 2. Bathymetric distribution of ostracodes off the north coast of Jamaica.

	E 1 (175')	E 3 (145')	E 5 (120')	M 1 (115')	PB 5 (110')	PB 2 (95')	E 10 (65')	E 16 (40')	
1010'	960'	810'	680'	510'	400'				

Species
Cytherella pandora Kornicker
Cytherelloidea precipua van den Bold
Cytherelloidea sp.
Argilloecia sp.
Propontocypris sp. 1
Propontocypris sp. 2
Triangulocypris laeva (Puri)
Triangulocypris sp.
Macrocyprina sp. aff. *M. maculata* (Brady)
Macrocypris sp.
Bairdia longisetosa Brady
Bairdia sp. aff. *B. fusca* Brady
Bairdia victrix Brady
Bairdia antillea van den Bold (rew?)
Bairdia sp. 4
Bairdia sp. 5
Bairdia sp. 2 (Cuba)
Bairdia dimorpha van den Bold
Paranesidea bradyi (van den Bold)
Paranesidea tuberculata (Brady)
Paranesidea sp. aff. *P. tuberculata* (Brady)
Paranesidea sp. aff. *P. fortificata* (Brady)
Paranesidea sp. (Colon harbor)
Anchistrocheles? sp. aff. *A*? *angulata* (Brady)
Triebelina sertata Triebel
Glyptobairdia coronata (Brady)
Parakrithe sp.
Krithe dolichodeira van den Bold
Cushmanidea sp.
Morkhovenia inconspicua (Brady)
Orionina serrulata (Brady)
Caudites howei Puri
Caudites nipeensis van den Bold
Radimella confragosa (Edwards) (rew?)
Radimella confragosa forma A
Radimella wantlandi (Teeter)
Quadracythere producta (Brady)
Quadracythere lankfordi Teeter
Jugosocythereis pannosa (Brady)
Coquimba sp.
Puriana sp. aff. *P. rugipunctata* (Ulrich and Bassler)
Hermanites hornibrooki (Puri)
Neocaudites nevianii Puri
Cativella sp.
Occultocythereis angusta van den Bold
Caribella? *yoni* (Puri)
Cytheretta pumicosa (Brady)
Loxoconcha fischeri (Brady)
Loxoconcha dorsotuberculata (Brady)
Loxoconcha magna Teeter
Loxoconcha rugosa van den Bold
Loxoconcha wilberti Puri
Loxoconcha sp.
Paracytheridea tschoppi van den Bold
Paracytheridea sp. aff. *P. altila* Edwards
Paracytheridea sp. aff. *P. hispida* van den Bold
Paracytheridea sp.
Cytheropteron sp.
Cytherura sp. aff. *C. johnsoni* Mincher
Cytherura sp.
Hemicytherura cranekeyensis Puri
Pseudocythere sp.
Pseudoceratina droogeri van den Bold
Bythoceratina sp.
Paradoxostoma sp.
Xestoleberis sp. 1
Xestoleberis sp. 2
Xestoleberis sp. 3
Uroleberis angulata (Brady)

escarpment | Lower Slope | Upper slope | Fore-reef terrace

in the depth distributions of the two subspecies of *Costa variabilocostata* off South America and that of the two subspecies of *Carinocythereis antiqua* (one from 0-90 m, the other from depths below 75 m) in the Adriatic (Breman, 1976) and in the Gulf of Biscaye (Carbonel, 1973).

The experience in the Gulf of Mexico and off South America shows, that est-

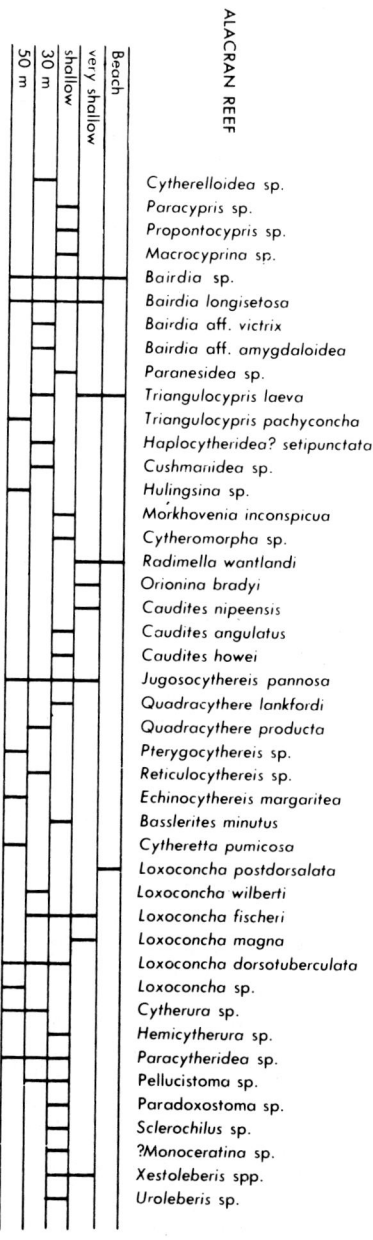

Table 3. Bathymetric distribution of ostracodes off the Alacrán reef.

183

ablishing a depth-zonation on a single traverse, or a few closely spaced traverses, has to be viewed with some scepticism if we want to extend the zonation laterally along the coast for larger distances. This will be even more the case with fossil formations where our information is generally much less complete.

Bathymetric zonation of the continental slope and deep oceanic regions appears possible, using the genus *Krithe* (Van Morkhoven 1972). Although his zonation is not confirmed by the results of the Hydrographer samples, due to a scarcity of *Krithe* species, the distribution of species of *Krithe* in Tertiary deep-water deposits of the Caribbean (Haiti, Jamaica, Trinidad) is very similar, and we find a general increase in size with deeper water. Also in both cases, we find a reversal in valve size, which in the Caribbean Neogene, takes place in what are obviously sediments deposited at considerable depth, perhaps corresponding to the 900–1000 m depth at which this change occurs in the Gulf of Mexico. The accompanying ostracode fauna is typically comprised of species of the genera *Bythocypris, Cardobairdia, Agrenocythere, Bradleya* and *Abbyssocythere* with some deepwater species of *Bairdia* and *Cytherella*. In slightly shallower water, species of *Trachyleberis, Trachyleberidea, Ambocythere, Byhocythere* and *Bythoceratina* are present.

References

Baker, J.H. & Hulings, N.C. 1966. Recent Marine Ostracod Assemblages of Puerto Rico. Publ. Inst. Mar. Sci., Texas 11:108–125.

Benda, W.K. & Puri, H.S. 1962. The distribution of foraminifera and Ostracoda off the Gulf Coast of the Cape Romano area, Florida. Trans. Assoc. Gulf Coast Geol. Soc. 12:303–336.

Benson, R.H. 1971. A new Cenozoic Deep-Sea Genus Abyssocythere (Crustacea, Ostracoda: Trachyleberididae), with descriptions of five new species. Smiths. Contr. Paleobiol. 7:1–20.

Benson, R.H. 1972. The Bradleya Problem, with description of two new Psychrospheric ostracode genera, Agrenocythere and Poseidonamicus (Ostracoda: Crustacea). Smiths. Contr. Paleobiol. 12, 138 pp.

Benson, R.H. & Coleman, G.L. II. 1963. Recent marine ostracodes from the eastern Gulf of Mexico. Univ. Kansas, pal. contr., Anthrop. 2.:1–52

Bold, W.A. van den. 1966a. New species of the ostracode-genus Ambocythere. Mag. Nat. Hist. ser. 13, 8 (1): 1–18.

Bold, W.A. van den. 1966b. Ostracoda from Colon Harbour, Panama. Carib. Jour. Sci. 6 (1/2):43–53.

Bold, W.A. van den. 1966c. Miocene and Pliocene Ostracoda from Northeastern Venezuela. Verh. Kon. Nederl. Akad. Wetensch., ser. 1, 23 (3):43 pp.

Bold, W.A. van den. 1970. The genus Costa (Ostracoda) in the Upper Cenozoic of the Caribbean Region. Micropal. 16 (1):61–75.

Bold, W.A. 1974a. Ornate Bairdiidae in the Caribbean. Geosci. and Man 6:29–40.

Bold, W.A. 1974b. Taxonomic status of Cardobairdia van den Bold, 1960, and Abyssocypris n. gen., two genera of deepwater ostracodes from the Caribbean Tertiary. Geosci. and Man 6:65–79.

Bold, W.A. van den. 1975a. Neogene biostratigraphy (Ostracoda) of southern Hispañola. Bull. Amer. Pal. 66 (286):549–625.

Bold, W.A. van den. 1975b. Distribution of the Radimella confragosa group (Ostracoda, Hemicytherinae) in the late Neogene of the Caribbean. Jour. Paleont. 49 (4):692–700.

Bold, W.A. van den. 1975c. Ostracodes from the late Neogene of Cuba. Bull. Amer. Pal. 68 (289):121–167.

Bold, W.A. van den. 1976. Distribution of species of the tribe Cyprideidini (Ostracoda, Cytherideinae) in the Neogene of the Caribbean. Micropal. 22 (2): 1-42.
Brady, G.S. 1866. On new or imperfectly known Species of Marine Ostracoda. Trans. Zool. Soc. London 5:359-393.
Brady, G.S. 1869. Description of Ostracoda (from Hong Kong, Nouvelle Providence, Saint Vincent, Golfe de Gascogne, Colon Aspinwall, Port-au-Prince), In: De Folin & Périer: Les Fonds de la Mer, vol. 1, pp. 113-176.
Brady, G.S. 1870. Description of Ostracoda (from Saint Vincent du Cap Vert, Strait Maggellan, Santiago de Cuba, Isles Lucayes, Vera Cruz and Carmen, Coast of Africa), In: de Folin & Périer: Les Fonds de la Mer, vol. 1, pp. 177-256.
Brady, G.S. 1880. Report on the Ostracoda dredged by HMS Challenger during the years 1873-76. Report on the Sci. rcs. of the voy. of HMS Challenger. Zool., vol. 1, pt. 3, pp. 1-184.
Breman, E. 1976. The distribution of ostracodes in the bottom sediments of the Adriatic Sea. Diss. Vrije Univ. Amsterdam, Krips Repr., Meppel, Netherlands. 165 pp.
Carbonel, P. 1973. Répartition des Carinocythereis dans le Golfe de Gascogne. Rev. Esp., Micropal., 5 (3):147-154.
Curtis, D. Malkin. 1960. Relation of environmental energy levels and ostracod biofacies in East Mississippi delta area. Bull. Amer. Assoc. Petr. Geol., 44 (4):pp. 471-494.
Drooger, C.W. & Kaasschieter, J.P.H. 1958. Foraminifera of the Orinoco-Trinidad-Paria Shelf. Rept. Orinoco Shelf Exp., vol. 4. Verh. Kon. Nederl. Acad. Wetensch. afd. Natuurk., ser. 1, vol. 22. Description of Ostracoda, pp. 88-92.
Howe, H.V. & Bold, W.A. van den. 1975. Mudlump Ostracoda. Bull. Amer. Paleont. 65 (282): 303-315.
Hulings, N.C, 1967. A review of the recent marine Podocopid and Platycopid Ostracodes of the Gulf of Mexico. Contr. Mar. Sci. 12:80-100.
Hulings, N.C. & Puri, H.S., 1964. The Ecology of shallow water ostracodes of the West Coast of Florida. Pubbl. Staz. zool. Napoli 33 suppl.:308-344.
Keij, A.J. 1954. Ostracoda: Identification and description of species, In: Van Andel & Postma: Recent sediments of the Gulf of Paria. Report Orinoco shelf Exp., vol. 1,. Verh. Kon. Nederl. Acad. Wetensch. afd. Natuurk., sert. 1, vol. 20, no. 5, pp. 218-231.
Keij, A.J. 1976. Note on Havanardia and Triebelina species (Ostracoda). Proc. Kon. Nederl. Acad. Wetensch., ser. B., 70 (1):36-44.
Keyser, Ditmar, 1975. Ostracodes of the mangroves of South Florida, their ecology and biology. Bull. Amer. Paleont. 65 (282):489-498.
Kontrovitz, Mervin. 1971. A new Holocene species of Echinocythereis (Ostracoda). Tulane Stud. Geol. 8 (3):166-168.
Morkhoven, F.P.C.M. van. 1972. Bathymetry of Recent marine ostracoda in the Northwest Gulf of Mexico. Trans. Gulf Coast Assoc. Geol. Soc. 22: 241-252.
Puri, H.S. 1960. Recent Ostracoda from the west Coast of Florida. Trans. Gulf Coast Assoc. Geol. Soc. 10:107-149.
Puri, H.S. 1967. Ecologic distribution of Recent Ostracoda. Proc. Sympos. on Crustacea, pt. 1, pp. 457-495.
Puri, H.S. & Hulings, N.C. 1957. Recent Ostracode facies from Panama City to Florida Bay area. Trans. Gulf Coast Assoc. Geol. Soc., vol. 7, p. 167-190, 11 text-figs., 2 tables.
Rothwell, W.T., Jr. 1949. Preliminary report on Ostracoda in bottom samples and cores of the norteastern Gulf of Mexico. Bull. Soc. Amer. 60 (12):1918 (abstract).
Sandberg, P.A., 1964a. Notes on some Tertiary and Recent brackish-water Ostracoda. Pubbl. staz. zool. Napoli 33 suppl.:496-514.
Sandberg, P.A., 1964b. The Ostracod genus Cyprideis in the Americas. Acta Univ. Stockh. Stockholm, Contr. in Geol. 12: 178 pp.
Sandberg, P.A., 1966. The modern ostracods Cyprideis bensoni n. sp., Gulf of Mexico, and C. castus, Baja California. Jour. Paleont. 40 (2):447-449.
Sandberg, P.A. & Plusquellec, P.L. 1974. Notes on the Anatomy and Passive Dispersal of Cyprideis (Cytheracea, Ostracoda). Geosci and man 6:1-26.
Swain, F.M., 1955. Ostracoda of San Antonio Bay, Texas. Jour. Paleont. 29 (4):561-646.

Swain, F.M. & Engel, P.L. 1954. Ostracoda, In: Study of nearshore Recent sediments and their environments in the northern Gulf of Mexico. Calif. Univ. Scripps Inst. Oceanogr., Amer. Pter. Inst. Res. Proj. 51, 3. Quart., Rep. 1954, pp. 7–8.

Teeter, J.W. 1975. Distribution of Holocene Marine Ostracoda from Belize. In: Wantland, K.F. and Pusey, W.C. III: Belize Shelf–Carbonate sediments, clastic sediments and ecology. Amer. Assoc. Petr. Geol., Studies in Geology 2: 400–499.

Tressler, W.L. 1949. Marine Ostracoda from Tortugas, Florida. Jour. Washint. Acad. Sci. 39 (10):335–343.

Tressler, W.L. 1954. Marine Ostracoda, In: R.S. Galtsoff (ed.), Gulf of Mexico, its origin, waters and marine life. Fish and Wildlife Serv., Fishery Bull. 89, 55:429–437.

Nomenclatorial Note

Reussicythere howei nom. nov.
Cythere reussi Brady, 1969, Fonds de la Mer. vol. 1, p. 153, pl. 18, fig. 9, 10.
Not *Cythere (Bairdia) reussi* Speyer, 1863, Ber. Ver. f. Naturk. zu Cassel, no. 13, p. 45, pl. 1, fig. 7 a–c.
Not *Cythere reussi* Brady, 1880, Challenger Report, vol. 1, p. 74, pl. 14, fig. 2 a–d.
Not *Cythere reussi* Prochaska, 1893, Rozpr. Ceska. Acad. Praze, vol. 2, H. 24, p.79, pl. 2, fig. 1 a–b (= *Cnestocythere lamellicostata* Triebel).
Reussicythere reussi (Brady) van den Bold, 1966b, p. 46, pl. 3, fig. 2 a–d, pl. 5, fig. 1 a–b; Teeter, 1975, p. 449, fig. 11r, 12c.
The new name is in honour of the late Dr. H.V. Howe, who pointed out the homonomy of the original name many years ago. I informed Teeter about this, but he did not change the name, and this is the first opportunity I have had for correcting the record.

Author's address:

4646 Bennett Drive, Baton Rouge, La. 70808, USA

Discussion

M. Ahmad: I wonder if you found any ecological differences in occurrence of the genus *Loxoconcha* and *Loxocorniculum?*

W.A. van den Bold: Tuberculate species of *Loxoconcha* (or *Loxocornuculum*) are strongly concentrated in the calcareous platform facies, altough they (and especially *L. fischeri*) are also found in a less calcareous environment and occasionally in deeper water. They are much less common in the Gulf of Mexico Province proper than in the Caribbean.

Sixth Intern. Ostracod Symposium, Saalfelden

PRELIMINARY NOTES ON THE DISTRIBUTION OF POLYCHAETA AND OSTRACODA OF THE AUSTRALIAN COAST

GESA HARTMANN-SCHRÖDER & GERD HARTMANN

Abstract

Between August 1975 and March 1976 the authors collected Ostracoda and Polychaeta along the Australian coast. The trip started in Derby in Northwestern Australia, continued along the West, South and East Coasts, and ended in Queensland on Heron Island, on the Barrier Reef.
 Although it is not yet possible to present a report on the taxomy of the animals collected, preliminary studies of the material allow a rough zonation of the fauna according to the above-mentioned coast lines, and a discussion of the ecological factors which influence the distribution of the ostracods and polychaeta. Both groups respond similarly to the influencing factors.
 The establishment of a complete lack of a cool-temperate fauna (Ostracoda, Polychaeta) in Southern Australia is considered an important result.

Introduction

During the Ostracod Symposium in Newark (Delaware, 1972) the authors of this paper reported on their studies of the ostracods of the coastlines of Angola, Namibia, South Africa and Mozambique, and compared their results with previous studies of the coasts of South America. Particular attention was paid to the cold water coasts of the southern continents, which are a notable feature of the western coasts of the Southern Hemisphere. The zonal distribution of ostracods and polychaetes was summarized in a figure which is again presented here (Fig. 1).
 In 1974 the final results of our African studies were published in a supplementary volume of the 'Mitteilungen des Hamburger Zoologischen Museums und Instituts'. The following year, 1975, funds were made available to carry out a comparative survey of the Australian coastline. Although our recent return from Australia in March 1976 makes it impossible to present a species list of polychaetes and ostracods collected from this continent, it is possible to give a rough picture of the distribution and zonation of populations from these animals.
 Our trip began in September from the northwest corner of Australia, Derby, and proceeded along the west, south and east coasts to end on the Barrier Reef in February. Altogether we established 54 stations and collected 196 samples. All samples were examined and sorted alive, and were then preserved in formalin for comparatitive examinations to be conducted later. The following hydrographical data were obtained at each sample station: temperature, salinity, oxygen-concentration, pH, calciumcarbonate concentration, coarseness of the sediment, and type of substrate. We travelled in a Volkswagen-VAN fitted out for field studies and covered a total distance of approximately 25000 km.

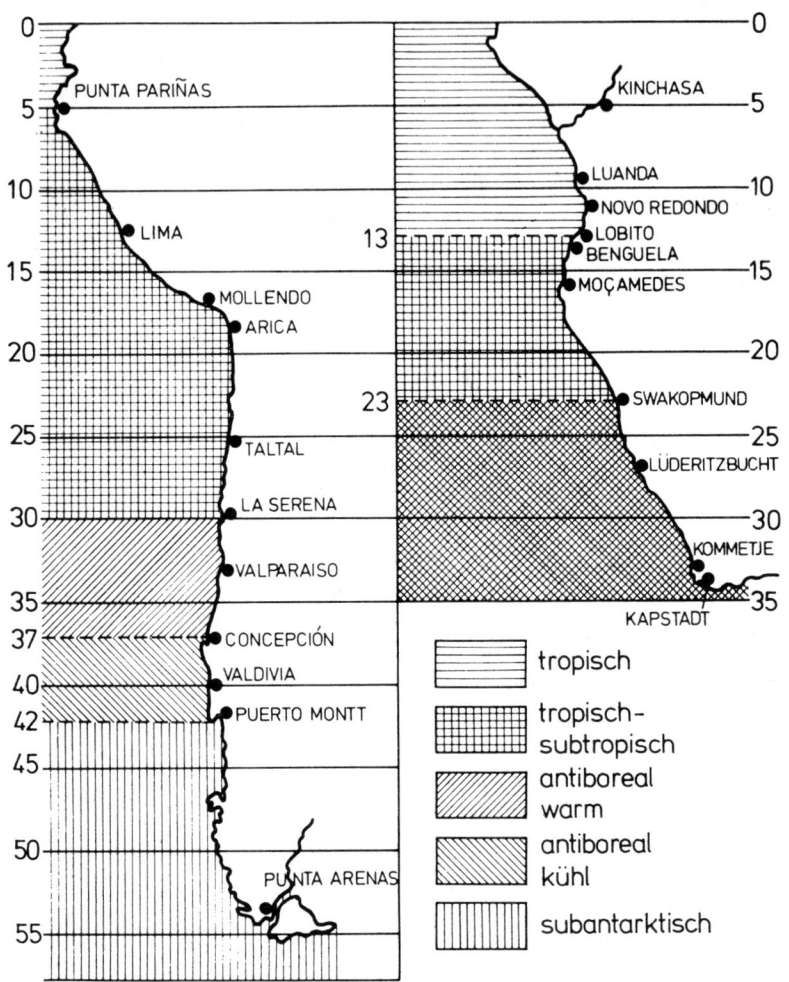

Fig. 1. Biogeographical zonation of the west coasts of Africa and South America. From Hartmann-Schröder & Hartmann (1974).

The zoogeographical zonation of the Australian coast based on Ostracoda and Polychaeta.

1. *The tropical zone of the West Coast*
(Fig. 2)

The north-west coast of the Australian continent, which has often been referred to as the 'Dampier Province' in earlier papers is a natural tropical coastline. Large

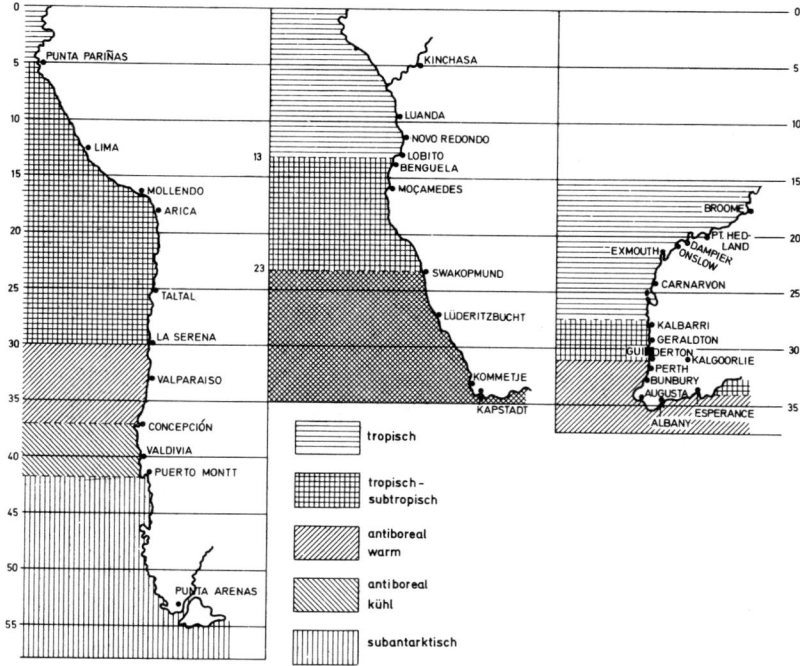

Fig. 2. Biogeographical zonation of the west coasts of Africa, South America and Australia. Original.

stands of mangroves (mainly *Avicennia marina*) are typical of the warm climate; a large variety of different tropical coastal types were particulary evident in the region around Broome.

The macrofauna of this coast is characterised by tropical species, and includes fiddler and ghost crabs of the genera *Uca*, respectively *Ocypode*, and polychaetes such as a dark green phyllodocid and a tubicolous onuphid.

Water temperature varied from 19.5°C (under mangroves) and 27°C (in open coastal waters) during the spring season of September. The salinity was rather high, never dropping below 36‰ even among mangroves, and was generally higher (in bays and mud flats) where it could reach as much as 42‰. None of the samples showed any brackish water influence, although this is no doubt an important feature during the rainy season.

The ostracod fauna is almost entirely represented by marine species. We found various species described by Kingma & Keij, for example *Tanella gracilis* (also known from South East Asia), *Neomonoceratina koenigswaldi*, *Actinocythereis scutigera*, *Cytherelloidea keiji*, *Mosella* and *Paracytheridea remanei* (previously described from the Red Sea). Species of the genus *Paradoxostoma* have their nearest relatives in the Red Sea, while those of the genus *Loxoconcha* have them on the East Coast of Africa (Hartmann 1964, Hartmann-Schröder & Hartmann 1974). We also found *Phlyctenophora zealandica*, but this was a characteristic inhabitant of mangrove swamps. Although we have not completed our studies of the coast, it is apparent that the species assemblage of the Tropical West Coast of Australia belongs to that of the Malayan Archipel and, therefore, to the Tropical Indo-West-Pacific.

The tropical fauna found between Derby and Broome was identical to that discovered in Port Hedland, Port Samson, Dampier, Onslow and Exmouth. At Carnarvon, a region generally regarded as belonging to the tropics, the littoral fauna begins to change, and hydrographical data no longer indicate purely tropical conditions. Most of the tropical species are still found, but in conjunction with new faunal members. This change is also apparent in the macrofauna. The genus *Ocypode* is now represented by a different species, some *Uca* no longer occur, and *Neopomatus ushakovi*, a typically tropical polychaete (see Hartmann-Schröder 1971), is replaced by *Mercierella enigmatica*.

Shark Bay, south of Carnarvon, probably belongs to the tropics. It has large stands of mangroves and living coral reefs are found off the coast. However, as we shall see later, the Australian mangroves are not reliable indicators of tropical waters. Stands of mangroves are also present in sheltered lagoons near Bunburry in the Lechenault-Inlet (34° S), and even at the southern tip of Australia near Welshpool (at 38°C). It is quite probable, however, that these stands are relict populations of warmer times when Australia was completely surrounded by trop-

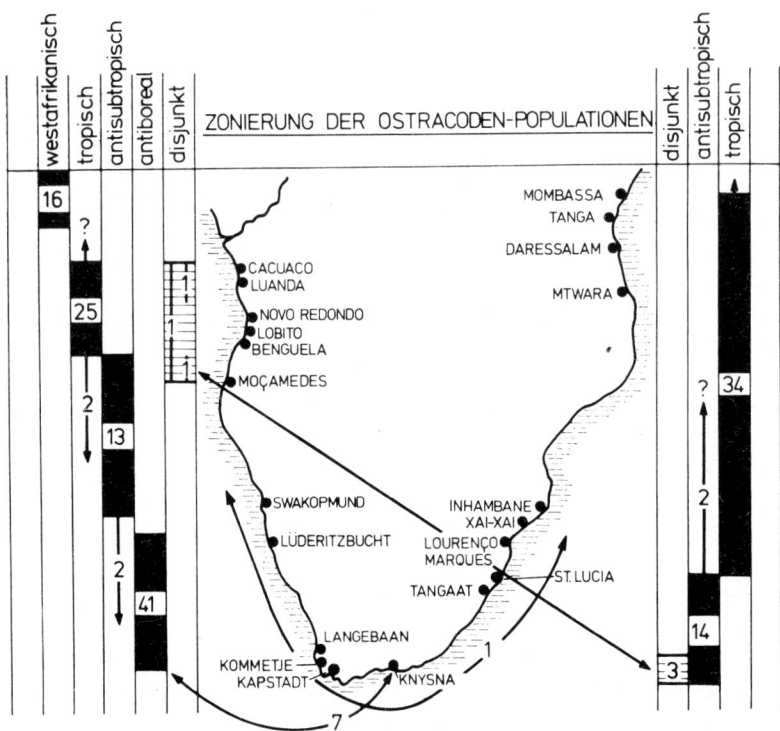

Fig. 3. Disjunct distribution of ostracods on the African coast. From Hartmann-Schröder & Hartmann (1974).

ical to subtropical waters. This situation can be compared to the results we obtained in previous studies of ostracods and polychaetes in South America and Africa. There we found indicators of warmer periods in some populations of ostracods, for tropical species occurred in disjunct distributions on the west and east coasts of both continents. It is possible that these distributions as well as the distribution in Australia are the result of the same geological event (Fig. 3).

With the exception of such relict populations we find completely different conditions when we compare the West Australian Tropics with the tropical west coasts of South America and Africa. Whereas in South America the tropics end at Punta Aguja, 7° S, and in Africa near Lobito, 13° S, we still find tropical conditions in Australia in Shark Bay which is at 26-27° S, i.e. 13-20° further south than in the two other continents. Thus the biogeographical zonation of coastal waters is different in Australia. Notable is the fact that there is neither any sign of an upwelling of cold water so characteristic of the west coasts of South America and Africa, nor influence of a cold equator-directed northward current (see (Fig. 2).

The tropical-subtropical transition zone of the West Coast
(Fig. 2)

South of Shark Bay, at Kalbarry, there was a considerable change in the ostracod and polychaete fauna. Hydrographical conditions, however, remain those of a warm coast. Ocean water temperatures of about 20°C were observed, and the salinity was measured to a maximum of 36‰. This was the first point where brackish water conditions appeared to persist throughout the year, and for the first time brackish water biotopes with a special fauna occur. It is significant that a species of *Cyprideis* was found. This is a typical brackish water indicator which is not present in the tropical region of the west coast. Besides *Cyprideis, Callistocythere* occurs and continues southwards where it is found in larger or smaller populations in all brackish environments. Of polychaetes we found an orbiniid, a capitellid and a small sabellid. The marine polychaete and ostracod faunas are very rich in species, especially in Paradoxostominae, but differ from the northern faunas in species and generic composition. Some tropical representatives were, however, still observed. This region, which we will call the Tropical-Subtropical Transition Zone, extends southwards to Guilderton (just nort of Perth) at 32° S. Mangroves are not longer present and *Uca* no longer occurs. *Ocypode* is still distributed, however, to the southern border of this zone.

3. *The warm-antiboreal zone of the West Coast*
(Fig. 2).

South of Guilderton a second change in the fauna takes place. This change, however, affects only the marine littoral fauna. The ostracod and polychaete species of the brackish water environments, now represented by large lagoons and so called 'Inlets', remain the same as in the tropical-subtropical transition zone. This is a well known phenomenon. Remane (1940) referred to it as one of his brackish water rules: "the fauna of brackish water biotopes is more extensively distributed than that of the adjacent marine waters".

Hydrographical conditions in this southern zone are comparable with those of the warm-antiboreal coast of South America. In South America (Fig. 2) the boundary of the warm-antiboreal zone is near 30° S. In contrast, in Africa boreal conditions start south of 23° S. It was not possible to distinguish between warm- and cold-antiboreal zones there. By comparison the west coast of Australia is obviously warmer. The transition from subtropical waters to antiboreal waters is 2° further south than in South America and 9° further south in Africa.

The composition of the ostracod fauna of the warm-antiboreal zone is influenced mainly by the occurrence of the large brown algae (*Ecklonia* and others, Figure 4, from Knox). *Paradoxostoma*, *Cytherura* s.l. and *Paracytheridea* and species of the *Procythereis*-group are prominent. The latter is a southern element. Among the calcareous algae species of *Aurila* and *Cytherura* occur in abundance. The whole tip of South-West-Australia east to Esperance (ca. 125° east) has a uniform fauna of Ostracoda.

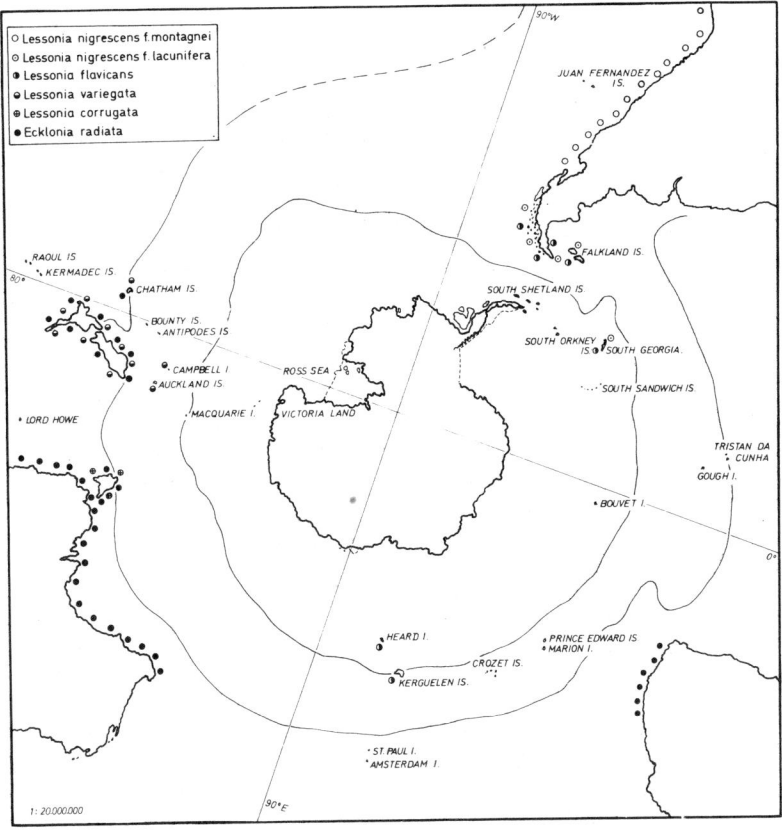

Fig. 4. Distribution of the Brown Algae *Macrosystis* and *Durvillea*. From Knox (1960).

The Polychaeta show a similar picture. For instance, the serpulid *Galeolaria caespitosa* forms large incrustations on the rocky shores throughout this zone. Another obvious species is a spionid that lives in the humid sand high above the intertidal region.

The hydrographical conditions are equally uniform. The early summer temperature of the ocean was between 18° and 19.6°C. Salinity was slightly lower than in the tropical-subtropical transition zone and lay between 35 and 36.4‰. A biotop peculiar to this part of the coast is that of inlets and lagoons. The rivers which feed the inlets come from large *Eucalyptus* forests. Their water is deep brown, thus indicating high acidity (humic acid). In one case, we measured a pH of 4.5, but usually it was between 5-6. Even in these inlets the salinity was still high, being polyhaline (25-30‰), and even in these waters brackish water ostracods and polychaetes, *Callistocythere, Macrocypris, Polydora,* the orbiniid and sabellid and others occurred. It is astonishing that the ostracod shells do not show any deficiency of calcium carbonate assimilation.

Our stations at Perth, Bunburry, Busselton, Margaret River, Augusta, Walpole, Denmark, Albany and Esperance belong to the Warm-Antiboreal Zone of Western Australia.

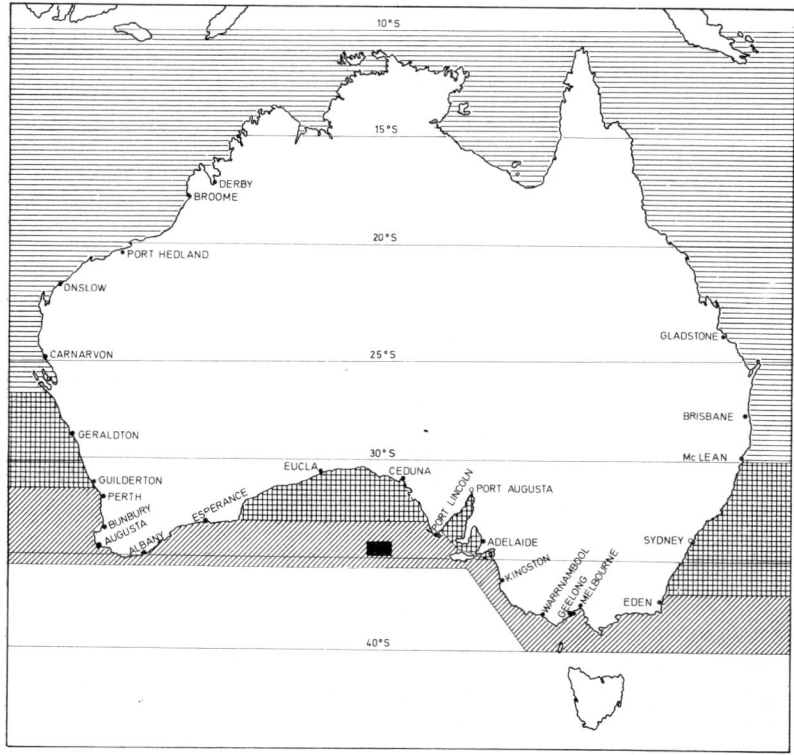

Fig. 5 Biogeographical zonation of the Australian coast. Original.

4. *The subtropical-tropical-zone of the Great Australian Bight*
(Fig. 5)

Not very much is known about the littoral fauna of the Great Australian Bight. Usually it is referred to the warm-temperate zone and united with the west Australian coast (Knox 1960), but the biogeographical situation is by far more complex. It is very difficult to reach the coast, and the climatic and geographical conditions are very rough. Despite millions of flies, a temperature of 46°C in the shade, and high sand dunes we managed to take a sample near Eucla (ca. 129° east). More intensive studies were possible at Ceduna, Port Lincoln, Port Augusta, Wallaroo, Adelaide and on Kangaroo Island.

Not all of these stations belong to the tropical-subtropical zone, for these conditions are only found in the inner parts of the Great Australian Bight, The Spencer and the St. Vincent Gulfs. The south tip of the Eyre Peninsula (Port Lincoln) and Kangaroo Island are situated in a cooler climate. Their fauna corresponds with that of the south-east tip of Australia.

Among the animals which we obtained at Eucla were some species already known from the warm-antiboreal coast around Esperance; but in addition there were others which we found later in the eastern part of the Australian Bight and which belong to the tropical-subtropical section of the south coast near Ceduna, Port Augusta and Adelaide. This eastern portion of the Australian Bight is the warmest part of the south coast. Closed stands of mangroves are again present, and there is a complete lack of brackish water biotopes.

These conditions indicate that the brackish water and the warm-boreal marine faunas of the south-west tip of the continent are completely isolated. We can conclude that the polychaetes and ostracods of this region belong to an isolated endemic fauna.

5. *The warm-antiboreal zone of the South East Coast*
(Fig. 5)

South of Adelaide we again have a warm-antiboreal fauna as is also found, as mentioned above, at the southern tip of the Eyre-Peninsula and on Kangaroo Island. We took samples near Kingston, Boatswain, Port Mcdonnel, Lake George, Warrnambool, Geelong, Foster, Wilsons Promontory, PortFranklin, Port Welshpool, Lakes Entrance, Mallacoota and Eden. The marine littoral fauna of ostracods and polychaetes and the fauna of the large brackish water biotopes, which are once more developed, are uniform. There are almost no affinities between the South East and the South West, and this is true both for the brackish and the marine fauna. A notable feature, moreover, is the first occurrence of warm water elements from the east coast of Australia, – developed, as one would expect, mainly in the eastern sections of this region. The mangroves surprisingly present in sheltered bays and estuaries are without doubt isolated relict stands – as suggested above. They contain a warm-boreal fauna and only a few subtropical-tropical species.

6. The subtropical-tropical transition zone of the East Coast

North of Eden the fauna changes again. Visible indicators are closed stands of mangroves which occur uninterrupted from here to the tropical North. We found the first rather short *Avicennia*-stands in estuaries near Merimbula and in the mouth of the Wagonga River near Narooma. They indicate the southern margin of a very large tropical-subtropical transition zone, extending from 36° S up to the Nambucca River at 30° S. The fauna of polychaetes and ostracods is characterised by a diminution of southern and an increase in tropical species – sampling from south to north. Nevertheless, it is easy to draw the limits of this region at 36° South and 30° South. Indicators of the northern boundary are, in particular, the brackish water polychaetes of the genera *Mercierella* and *Neopomatus*. Going northward the first occurrence of the tropical species *Neopomatus ushakovi* is in the Nambucca River.

The brackish water fauna is rich in species but we are as yet unable to define its zoogeographical provenance. The entire transition zone has warm water conditions. The temperature of open ocean was (in summer) between 21° and 27.5°C, the salinity was between 35-36.5‰. Near Mclean at the Clarence River species of the tropical genus *Uca* (fiddler crab) occurred again – after an interim of 7 month of travelling and sampling. From there on northwards only tropical conditions occur. Here again we found the dark green phyllodocid and the big tube dwelling onuphid of the tropical West, which are probably distributed all along the northern coast. We finished our studies with a special reef-programme on Heron Island, Great Barrier Reef.

Summary

Although we have not yet studied our material in detail, we feel that we can make the following statement:

The zoogeographical zonation of the Australian coast is completely different from that of the other Southern Continents of Africa and south America. There is no cold water current along the west and south coasts and no upwelling of deep water occurs. Thus the climate conditions of the Australian coasts are warmer than those of Africa and South America. Relict populations of mangroves all over Australia indicate a warmer climate in the past. This is also true for the southern tips of Africa and South America, as disjunct populations of warm water elements on their west and east coasts indicate.

We would like to express our thanks to all those who helped us to accomplish our programme and especially to our colleagues from the Western Australian Museum in Perth and the Autralian Museum in Sydney. We have to thank Pat Hutchings from the Australian Museum in Sydney for arranging the transport of our material and many other things. John Neale and Harm Gross went over the manuscript and transformed our English into understandable language. Many thanks also to them.

References

Hartmann, G. 1964. Zur Kenntnis der Ostracoden des Roten Meeres. Kiel. Meeresforsch. 20:35-127.

Hartmann-Schröder, G. 1971. Zur Unterscheidung von Neopomatus Pillai und Mercierella Fauvel (Serpulidae, Polychaeta). (Mit neuen Beiträgen zur Kenntnis der Ökologie und der Röhrenform von Merciella enigmatica Fauvel). Mitt. Hamburg. Zool. Mus. Inst. 67:7-27.

Hartmann-Schröder, G. & Hartmann, G. 1962. Zur Kenntnis des Eulitorals der chilenischen Pazifikküste und der argentinischen Küste Südpatagoniens unter besonderer Berücksichtigung der Polychaeten und Ostracoden. Mitt. Hamburg. Zool. Mus. Inst., Suppl.-Bd. zu 60:1-270.

Hartmann-Schröder, G. & Hartmann, G. 1965. Zur Kenntnis des Sublitorals der chilenischen Küste unter besonderer Berücksichtigung der Polychaeten und Ostracoden. (Mit Bemerkungen über den Einfluß sauerstoffarmer Strömungen auf die Besiedlung von marinen Sedimenten.) Mitt. Hamburg. Zool. Mus. Inst., Suppl.-Bd. zu 62:1-384.

Hartmann-Schröder, G. & Hartmann, G. 1974. Zur Kenntnis des Eulitorals der afrikanischen Westküste zwischen Angola und Kap der Guten Hoffnung und der afrikanischen Ostküste von Südafrika und Moçambique unter besonderer Berücksichtigung der Polychaeten und Ostracoden. Mitt. Hamburg. Zool. Mus. Inst., Suppl.-Bd. zu 69:1-520.

Knox, G. A. 1960. Littoral ecology and biogeography of the southern oceans. Proc. Roy. Soc., B 152:577-624.

Remane, A. 1940. Einführung in die zoologische Ökologie der Nord- und Ostsee. Die Tierwelt der Nord-und Ostsee 34:1-238.

Authors' address:

Martin-Luther-King-Platz 3, 2 Hamburg 13, B.R.D.

PRELIMINARY ACCOUNT OF THE ECOLOGY OF OSTRACODS ON THE ROCKY SHORE OF HELGOLAND*

ULRIKE SKAUMAL

Abstract

The ecology of recent Ostracoda from the North-Sea in the vicinity of Helgoland is discussed. Here, only the rocky shores have been taken into consideration. The influence of some ecofactors on the distribution of nine well-known species of Ostracoda is shown.

1. Introduction

The present study is part of an investigation dealing with spatial and temporal distribution patterns of ostracods on rocky shores of Helgoland. Here, spatial distribution and its causes are reported.

2. Study Area

2.1. Location and Description

The island of Helgoland is situated in the southern North Sea (German Bay) approximately 50 km from the continent.

It arose when parallel strata of rock of various degrees of hardness were uplifted during the Tertiary by vaulting of a salt dome. These strata were further subjected to isostatic uplift and erosion. As a result, platforms with several degrees of elevation were produced and those along the shoreline are periodically inundated by the tide. Today, the latter are covered with boulders of various size, blocks of concrete, and sand. These form a very suitable substrate for the attachment and growth of marine algae (Gellert 1958).

Samples for this investigation were collected from the littoral of the NE rocky shore. Zones of the littoral were distinguished as suggested by Stephenson & Stephenson (1972). This study site could be divided along the shore line into five localities which contain different habitat structures (Table 1, Fig. 1).

---

* Teil einer Dissertation, angefertigt am Fachbereich Biologie der Universität Hamburg.

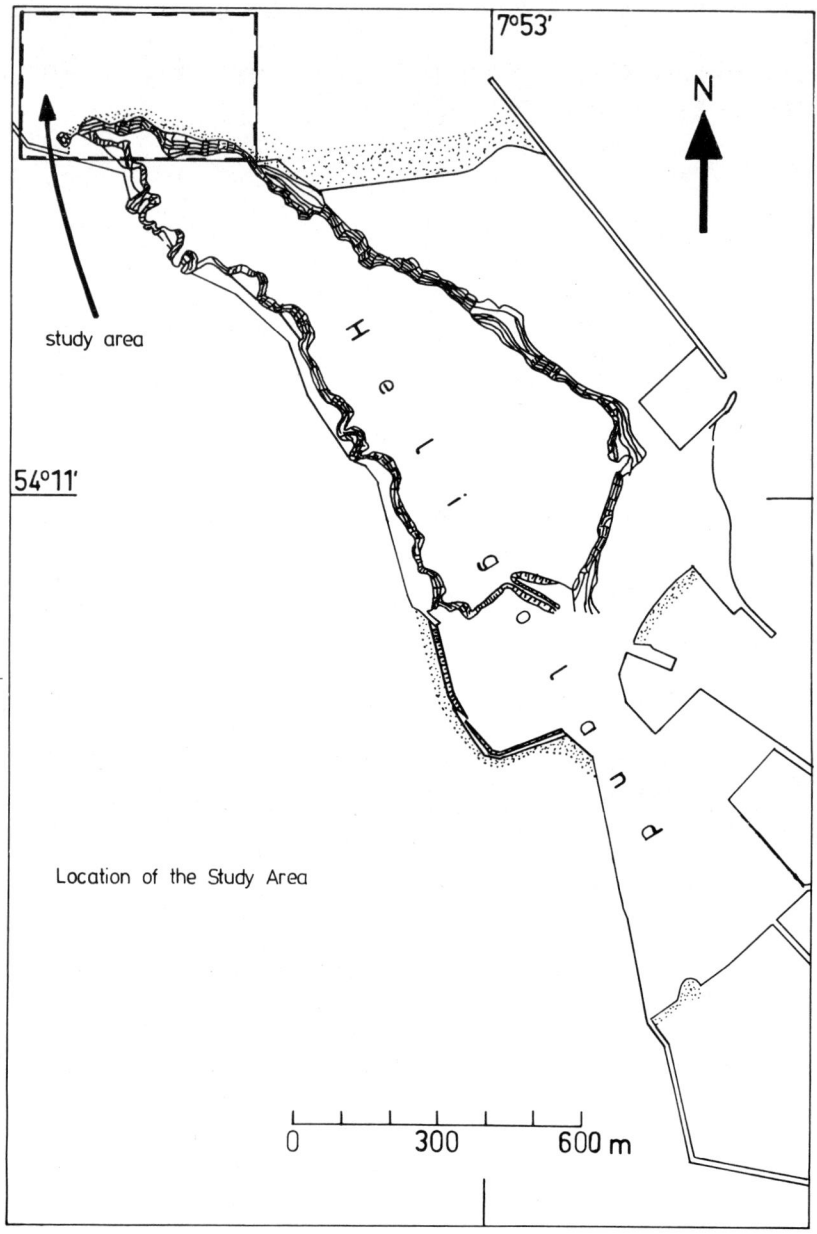

Fig. 1. Location of the study area.

Table 1. Description of the study site

Locality	Description
I Rock Beach	sand; debris; stony bank covered with *Fucus*
II Large Rockpool	sand; debris; stony bank covered with *Fucus*; salt-water pool
III Mussel Bank	sand; debris; stones covered with *Enteromorpha*; cliffs and concrete bunker covered with *Fucus*; sediment between stone layers; mussel bank
IV Sandy Bay	sand; stones covered with *Enteromorpha*
V Cliffs	surface area of rock layers covered with *Enteromorpha*; cliffs covered with *Fucus*; rock channels connected with the sea occasionally containing *Cladophora, Chaetomorpha*, etc.; sediment between stone layers; small accumulations of debris; *Laminaria*-holdfasts

2.2. *Physical characteristics*

2.2.1. Water-temperature and salinity
Water-temperature ranged from 2.1 to 18.1°C but varied greatly and was correlated to air-temperature. Salinity ranged from 29.6 to 31.5‰ and was thus nearly euhaline.

2.2.2. Substrata
The most important substrate was the various species of algae. Sand and debris were alternative habitats. Special attention was given to the former substrate 'algae', because of its abundance and for other qualitative reasons. For example, the morphology of the thalli strongly. influences certain important features of the in-

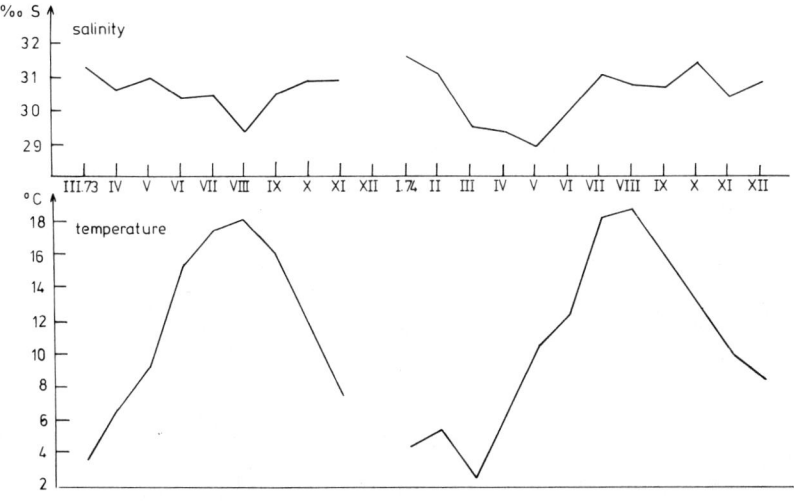

Fig. 2. Temperature and salinity in the study area.

dividuals' survival such as probability of desiccation, mechanical shelter against wave-action, amount and composition of food, and the chemo-physical qualities of the habitat (Wieser 1951).

2.2.3. Exposure

Depending on the substrate, wave-action of various intensities can prevent ostracods from attaching themselves. Because of this problem, the degree of exposure to turbulence should be determined. However, suitable methods for quantitative determination of this phenomenon have not yet been devised and, therefore, it became-necessary to use the subjective terms strong, intermediate and low.

A fairly clear relation between exposure and habitat structure was evident. Exposure is most severe on unsheltered sand and debris and one may occasionally notice extensive rearrangement of the substrate.

An intermediate intensity of exposure is found on *Fucus*. A flat thallus as well as a slimy surface, when subjected to strong wave-action, certainly impede settlement by ostracods.

A minimum exposure has been observed in *Laminaria*-holdfasts, in sand and debris between layers of rock, and in fine algae.

3. Methods

Samples were collected by hand from March 1973 until July 1974 as described by Wieser (1951). Afterwards, algae and their holdfasts were dried at approximately 100°C for about 24 hours. They were then weighed.

Quantitative information on the distribution of ostracods was obtained by calculating for each sample the number of individuals/100 g of dry algae or the number of individuals/10 cm² of sand.

Determination of species according to Sars (1928), Klie (1938) and Hartmann & Puri (1974).

4. Results

Nine species of ostracods were found and these can be arranged according to abundance as follows: *Hirschmannia viridis* (O.F. Müller 1785); *Paradoxostoma variabile* (Baird 1850); *Hemicytherura cellulosa* (Norman 1865); *Paradoxostoma* sp.; *Semicytherura nigrescens* (Baird 1838); *Heterocythereis albomaculata* (Baird 1850); *Hemicythere villosa* (G.O.Sars 1865); *Elofsonella concinna* (Jones 1856); *Eucytheridea punctillata* (Brady 1865). Only the first three species occured regular as adults and/or instars in at least one sample. The remainder were only found sometimes.

4.1. *Horizontal distribution and its causes (Table 2)*

Ostracod abundance is significantly lower in localities I and II than in the rest of the study area. Maximum abundance was found in locality V. The scarcity of ostracods in localities I and II is, presumably, caused by the nearly total absence of a suitable substrate (see Table 1). These areas are largely composed of sand and debris, and as a result of the almost continual rearrangement of materials characteristic of this substrate, they are nearly devoid of algae, with the exception of occasional beds of Fucus. Furthermore, H_2S production could be observed throughout nearly every month. The possibility of immigration was thus very poor.

Localities I and II are nearly identical not only with regard to their outward appearence but also with regard to the distribution of ostracods. In locality I *Hirschmannia viridis, Paradoxostoma variabile,* and *Paradoxostoma sp.* were observed. Locality II contained small numbers of the other species as well. Additionally, this locality is characterised by a somewhat greater abundance of those individuals belonging to the first three species. In either locality, the samples containing more than 90 individuals per 100 g of algae were of fine algae from the infralittoral zone.

Table 2. Horizontal distribution of species

Species	Locality	Substrate
Hirschmannia viridis	I-V	sand; *Enteromorpha*; fine algae*; *Fucus*; *Laminaria*-holdfasts
Paradoxostoma variabile	I-III, V	fine algae; *Fucus*; ?*Laminaria*-holdfasts
Paradoxostoma sp.	I-III, V	fine algae; *Fucus*; ?*Laminaria*-holdfasts
Hemicytherura cellulosa	II-V	*Enteromorpha*; fine algae; *Fucus*; *Laminaria*-holdfasts
Semicytherura nigrescens	II, V	fine algae; *Fucus*, *Laminaria*-holdfasts
Heterocythereis albomaculata	II-V	sand; ?*Enteromorpha*; *Fucus*; *Laminaria*-holdfasts
Hemicythere villosa	II, IV, V	sand; ?*Enteromorpha*
Elofsonella concinna	II, V	sand; *Laminaria*-holdfasts
Eucytheridea punctillata	II	sand

* fine algae = *Cladophora* spp., *Chaetomorpha* spp., etc.

Localities III and V are similar to each other. This is especially evident when the cliffs of locality III (which occasionally become dry) and the cliffs of locality V are compared. Only in these two localities was it possible to find an average of more than one individual per 100 g of *Fucus*. Also, it is here that ostracods were found in greatest abundance; however, on *Laminaria*-holdfasts only. The fronds were bare of ostracods. There is reason to presume that this is influenced by wave-action (Reys 1963; Whatley & Wall 1975).

The upper zone of locality III is comparable to localities I and II in that there are no algae (except for occasional *Fucus vesiculosus*). For this reason only few ostracod specimens were collected. During particulary low tides, a small area similar to the cliffs of locality V can be reached. This area, partly consisting of layers of rock and covered with fine algae, is much more favourable to the occurence of ostracods and large numbers are obtained here. The reason for this phenomenon along the cliffs will be returned to below. The only species which was not found there is *Semicytherura nigrescens*. The reason for this is not clear.

Locality IV, the sandy bay, proved to be an unsuitable habitat for ostracods. Virtually the only substrate available is sand, and as already noted its mobility is probably too great for the surface or the interstitial habitat to be tenable (Whatley & Wall 1969, Williams 1969). This is the only locality from which the two species of *Paradoxostoma* are absent. Species of this genus are probably plant suckers and their absence here results from the lack of suitable algae. *Enteromorpha* is found, but only as a thin layer in the supralittoral fringe, and during low tide it becomes too dry for occupation by ostracods.

In locality V all the species (except *Eucytheridea punctillata*) were found and in their greatest abundance too. During the entire sampling period, more individuals were collected here than in all the remaining localities taken together. This great abundance is made possible by a substrate which affords protection against exposure, both in the form of fine algae and sand which collects between layers of rock and *Laminaria*-holdfasts, and which perhaps also offers a suitable diet.

In conclusion it appears that important (and perhaps the most important) factors influencing the horizontal distribution of ostracods are the substrate and the exposure to wave action.

4.2. *Vertical distribution and its causes*

In the supralittoral zone, which remains dry for the longest period of any of the vertical zones, no ostracods could be found. The upper limit of their distribution is in the upper portion of the next seaward zone, the upper littoral zone. Here, the following species are encountered first: *Hirschmannia viridis, Paradoxostoma variable, Paradoxostoma* sp., *Hemicytherura cellulosa,* and *Hemicythere villosa*. *Semicytherura nigrescens* and *Heterocythereis albomaculata* have their upper limit in the midlittoral zone, and seldom occur in the upper littoral zone. Further seawards the latter two species appear more consistently, probably because the substrate is exposed to air for periods too short to result in desiccation. Occasionally, extreme low tide levels in combination with appropriate atmospheric conditions expose large areas of the 'lower' infralittoral. Whatley & Wall (1975) suppose that this is a zone where surf action is only operative during the coincidence of extreme low tides and extreme storm. Samples collected here usually contain far more ostracods per unit of algae or sand than samples from more regularly exposed shores. This means that the lower a sampling site is situated with respect to the sea-level, the more specimens it is likeley to contain (Table 3). This observation strongly supports the hypothesis that vertical distribution depends on duration of dehydration (Colman 1940, Wieser 1959) and on surf action (Whatley & Wall 1975).

Heterocythereis albomaculata generally did not appear higher up than the infralittoral fringe. Substantial numbers of this species were only collected in deeper zones, which suggests that this species may be extremely sensitive to desiccation. Also, *H. albomaculata* is a relatively stenohaline species (Elofson 1942, Neale 1964) and is here as elsewhere, only reported from euhaline water. In the light of this observation, this restriction to zones usually covered with water is understandable for in other zones salinity may strongly fluctuate with rain or evaporation. The distribution of this species is also related to that of a suitable substrate, for it was almost only found in *Laminaria*-holdfasts or in sand between layers of rock.

Table 3. Vertical distribution of ostracoda (calculated as the average over the sampling period of the no. of species/100 g dried algae for locality V)

Zone	Enteromorpha	Fucus	fine algae
upper littoral zone	27.4	1.5	47.8
'upper' midlittoral zone	27.7	2.0	1290.18
'lower' midlittoral zone	–	1.0	513.2
infralittoral fringe	–	4.35	1983.2
infralittoral zone	–	3.1	3243.1

The lower limits to the distribution of all the species are below the low water mark and probably extend into that portion of the 'phytal'-zone of the infralittoral which always remains under water.

All of the above supports the statement of Hartmann in Bronn (1976) that the fauna of the littoral zone is poor and is comprised of eurytopic representatives of the infralittoral fauna which consists of more species than those considered here. Examples of such additional forms would include *Leptocythere*, *Loxoconcha*, and *Cytherois* species.

In the supralittoral fringe, which remains dry for relatively extensive periods, the only substantial beds of algae consist of *Enteromorpha*. Here, ostracods were seldom collected because this algae becomes rather desiccated. In the midlittoral zone colonisation of each of the three species of *Fucus* (*F. vesiculosus*, *F. platycarpus*, *F. serratus*) is similarly sparse because of their comparably poor qualities as a habitat, i.e. a flat thallus and slimy surface and because of wave-action.

In contrast, algae such as *Cladophora* and *Chaetomorpha* represent better habitats because of their more finely packed thalli. They thus influence the quantitative distribution of ostracods to a greater extent.

In the infralittoral ostracod abundance is high within the holdfasts of the main algal species there (*Laminaria saccharina* and *L. digitata*) because of the protection offered against exposure. The phylloids were bare of ostracods, probably for the same reason as in *Fucus*.

Temperature as such does not appear to influence distribution of the species. Life cycles of some species, however, greatly depend on this parameter. Similarly, salinity is generally unimportant although data are scarce. In conclusion, the duration of exposure to air, substrate, and exposure to wave action appear to be the major factors influencing the vertical distribution of ostracods.

5. Summary

1. The horizontal distribution of ostracod abundance increases in general from locality I to V, the exception being locality IV. Locality V is the most densely populated area. Important physical factors affecting the horizontal distribution are the type of substrate and the extent of exposure to wave-action. The highest concentrations of ostracods are found in fine algae, sand, and debris from between layers of rock or from *Laminaria*-holdfasts.
2. The vertical distribution extends from the upper or midlittoral to the infralittoral zone which extends beyond the area of study. The concentration of ostracods within a sample increases regulary with the distance of a sampling site from the upper limit of a species distribution. The distribution pattern is strongly related to the exposure to air, the type of substratre, and exposure to wave-action.

Acknowledgements

Part of the investigation was done in the 'Biologische Anstalt Helgoland, Meeresstation Helgoland'. I wish to thank all the members of the staff, especially those of the 'Arbeitsgruppe Gastforschung', for assistance in the course of the work. Further, I wish to extend my thanks to Dr. H. Gross, who translated the paper.

References

Colman, J. 1940. On the faunas inhabiting intertidal seaweeds. J.Mar. Biol. Assoc. Plymouth 24:129-199.
Elofson, O. 1941. Zur Kenntnis der marinen Ostracoden Schwedens. Zool. Bidr. Uppsala 19:215-534.
Gellert, J. F. 1958. Grundzüge der Physischen Geographie von Deutschland, Bd. I. VEB Dt. Verl. d. Wissenschaften: 362-372.
Hartmann, G. 1976. Ostracoda. In: Gruner, H.-E. (Ed.), Dr. H.-G. Bronns Klassen und Ordnungen des Tierreiches, 5. Bd., I. Abt., 2. Buch, IV. Teil, 3. Lfg. VEB Gustav Fischer: pp. 569-786. Jena.
Hartmann, G. & H. S. Puri. 1974. Summary of Neontological and Paleontological Classification of Ostracoda. Mitt. Hamburg. Zool. Mus. Inst. 70:7-73.
Klie, W. 1938. Ostracoda. In: Dahl, P. Tierwelt Deutschlands 34(3):148ff.
Neale, J. 1964. Some Factors Influencing the Distribution of Recent British Ostracoda. Pubbl. Staz. Zool. Napoli 33 suppl.: 247-307.
Reys, S. 1963. Ostracodes des Peuplements Algaux de l'Etage Infralittoral de Substrat Rocheux. Rec. Trav St. Mar. End. Bull. 28(43):33-47.
Sars, G. O. 1928. An account of the Crustacea of Norway. Vol. IX, Ostracoda. 271 pp. Oslo.
Stephenson, T. A. & A. Stephenson. 1972. Life between tidemarks on rocky shores. W. H. Freeman & Company. San Francisco. 425 pp.
Whatley, R. C. & D. R. Wall. 1969. A preliminary account of the ecology and distribution of recent Ostracoda in the Southern Irish Sea. In: Neale, J. W. (ed.), The Taxonomy, Morphology, and Ecology of Recent Ostracoda. Oliver & Boyd: pp. 268-298. Edinburgh.

Whatley, R. C. & D. R. Wall 1975. The relationship between Ostracoda and Algae in littoral and sublittoral marine environments. Bull. Amer. Paleont. 65 (282):173-203.

Wieser, W. 1951. Untersuchungen über die algenbewohnende Mikrofauna mariner Hartböden I. Österr. Zool. Zschr. 3 (3/4):425-480.

Wieser, W. 1959. Zur Ökologie der Fauna mariner Algen mit besonderer Berücksichtigung des Mittelmeeres. Int. Rev. ges. Hydrobiol. 44:137-179.

Williams, R. 1969. Ecology of the Ostracoda from selected marine intertidal localities on the Coast of Anglesey. In: Neale, J. W. (ed.), The Taxonomy, Morphology, and Ecology of Recent Ostracoda. Oliver & Boyd: pp. 299-328. Edinburgh.

Author's address:

Zool. Institut und Zool. Museum, Martin-Luther-King-Platz 3, 2000 Hamburg 13, B.R.D.

Discussion

Athersuch: Did any of the algae mentioned bear epiphytes?

Skaumal: I did not see any.

Whatley: Your fauna is very familiar to me, having studied a number of littoral phytal faunas in Britain. The absence of such important species in your fauna, such as *Cythere lutea* and *Loxoconcha rhomboidea* is unusual. Do you have any explanation for their apparent absence in Helgoland? It is gratifying to note that your ecological conclusions conform, almost exactly, to those of Wall and I (Whatley & Wall 1975).

Skaumal: In *Laminaria*-holdfasts from greater depth – they were collected by boat – I found some individuals belonging to *Loxoconcha* sp. Species like these did not occur in the main study area. This is why I think that factors influencing their apparent absence in the study area are closely related to depth. I do not know whether the species were *L. rhomboidea* but probably it is the same with their distribution. For the absence of *Cythere lutea* I do not have any explanation.

Sixth Intern. Ostracod Symposium, Saalfelden

ECOLOGY AND ZOOGEOGRAPHY OF RECENT BRACKISH-WATER OSTRACODA (CRUSTACEA) FROM SOUTH-WEST FLORIDA

DIETMAR KEYSER

Abstract

Die ökologische und zoogeographische Verbreitung von 36 rezenten Brackwasserarten der Ostracoden aus Südwest-Florida wurde untersucht. An ökologischen Parametern wurden folgende Faktoren berücksichtigt: Salzgehalt, Substrat, Wassertemperatur, pH-Wert und Wassertiefe. Mehrere Ostracodenvergesellschaftungen werden hinsichtlich ihrer ökologischen Präferenzen dargestellt. Die hier vorliegenden Fauna wird mit ähnlichen aus dem Golf von Mexico, der Karibischen Region, sowie von der Mittelamerikanischen Pazifikküste verglichen.

Introduction

The Gulf of Mexico, especially the west coast of Florida, is fairly well known in respect of its ostracod fauna. Despite this knowledge, there is a lack of information about typical brackish water Ostracoda and of the ecology of living ostracods. Many of the existing publications deal with the ecology of these animals, but because the research was done only by using the dead shells they cannot give valid data for the ecology of the species concerned. The present study tries to rationalize this situation.

A second problem is the evaluation of the ostracod fauna in the Gulf of Mexico in zoogeographical connotations. The taxonomic situation has been very confused over a long period of time. Only during the past decade a number of workers have attempted to clarify the situation (Van den Bold div., Morales 1966, King & Kornicker 1970, Keyser 1976b), and it is now possible to attempt a comparison of the Ostracoda of the Gulf Coast with other faunas, especially those of the Caribbean and the Pacific.

Methods

The material was collected between 1969 and 1971, mainly in the mangrove region of the Everglades National Park in southwestern Florida. One hundred and eighty-three samples from 120 stations have been examined. The samples were collected with a handnet of 240 µm mesh. Salinity, temperature and conductivity were measured with a portable conductivity-meter. The salinity tolerance was classified using the 'Venice-System'; pH was determined with the use of a portable pH-meter. The substrate was divided into sand, mud, peat, silt and shell sand by in situ observation. The grain size was determined at three stations and the depth was measured with a rod. The taxonomy and the ecological boundaries of each species have been given previously (Keyser 1976a, b).

The species encountered were:

aff. *Aglaiocypris eulitoralis* forma *floridensis* Hartmann 1974; *Actinocythereis subquadrata* Puri 1960; *Aurila amygdala* (Stephenson 1944); *Candona annae* Mehes 1913; *Candona ? balatonica* Daday 1894; *Cypretta brevisaepta* Furtos 1934; *Cypria pseudocrenulata* Furtos 1936; *Cyprideis beaveni* Tressler & Smith 1948; *Cyprideis salebrosa* Van den Bold 1963; *Cypridopsis okeechobei* Furtos 1936; *Cytheridella alosa* (Tressler 1939); *Cytheromorpha paracastanea* (Swain 1955); *Cytherura elongata* Edwards 1944; *Cytherura sandbergi* Morales 1966; *Cytherura* sp. aff. *C. forulata* Edwards 1944; *Darwinula furcabdominis* Keyser 1976; *Darwinula stevensoni* (Brady & Norman 1889); *Dolerocypria fastigata* Keyser 1976; *Haplocytheridea setipunctata* (Brady 1869); *Heterocypris punctata* Keyser 1976; *Leptocythere darbyi* Keyser 1976; *Limnocythere floridensis* Keyser 1976; *Loxoconcha matagordensis* Swain 1955; *Mungava marthapuriae* Keyser 1976; *Neocaudites nevianii* Puri 1960; *Paracytheroma stephensoni* (Puri 1954); *Parapontoparta subcaerulea* Keyser 1976; *Perissocytheridea brachiforma* forma *inferior* Swain 1955; *Perissocytheridea cribrosa* (Klie 1933); *Pontoparta hartmanni* Keyser 1976; *Radimella floridana littorala* (Grossman 1965); *Reticulocytheries floridana* Puri 1960; *Reticulocythereis purii* Keyser 1976; *Thalassocypria gesinae* Keyser 1976; *Thalassocypria vavrai* Keyser 1976; *Xestoleberis mixohalina* Keyser 1976.

Factors controlling distribution

Salinity

The salinity seems to exert considerable influence upon the distribution of the ostracods. Fig. 1 shows that there is a boundary between the α- and β-oligohaline zones. The limnic species *Candona ? balatonica* and *Cypridopsis okeechobei* for example, are limited by this boundary. *Cyprideis salebrosa* seems to tolerate the higher salinity and cannot be called a limnic species at this time (Van den Bold 1963). *Cytheridella alosa*, on the other hand, is a true limnic species which, however, is often washed into the β-oligohaline zone.

The next group, with a maximum in the β-oligohaline, is made up by 5 species, *Mungava marthapuriae*, *Candona annae*, *Darwinula stevensoni*, *Darwinula furcabdominis* and *Cypretta brevisaepta*. They often tolerate a salinity of more than 10‰.

Some limnic Ostracoda, such as *Cypria pseudocrenulata*, *Heterocypris punctata*, *Limnocythere floridensis*, *Dolerocypria fastigata* and *Pontoparta hartmanni* are more euryhaline and live mainly in the α-oligohaline and β-mesohaline zones where there is a low level of competition.

Thalassocypria gesinae, *Parapontoparta subcaerulea* and *Thalassocypria vavrai* are euryhaline species. It seems that their maximum population density is influenced by the absence of each other. *T. gesinae* prevails in the β-mesohaline, *P. subcaerulea* in the α-mesohaline and *T. vavrai* in the lower (β) polyhaline zone. In the upper (α) polyhaline *Aff. Aglaiocypris eulitorales* is dominant. This succession of different Cypridacea may be controlled by competition.

Cyprideis beaveni, *Perissocytheridea brachyforma*, *Radimella floridana* littorala and *Cytherura sandbergi* represent the euryhaline Cytheracea. Their maximum density lies in the mesohaline zone which experiences a strong tidally controlled salinity range (Den Hartog 1964). This region has the lowest amount of competition. *Perissocytheridea cribrosa* might belong in this group, but it prefers the polyhaline zone.

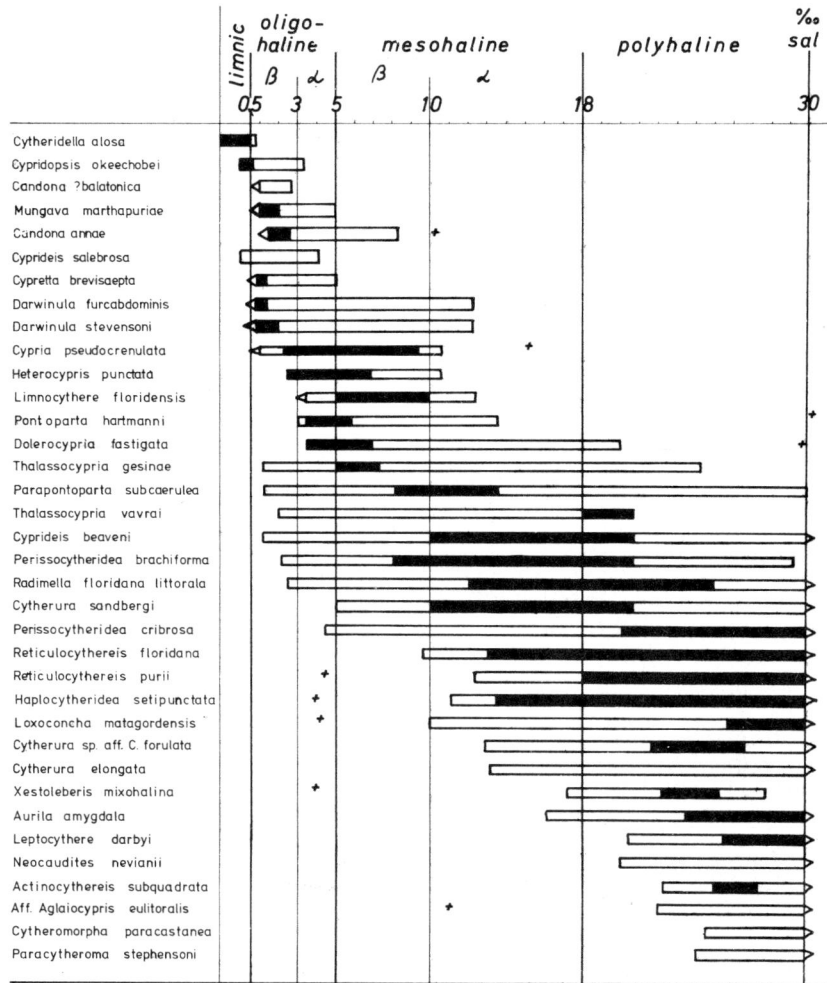

Fig. 1   Species distribution in the mixohaline waters of South-West Florida

☐ Total range   ■ Range of maximal abundance

▷ not studied range   + single findings

The group of thalassogenious species which are enabled to tolerate the polyhaline and α-mesohaline consists of seven species: *Haplocytheridea setipunctata, Cytherura* sp. aff. *C. forulata, Loxoconcha matagordensis, Xestoleberis mixohalina, Cytherura elongata, Reticulocythereis floridana* and *Reticulocythereis purii*.

The maximum population density of the remaining seven species lies probably in the euhaline, which has not been considered in the present study. These species are: *Actinocythereis subquadrata, Aurila amygdala, Cytheromorpha paracastanea, Leptocythere darbyi, Paracytheroma stephensoni, Neocaudites nevianii* and aff. *Aglaiocypris eulitoralis*.

	shell sand	sand	sandy mud	mud	peat	silt	organic debris
Cytheridella alosa		xxxx					+
Cypridopsis okeechobei		+					xxxx
Candona ?balatonica							xxxx
Mungava marthapuriae		+					xxxx
Candona annae							xxxx
Cyprideis salebrosa		xxxx					
Cypretta brevisaepta							xxxx
Darwinula furcabdominis							xxxx
Darwinula stevensoni							xxxx
Cypris pseudocrenulata	+	xxxx	+	+	+		+
Heterocypris punctata					xxxx		+
Limnocythere floridensis						xxxx	+
Pontoparta hartmanni		xxxx					+
Dolerocypria fastigata		xxxx					+
Thalassocypria gesinae							xxxx
Parapontoparta subceaerulea	xxxx			+	xxxx		+
Thalassocypria vavrai		xxxx		+			xxxx
Cyprideis beaveni		+	+	xxxx	+		+
Perissocytheridea brachiforma			xxxx	+			
Radimella floridana littorala	xxxx	xxxx			xxxx		
Cytherura sandbergi		xxxx			+		+
Perissocytheridea cribrosa			xxxx	+			
Reticulocythereis floridana		xxxx	+				
Reticulocythereis purii		xxxx					
Haplocytheridea setipunctata		xxxx				+	
Loxoconcha matagordensis	xxxx				xxxx		
Cytherura sp.aff.*C.forulata*	xxxx	xxxx					
Cytherura elongata		xxxx				+	
Xestoleberis mixohalina	xxxx				xxxx		+
Aurila amygdala	xxxx						
Leptocythere darbyi	xxxx			+			
Neocaudites nevianii	xxxx	xxxx					
Actinocythereis subquadrata	xxxx	xxxx				+	
Aff. *Aglaiocypris eulitoralis*	xxxx						
Cytheromorpha paracastanea	xxxx	xxxx					
Paracytheroma stephensoni		+	+			xxxx	

Fig. 2. Species distribution in correlation to the substrate.
xxxx   main range   + single findings

Substrate

After salinity, the substrate is the most important factor controlling the distribution of Ostracoda. Generally the classification of the substrate is very difficult, because the substrate types are usually obscured by some other ingredients. Especially in the brackish water region, with its quick changing currents, a complete description of the substrate is nearly impossible in terms of microhabitat of the Ostracoda. Nevertheless, it seems worthwhile to estimate the nature of the substrate in general terms and to try to compare the living populations of Ostracoda found on these different substrates.

Ostracods do not live on sterile sand or shell sand (Hulings & Puri 1964). On exposed rock, in the running waters of the channels, the small ostracod population is restricted to minor sandfilled pits. Fig. 2 shows the prevailing substrate types for the different species. To establish this table the number of specimens per sample was used, not the number of samples in which the ostracod species was present.

There is a striking difference between the limnic and marine species, the latter prefer sand or shell sand while the limnic forms were found more on organic detritus which was more abundant in the limnic region. However, both limnic Cytheracea, *Cytheridella alosa* and *Cyprideis salebrosa* obviously favoured sandy ground, which might be due either to their creeping movement or to their feeding as scavengers.

The brackish water ostracods are found on the whole range of substrates in this zone. The swimming forms were mostly collected over sandy bottoms; the creeping ones mostly on peat or sandy mud bottoms. Polyhaline species such as *Cytherura, Reticulocythereis* and *Haplocytheridea,* live mainly on sand, whereas *Perissocytheridea* prefers sandy mud. Typical mud favouring forms are *Cyporideis beaveni* and *Heterocypris punctata. Cyprideis* being only found in great numbers on mud. *Parapontoparta* as well as *Thalassocypria vavrai* were also often collected on this substrate. Peat and shell sand give rise to a characteristic fauna, which consists of *Radimella floridana* littorala, *Loxoconcha matagordensis* and *Xestoleberis mixohalina*. Silt has proved to contain only a low number of such species as *Haplocytheridea setipunctata, Actinocythereis subquadrata* and *Paracytheroma stephensoni* in the brackish water zone. A fairly rich fauna lives on silt in the euhaline. Shell sand seems to contain the same species which live on sand.

No interdependence can be stated between the phytal and the present ostracod species. *Loxoconcha* and *Xestoleberis* flourished during the summer vegetation of *Caulerpa* and *Acetabularia,* but were present also at other times when no vegetation was observable. They obviously did not depend on the plants. Whether or not *Cypretta* and *Cypridopsis* need plant life could not be defined. They were more numerous in the vegetation, but no definite answer is possible at this time, for the amount of decaying material in these areas was also very high.

Temperature (Fig. 3)

The temperature during the collection interval varied between 13° and 34°C. Ostracods were mainly found in the 20° to 30°C range. Stenohaline marine ostracods

Fig. 3. Temperature range of each species

```
                                       15      20      25      30      35
                                    °C
```

Species	15	20	25	30	35
Cytheridella alosa			xxxxxxxxxxxxxxx		
Cypridopsis okeechobei			xxxxxxxxxxxxxxxxxxxxx		
Candona ?balatonica				xxxxx	
Mungava marthapuriae				xxxxx	
Candona annae			xxxxxxxxxxxxxxxxxxxx		
Cyprideis salebrosa				x·	
Cypretta brevisaepta			xxxxxxxxxxxxxxxxx		
Darwinula furcabdominis			xxxxxxxxxxxxxxxxxxx		
Darwinula stevensoni			xxxxxxxxxxxxxxxxxxx		
Cypria pseudocrenulata			xxxxxxxxx·xxxxxxxxxxxx		
Heterocypris punctata			xxxxxxxxxxxxxxx·xxxxxxxxxxxxx		
Limnocythere floridensis			xxxxxxxxxxxxxxxxxxxxx		
Pontoparta hartmanni		--------------	xxxxxxxxxx		
Dolerocypria fastigata		xxxxxxxxxxxxxxxxxxxxxxxxxxxxxxxxx			
Thalassocypria gesinae		xx·xxxxxxxxxxxxxxxxxxxxxxxxx			
Parapontoparta subcaerulea		xxxxxxxxxxxxxxxxxxxxxxxxxxxxxxx			
Thalassocypria vavrai		xxxxxxxxxxxxxxxxxxxxxxxxxxxxxx			
Cyprideis beaveni		xxxxxxxxxxxxxxxxxx·xxxxxxxxxxxxx·xxxx			
Perissocytheridea brachiforma		----xxxxxxxxxxxxxxxxxxxxxxxxxxx			
Radimella floridana littorala		xxxxxxxxxxxxxx·xxxxxxxxxxxxxxxxx			
Cytherura sandbergi		--------- xxxxxxxxxxxxxxxxxxxxxxxxxxxxx			
Perissocytheridea cribrosa		xxxxxxxxxxxxxxxxxxxxxx·xxxxxxxxx			
Reticulocythereis floridana		--------xxxxxxxxxxxxxxxxxxxxxxxxxxxxx			
Reticulocythereis purii		--------xxxxxxxxxxxxxxxxxxxxxxx·xxxxxxxx			
Haplocytheridea setipunctata		-------xxxxxxxxxxxxxxxxx·xxxxxxxxxxxx			
Loxoconcha matagordensis		----xxxxxxxxxxxxxxxxxxxxxxxxx			
Cytherura sp.aff. *C. forulata*		xxxxxxxxxxxxxxxxxxxxxxx·xx			
Cytherura elongata			xxxxxxxxxxx		
Xestoleberis mixohalina		--------xxxxxxxxxxxxxxxx ----------			
Aurila amygdala		------------------- xxxxxxxxxxxxxxxx			
Leptocythere darbyi			xxxxxxxxxx		
Neocaudites nevianii			xxxxx		
Actinocythereis subquadrata		----------------xxxxxxxxxxxxxxxx			
Aff. *Aglaiocypris eulitoralis*			xxxxxxxxxxxx		
Cytheromorpha paracastanea			xxxxxxxxxxxxx		
Paracytheroma stephensoni			xxxxxxx		

xxx determined range
--- uncertain range

are also stenothermal, they were found mainly at 25° to 30°C and sometimes up to 34°C. Euryhaline ostracods seem to be also eurythermal. Limnic forms show a maximum temperature limit at 30°C.

Definite statements concerning the optimum temperatures cannot be given. Only laboratory experiments can give reliable data, as done for *Aurila conradi* (= *?Radimella floridana*) by Kornicker & Wise (1960). They could show that this species is immobilized below 6°C and can survive 0°C for 24 hours. Its activity was limited by an upper temperature of 36°C. These temparature limits show the range of water temperatures the brackish water ostracods have to withstand in these shallow tidal areas. This preadaptation might be one cause which separates the mixohaline and the marine animals (see also Den Hartog 1964).

Hydrogen-ion-concentration (pH)

During the investigation the pH did not seem to have any influence on the distribution. The ostracods were found between 6.5 and 9.2 pH. In the limnic zone the difference was smaller. In the marine and polyhaline zone the animals were collected at a pH below 8, with the one exception of *Aurila amygdala* at 8.8 pH.

Results

The main factor controlling the local distribution of the investigated Ostracoda was the salinity. This agrees with the findings of several authors (Hartmann 1956, 1957, 1962, Morales 1966, King & Kornicker 1970).

The second most important factor in controlling their distribution was found to be the substrate. Temperature seems to have only a minor effect on the distribution in this geographically small area.

Current strength influenced mainly the swimming forms, but only in small areas. Water depth, pH and turbidity showed no apparent influence on the ostracod fauna in the eulitoral zone.

Associations (Fig. 4)

Using the salinity data for each species, 5 ostracod-associations can be recognized in the brackish water zone of southwestern Florida. The substrate preference of each species is also given:

1) Limnic-Oligohaline assemblage; composed mainly of Cypridacea, with only two Cytheracea occurring:

Cytheridella alosa	(sand)
Cypridopsis okeechobei	(organic debris/phytal)
Candona ?balatonica	(organic debris)
Mungava marthapuriae	(organic debris)
Candona annae	(organic debris)
Cyprideis salebrosa	(sand)
Cypretta bevisaepta	(organic debris)
Darwinula furcabdominis	(organic debris)

The latter two species also occur in assemblage 2.

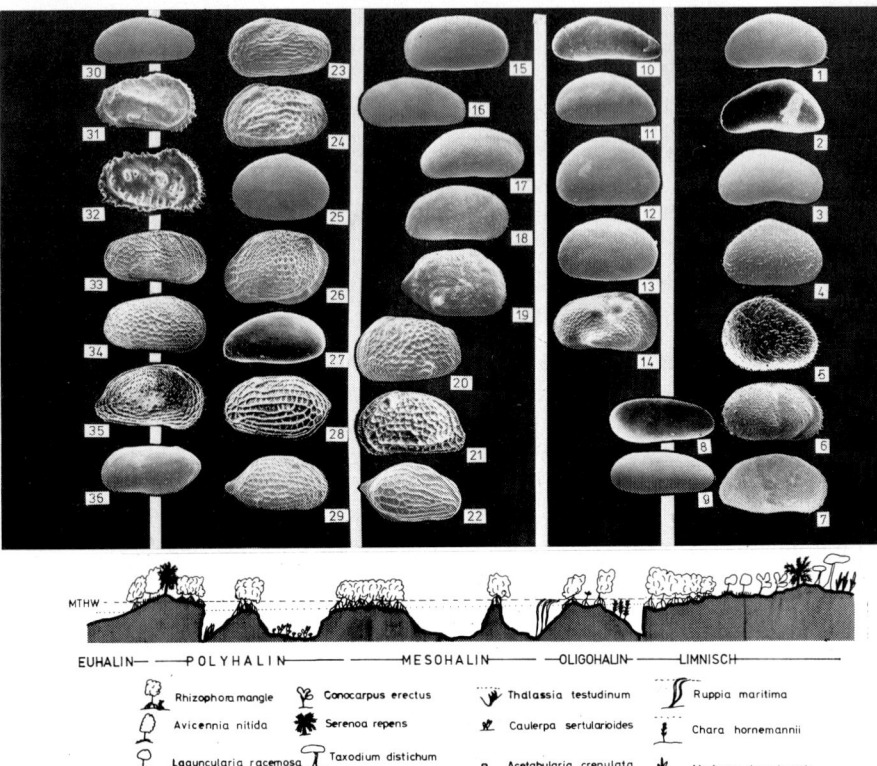

Fig. 4. Ostracode assemblages and vegetation distribution in the whitewater Bay (Florida)

A) *Limnic-Oligohaline Assemblage*
1. Mungava marthapuriae
2. Candona ?balatonica
3. Candona annae
4. Cypridopsis okeechobei
5. Cypretta brevisaepta
6. Cytheridella alosa
7. Cyprideis salebrosa
8. Darwinula stevensoni
9. Darwinula furcabdominis

B) *Oligo-Mesohaline Assemblage*
10. Dolerocypria fastigata
11. Pontoparta hartmanni
12. Cypria pseudocrenulata
13. Heterocypris punctata
14. Limnocythere floridensis

C) *Euryhaline Assemblage*
15. Thalassocypria gesinae
16. Thalassocypria vavrai
17. Parapontoparta subcaerulea

18. Cyprideis beaveni
19. Perissocytheridea brachiforma
20. Perissocytheridea cribrosa
21. Radimella floridana littorala
22. Cytherura sandbergi

D) *Meso-Polyhaline Assemblage*
23. Reticulocythereis purii
24. Reticulocythereis floridana
25. Haplocytheridea setipunctata
26. Loxoconcha matagordensis
27. Xestoleberis mixohalina
28. Cytherura sp. aff. C. forulata
29. Cytherura elongata

E) *Poly-Euhaline Assemblage*
30. Aff. Aglaiocypris eulitoralis
31. Neocaudites nevianii
32. Actinocythereis subquadrata
33. Leptocythere darbyi
34. Cytheromorpha paracastanea
35. Aurila amygdala
36. Paracytheroma stephensoni

2) Oligo-Mesohaline assemblage:

Cypria pseudocrenulata	(independent of substrate)
Heterocypris punctata	(mud)
Limnocythere floridensis	(peat)
Pontoparta hartmanni	(sand)
Dolerocypria fastigata	(sand)

All these species except *Cypria pseudocrenulata* were found alone. *Dolerocypria fastigata* was also found only together with *Pontoparta hartmanni*. This is obviously a function of substrate selection.

3) Euryhaline assemblage; composed of the euryhaline ostracods which favour the mesohaline zone with large and quick salinity changes.

Here associations are encountered which characterize a very small biotope. The combination of *Cyprideis beaveni* and *Thalassocypria vavrai* for example, indicates a pure mud substrate in a stagnant water region and *Perissocytheridea brachiforma*, *Cytherura sandbergi* and *Radimella floridana* littorala are characteristic of sandy mud in open tidal waters. A substrate rich in debris yields *Thalassocypria gesinae* and *Parapontoparta subcaerulea* in association.

The Euryhaline assemblage consists of the following species:

Thalassocypria gesinae	(sand)
Parapontoparta subcaerulea	(peat)
Thalassocypria vavrai	(sand)
Cyprideis beaveni	(independent of substrate)
Perissocytheridea brachiforma	(sandy mud)
Radimella floridana littorala	(shell sand, peat)
Cytherura sandbergi	(sand)
Perissocytheridea cribrosa	(sandy mud)

The latter three species are often encountered in the following assemblage.

4) Meso-Polyhaline assemblage:

Reticulocythereis floridana	(sand)
Reticulocythereis purii	(sand)
Haplocytheridea setipunctata	(sand)
Loxoconcha matagordensis	(shellsand)
Cytherura sp. aff. *C. forulata*	(sand, shellsand)
Cytherura elongata	(sand, shellsand)
Xestoleberis mixohalina	(peat)

Here again characteristic combinations are observed. *Radimella floridana* littorala, *Loxoconcha matagordensis* and *Xestoleberis mixohalina* are typically found in the zone of seasonal vegetation or on shell sand or peat in open waters. Both species of *Reticulocythereis* often occur together, as do the three species of *Cytherura*. An explanation for these phenomena cannot be given at this time.

5) Poly-Euhaline assemblage:

Aurila amygdala	(shell sand)
Leptocythere darbyi	(shell sand)
Neocaudites nevianii	(shell sand)
Actinocythereis subquadrata	(sand, shell sand)
Aff. Aglaiocypris eulitoralis	(shell sand)
Cytheromorpha paracastanea	(shell sand)
Paracytheroma stephensoni	(silt)

Zoogeographical distribution (Fig. 5)

The zoogeographical comparison of the study area with other parts of middle America is difficult due to taxonomic uncertainties (Swain 1955, Curtis 1960).

The present results have been compared with two studies from the Gulf of Mexico (Morales 1966, King & Kornicker 1970). The Caribbean Sea area is fairly well known in respect of its fossil and subfossil Ostracoda due to the work of Van den Bold (div.). Recent species were mentioned by Klie (1933, 39) and Triebel (1961, 62). The Pacific coast is well known only in South California in the USA and Baja California and the Gulf of California in Mexico (Juday 1907, Benson 1959, Benson & Kaesler 1963, McKenzie & Swain 1969, Swain & Gilby 1974). The Ostracoda of the further south lying coast in the area are known merely from publications by Hartmann (1955, 1956, 1957, 1959) from El Salvador and Coryell & Fields (1937) from Panama. Ishizaki & Gunther (1974) recently published a paper on the Cytheruridae of Panama. Allison & Holden (1971) dealt with the fauna of Clipperton Island and Triebel (1956) as well as Pokorny (1968 a, b, c, 1969, 1970) described ostracods from the Galapagos.

Distribution in the Gulf of Mexico (Florida-Texas-Mexico)

The ostracod fauna of the Gulf of Mexico is fairly uniform. This is surprising because the studies with which the present work is compared (Morales 1966, King & Kornicker 1970) were carried out in different environments.

Ten corresponding species are common to all three areas:

Aurila amygdala, Radimella floridana (littorala), *Cytherura elongata, Cytherura* sp. aff. *C. forulata, Cytherura sandbergi, Haplocytheridea setipunctata, Leptocythere darbyi* (? *nikraveshae*), *Loxoconcha matagordensis, Paracytheroma stephensoni,* and *Perissocytheridea brachiforma.*

Three further species may be common, but taxonomic uncertainties were present:

Actinocythereis subquadrata	= ?(*A. triangularis* sensu Morales),
Cytheromorpha paracastanea	= ?(*C. warneri* sensu King & Kornicker), and
Xestoleberis mixohalina	= ?(*X.* sp. sensu King & Kornicker, *X. rigbyi* sensu Morales).

Two species are found only in Mexico and Florida:
Neocaudites nevianii, and *Perissocytheridea cribosa.*

Species	limnic	mixo-haline	eu-haline	Florida	Texas	Mexico	Caribbean	Pacific
Actinocythereis subquadrata			▨					
Acuticythereis sp. A								
Acuticythereis sp. B								
Aurila amygdala								
Bairdia bradyi								
Basslerites minutus								
Campylocythere laevissima			▨					
Cativella dispar								
Cobanocythere subterranea								
Cobanocythere labiata								
Cyprideis beaveni		▨						
Cyprideis bensoni								
Cyprideis castus								
Cyprideis mexicana								
Cyprideis pacifica		▨						
Cyprideis salebrosa	▨							
Cyprideis torosa		▨						
Cytherella sp. aff.C.harpago			▨					
Cytherelia mejanguerensis								
Cytheridella alosa	▨							
Cytheromorpha paracastanea			▨					
Cytheromorpha warneri								
Cytherura elongata								
Cytherura sp. aff.C.forulata								
Cytherura ostilicola								
Cytherura palacii			▨					
Cytherura radialirata								
Cytherura sandbergi		▨						
Cytherura sp.								
Cytherura swaini								
Elofsonella salvadoriana		▨						
Haplocytheridea bradyi								
Haplocytheridea setipunctata			▨					
Hemicytherura cranekeyensis								
Hulingsina ashermani								
Hulingsina sp. aff.H.rugipustulosa								
Hulingsina sandersi								
Leptocythere darbyi			▨					
Leptocythere nikraveshae								
Limnocythere floridensis	▨							
Limnocythere sanctipatricii			▨					
Limnocythere sp.								
Loxoconcha lapidiscola								
Loxoconcha matagordensis		▨						
Loxoconcha purisubrhomboidea								
Loxoconcha sp. aff.L.sarasotana								
Loxoconcha schusterae								
Megacythere johnsoni								
Neocaudites nevianii			▨					
Orionina bradyi								
Paijenborchella mediterranea								
Palaciosa vandenboldi								
Paracytheridea troglodyta		▨						
Paracytheridea vandenboldi			▨					
Paracytheroma costata								
Paracytheroma magna								
Paracytheroma stephensoni								
Paracytheroma undulimarginata		▨						
Paradoxostoma salvadorianus								
Paradoxostoma ? sp.								
Parvocythere dentata								
Pellucistoma magniventra								
Pericythere foveata	▨							
Perissocytheridea brachiforma								
Perissocytheridea cribrosa								
Perissocytheridea dentatomarginata								
Perissocytheridea excavata								
Perissocytheridea meyerabichii		▨						
Perissocytheridea punctata								
Perissocytheridea rugata								
Perissocytheridea swaini								
Pumilocytheridea ayalai								
Radimella floridana			▨					
Reticulocythereis floridana								
Reticulocythereis multicarinata								
Reticulocythereis purii								
Sclerochilus centroamericanus								
Tanella gracilis			▨					
Triebelina gierloffi								
Xestoleberis eulitoralis		▨						
Xestoleberis mixohalina								
Xestoleberis rigbyi								
Xestoleberis sp.			▨					

FIG. 5 Zoogeographical comparison of the Ostracode-fauna

Van den Bold reported both species (*Neocaudites* 1966b from Venezuela, *Perissocytheridea* 1963 from Trinidad), Klie (1933, 39) only *Perissocytheridea,* from the coast of South America. Questionable is the conspecifity of *Perissocytheridea cribrosa* and *P. bicelliformis* mentioned by Swain (1955) and Engel & Swain (1967). The absence of *Neocaudites nevianii* from Texas is striking. It is probably a true tropical species.

Naturally the ostracod fauna of Florida corresponds most closely to that of Texas, since the environment of the two study areas is more comparable and the geographical situation is also similar. A comparison between the Cypridacea of the brackish water zone based on the figured shell characteristics by King & Kornicker (1970) seems to reveal the same genera, although a definite assignment is not possible due to lack of information of the softparts:

Aglaiocypris sp.	? = *Thalassocypria* sp.
Astenocypris sp.	? = *Parapontoparta* sp.
Cypridopsis sp.	= *Cypridopsis* sp.
Potamocypris sp.	= *Potamocypris* sp.A
Potamocypris smaragdina	= *Potamocypris* sp.B

The same is true for *Cyprideis beaveni* and unfortunately questionable between *Limnocythere floridensis* and *L.* sp. as well as *Reticulocythereis floridana* and *R. multicarinata*.

Species common between Texas and Mexico are:

Megacythere johnsoni, Perissocytheridea excavata, and *Perissocytheridea rugata.*

In summary, the ostracod fauna of the brackish water zone is very homogenous. Unfortunately, only the fossil fauna of the Caribbean is known from the works of Van den Bold and, therefore, little can be said about the distribution of the Recent Ostracoda. The fossil findings, however, indicate a partly communality between the ostracod faunas of the Gulf of Mexico and the Caribbean.

Gulf of Mexico-Pacific coast of Middle America

The brackish water fauna of Florida at the specific level is not similar to the ostracod fauna which Hartmann (1957) described from El Salvador although the genera are the same. No species which Hartmann described was found in the Gulf of Mexico. Van den Bold (1964), however, encountered three species from El Salvador in his samples from the Caribbean: *Loxoconcha lapidiscola, Palaciosa vandenboldi* and *Perissocytheridea punctata*. There seem to be a few species common between the Caribbean and the Pacific. Only one species has so far been found in the Pacific as well as in the Caribbean and the Gulf of Mexico, that is *Cytherura sandbergi* (Ishizaki & Gunther 1974, Benson & Kaesler (?) 1963, Morales 1966, King & Kornicker 1970). The connections between the Gulf of Mexico and the Pacific are almost non-existent. The ostracod faunas of these two regions, which have been physically separated since the Miocene have become widely divergent due to separate evolutions.

In this connection it is interesting that Eibl-Eibesfeld (Triebel 1956) found two species on the Galapagos Islands which are conspecific with ostracods from Florida. These species are *Cyprideis beaveni* (= *C. stenophora*) and *Xestoleberis arcturi*. This might be related to the extinction of the tropical fauna at the west coast of the continents with their cold currents in equatorial directions and the survival on islands along these coasts (Ekman 1953). Hartmann (1974) explained this by the extension of the cold Pleistocene currents along the west coast of the continents, which resulted in the extinction of the coastal tropical fauna, while the island were mostly not influenced by these currents, enabling the fauna to survive. Whether or not this can explain the occurrence of the same species on the Galapagos Islands and in Florida remains to be demonstrated by further studies.

Mangrove dependence

A typical mangrove fauna could not be found at the specific level. A correspondence could only be observed at the generic level. This fauna does not seem to depend on mangroves but is typical of a tropical brackish water flat (Bate 1971).

Summary

The ecology and zoogeographical distribution of 36 ostracod species collected in the brackish water region of South West Florida has been investigated. 183 samples from 120 stations were studied. The ecologic parameters recorded were; salinity, substrate, water temperature, pH, water depth, turbidity and currents.
1. The most important factor controlling the distribution of ostracods in this region is the salinity.
2. Five assemblages could be recognized in respect of their salinity preferences:
Limnic-Oligohaline assemblage
Oligo-Mesohaline assemblage
Meso-Polyhaline assemblage
Poly-Euhaline assemblage
Euryhaline assemblage
3. Typical for the brackish water zone is the occurrence of the Thalassocyprididae. *Heterocypris punctata, Cyprideis beaveni, Perissocytheridea brachiforma* and *Perissocytheridea cribrosa* are also true brackish water ostracods. *Cyprideis salebrosa* has been classified as a limnic species with oligohaline tolerance.
4. The next most important factor controlling the distribution is the substrate. It was classified into shell sand, sand, mud, sandy mud, peat, silt and organic debris.

The Cypridacea of the Limnic-Oligohaline assemblage prefer substrates which are rich in debris, the Cytheracea more sandy ground. The explanation is probably their different feeding habits. The ostracods of the Poly-Euhaline assemblage prefer sand and shell sand. *Cyprideis beaveni* and *Cypria pseudocrenulata* could not be correlated with a special substrate.
5. Water temperature has only a minor effect on the distribution. Extreme temperatures are a limiting factor. Only *Leptocythere darbyi* shows a seasonal fluctuation which is correlated with temperature.
6. Turbidity, currents, pH and water depth do not reveal any influence on the distribution of the ostracods.
7. The zoogeographical comparison of the investigated brackish water fauna with other known faunas of the Gulf of Mexico showed a good accordance at specific level (16 species). 6 species have been reported from the Gulf of Mexico and the

Caribbean. Only one species (*Cytherura sandbergi*) is reported live from both sides of the middle American landbridge. Communality can be found only at generic level (8 genera) between the Pacific and the Gulf of Mexico/Caribbean.
8. A connection between the mangrove vegetation and the distribution of ostracod species could not be found.

Acknowledgments

Thanks are due to Prof. Dr. G. Hartmann, Zoologisches Institut und Museum Universität Hamburg and Dr. Harbans S. Puri, Tallahassee, Florida, for the initiation of this subject and the most valuable help during the study. Gratitude is also expressed to the Florida Bureau of Geology and the Everglades National Park for help during the collection of the material.

References

Allison, E. C. & J. C. Holden. 1971. Recent Ostracodes from Clipperton Island, Eastern Tropical Pacific. Trans.S.Diego Soc. nat. Hist. 16:165-214.
Bate, R. H. 1971. The distribution of recent Ostracoda in the Abu Dhabi Lagoon, Persian Gulf. Bull. Centre Rech. Pau SNPA, suppl. 5:239-256.
Benson, R. H. 1959. Ecology of recent Ostracodes of the Todos Santos Bay Region, Baja California, Mexico. Paleont. Univ. Kans. 23:1-80.
Benson, R. H. & Kaesler. 1963. Recent marine and lagoonal ostracodes from the Estero de Tastiota Region, Sonora, Mexico (Northeastern Gulf of California). Paleont. Contr. Univ. Kans. 2:1-34.
Bold, W. A. Van den. 1957. Oligo-Miocene Ostracoda from southern Trinidad. Micropaleontology 3:231-254.
Bold, W. A. Van den. 1963. Upper Miocene and Pliocene Ostracoda of Trinidad. Micropaleontology 9:361-424.
Bold, W. A. Van den. 1964. Nota preliminar sobre los Ostracodos del Mioceno-Reciente de Venezuela. Escuelo de Geologia, Univ, Centr. de Venezuelas, GEOS-Caracas 11:7-18.
Bold, W. A. Van den. 1966a. Ostracoda from Colon Harbour, Panama. Caribb.J.Sci. 6:43-64.
Bold, W. A. Van den. 1966b. Miocene and Pliocene Ostracoda from northeastern Venezuela. Verh.K.Akad.Wet. 23:5-42.
Bold, W. A. Van den. 1972. Ostracoda of the La Boca Fromation, Panama Canal Zone. Micropaleontology 18:410-442.
Coryell, H. N. & S. Fields. 1937. A Gatun Ostracod fauna from Panama Am.Mus.Novit. 956:1-18.
Curtis, D. M. 1960. Relation of environmental energy levels and ostracod biofacies in East Mississippi Delta area. Bull.Am.Ass.Petrol.Geol. 44:471-494.
Den Hartog, C. 1964. Typologie des Brackwassers. Helgoländer wiss. Meeresunters. 10:377-387.
Ekman, S. 1953. Zoogeography of the Sea. Sidgwick and Jackson Ltd., London.
Engel, P. L. & F. M. Swain. 1967. Environmental relationships of recent Ostracoda in Mesquite, Aransas and Copano Bays, Texas Gulf Coast. Trans. Gulf Coast Ass.Geol.Soc. 17:408-427.
Hartmann, G. 1955. Neue marine Ostracoden der Familie Cyprididae und der Subfamilie Cytherideinae der Familie Cytheridae aus Brasilien. Zool.Anz. 154:109-127.
Hartmann, G. 1956. Zur Kenntnis des Mangrove-Estero-Gebietes von El Salvador und seiner Ostracoden Fauna. Kieler Meeresforsch. 12:219-248.

Hartmann, G. 1957. Zur Kenntnis des Mangrove-Estero-Gebietes von El Salvador und seiner Ostracoden Fauna. Kieler Meeresforsch. 13:134-159.

Hartmann, G. 1959. Zur Kenntnis der lotischen Lebensbereiche der pazifischen Küste von El Salvador unter besonderer Berücksichtigung seiner Ostracodenfauna. Kieler Meeresforsch. 15:187-241.

Hartmann, G. 1962. In: Hartmann-Schröder, G. & G. Hartmann, Zur Kenntnis des Eulitorals der chilenischen Pazifikküste und der Küste Südpatagoniens unter besonderer Berücksichtigung der Polychaeten und Ostracoden. Mitt.Hamb.Zool.Mus.Inst., suppl. 60:1–270.

Hartmann, G. 1974. Zur Kenntnis des Eulitorals der afrikanischen Westküste zwischen Angola und Kap der guten Hoffnung und der afrikanischen Ostküste von Südafrika und Mocambique unter besonderer Berücksichtigung der Polychaeten und Ostracoden. Teil III, Die Ostracoden des Untersuchungsgebietes. Mitt.Hamb.Zool.Mus.Inst.suppl. 69:229–520.

Hulings, N. C. & H. S. Puri. 1964. The ecology of shallow water Ostracods of the west coast of Florida. Pubbl.Staz.Zool.Napoli 33:308–344.

Ishizaki, K. & F. J. Gunther. 1974. Ostracoda of the Family Cytheruridae from the Gulf of Panama. Tohoku Univ., Sci.Rep., 2nd ser. 45:1–50.

Juday, O. 1907. Ostracoda of the San Diego Region, II: Littoralforms. Univ.Calif.Public.Zool. 3:153–156.

Keyser, D. 1976a. Ostracoden aus den Mangrovegebieten von Südwest-Florida. Abh.Verh.naturwiss.Ver.Hamburg (NF) 18/19:255–290.

Keyser, D. 1976b. Brackwasser-Cytheracea aus Südflorida (Crustacea: Ostracoda, Podocopa). Abh.verh.naturwiss.Ver.Hamburg (NF) 20 (in press).

King, C. E. & L. S. Kornicker. 1970. Ostracoda in Texas Bays and Lagoons: An ecologic study. Smithsonian Contr.Zool. 24:1–92.

Klie, W. 1933. Zoologische Ergebnisse einer Reise nach Bonaire, Curacao und Aruba im Jahre 1930, No. 5. Süßwasser und Brackwasser Ostracoden von Bonaire, Curacao und Aruba. Zool.Jb. 64:369–390.

Klie, W. 1939a. Ostracoden aus den marinen Salinen von Bonaire, Curacao und Aruba. Capita Zoologica 8:3–19.

Klie, W. 1939b: Brackwasserostracoden aus Nordbrasilien. Zool.Jb.Syst. 72:359–372.

Kornicker, L. S. & C. D. Wise. 1960. Some environmental boundaries of a marine ostracode. Micropaleontology 6:393–398.

McKenzie, K. G. & F. M. Swain. 1967. Recent Ostracoda from Scammon Lagoon, Baja California. J.Paleont. 41:281–305.

Morales, F. G. A. 1966. Ecology, distribution and taxonomy of recent Ostracoda of the Laguna de Terminos, Campeche, Mexico. Univ.National Autonoma de Mexico, Inst.Geol.Bull. 81:1–102.

Pokorny, V. 1968a. Radimella gen.n., a new species of the Hemicytherinae (Ostracoda, Crustacea). Acta Univ. Carol. 4:359–373.

Pokorny, V. 1968b. Baltraella baltraensis gen.n., sp.n. (Bythocytheridae, Ostracoda) from Recent Deposits of the Galapagos Islands. Cas.min.geol. 13:385–390.

Pokorny, V. 1968. Havanardia g.nov., a new genus of the Bairdiidae (Ostracoda, Crust.). Vestnik Ustredniho Ustavu Geol. 43:61–63.

Pokorny, V. 1969. The genus Radimella Pokorny, 1969 (Ostracoda, Crustacea) in the Galapagos Islands. Acta Univ. Carol. 4:375–389.

Pokorny, V. 1970. The genus Caudites Coryell & Fields, 1937 (Ostracoda, Crust.) in the Galapagos Islands. Acta Univ. Carol. 4:267–302.

Swain, F. M. 1955. Ostracoda of San Antonio Bay, Texas. J.Paleont. 29:561–646.

Swain, F. M. & J. M. Gilby. 1974. Marine Holocene Ostracoda from the Pacific coast of North and Central America. Micropaleontology 20:257–352.

Triebel, E. 1956. Brackwasserostracoden von den Galapagos-Inseln. Senckenberg. biol. 37:447–467.

Triebel, E. 1961. Süwasser-Ostracoden von den Karibischen Inseln; 1. Cypridiini. Senckenberg biol. 42:51–74.

Triebel, E. 1962. Süßwasser-Ostracoden von den Karibischen Inseln; 2. Xenocypris n. g. Senckenberg biol. 43:47–63.

Author's address:

Zool. Institut und Zool. Museum, Martin-Luther-King-Platz 3, 2000 Hamburg 13, B.R.D.

Sixth Intern. Ostracod Symposium, Saalfelden

OSTRACODA FROM BOTTOM CORES OFF THE COAST OF MONTENEGRO

ANA SOKAČ

Abstract

Recent ostracoda from bottom cores off the coast of Montenegro were studied. Six samples from the localities Kotor, Budva, Bar and Ulcinj were taken into account for this investigation. With respect to the ostracod and sediment contents similarities and differences between the samples analysed are discussed.

Introduction

The rich, well preserved ostracod fauna in bottom core samples collected off the Coast of Montenegro (Sokač 1975) is partly considered again in order to establish the relationship between the sediments and the ostracod fauna. For this purpose, granulometric and differential thermal analyses have been carried out on the samples from the sites Kotor-3, Budva-2, Budva-3, Bar-1, Bar-3, and Ulcinj-4.

Previous work

Recently a few important studies have been published on the distribution of ostracoda from the bottom cores in the Adriatic Sea. The ostracod fauna north of the line Gargano-Zadar was investigated by Bonaduce, Ciampo & Masoli (1975). The systematic part is particularly detailed, and the correlations between the ostracod fauna and environmental factors are given. Breman (1975) described ostracods from the entire area of the Adriatic Sea, except the Yugoslav and Albanian territorial waters. Some new species from the bottom cores of the Adriatic Sea have been described by Breman (1976) as well. Uffenorde (1975) discussed the dynamics of recent marine benthonic ostracod assemblages in the Limski Kanal.

The present paper is a continuation of the research on the microfauna from the bottom cores off the coast of Montenegro, carried out by Sokač (1975, an exhaustive bibliography is presented).

Material and methods

The material described in this work consists of 6 grab-samples (Sokač 1975, p. 112, Fig. 1), which were taken with a Paterson-grab sampler. Each sample was collected from 20 sq.cm of the sediment surface and usually included 2-3 cm of the uppermost layer. The water temperature near the bottom was measured with a Nansen apparatus.

Two sedimentary methods were used. The sediments were examined by granulometric and differential thermal analyses (DTA). The granulometric analyses

Genera and species	KOTOR - 3	BUDVA - 2	BUDVA - 3	BAR - 1	BAR - 3	ULCINJ - 4
Polycope reticulata G.W. MÜLLER	●					
Polycope sp.	●					
Cytherella alvearium BONADUCE, CIAMPO & MASOLI			●			
Cytherella sp.				●		
Cytherelloidea sordida G.W. MÜLLER					●	
Darwinula stevensoni (BRADY & ROBERTSON)						●
Bairdia longevaginata G.W. MÜLLER				●	●	
Candona neglecta SARS				●		●
Candona sp.						●
Propontocypris setosa (G.W. MÜLLER)	●		●			●
Ilyocypris bradyi SARS						●
Leptocythere ramosa (ROME)			●			
Callistocythere adriatica MASOLI			●			
Aurila convexa (BAIRD)					●	
A. interpetis ULICZNY				●	●	
A. prasina BARBEITO-GONZALES					●	
A. speyeri (BRADY)				●	●	●
A. aff. convexa (BAIRD)						●
Aurila sp.				●		
Urocythereis sp.		●	●		●	
Heterocythereis albomaculata (BAIRD)				●		
Bosquetina pectinata (BOSQUET)	●					
Acantocythereis histrix (REUSS)			●			
Carinocythereis antiquata (BAIRD)	●	●	●	●		●
Hiltermannicythere turbida (G.W. MÜLLER)	●	●	●	●	●	●
Costa edwardsii (ROEMER)	●					
Pterygocythereis jonesi (BAIRD)	●		●			●
P. ceratoptera (BOSQUET)			●			
Cytheretta subradiosa (ROEMER)		●	●	●		●
Cytheretta sp.		●		●		●
Cytheridea neapolitana KOLLMANN	●		●	●		●
Cyprideis torosa (JONES)				●		
Pontocythere elongata (BRADY)		●	●	●		●
Neocytherideis fasciata (BRADY & ROBERTSON)		●		●		●
N. subspiralis (BRADY, CROSSKEY & ROBERTSON)					●	
Neocytherideis sp.		●				●
Krithe sp.	●					
Semicytherura incongruens (G.W. MÜLLER)		●		●		
S. acuminata (G.W. MÜLLER)					●	●
S. sulcata (G.W. MÜLLER)		●				●
S. inversa (SEGUENZA)			●			
Semicytherura sp.						●
Tetracytherura angulosa (SEGUENZA)			●		●	
Paracytheridea sp.			●	●		
Pseudocytherura calcarata (SEGUENZA)			●			
Pseudocytherura sp.				●		
Cytheropteron rotundatum (G.W. MÜLLER)			●			
Loxoconcha agilis RUGGIERI	●		●			
L. gibberosa TERQUEM				●		
L. napolitana PURI		●	●			
L. rhomboidea (FISCHER)		●	●		●	
L. parallela G.W. MÜLLER					●	
L. tumida BRADY		●				●
L. rubritincta RUGGIERI		●				
L. stellifera G.W. MÜLLER						●
Loxoconcha sp.		●		●		●
Limnocythere inopinata (BAIRD)						●
Monoceratina sp.	●					
Paradoxostoma simile G.W. MÜLLER	●				●	
Xestoleberis communis G.W. MÜLLER		●	●	●	●	●
X. dispar G.W. MÜLLER			●	●	●	
X. plana G.W. MÜLLER					●	

Fig. 1. Tabular review of the ostracod genera and species from bottom cores off the Coast of Montenegro.

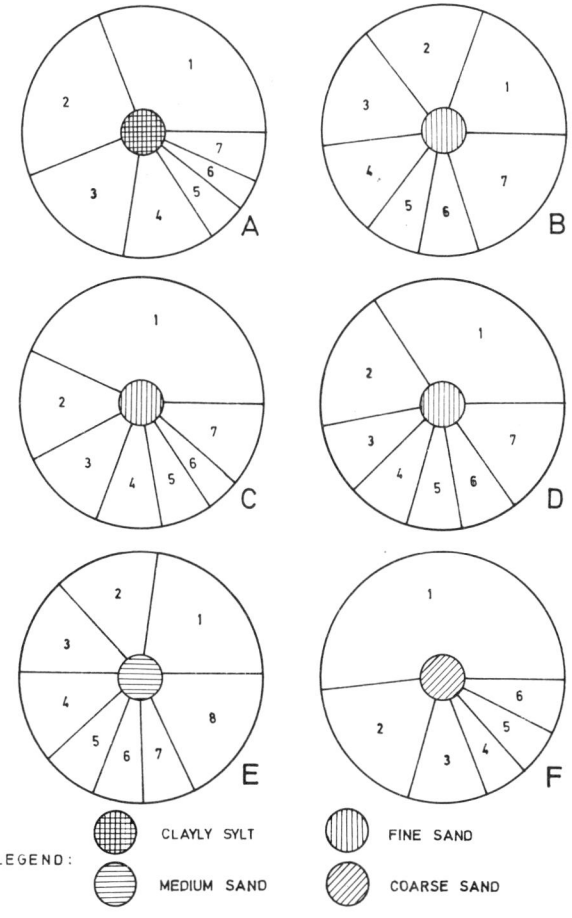

Fig. 2. Cyclodiagrams of the samples from bottom cores off the coast of Montenegro.
A. Kotor-3, 1. *Cytheridea*, 2. *Pterygocythereis*, 3. *Costa*, 4. *Bosquetina*, 5. *Carinocythereis*, 6. *Krithe*, 7. Other genera in order of abundancy: *Loxoconcha, Polycope, Propontocypris, Paradoxostoma, Monoceratina*,
B. Ulcinj-4, 1. *Pontocythere*, 2. *Loxoconcha*, 3. *Semicytherura*, 4. *Cytheridea*, 5. *Neocytherideis*, 6. *Carinocytherideis*, 7. Other genera: *Cytheretta, Hiltermannicythere, Xestoleberis, Pterygocythereis, Aurila, Propontocypris, Ilyocypris, Limnocythere, Candona, Darwinula*,
C. Budva-2, 1. *Pontocythere*, 2. *Neocytherideis*, 3. *Loxoconcha*, 4. *Cytheretta*, 5. *Carinocythereis*, 6. *Semicytherura*, 7. Other genera: *Xestoleberis, Hiltermannicythere, Urocythereis*,
D. Bar-1, 1. *Pontocythere*, 2. *Xestoleberis*, 3. *Loxoconcha*, 4. *Neocytherideis*, 5. *Cytheretta*, 6. *Aurila*, 7. Other genera: *Semicytherura, Carinocythereis, Hiltermannicythere, Cytherella, Cytheridea, Bairdia, Heterocythereis, Paracytheridea, Pseudocytherura, Candona, Cyprideis*,
E. Budva-3, 1. *Pontocythere*, 2. *Loxoconcha*, 3. *Carinocythereis*, 4. *Aurila*, 5. *Semicytherura*, 6. *Cytheridea*, 7. *Pterygocythereis*, 8. Other genera: *Hiltermannicythere, Leptocythere, Xestoleberis, Cytheropteron, Cytheretta, Cytherella, Acantocythereis, Paracytheridea, Callistocythere, Propontocypris, Tetracytherura, Urocythereis, Pseudocytherura*,
F. Bar-3, 1. *Aurila*, 2. *Xestoleberis*, 3. *Loxoconcha*, 4. *Pontocythere*, 5. *Bairdia*, 6. Other genera: *Hiltermannicythere, Tetracytherura, Semicytherura, Neocytherideis, Paradoxostoma, Urocythereis, Cytherelloidea*.

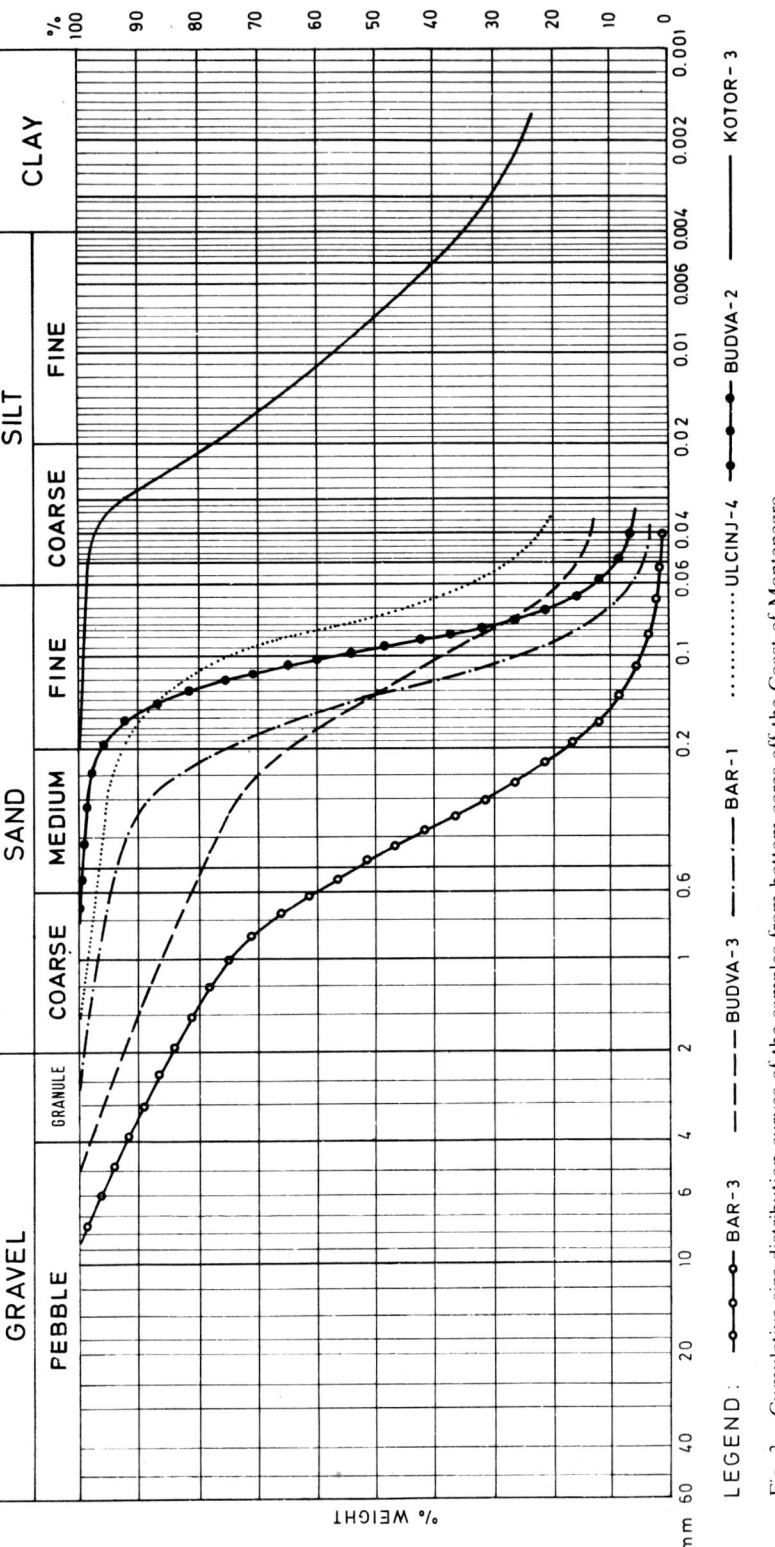

Fig. 3. Cumulative size distribution curves of the samples from bottom cores off the Coast of Montenegro.

of the samples Budva-2, Budva-3, Bar-1, Bar-3, and Ulcinj-4 were carried out with a sieve-method, the sample Kotor-3 with the help of a sedimentary balance, and the sieve-method. Differential thermal analyses were carried out, for easier determination, on the fraction of 0.1 mm, with the exception of the sample Kotor-3, where, because of the treatment on the sedimentary balance, it was possible to take a fraction smaller than 0.04 mm.

Discussion

This micropaleontological investigation of the samples from the bottom cores taken from the sites Kotor-3, Budva-3, Bar-1, Bar-3 and Ulcinj-4 has added several ostracod taxa to the list presented in a previous study, and some minor corrections have been made (Sokač 1975). In this material, 35 genera and 62 species have been recorded (Fig. 1). The analyses of the ostracod assemblages will deal specially with those general which are percentually more significant. The percentages of the genera present are presented in cyclodiagrams (Fig. 2).

Three essential ecological factors (substrate, depth, and temperature) will be considered in relation to ostracod fauna.

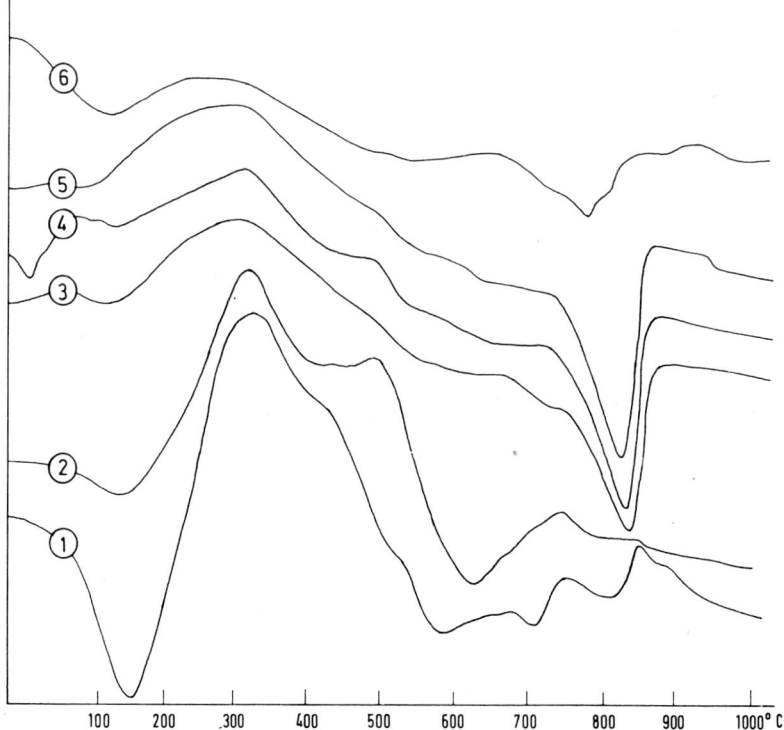

Fig. 4. DTA curves of the samples from bottom cores off the Coast of Montenegro.
1. Kotor-3, 2. Ulcinj-4, 3. Bar-1, 4. Budva-3, 5. Bar-3, 6. Budva-2.

The results of granulometric analyses (Fig. 3) have made it possible to distinguish the following types of bottoms:
a) clayey sand
b) fine sand
c) medium sand
d) coarse sand

Mineral contents have been established by differential thermal analyses. The results of these analyses indicate that the sediments analysed are not rich in clay minerals because the endothermal effects shown by the DTA curves demonstrate mainly the peaks of calcite (Fig. 4). Because of the high amounts of organisms present, it can be assumed that the calcite is partly of biogenic origin. Thus, it is not possible to distinguish between the calcite of lithogenic and biogenic origin.

The samples will be analysed according to the types of sediment.

Clayey sand

To this category belongs the sample Kotor-3 (depth 34 m, temperature 15.32°C). The granulometric composition of the sediment is: 2% fine sand, 63% silt, and 35% clay (Fig. 3). The DTA curves indicate a mixture of clay minerals (illyte, montmorillonite, caolinite), some calcite and dolomite (Fig. 4).

In ostracod fauna assemblage, the species *Cytheridea neapolitana, Pterygocythereis jonesi, Costa edwarsii, Bosquetina pectinata, Carinocythereis antiquata* and *Krithe* sp. are present in significant percentages (Fig. 2/ A, Plate 1, Fig. 1). Other characteristic, but less abundant species are *Loxoconcha agilis, Polycope reticulata, Polycope* sp., *Propontocypris setosa, Paradoxostoma simile* and *Monoceratina* sp.

Fine sand

The following samples belong to this type of bottom category:
1. Ulcinj-4 (depth 26 m, temperature 18.85° C). The sediment is composed of 3% coarse sand, 4% medium sand, 58% fine sand, 35% silt and clay (Fig. 3). The DTA curve for this sample demonstrates many effects in some temperature intervals, which indicates benthonite (illyte-montmorillonite). It is important to note that calcite is lacking in this sample of marine sediment, while in other samples it is almost always present, and the magnesite from the minerals of the carbonate group appears (temperature interval 560-680°C). This feature distinguishes this sample from the others.

In the sample Ulcinj-4, the association of ostracod consists of the following species: *Pontocythere elongata, Loxoconcha tumida, L. stellifera, Loxoconcha* sp., *Semicytherura acuminata, S. sulcata, Semicytherura* sp., *Cytheridea neapolitana, Neocytherideis fascinata, Neocytherideis* sp., *Carinocythereis antiquata, Cytheretta subradiosa, Cytheretta* sp., *Hiltermannicythere turbida, Xestoleberis communis, Pterygocythereis jonesi, Aurila speyeri, A.* aff. *convexa, Propontocypris setosa.* Species which occur in freshwater and in oligohaline environments, as for example *Limnocythere inopinata, Darwinula stevensoni, Candona neglecta* and *Ilyocypris bradyi,* are also present.

2. Budva-2 (depth 10 m, temperature 21.10°C). The sediment consists of 3% me-

Plate 1.
1. Ostracod assemblage from the bottom core of the sample Kotor-3.
2. Ostracod assemblage from the bottom core of the sample Ulcinj-4.
Photomicrographs enlarged: 20 ×. Taken by: N. Rendulić.

Plate 2.
1. Ostracod assemblage from the bottom core of the sample Budva-2.
2. Ostracod assemblage from the bottom core of the sample Bar-1.
Photomicrographs enlarged: 20×. Taken by: N. Rendulić.

dium sand, 86% fine sand, and 13% silt and clay (Fig. 3). Except for some calcite, differential thermal analysis did not show the peaks of other minerals (Fig. 4).

The ostracod fauna assemblage includes the following species: *Pontocythere elongata, Neocytherideis fascinata, Neocythereis* sp., *Loxoconcha napolitana, L. rhomboida, L. tumida, L. rubritincta, Loxoconcha* sp., *Cytheretta subradiosa, Cytheretta* sp., *Carinocythereis antiquata, Semicytherura incongruens, S. sulcata, Xestoleberis communis, Hiltermannicythere turbida* and *Urocythereis* sp.

3. Bar-1 (depth 7 m, temperature 21.15°C). The sediment contains 5% coarse sand, 20% medium sand, 69% fine sand, and 6% silt (Fig. 3). As in the previous sample, the DTA curve demonstrated only some calcite (Fig. 4).

The association of ostracod fauna contains the following forms: *Pontocythere elongata, Xestoleberis communis, X. dispar, Loxoconcha gibberosa, L. rhomboidea, Neocytherideis fascinata, Cytheretta subradiosa, Cytheretta* sp., *Aurila prasina, Aurila* sp., *Semicytherura incongruens, S. sulcata, Carinocythereis antiquata, Hiltermannicythere turbida, Cytherella* sp., *Cytheridea neapolitana, Bairdia longevaginata, Heterocythereis albomaculata, Paracytheridea* sp., *Pseudocytherura* sp., *Candona neglecta* and *Cyprideis torosa*.

The analyses of ostracod fauna revealed great similarity between the samples Budva-2 (Fig. 2/C, Plate 2, Fig. 1) and Bar-1 (Fig. 2/D, Plate 2, Fig. 2). This is revealed, in the first place, by the fact that the genus *Pontocythere* is present in the highest percentage. Then follow the genera *Neocytherideis, Loxoconcha,* and *Cytheretta*. A higher percentage of the genus *Xestoleberis* was recorded in the sample Bar-1. The similarity in contents as well as the percentage of single forms, however, is obvious. The ostracod fauna of the sample Ulcinj-4 (Fig. 2/ B, Plate 1, Fig. 2) differs markedly. Here too, the genus *Pontocythere* is present in the highest percentage the number are markedly lower than those found in the samples Budva-2, and Bar-1. At the same time, the content differs markedly. In addition to *Pontocythere*, the genera *Loxoconcha, Semicytherura, Cytheridea, Neocytherideis,* and *Carinocytherideis* are present in significant numbers.

Medium sand

The sample Budva-3 (depth 27 m, temperature 18.25°C) belongs to this category. The sand contains 8% gravel, 10% coarse sand, 17% medium sand, 46% fine sand, 19% silt (Fig. 3). Differential thermal analyses have shown a certain amount of calcite. No evidence of other minerals exists.

The ostracod fauna includes the following species: *Pontocythere elongata, Loxoconcha agilis, L. napolitana, L. rhomboiea, Carinocythereis antiquata, Aurila interpretis, A. speyeri, Semicytherura incongruens, S. inversa, Cytheridea neapolitana, Pterygocythereis jonesi, P. ceratoptera, Hiltermannicythere turbida, Leptocythere ramosa, Xestoleberis communis, X. dispar, Cytheropteron rotundatum, Cytheretta subradiosa, Cytherella alvearium, Acantocythereis histrix, Paracytheridea* sp., *Callistocythere adriatica, Propontocypris setosa, Tetracytherura angulosa, Urocythereis* sp. and *Pseudocytherura calcarata*.

It is possible to compare partly the ostracod assemblage in the sample Budva-3 (Fig. 2/ E, Plate 3, Fig. 1) with Ulcinj-4. This could be due to the similar depths

Plate 3.
1. Ostracod assemblage from the bottom core of the sample Budva-3.
2. Ostracod assemblage from the bottom core of the sample Bar-3.
Photomicrographs enlarged: 20×. Taken by: N. Rendulić.

and temperature. However, the ostracod assemblage in the sample Budva-3 is characterized by relatively higher percentages of the genera *Aurila* and *Pterygocythereis*, which are rather scarce in the sample Ulcinj-4. On the other hand, the genus Neocytherideis has not been recorded in the sample Budva-3. In addition to the genera *Aurila* and *Pterygocythereis*, the genera *Pontocythere, Loxoconcha, Carinocythereis, Semicytherura* and *Cytheridea* are numerically important.

Coarse sand

Bar-3 (depth 49 m, temperature 15.20°C) has the following granulometric content: 25% gravel, 25% coarse sand, 42% medium sand, 6% fine sand, and 2% silt (Fig. 3). This sample shows no traces of other minerals except for some calcite. (Fig. 4).

In this sample, the following ostracod species have been found: *Aurila convexa, A. prasina, A. speyeri, Xestoleberis communis, X. dispar, X. plana, Loxoconcha rhomboidea, L. paralella, Pontocythere elongata, Bairdia longevaginata, Hiltermannicythere turbida, Tetracytherura angulosa, Semicytherura incongruens, Neocytherideis fascinata, Paradaxostoma simile, Urocythereis* sp. and *Cytherelloidea sordida*.

The high percentage of the genera *Aurila, Xestoleberis,* and *Loxoconcha* is characteristic of the sample Bar-3 (Fig. 2/F, Plate 3, Fig. 2). Than follow the genera *Pontocythere, Bairdia,* and others. It is known that the genus *Aurila* is confined to medium sand substrate, and here it has frequently been found in coarse sand. The genera *Xestoleberis* and *Loxoconcha* are usually found in nearshore environments. The majority of the genera and their respective species found in this sample occur on medium sand substrates.

Conclusions

The discussion on the samples of sediments from the bottom cores off the Coast of Montenegro reported in this paper includes information on the content of the sediments, depth, temperature of the sea bottom, and association of ostracod fauna. This is the first time that differential thermal analysis has been applied to investigations of ostracod fauna in order to determine the mineral content of the sediments.

The analysis of the samples of fine sand (Ulcinj-4, Budva-2 and Bar-1) has shown that samples of the same type of sediment, of similar granulometric composition but of different mineral content, have different associations of ostracod fauna. For example, samples Budva-1 and Bar-1, with similar mineral content, have similar ostracod fauna, but differ from the sample Ulcinj-4 with a different mineral content. This indicates that the relation of the ostracod fauna to the sediment does not only depend on the granulometric composition but also on the mineral content.

Acknowledgements

I would like to take this opportunity to express my gratitude to my colleague V. Babić who did the differential thermal analyses, and to Dr. J. Tišljar for granulometric analyses. I also wish to thank N. Rendulić who picked out ostracods and took microphotographs.

References

Bonaduce, G. Ciampo, G. & Masoli, M. 1975. Distribution of Ostracoda in the Adriatic Sea. Pubbl. Staz. Zool. Napoli 40:304 pp.
Breman, E. 1975. The distribution of Ostracodes in the bottom sediments of the Adriatic Sea. Diss. Vrije Universiteit Amsterdam, 165 + XX pp.
Breman, E. 1976. Five ostracode species from Adriatic deep-sea sediments. Koninkl. Nederl. Akad. v. Wetensch., ser. B 79/1:9-11.
Sokač, A. 1975. Mikrofauna sedimenata morskog dna iz Crnogorskog primorja. Studia Marina 8: 111–120.
Uffenorde, H. 1975. Dynamics in Recent marine benthonic ostracode assemblages in the Limski Kanal (Northern Adriatic Sea). Bull. Amer. Paleont. 65 (282):13 textfig.

Author's address:

Rudarsko-geološko-naftni fakultet, Pierottijeva 6, 41000 Zagreb, Yugoslavia

Discussion

Uffenorde: Did you study the ratio between living and dead ostracods in your grab samples?
Did you have any information on the sedimentation rate at your sampling stations, to be sure that you were dealing with fairly recent faunas that can be compared with ecological data? Along the Yugoslavian coast of the open Adriatic Sea you often find very low sedimentation rates and the faunas you get in a sample are often not in an equilibrium with their Recent environment.
Sokač: I could not get any data on sedimentation rates in the investigated area. I had the samples from the bottom cores and I did not find any differences between ostracod species which could tell me that they were not from the corresponding present time environments. I never said in my paper that I was dealing with biozenose, and it is possible that these samples belong to recent and subrecent faunas.

Sixth Intern. Ostracod Symposium, Saalfelden

DELIMITATION OF THE ANTARCTIC CONVERGENCE BY CLUSTER ANALYSIS AND ORDINATION OF BENTHIC MYODOCOPID OSTRACODA

ROGER L. KAESLER, PATRICK S. MULVANY & LOUIS S. KORNICKER

Abstract

The purpose of our research was to determine the influence of the Antarctic Convergence on the distribution of benthic myodocopid Ostracoda. Eighty-eight species of myodocopid Ostracoda have been reported from benthic samples of the American Quadrant of the southern oceans. Only species that occurred in five or more samples were used in the study: *Doloria pectinata, Anarthron dithrix, Philomedes orbicularis, Scleroconcha gallardoi, Empoulsenia pentathrix,* and *Skogsbergiella macrothrix*. These six species occured at 64 stations. R-mode cluster analysis showed two groups of species that commonly co-occur: 1. *D. pectinata, A. dithrix,* and *S. macrothrix;* and 2. *P. orbicularis, S. gallardoi,* and *E. pentathrix*. Q-mode cluster analysis produced fourteen clusters, almost all of which contained stations from only one biogeographic region. Nonmetric multidimensional scaling was computed in both two and three dimensions. These ordination techniques clearly separated Antarctic from Subantarctic stations. Moreover, stations from the South Pacific and South Atlantic were differentiated to some extent. On the basis of these results, we judge that the distribution of these six benthic myodocopid Ostracoda is influenced by the position of the Antarctic convergence. Therefore, their distribution may be useful in determining the position of the convergence. If similar results should be obtained with podocopid ostracodes, whose remains are more likely to be preserved in ancient sediments, the study of podocopids from cores obtained in the vicinity of the present convergence might reveal the position of the convergence in the geological past.

Introduction

The Antarctic Convergence, marking the boundary between the Antarctic and Subantarctic regions of the southern oceans (Figure 1), is one of the major oceanographic features of the world and is the most important feature of the southern oceans. It is marked by a relatively sudden northward increase in temperature of surface water and results from the convergence of currents carrying relatively warm Subantarctic Upper Water (8° to 9°C) with those carrying cold, nutrient-enriched Antarctic Intermediate Water (3° to 7°C) (Sverdrup et al., 1942, pp. 139, 140, 941, 942). The colder, more dense Antarctic Intermediate Water sinks to a depth of 800 m to 1200 m below which the Antarctic Convergence is expected to have little effect on the benthic organisms.

The purpose of our research was to study the effect of the Antarctic Convergence on the distribution of benthic myodocopid Ostracoda from Antarctic and Subantarctic regions. We demonstrate that the myodocopid fauna of the shelf and slope is strongly affected by the descending cold Antarctic Intermediate Water. Podocopid ostracodes are much more likely than myocopid ostracodes to be found in cores because of their generally more robust carapaces. If living species of podocopids can be shown to be affected by the Antarctic Convergence simi-

235

larly to the myodocopids, the podocopids could be used to trace the position of the Antarctic Convergence relative to the cores through geological time.

Materials and methods

Kornicker (1975) has recently described 122 species of myodocopid Ostracoda from the southern oceans and has discussed their biotic zonation from extensive benthic collections between Antarctica and 35° S latitude in all quadrants of the southern oceans. For our quantitative study, it has been necessary to limit somewhat both the number of samples and the number of species. The vast size of the original data matrix made multivariate analysis of it prohibitively expensive. Moreover, many samples contained only a single species, and some species occurred in very few samples. The effect of rare species and nearly barren samples is to obscure results of cluster analysis and ordination. In order to limit the number of samples, we chose to study only the American Quadrant, which has been more thoroughly sampled than other areas. Of the 88 species that occurred in ben-

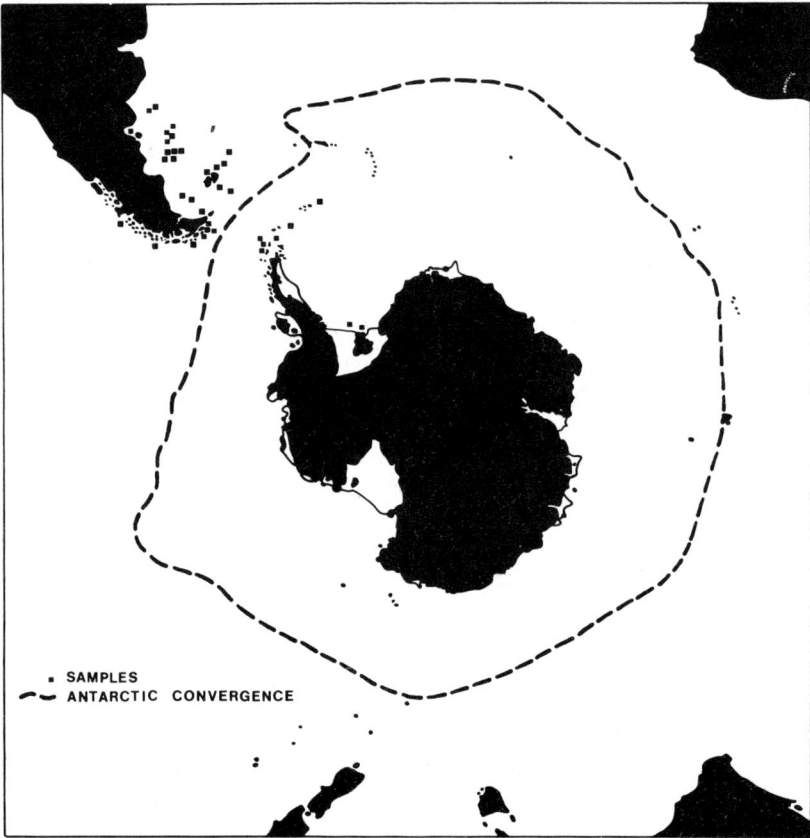

Fig. 1. Map of southern oceans showing Antarctic Convergence and location of stations from which samples used in the study were collected.

thic samples of the American Quadrant, only those that occurred in five or more samples were used in the study: *Doloria pectinata, Anarthron dithrix, Philomedes orbicularis, Scleroconcha gallardoi, Empoulsenia pentathrix,* and *Skogsbergiella macrothrix.* At least one of these six species occurred in 64 samples, the locations of which are shown in Figure 1. Note that many of the samples were quite close together and several may be represented by a single square in Figure 1. For each sample the data consisted of presence-absence information for each of the six species, and many of the samples had identical species lists and were thus perfectly similar to each other for the purposes of the study.

Two quantitative methods were chosen for analysis of the data: cluster analysis and nonmetric multidimensional scaling. Cluster analysis is a means of quantitatively developing a classification of the entities under study (Sokal & Sneath 1963, Kaesler 1966, Sneath & Sokal 1973). Q-mode cluster analysis, in an ecological or paleoecological context, develops a classification of samples on the basis of their contained species. The similar samples grouped together by cluster analysis are said to have come from the same biotope. R-mode cluster analysis is used to make a classification of species, not taxonomically, but on the basis of their presence or absence in samples from the study area. The groups of similar species are referred to as communities, portions of communities, assemblages, or biofacies.

Literally dozens of methods cluster analysis have been developed (Williams 1971). For ease of computation we have chosen an agglomerative, hierarchical clustering procedure, that is, one that 1) builds clusters by adding individual stations (or species, in R-mode cluster analysis) rather than by dividing the whole set of stations (or species) into subsets on the basis of differences between groups of stations (or species) and that 2) arranges clusters and subclusters in a hierarchy so that all stations (or species) eventually join each other at a low level of similarity.

Similarities between stations and species were expressed as Jaccard's coefficients (Kaesler 1966). The chioce of Jaccard's coefficient was entirely arbitrary. Numerous coefficients of similarity are available; Jaccard's coefficient is conceptually straightforward, easy to compute, and highly correlated with many other similarity coefficients (Kaesler et al., in preparation).

The use of cluster analysis is not without disadvantages. The clustering technique used, the unweighted pair-group method with arithmetic averages (UPGMA), constructs hyperspherical clusters regardless of whether such clusters exist in the actual configuration of the data in multidimensional space. Moreover, due to averaging of similarities during clustering, considerable distortion of similarities may result, particularly between samples in different clusters. Thus, because of the need to construct hyperspherical clusters and to average similarities, the similarities shown in the dendrogram may be quite different from the similarities in the original matrix of Jaccard's coefficients. Fortunately, it is possible to assess the amount of distortion by using the coefficient of cophenetic correlation (Sneath & Sokal 1973), which is a product-moment correlation coefficient computed between corresponding elements of the original similarity matrix and a matrix of the similarities shown on the dendrogram. A coefficient greater than

0.75 is usually taken to indicate that the dendrogram is a good representation of the similarities, although, of course, individual pairs of stations with low similarities may still have strongly distorted similarities of the dendrogram.

The advantage of ordination techniques over cluster analysis is that '... they allow one to examine a scatter diagram displaying the structure of the data without having to first assume that clusters are present' (Rohlf 1970, p. 60). Nonmetric multidimensional scaling (Kruskal 1964a,b, Green & Carmone 1970, Rohlf 1972) is an ordination method that is particularly well suited for presence-absence data that are so common in ecology and paleoecology. Moreover, it has the special advantage that '... it seems better than principal component analysis in giving balance between the large intercluster distances and the fine differences between members of a given cluster' (Sneath & Sokal 1973, p. 250).

In an ecological and paleoecological context, nonmetric multidimensional scaling is a means of representing samples as points in a scatter diagram in such a way that the distance between the points corresponds to the dissimilarities in faunal composition of the samples. One starts with an initial configuration, in our study a three-dimensional principal coordinate analysis. The distances in the initial configuration are plotted against the computed distances between all samples on what is called a Shepard diagram, and a monotone function is fitted to the scatter diagram. The departure of the plotted points from the monotone function is called the stress (Kruskal 1964a). If the stress is not sufficiently small, the initial configuration is changed slightly to improve the goodness of fit; and the stress is recomputed. The process continues until the computer reaches a predetermined criterion and stops. In our study, iterations stopped when the mean stress ratio over the previous five iterations was less than 0.999, indicating negligible improvement in the stress with successive iterations.

Results and discussion

Figure 2 shows the results of Q-mode cluster analysis of the sixty-four samples used in the study. Samples within each of the 14 labeled groups were identical in faunal composition. The cluster analysis effectively separated samples from the Antarctic region from those collected north of the Antarctic Convergence.

Cluster B, comprising all but three of the samples collected north of the Antarctic Convergence, contained thirty-six samples belonging to seven identical groups. All but three samples from north of the Subantarctic region were grouped into Cluster A, which contained twenty-one samples belonging to four identical groups. All but one of the identical groups contained samples from both the Subarctic region as well as the South Atlantic or South Pacific oceans. The deepest sample in Cluster B was from the South Pacific Ocean west of Chile and was collected at a depth of more than 1200 m. Most of the other samples were from depths less than 200 m, and all others were from above the layer of Antarctic Intermediate Water extending north from the Antarctic Convergence at a depth of 800 to 1200 m.

Cluster C contained twenty-eight samples belonging to seven identical groups. Only three Subantarctic samples joined this cluster, all of them in the same

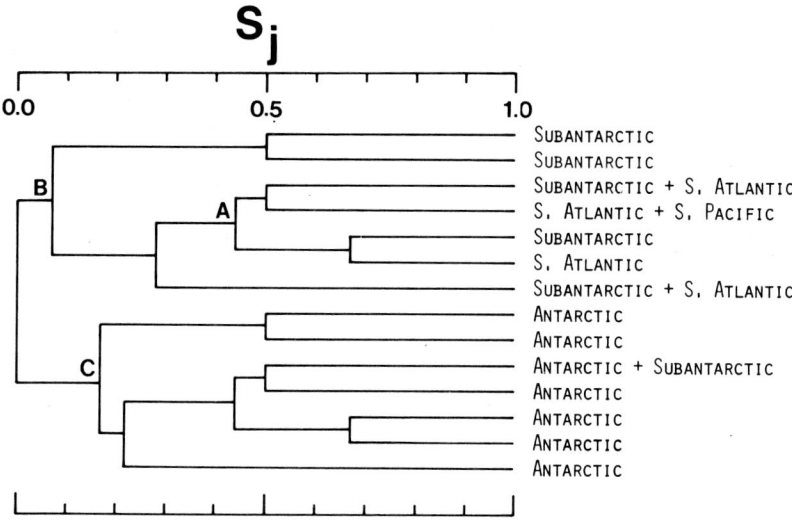

Fig. 2. Q-mode cluster analysis (UPGMA) of sixty-four samples belonging to 14 groups; samples within each of the fourteen labeled groups are identical; $r_{CC} = 0.942$.

identical group. Two of those samples were collected at depths greater than 1000 m and were, thus, more a part of the Antarctic region than of the Subantarctic region because of the cold temperature of the water at that depth. Most other samples were collected at depths of less than 200 m. The substructure within Cluster C is not readily interpretable because samples from several subregions of the Antarctic region are often grouped together.

The results of Q-mode cluster analysis indicate that the distributions of the six species of myodocopid Ostracoda used in the study were strongly influenced by the Antarctic Convergence. Only one Subantarctic station was misclassified. The lack of interpretable substructure within the cluster of Antarctic samples and the mixing of Subantarctic samples with those from more northerly localities suggest that the ostracodes and their biotopes are very broadly distributed within the Antarctic and Subantarctic regions. This result agrees with the interpretation of Kornicker (1975), who found biofacies to be broadly distributed in a circumpolar fashion in the Antarctic region and more narrowly in Subantarctic and more northerly regions.

Figure 3 shows the results of two-dimensional nonmetric multidimensional scaling. Here the fourteen groups of identical samples are ordinated rather than being classified as in the previous figure. The ordination shows samples from the Antarctic region grouped on the right side of the scatter diagram. The sample group that contains both Antarctic and Subantarctic samples is the same group that clustered with Antarctic samples in Figure 2, Cluster C. Recall that two of the Subantarctic samples in that sample group were from deep, Antarctic Intermediate Water north of the Antarctic Convergence. The otherwise clearcut separation of Antarctic from Subantarctic sample groups indicates the effect of the Antarctic Convergence on the distribution of the ostracodes. The strong overlap

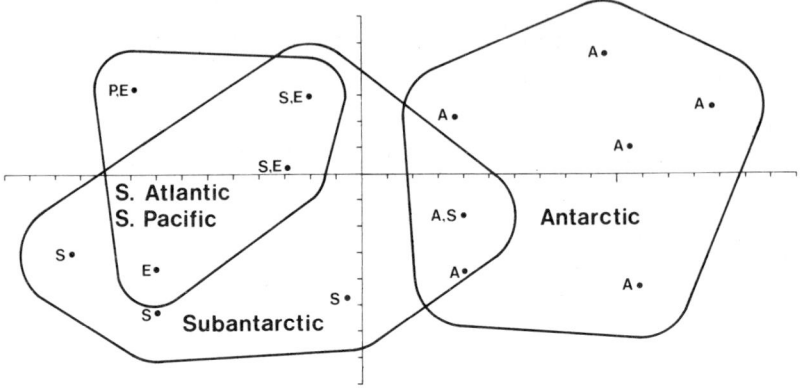

Fig. 3. Results of two-dimensional nonmetric multidimensional scaling of sixty-four samples belonging to fourteen groups of identical samples; A indicates Antarctic region; S, Subantarctic; P, South Pacific Ocean; E, South Atlantic Ocean. Stress equals 0.297.

of Subantarctic samples with those from more northerly localities is consistent with the lack of an oceanographic northern boundary to the Subantarctic region.

The results of three-dimensional nonmetric multidimensional scaling (Figure 4) further emphasize the similarity among samples from the South Atlantic and South Pacific oceans and the dissimilarity of most Antarctic samples from the mixed Antarctic-Subantarctic group of samples.

Figure 5 shows the results of R-mode cluster analysis of a matrix of Jaccard's coefficients that expressed the similarity of distribution of the six species studied. The similarities are, in general, very low. The two species in Cluster A, *Anarthron dithrix* and *Skogsbergiella macrothrix*, have the closest similarity of distribution and are both much more commomn in the South Atlantic and South Pacific oceans than in either the Subantarctic or Antarctic regions. Cluster B includes the three species most commonly found north of the Antarctic Convergence, and

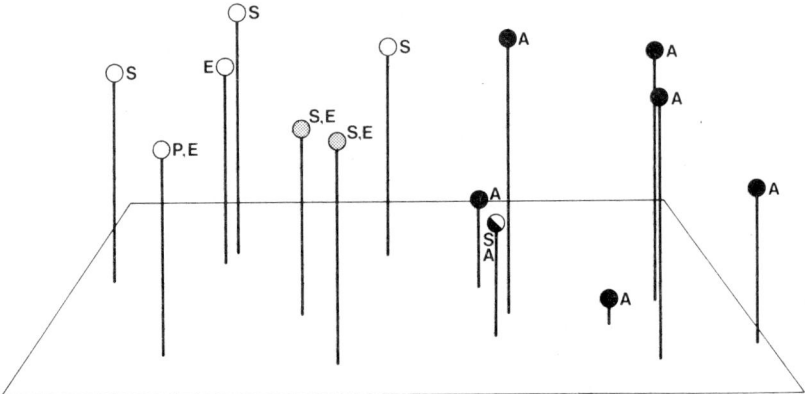

Fig. 4. Results of three-dimensional nonmetric multidimensional scaling of sixty-four samples belonging to fourteen groups of identical samples; A indicates Antarctic region; S, Subantarctic; P, South Pacific Ocean; E, South Atlantic Ocean. Stress equals 0.236.

Table 1. Species studies, number of samples in which each occurred, and proportion of occurrences from the South Atlantic and South Pacific oceans (S.A.), Subantarctic region (SUB), and Antarctic region (ANT).

Species	N	S.A.	SUB	ANT
Skogsbergiella macrothrix	21	0.81	0.19	
Anarthron dithrix	15	0.80	0.20	
Doloria pectinata	15	0.13	0.87	
Empoulsenia pentathrix	10		0.30	0.70
Scleroconcha gallardoi	9			1.00
Philomedes orbicularis	19			1.00

Cluster C contains the typically Antarctic species. Table 1 shows the number of samples in which each species occured and the proportion of its occurences in each of the three biogeographic regions. Note that samples from the South Atlantic and South Pacific oceans comprise 0.81 and 0.80, respectively, of occurrences of *S. macrothrix* and *A. dithrix,* with joint occurrences accounting for their relatively high similarity in Figure 5. Eighty-seven percent of the occurrences of *Doloria pectinata* were from Subantarctic samples, but *Scleroconcha gallardoi* and *Philomedes orbicularis,* on the other hand, were limited to samples from the Antarctic region. Their very low similarity to each other suggests that they are unlikely to be found together in spite of their strong preference for Antarctic waters. Neither water depth nor geographic location can account for the differences of distribution of the two species.

Kornicker (1975) recognized a *Skogsbergiella-Empoulsenia* biofaces in the Antarctic region completely surrounding Antarctica and extending north at least as far as 40° S latitude on both the east and west sides of South America. In our study we have never found *Skogsbergiella* and *Empoulsenia* together, but Kornicker was dealing with four species of *Skogsbergiella* plus a number of specimens of indeterminate species and five species of *Empoulsenia* plus specimens of indeterminate species. Our result does not contradict his because our study was based

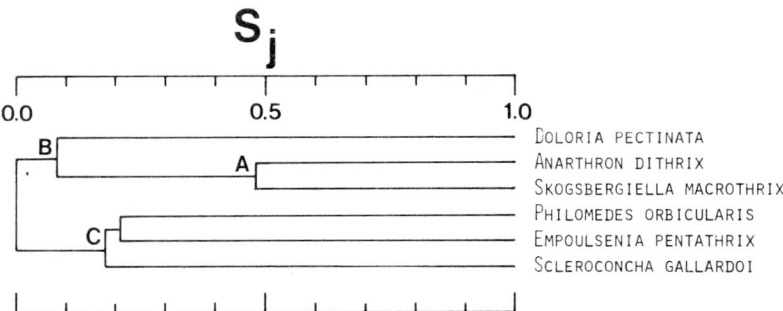

Fig. 5. R-mode cluster analysis (UPGMA) of six species of myodocopid Ostracoda based on their presence or absence in 64 samples from the American Quadrant of the southern oceans; $r_{CC} = 0.999$.

on the more common species and included only species from the American Quadrant.

With six species of myodocopids in the study, sixty-three different assemblages of Ostracoda were possible ($2^6 = 64$ minus 1 for the 'barren assemblage'). In sixty-four samples, only fourteen of these possible assemblages were realized, ranging in frequency from one to twelve samples. Table 2 shows the frequency of occurrence of the fourteen realized assemblages. The presence of five of the six possible monospecific assemblages does not necessarily imply inadequate sampling because the species list included three dominantly Antarctic ones and three from Subantarctic or more northerly regions. This increases the likelihood of finding mono- or dispecific assemblages and greatly reduces the likelihood of finding assemblages with more than three species. Indeed, none of the latter were found. The presence of five dispecific assemblages out of fifteen possible can be explained similarly. Since the ostracodes came from two biogeographic regions rather than one, only six dispecific assemblages are possible because of the unlikelihood of Antarctic species occurring in samples from north of the Antarctic Convergence.

Table 2. Species composition of the fourteen realized assemblages and the number of samples that contained each assemblage. *Skogsbergiella macrothrix*, Sk; *Anarthron dithrix*, An; *Doloria pectinata*, Do; *Empoulsenia pentathrix*, Em; *Scleroconcha gallardoi*, Sc; and *Philomedes orbicularis*, Ph;.

Number of samples	Sk	An	Do	Em	Sc	Ph
1	×	×	×			
1		×	×			
1				×	×	
2				×	×	×
2					×	×
2	×		×			
3				×		×
3		×				
4					×	
4				×		
7	×					
11			×			
11	×	×				
12						×

Conclusions

Q- and R-mode cluster analysis and nonmetric multidimensional scaling have been used to demonstrate the effect of the Antarctic Convergence on the distribution of six common species of myodocopid Ostracoda from sixty-four benthic samples from the shelf and slope of the American Quadrant of the southern oceans. The Antarctic Convergence has a pronounced effect on the distribution of the species studied above a depth of about 800 m, but below that depth the Antarctic Intermediate Water and perhaps its Antarctic fauna underlie the Subantarctic region. Two species were restricted to samples from the Antarctic region, south of the Antarctic Convergence, and two species had at least eighty percent of their occurrences in samples from the South Atlantic or South Pacific oceans. Three species were found exclusively north of the Antarctic Convergence, one of them, *Doloria pectinata,* with eighty-seven percent of its occurrences in samples from the Subantarctic region.

The myodocopid species studied could clearly be used to locate the position of the Antarctic Convergence with respect to the stations at which the samples were collected. Unfortunately, myodocopid ostracodes are not usually well preserved in cores. However, if a few characteristic species of podocopid ostracodes could be shown to be affected by the Antarctic Convergence similarly to the myodocopids, specimens from cores could be used to trace the position of the Antarctic Convergence through the geological past.

Acknowledgments

We are grateful to R. M. Peterson for his critical review of the manuscript and to Dr. Neal Roth, Smithsonian Institution, who was consulted during the initial phase of the work. All computations were done at The University of Kansas Computation Center using the NT-SYS system of computer programs written by F. James Rohlf and his associates. Research was supported by the Wallace E. Pratt Research Fund provided to the Department of Geology, The University of Kansas, by the Exxon Corporation.

References

Green, P.E. & Carmone, F.J. 1970. Multidimensional scaling and related techiques in marketing analysis.Allyn and Bacon, Inc., Boston 203 pp.
Kaesler, R.L. 1966. Quantitative re-evaluation of ecology and distribution of Recent Foraminifera and Ostracoda of Todos Santos Bay, Baja, California, Mexico, Univ. Kansas Paleontological Contr., Paper 10:1 1–50.
Kaesler, R.L., Crossman, J.S., Toole, T.D. & Urban, R.D. in prep. An evaluation of some similarity and distance coefficients used in cluster analysis.
Kornicker, L.S. 1975. Antarctic Ostracoda (Myodocopina). Smithsonian Contrib. to Zoology 163: 1–720.
Kruskal, J.B. 1964a. Multidimensional scaling by optimizing goodness of fit to a nonmetric hypothesis. Psychometrika 29:1-27.
Kruskal, J. B. 1964b. Nonmetric multidimensional scaling: a numerical method. Psychometrika 29: 115–129.
Rohlf, F.J. 1970. Adaptive hierarchical clustering schemes. Systematic Zoology 19:58–82.
Rohlf, F.J. 1972. An empirical comparison of three ordination techniques in numerical taxonomy. Systematic Zoology 21:271–280.
Sneath, P.H.A. & Sokal, R.R. 1973. Numerical taxonomy. W.H. Freeman & Company, San Francisco, 573 pp.

Sokal, R.R. & Sneath, P.H.A. 1963. Principles of numerical taxonomy. W.H. Freeman & Company, San Francisco, 359 pp.
Sverdrup, H.U., Johnson, M.W. & Fleming, R.H. 1942. The Oceans. Their physics, chemistry, and general biology. Prentice-Hall, Inc., Englewood Cliffs, New Jersey, 1087 pp.
Williams, W.T. 1971. Principles of clustering. Annual Review of Ecology and Systematics 2:303–326.

Authors' addresses:

R.L. Kaesler & P.S. Mulvany, Dept. of Geology and Museum of Invertebrate Paleontology, The University of Kansas, Lawrence, Kansas 66045, USA.

L.S. Kornicker, National Museum of Natural History, Smithsonian Institution, Washington, D.C. 20560, USA.

Discussion

Sohn: How far back in geologic time do you expect to find the Antarctic Convergence?

Kaesler: Greg, I don't know the answer, but I understand that study of cores has shown that the Antarctic Convergence has been in existence for at least 20 million years.

Sixth Intern. Ostracod Symposium, Saalfelden

LIFE HISTORY PATTERNS OF THE MYODOCOPID OSTRACOD EUPHILOMEDES PRODUCTA POULSEN, 1962

JAMES H. BAKER

Abstract

Euphilomedes producta (Poulsen 1962) occurs from shallow waters off British Columbia, Canada, to the deeper waters of southern California, United States. Off southern California, *E. producta* is the numerical dominant in a recurrent group of six ostracod species.

Data from 12 monthly samples indicate that *E. producta* juveniles pass through five instar stages in a two year period. Reproduction is continuous throughout the year, but shows a distinct 'periodic' high and low.

Using a modified version of the egg-ratio method to determine birth rate ($\bar{b}$), death rate ($\bar{d}$), and per capita rate of change (r), predatory pressure appeared to cause major population lows with lack of food causing small reductions. Statistical comparison of adult males and females with Instar V males and females appeared to substantiate postadult molting.

Introduction

In a recurrent groups analysis of benthic myodocopid ostracods of southern California, Baker (in press) identified two recurrent groups and a group of ostracods not belonging to any recurrent group. *Euphilomedes producta* (Poulsen 1962) was the numerical dominant in Recurrent Group No. 1 of six ostracod species.

Numbers of individuals/m^2 regressed on depth, water temperature, salinity, percent calcium carbonate (sediment), percent organic nitrogen (sediment), and median diameter of the sediment failed to sufficiently explain the distributional patterns of *E. producta* and the other species. Therefore, using the relative numbers of instars and adults and the egg-ratio method of Edmondson (1960), the effects of predation and food availability on the distribution of *E. producta* was examined. Observations on the life history and postadult molting are also included.

Distribution

Euphilomedes producta has been reported from Departure Bay, Cape Keppel, and Brandon Island, British Columbia (Lucas 1931, Poulsen 1962); Puget Sound, Washington (Lie 1968); and from Pt. Conception to San Diego, California (Baker 1975). It has been found at a depth range of 13 to 401 m in sediments of a sand-silt mixture. In southern California, the temperature range was 9.4 to 19.2° C and salinity was 24.7 to 34.2‰. Ranges of other environmental parameters off southern California are also listed in Baker (1975, p. 98).

Percent calcium carbonate (sediment) was highly correlated with *E. producta*, but only accounted for 2.9% of the variability in its abundance (Baker 1975).

Based upon the depth distribution of 16,975 specimens from 224 stations, *E. producta* occurred in greatest numbers at 60 to 90 m where the temperature of the bottom sediments varied from 10.5 to 12.0°C. In Puget Sound, *E. producta* occurred in greatest numbers at temperatures of 10.5 to almost 12.0°C (Lie 1968). However, in southern California, temperature was not correlated to variability in abundance of *E. producta*.

Studies of Lie (1968) and Baker (in press) indicated a seasonal trend in numbers with maxima in July to August in Puget Sound and maxima in February and September in southern California, respectively.

Study area

The area under study is the narrow shelf from San Diego to Pt. Conception, California, ranging in width from 1.6 to 19.2 km and averaging about 7.2 km wide (Allan Hancock Foundation 1965) (AHF). The shelf presents a complex environment with respect to both the sea floor itself and to its effects on the overlying water. More detailed description of the study area and the physical and chemical properties may be found in Emery (1960) and AHF (1965).

Methods and materials

This study is an outgrowth of a much larger analysis of the benthic myodocopid ostracods from San Diego to Pt. Conception, California. Methods of sample collection and preparation are presented in detail in AHF (1965) and Baker (1975).

In 1957, 107 mainland shelf samples were collected that contained *Euphilomedes producta* and where each month was represented by at least one sample. From this group, 12 samples, one for each month, were chosen on the basis of the following criteria: (1) a depth of about 60 meters, (2) closely spaced to each other, and (3) containing at least 100 individuals of *E. producta* in each sample. Two stations, 5108 (June) and 5325 (October), had less than 100 individuals; 20 and 27, respectively. One other station in June was from a depth of 15.2 m and had only three individuals. In October, the other two stations had only four and one individuals, respectively. This indicates that abundances were low during these months. At a depth of 60 m, temperature and salinity exhibit less variability. For the samples chosen, salinity varied from 33.53 to 33.89‰ and the temperature range was 9.8 to 12.9°C.

Each sample containing greater than 100 individuals was placed in a square petri dish divided into 36 squares. The specimens were stirred several times in both directions in order to obtain an equal distribution of both sexes and all size classes. Then using a table of random numbers, each square was assigned a number from 1 to 36. Beginning with square 1, all individuals were picked from successive squares until 100 had been chosen. If 100 individuals were chosen before all specimens had been picked from a square, the remainder in that square would be included.

These chosen individuals were then measured (length and height) and sexed. Indications of food material in gut, incidence of parasitism or commensalism, and in process of molting was also noted. Adult females were sorted into categories based upon the presence, state of development, and location of eggs.

Developmental history

Using the length-height plotting method of Kesling (1951), the larvae key of Kornicker (1969), and Przibram's theoretical growth factor for a linear dimension (1.26), *Euphilomedes producta* was determined to have five instar stages, of which only the last three were found in these samples. Elofson (1941, 1969) found *Philomedes globosa* to have only five juvenile stages and Kornicker (1969) concluded that the Philomedinidae as a group must also undergo five juvenile instar stages. Juvenile *E. producta* comprised 49.3% of the total individuals counted with Instars III, IV, and V represented by 0.9%, 19.9%, and 28.5% of the total, respectively. Earlier instars were lost because the samples had been washed over a 0.7 mm screen.

Adult *E. producta* were divided into males and females, and females were further separated into categories on the basis of egg development and location. The largest category was ovigerous females (O) with 20.5% of the total. Next were females with unextruded eggs (P) with 10.9% and all other categories totaled 10.0% of the total individuals counted. Adult females comprised 42.4% of the total and adult males only 8.2%.

Ovigerous females were found with three types of eggs. The first consisted of what appeared to be eggs that had just been laid and no differentiation of the embryo could be noted. Eggs of the second group had begun producing limb buds, with the first and second antennae showing the greatest development. Both the first and second groups had an average size of 0.28 mm. The third groups of eggs were larger, average of 0.35 mm, and the first and second limbs had become more segmented. These embryos were still enclosed in the egg membrane and appeared to be very near the time of release.

The average number of eggs per ovigerous female varied from 14 in March to 16 in July with an overall average of 15. Elofson (1941, 1969) found *P. globosa* had 8 to 16 eggs in the brood pouch and Kornicker (1975) found some Antarctic Philomedidae to produce up to 32 eggs per clutch.

Measured adult female *E. producta* had a length of 1.64 mm or greater. Two adult (based upon the 2nd antennae endopodite) male *E. producta* with lengths of 1.54 and 1.68 mm, respectively, were found. However, males generally reached maturity at a length of 1.75 mm. Females of Instar V varied in length from 1.36 to 1.63 mm, with Instar V males as long as 1.68 mm. Instars III through IV had a length range of 0.82 to 1.00 and 1.01 to 1.35 mm, respectively. Thus, Instars I and II, if present, would have a length range of 0.52 to 0.64 and 0.65 to 0.81 mm, respectively.

The sex ratio of *E. producta* Instars III, IV, and V was found to be essentially 1:1. Kornicker (1969) found the average percent of males in Instar IV of *Spinacopia sandersi* Kornicker, 1969, to be 54.5 or essentially one male to one female. One male for every five females was the ratio for adult *E. producta*. Scarcity of adult males is probably a combination of their shorter life spans and longer forays

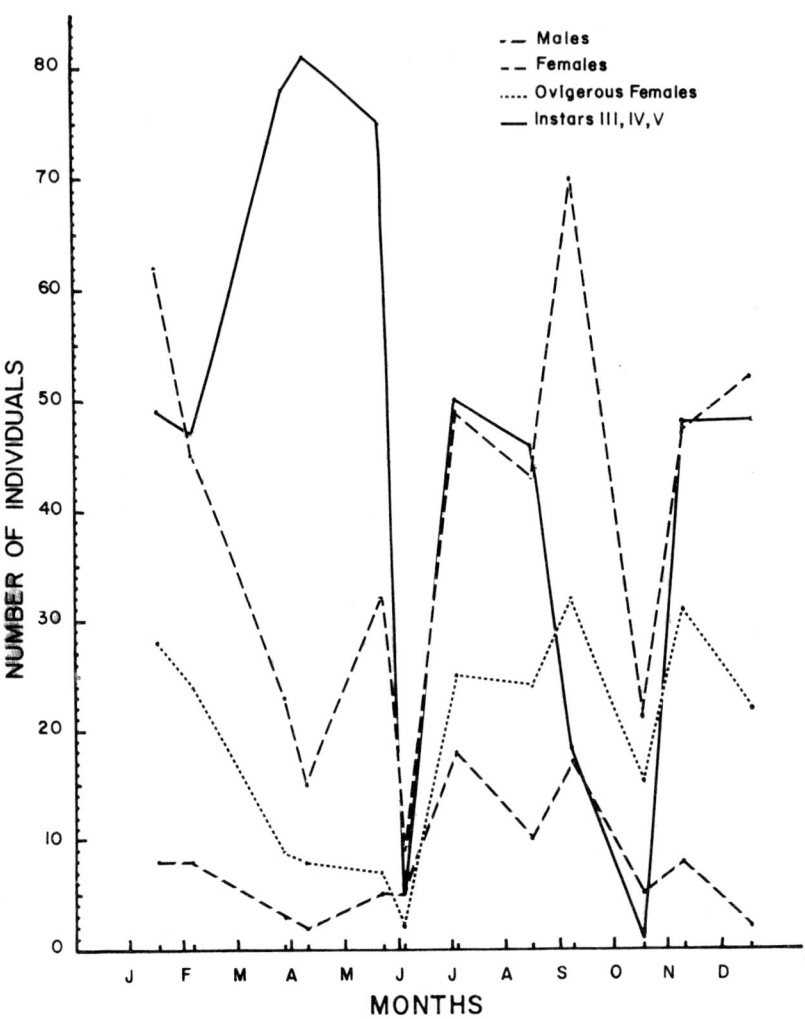

Fig. 1. Monthly distribution of adult males, total adult females, ovigerous females, and total number of juveniles of *Euphilomedes producta*.

into the plankton for purposes of reproduction (Müller 1894, Skogsberg 1920, Elofson 1941, 1969, Baker 1975).

Using Wilcoxon's signed-ranks test, the differences in numbers of males and females of Instars IV ($Z = -1.30$, $K = 9$, $p + 0.19$) and V ($Z = -0.61$, $K = 10$, $p = 0.54$) were found not to be significant. This supports the validity of the 1:1 sex ratio. Numbers of adult females were found to be significantly different from numbers of adult males ($Z = -3.06$, $K = 12$, $p < 0.001$), which emphasizes that something different is affecting adult males.

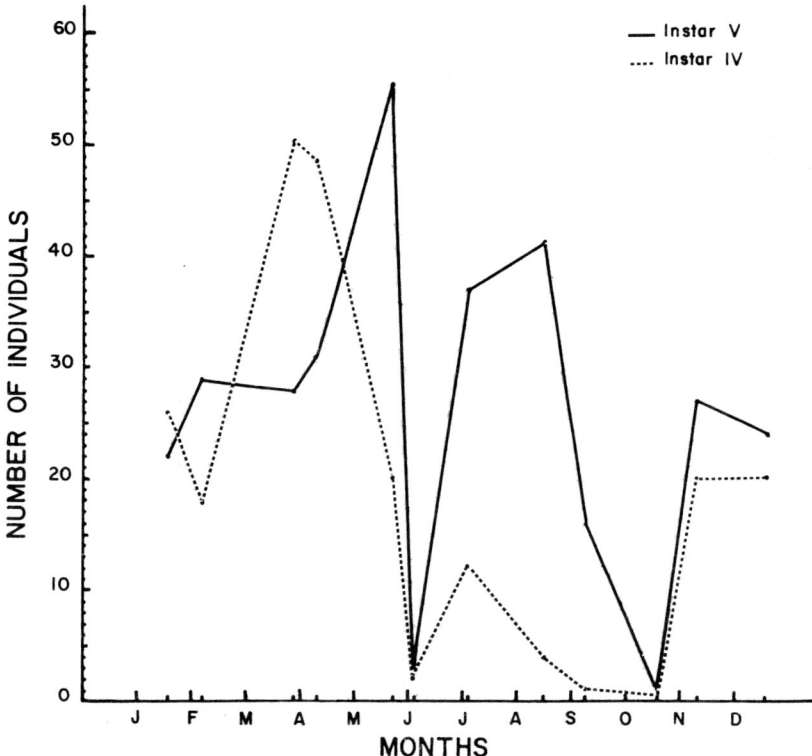

Fig. 2. Monthly distribution of Instars IV and V of *Euphilomedes producta*.

Seasonal distribution

The monthly distribution of adult males, total adult females, ovigerous females, and total number of juveniles is illustrated in Figure 1. Based upon the distribution of ovigerous females, reproduction appears to occur year round. However, a periodic high and low is also exhibited. If the total number of juveniles are divided into Instars IV and V and plotted separately, a distinct lag in peaks is noted (Figure 2). Instar V peaked later than Instar IV, which is to be expected.

Thus, *Euphiomedes producta* appears to reproduce all year long, but with a distinct periodicity. Lie (1968) found that in Puget Sound the maximum was in July to August while the high in southern California extends from July to the following January. Kornicker (1975) found that some of the Antarctic species also reproduce year round.

Juveniles of *E. producta* appeared to take approximately two years from time of leaving the brood pouch until reaching maturity. Figure 2 indicates that both Instars IV and V have a duration of about 120 days each. Therefore, Instars I, II, and III probably developed during the previous year. Elofson (1941, 1969) found that *Philomedes globosa* took 157 days each to complete Instars IV and V. Instars

I, II, and III took 150 to 170, >50, and >40 days, respectively. Once maturity is reached, both male and female *E. producta* join the breeding population which began to increase in July. Hulings (1969) found *Parasterope pollex* Kornicker 1967 (Bowman & Kornicker 1967), to live for only one year, including instar and adult stages.

Examination of ovigerous females showed the presence of eggs in the ovaries indicating that the females reproduce at least twice. An attempt was made to determine if the female had to mate again to produce a second batch of eggs or if sperm were retained from the first mating. However, neither a seminal receptacle nor sperm could be located by examination with a light microscope. Sectioning, which was not feasible at this time, may answer the question. Elofson (1941, 1969) and Kornicker (1975) found some ostracods produced at least two broods, but Hulings (1969) found *P. pollex* to reproduce only once.

Population analysis

Often life-table data are difficult or impossible to collect, and field data do not easily provide population statistics such as per capita rate of change (r), birth rate ($\hat{b}$), and death rate ($\hat{d}$). The egg-ratio method of Edmondson (1960), modified by van Dolah, Shapiro & Rees (1975), was used to delineate these statistics for the sexually producing amphipod, *Gammarus palustris* having a sex ratio of 1:1.

Euphilomedes producta reproduces sexually, has a 1:1 sex ratio, produces eggs, and broods the eggs for a period of time as does *G. palustris*. Therefore, the egg-ratio method appeared applicable to *E. producta*. In order to use this method, certain assumptions had to be made. Extrapolating from the work of Elofson (1941, 1969) on *Philomedes globosa* and work of van Dolah et al. (1975), *E. producta* was assumed to have an egg development time of 158 days. Because Instars I, II, and a portion of III are missing from the samples, only the adult females, juvenile females of Instars IV and V, and one-half of the eggs are used to represent the total population. Even using only those individuals represented in the samples without considering females of other stages, the general trend of the population as related to birth rate, death rate, and per capita rate of change can be determined.

Population parameters calculated for *E. producta* are illustrated on a monthly basis in Figure 3. Average brood size is believed to be correlated to food availability with low values indicating food limitation and high values the opposite (Hall 1964, van Dolah et al. 1975). Van Dolah et al. (1975) further observed that $\hat{b}$ and $\hat{d}$ generally rose and fell with nearly identical values when predation pressure was affecting the population. Food limitation appeared to cause $\hat{b}$ and $\hat{d}$ to progress with distinctly different values over time.

The decrease in adult numbers in January (see Figure 1) appears to be the result of predatory pressure. A corresponding high in numbers of juveniles and average brood size (Figure 3) does not indicate food to be limiting. Brooks and Dodson (1965b), Cooper (1965), Archibald (1975), and others have demonstrated that many fish predators are size-selective for large body sizes. Therefore, predators would select adults over juveniles.

The adult numbers continue to drop and reach the low point of the year in

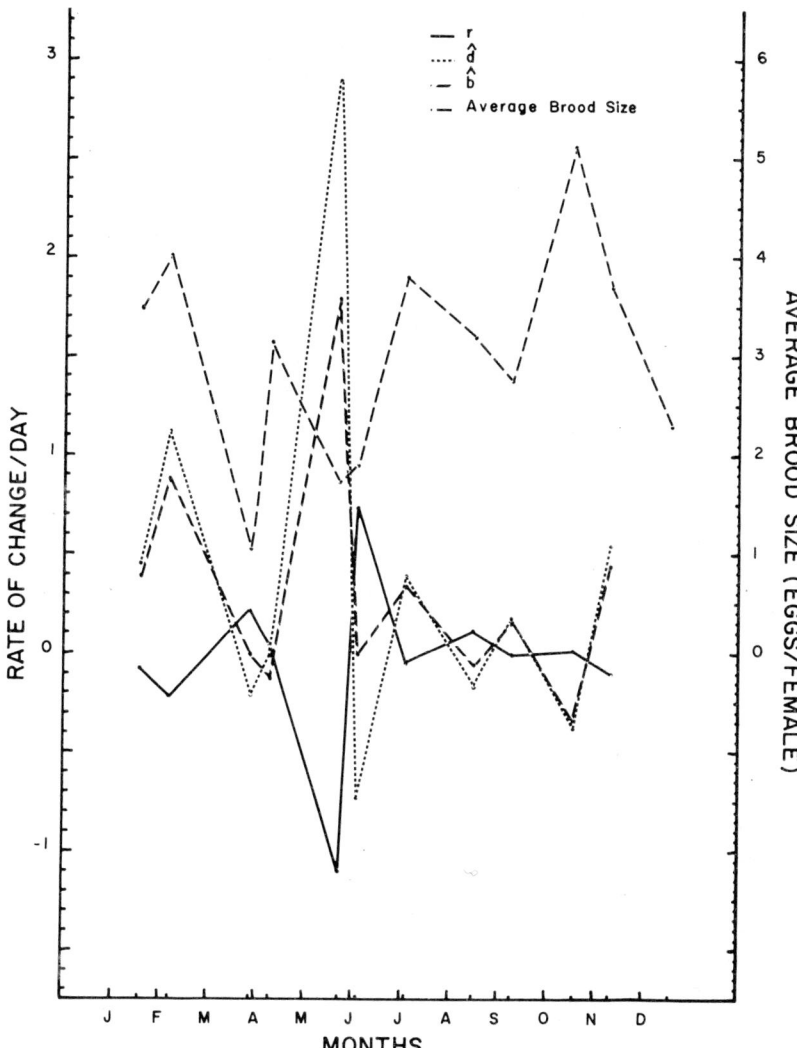

Fig. 3. Monthly distribution of the population parameters for *Euphilomedes producta*.

April. Average brood size also drops which tends to imply that food supply becomes limiting and predatory pressure decreases as indicated by the tendency of $\hat{b}$ and $\hat{d}$ to be opposing. Baker (1975) found *E. producta* to be a detritivore and, as such, food should seldom become limiting.

Death rate reaches the highest point in May followed closeley by a high birth rate (Figures 1 and 3). Average brood size also decreases. Numbers of adults and juveniles both decrease drastically. Even though the decrease in June may be a function of small sample size, the decrease appears to be real.

Adult numbers then increase during the remainder of the year. Other prey would become abundant during this time, thus relieving the pressure from *E. producta*. However, predatory pressure does appear to cause the population decreases in August and October. Again, the October low may be a function of small sample size, but the population decrease is real. Baker (1975) discusses known predators of ostracods and possible predators in the study area.

Molting

The latest study concerning molting is that of Kornicker (1975). He concluded that if postadult molting occurred, it was rare.

Males reached maturity at a length of 1.75 mm with the largest being 2.1 mm in length. Two adult male *Euphilomedes producta* were found which measured 1.54 and 1.68 mm, respectively. An explanation for the maturity of these two individuals at a smaller length is not readily apparent. Adult females measured 1.63 to 2.03 mm in length. Following the discussion of Kornicker (1975, p. 687), dispersion of carapace lengths within the adult females was compared to dispersion in Instar V females using the *t*-distribution. Dispersion within adult females was highly significant ($p \leq 0.001$) and dispersion within Instar V females was not significant. Dispersion within adult males was also highly significant ($p < 0.001$) and dispersion within Instar V males also was not significant.

Again, using the Wilcoxon's signed-ranks test, total numbers of individuals of Instar IV were compared to Instar V for each of the 12 months. Instar V was also compared to total adults. The difference between Instar IV and V was not significant ($Z = -1.49, K = 12, p = 0.136$). However, the difference between Instar V and adults was significant ($Z = -2.43, K = 12, p < 0.02$). This appears to lend support to the contention that the adult group consists of several size classes.

Appendages of adults were checked to see if any were undergoing molting as mentioned by Kornicker (1975) and none were found. However, there appears to be more than one size class within adult males and females indicating the occurrence of postadult molting in *E. producta*.

Conclusions

1. *Euphilomedes producta* undergoes five instar stages and reaches maturity in about two years after leaving the brood pouch.
2. Reproduction appears to occur all year long, but does show a periodic high and low.
3. Predatory pressure appears to be the primary factor affecting the distribution of *E. producta* with food availability causing minor limitations.
4. Dispersion within adult male and female lengths was significantly different than within Instar V lengths, implying postadult molting in both males and females.

Acknowledgments

Material for this study was kindly provided by Dr. John S. Garth, Allan Hancock Foundation, University of Southern California. Ms. Kay T. Kimball helped with statistical procedures and Mr. J. Thomas Ivy offered many helpful suggestions. Drs. Louis S. Kornicker and C. A. Bedinger, Jr., Mr. J. Thomas Ivy, and Ms. Kay T. Kimball critically read the manuscript. Mrs. Jeanne N. Rossman typed the manuscript.

References

Allan Hancock Foundation. 1965. An oceanographic and biological survey of the southern California mainland shelf. Calif. St. Wat. Qual. Control Bd. 27:1-232.

Archibald, C. P. 1975. Experimental observations on the effects of predation by goldfish (Carassius auratus) on the zooplankton of a small saline lake. J. Fish. Res. Board Can. 32 (9):1589-1594.

Baker, J. H. 1975. Distributions, ecology, and life histories of selected Cypridinacea (Myodocopida, Ostracoda) from the southern California mainland shelf. Doctoral Dissertation, University of Houston XVII + 185 pp.

Baker, J. H. In press. Distributional, ecological, and life history patterns of myodocopid Ostracoda (Myodocopina: Cypridinacea) from southern California: An introduction.

Bowman, T. E. & L. S. Kornicker. 1967. Two new crustaceans: the parasitic copepod Sphaeronellopsis monothrix (Choniostomatidae) and its myodocopid ostracod host Parasterope pollex (Cylindroleberidae) from New Zealand coast. Proc. U. S. natn. Mus. 123 (3613):1-28.

Brooks, J. L. & S. I. Dodson. 1965. Predation, body size and composition of plankton. Science 150 (3692):28-35.

Cooper, W. E. 1965. Dynamics and production of a natural population of a fresh-water amphipod, Hyalella azteca. Ecol. Monogr. 35(4):377-394.

Cushman, J. A. 1906. Marine Ostracoda of Vineyard Sound and adjacent waters. Proc. Boston Soc. nat. Hist. 32:359-385.

Edmondson, W. T. 1960. Reproductive rates of rotifers in natural populations. Memorie Ist. ital. Idrobiol. 12:21-77.

Elofson, O. 1941. Zur Kenntnis der marinen Ostracoden Schwedens mit besondered Berucksichtizung des Skageraks. Zool. Bidr. Upps. 19:215-534.

Elofson O. 1969. Marine Ostracoda of Sweden with special consideration of the Skagerrak. English translation. Israel Program for Scientific Translations Ltd., Jerusalem. Translated by A. Mercado. U. S. Dept. of Commerce, Clearing-House for Federal Scientific and Technical Information, Springfield, Virginia. V + 285 pp.

Emery, K. O. 1960. The sea off southern California, A modern habitat of petroleum. John Wiley and Sons, Inc., New York. IX + 366 pp.

Hall, D. J. 1964. An experimental approach to the dynamics of a natural population of Daphnia galeata mendotae. Ecology 45 (1):94-112.

Hulings, N. C. 1969. The ecology of the marine Ostracoda of Hadley Harbor, Massachusetts, with special reference to the life history of Parasterope pollex Kornicker, 1967, pp. 412-422. In: J. W. Neale (ed.), The taxonomy, morphology, and ecology of Recent Ostracoda. Oliver and Boyd, Edinburgh. IX + 553 pp.

Kesling, R. V. 1951. The morphology of ostracod molt stages. Illinois Biol. Monogr. 21(1-3):1-325.

Kornicker, L. S. 1967. A study of three species of Sarsiella (Ostracoda: Myodocopa). Proc. U. S. natn. Mus. 122(3594):1-46.

Kornicker, L. S. 1969. Morphology, ontogeny, and interspecific variation of Spinacopia a new genus of myodocopid ostracod (Sarsiellidae). Smithson. Contrib. Zool. 8:1-50.

Kornicker, L. S. 1975. Antarctic Ostracoda (Myodocopina). Smithson. Contrib. Zool. 163:1-720.
Lie, W. 1968. A quantitative study of benthic infauna in Puget Sound, Washington, U.S.A., in 1963-1964. Fiskdir. Skr., Sene Havunelersokelser 14(5):1-556.
Lucas, V. Z. 1931. Some Ostracoda of the Vancouver Island region. Contrb. Can. Biol. Fish. 6(17, N.S.):397-416.
Müller, G. W. 1894. Die Ostracoden des Golfes von Neapel und der angrenzenden Meeresabschnitte. Fauna Flora Golf. Neapel 21:1-404.
Poulsen, E. M. 1962. Ostracoda-Myodocopa. Part. 1. Cypridiniformes-Cypridinidae. Dana Rep. 57:1-414.
Skogsberg, T. 1920. Studies on marine ostracods. Part I (Cypridinids, Halocyprids, and Polycopids). Almquist and Wiksells Boktryckeri, Upsala. 784 pp.
Van Dolah, R. F., L. E. Shapiro & C. P. Rees. 1975. Analysis of an intertidal population of the amphipod Gammarus palustris using a modified version of the egg-ratio method. Mar. Biol. 33(4):323-330.

Author's address:

Southwest Research Institute, 3600 Yoakum Blvd, Houston, Texas 77006, USA

Discussion

Whatley: What was the earliest stage at which you could differentiate heteromorphs. and how did you do this?
Baker: The earliest stage found in this study was Instar III. Heteromorphy could be differentiated, even at this stage, by the endopodite of the 2nd antennae.

Sixth Intern. Ostracod Symposium, Saalfelden

DISTRIBUTION PATTERNS OF PELAGIC OSTRACODA OF THE PERU CURRENT SYSTEM (CRUSTACEA; OSTRACODA: HALOCYPRIDIDAE)

JOHANNES M. MARTENS*

Abstract

The halocyprid Ostracoda of the Expedition 'MARCHILE I' were studied. The samples were collected between 30°S and 42°S off the Chilean coast by a series of quantitative surface and oblique hauls. The distribution of adult and juvenile specimens was compared to the horizontal distribution of water masses, as defined by temperature-salinity curves and by dissolved oxygen concentrations. Five species seemed to be correlated with specific water masses:
1. *Conchoecia* (*Orthoconchoecia*) aff. *haddoni* and Subantarctic Surface Water (= Antiboreal Water).
2. *Conchoecia* (*Spinoecia*) aff. *porrecta* and Subtropical Surface Water.
3. *Conchoecia* (*Conchoecia*) *lophura* and Low Salinity Water.
4. *Conchoecia* (*Pseudoconchoecia*) *serrulata* and Subantarctic Subsurface Water.
5. *Conchoecia* (*Conchoecetta*) *giesbrechti giesbrechti* and Equatorial Subsurface water (= Gunther Current).

Introduction

Information on the occurrence of planktonic Ostracoda in the South East Pacific Ocean is very limited. Collections made by the Danish DANA Expeditions are located too far north near the equator (Poulsen 1973). Müller (1891) made the only recorded sample in the Peru Current Area. This was located near Antofagusta/Chile (25°S; 72°W) and contained only two species: *Conchoecia oblonga* Müller, 1891 (non Claus, 1890) = *giesbrechti giesbrechti* Müller, 1906, and *Conchoecia striata* Müller, 1891 (non Claus, 1890) = *striola striola* Müller, 1906. Therefore, it was of great importance to study a series of zooplankton samples containing several ostracod species collected during the oceanographic expedition MARCHILE I off the Chilean coast. This paper is a preliminary report on the horizontal distribution of some common species. Notes on vertical movements, larval development, and taxonomy will be published in the near future.

Material and mothods

During the southern summer 1960 (20th February to 29th March) the Expedition MARCHILE I was carried out to obtain oceanographic and biological data on a coastal belt of 120 miles off the Chilean coast (Sievers 1960). The stations were

---

* This study is based in part on a dissertation submitted in partial fulfilment of the requirements for a Doctor's degree at the University of Hamburg, Department of Biology, D-2000 Hamburg, West Germany.

located between 29°57'S and 42°S latitude on 14 transects perpendicular to the coast line. Each transect had a length of either 60 miles or 150 miles, with from seven to ten stations. At 118 stations oceanographic data were collected at standard depths from 0 m to 1000 m (see Brandhorst, 1971). Temperature, salinity, oxygen and nitrogen data were recorded on samples collected by means of Nansen bottles. At each of the the 90 plankton stations*, one 10 minute superficial horizontal haul and another 15 minute oblique one (from approximately 150-0m) were made. A net similar to Discovery Type No. 70 of 2.95 m length with an aperture of 0.70 m and a gauze of 22-25 meshes per linear cm was employed (Fagetti & Fischer, 1964: 141 ff). The plankton samples were divided into subsamples and sorted quantitatively (Fagetti & Fischer, 1964). Some papers have already been published on the results of different zooplankton groups such as Chaetognaths (Fagetti, 1968), Medusae (Kramp, 1966) and Copepods (Björnberg, 1973).

The taxonomic investigation of the pelagic ostracods, especially their soft parts, was carried out by means of an Interference Contrast Microscope and a scanning Electron Microscope. The majority of the 3500 specimens were juvenile individuals, which had not been previously identifiable. Therefore, a method for identification of juvenile halocyprids was developed (Martens in peparation). This is done by means of mathematical growth relationships and number of furcal claws.

Since the juveniles were to a large extent identifiable, only two percent of all investigated specimens could not be classified. More than 30 species were found. The precise number is not yet definite due to taxonomical difficulties in the *Conchoecia rotundata*-species group (see Angel 1972: 214 and Angel & Fasham, 1975: 712). The great majority belonged to the Halocyprididae. Except for a single male of *Anarthron reticulata* (Hartmann, 1965) all the species recorded are truly pelagic forms.

Hydrographic situation

The area under study is dominated by the northward flowing Peru Current as part of the anticyclonic circulation in the South Pacific Ocean (Wyrtki, 1965). At the surface two branches are distinguishable: 1) a coastal one of approx. 200 m maximum depth, the 'Peru Coastal Current' (after Gunther 1936), characterized by several upwelling areas (Brandhorst, 1971) and 2) one west of 82°W of approx. 600 m maximum depth (Wyrtki 1963), the 'Peru Oceanic Current' (after Gunther 1936). These two branches are usually split in the southern summer February-March by a flow to the south, the 'Peru Countercurrent' of approx. 600 m depth (Wooster & Gilmartin, 1961). Also moving southward is a subsurface current directly along the coast at a depth of 100-400 m, the 'Peru-Chile Undercurrent' (after Wooster & Gilmartin 1961), also known as 'Gunther Current' (after Brandhorst 1959).

It was, therefore, pertinent to establish if the species distribution was in any way related to the water masses. Considering these water masses one first finds the Subantarctic Surface Water, a large, very wide but shallow body (Fig. 1). In

---
* For benthos stations see Hartmann-Schröder & Hartmann (1965).

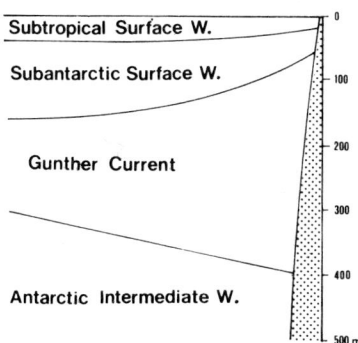 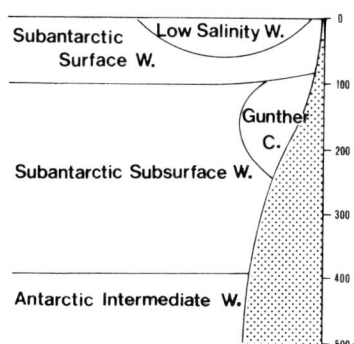

Fig.1 Vertical distribution of water masses of the Peru Current System.

this area it is identical with the Peru Coastal Current with a salinity of 33.8-34.2‰ and a temperature between 10° and 17°C. In the south this water mass appears at the surface, but north of 35°S it sinks to a depth of 50 m.

At these latitudes, the Subantarctic Surface Water is covered by the Subtropical Surface Water (Fig. 1) in the form of the southward flowing Peru Countercurrent. It is thus primarily oceanic in nature. The salinity is above 34.3‰ and the temperature between 17° and 20°C.

Also flowing south, but only along the coast, is a water mass of tropical origin, the Equatorial Subsurface Water formed by the Gunther Current (Fig. 1). Its salinity is between 34.4-34.9‰ and temperature between 7° and 13°C. The principal difference, however, is its extremely low amount of dissolved oxygen. This being at its center lower than 0.25 ml/l O_2, whereas all of the other water masses have a O_2 content between 2.0 and 5.0 ml/l (See also Reid, 1973). Under certain meteorological conditions this current reaches the surfaces and forms coastal upwelling areas.

In the entire area below 400 m is a uniform water body, the Antarctic Intermediate Water (S 34.2-34.4‰; T 3-7°C) which has no influence on the plankton hauls. In the northern section (29°S-37°S) this is covered by a wide layer of water from the Gunther Current. In the southern section, as the Gunther Current diminishes, there appears a transitional water mass of Antarctic Intermediate and Subantarctic Surface Water with temperatures from 7-10°C and a salinity from 34.0-34.4‰, which will be named the Subantarctic Subsurface Water (Fig. 1).

Finally there remains to be mentioned the influence of freshwater from inlets along the coast, which forms a long, shallow tongue above the Subantarctic Surface Water with temperatures between 10-19°C and salinity of 33.6‰ at the southern stations and of 34.0‰ at the northern ones (34°S). This will be called Low Salinity Water.

Distribution of Ostracoda

For each of the ostracod species a distribution map was prepared showing its hor-

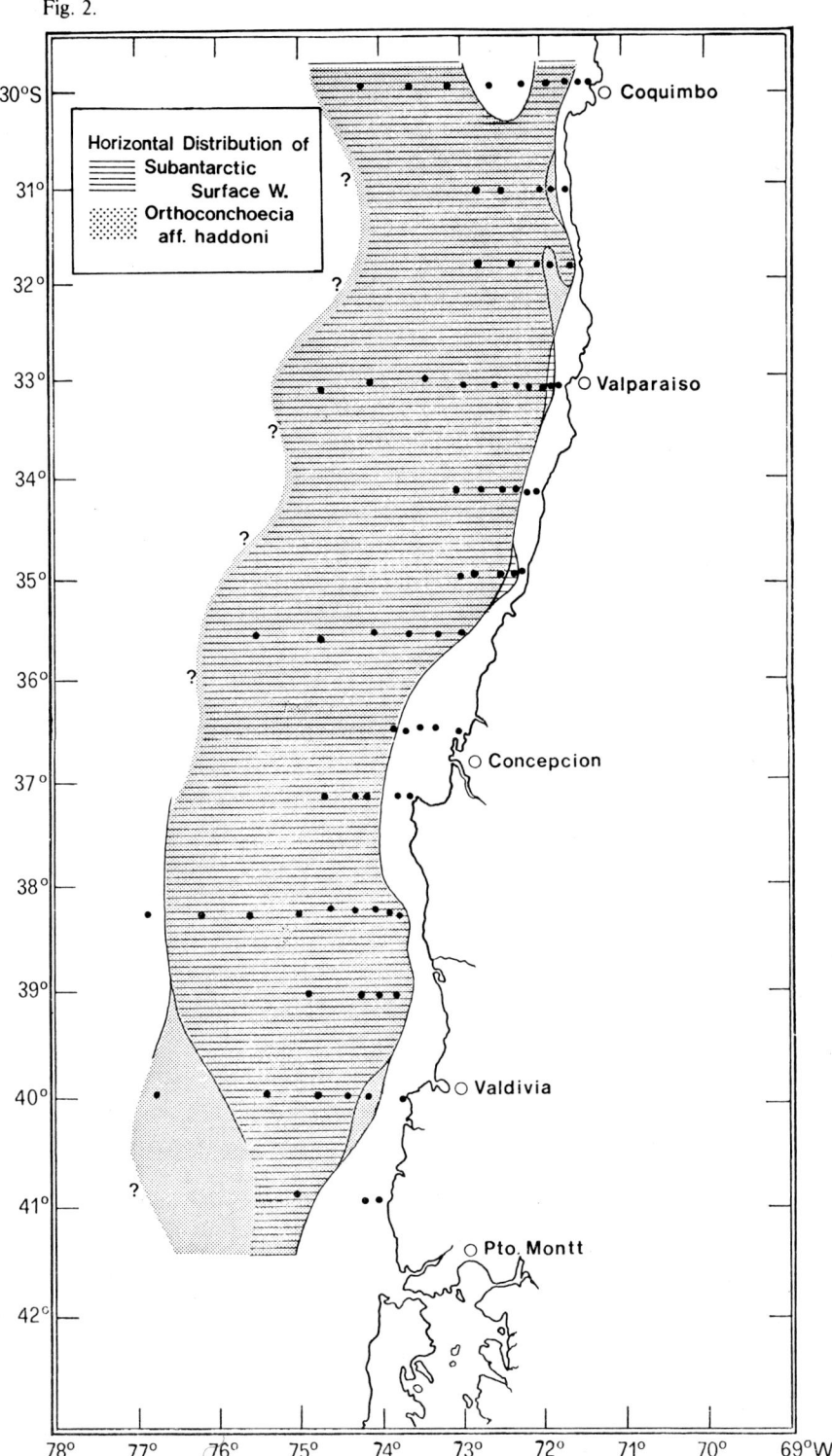

Fig. 2.

Fig. 3.

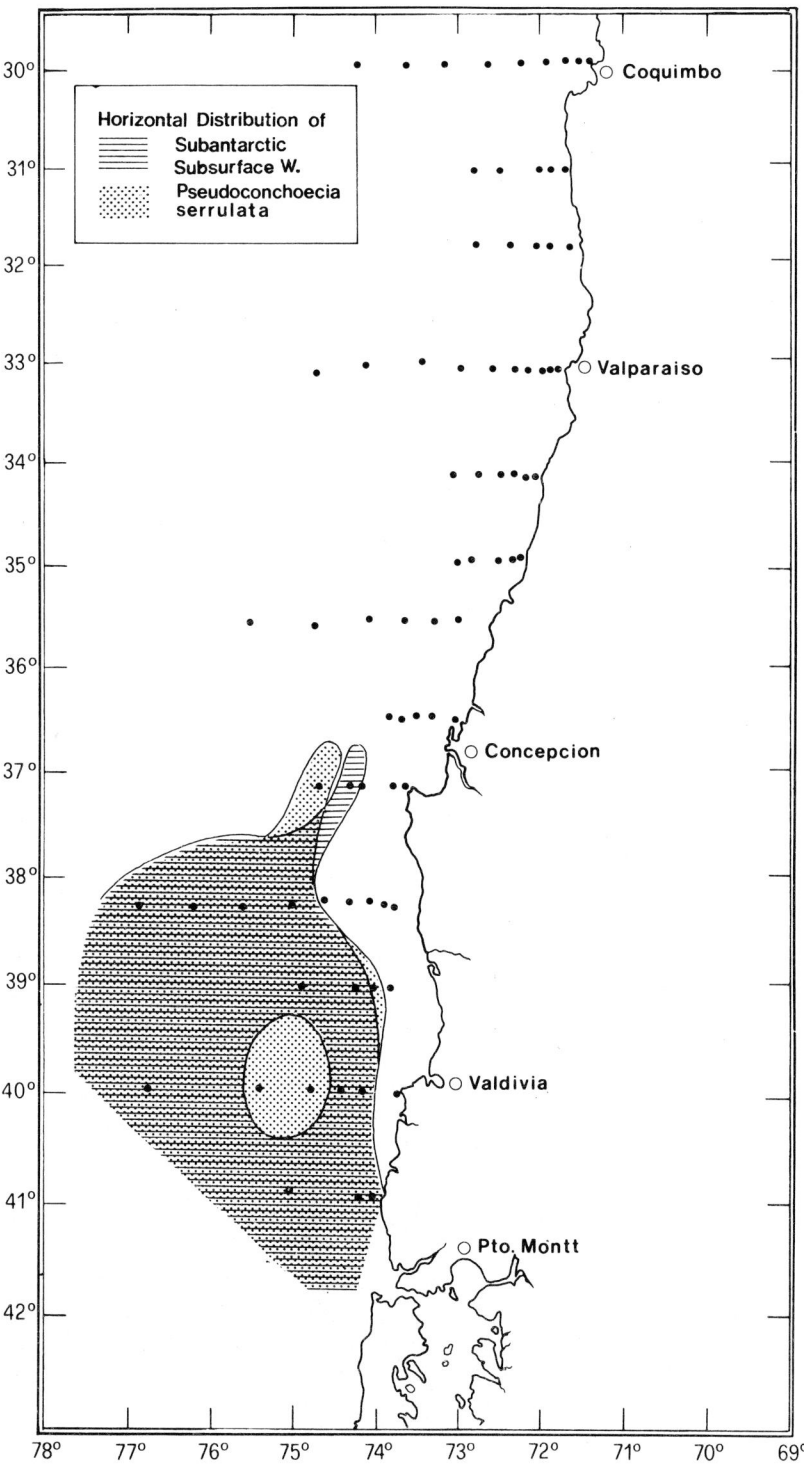

Fig. 4.

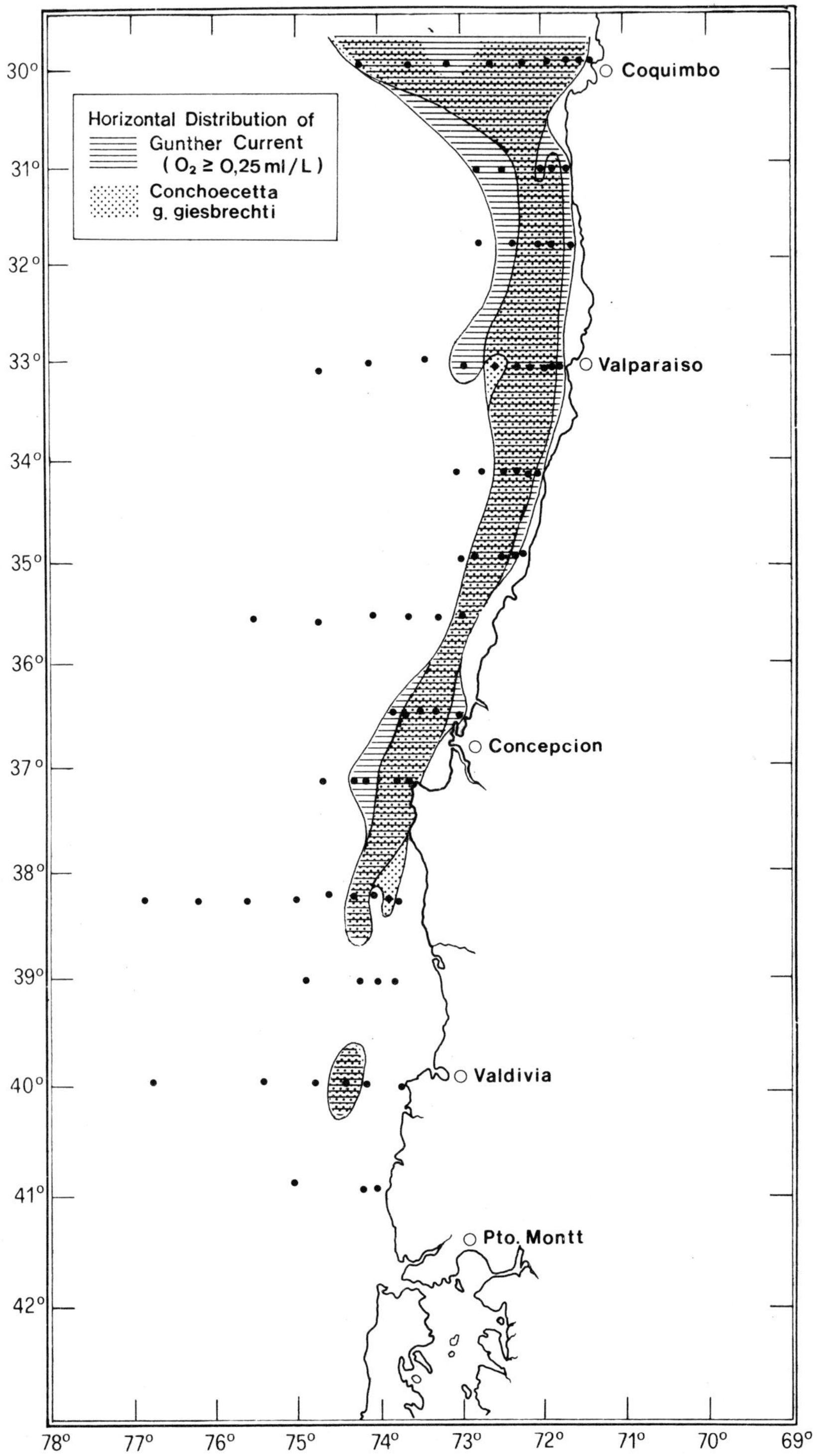
Horizontal Distribution of
Gunther Current
(O₂ ≥ 0,25 ml / L)
Conchoecetta
g. giesbrechti
Coquimbo
Valparaiso
Concepcion
Valdivia
Pto. Montt
30°
31°
32°
33°
34°
35°
36°
37°
38°
39°
40°
41°
42°
78°
77°
76°
75°
74°
73°
72°
71°
70°
69°

izontal components. As the majority of hauls were oblique it was difficult to ascertain an exact vertical pattern for each species. In the cases of five of the most frequent species a remarkable correlation could be established between these horizontal distributions and specific water masses.

The distribution of *Conchoecia* (*Orthoconchoecia*) aff. *haddoni* corresponded with that of the Subantarctic Surface Water (Fig. 2). *Conchoecia* (*Spinoecia*) aff. *porrecta* coincided with the Subtropical Surface Water. The Equatorial Subsurface Water or Gunther Current corresponded to the distribution of *Conchoecia* (*Conchoecetta*) *g. giesbrechti* Müller, 1906 (Fig. 3). *Conchoecia* (*Pseudoconchoecia*) *serrulata* Claus, 1874 is found in the Subantarctic Subsurface Water (Fig. 4) and *Conchoecia* (*Conchoecia*) *lophura* is typical of the Low Salinity Water from the fjords.

It thus appears that in the Peru Current Area a series of Halocyprididae are typical of well-defined water masses; that is, the hydrographical differences between these water bodies seem to be great enough to cause distributional boundaries. The use of these ostracod species as indicators of the various water masses might aid further research in this area.

Acknowledgements

I would like to thank Prof. Dr. G. Hartmann (Hamburg) for making the samples available for this study and H. Sievers (Instituto Hidrografico de la Armada, Valparaiso, Chile) for hydrographic data.

References

Angel, M. V. 1972. Planktonic oceanic Ostracods – historical, present and future. Proc. Roy. Soc. Edinburgh (B) 73 (22):213-228.

Angel, M. V. & Fasham, M. J. R. 1975. Analysis of the vertical and geographic distribution of the abundant species of planktonic Ostracods in the north-east Atlantic. J. mar. biol. Ass. U. K. 55:709-737.

Björnberg, T. K. S. 1973. The planktonic copepods of the Marchile I Expedition and of the 'Eltanin' cruises 3-6 taken in the SE Pacific. Bol. Zool. Biol. Mar. (N.S.) 30:245-394.

Brandhorst, W. 1959. Relationship between the hake fishery and a southerly sub-surface return flow below the Peru current off the Chilean coast. Nature 183:1832-1833.

Brandhorst, W. 1971. Condiciones oceanograficas estivales frente a la Costa de Chile. Revista Biol. Mar. 14(3):45-84.

Fagetti, E. 1968. Quetognatos da la expedición 'Marchile I' con observationes acerca de posible valor de algunas especies como indicadores de las masas de agua frente a Chile. Revista Biol. Mar. 13(2):85-171.

Fagetti, E. & Fischer, W. 1964. Resultados cuantitativos del zooplancton colectado frente a la costa Chilena por la Expedition 'Marchile I'. Montemar (Revista Biol. Mar.) 11(4):137-193.

Gunther, E. R. 1936. A report on oceanographical investigations in the Peru Coastal Current. Discovery Rep. 13:107-276.

Hartmann-Schröder, G. & Hartmann, G. 1965. Zur Kenntnis des Sublitorals der chilenischen Küste unter besonderer Berücksichtigung der Polychaeten und Ostracoden. Mitt. Hamb. zool. Mus. Inst. 62 (Suppl.):1-384.

Kramp, P. 1966. A collection of Medusae from the coast of Chile. Vidensk. Medd. Dansk naturh. foren. 129:1-38.
Müller, G. W. 1891. Über Halocypriden. Zool. Jb. Syst. 5:253-280.
Poulsen, E. M. 1973. Ostracoda-Myodocopa Part III B Halocypriformes Halocypridae Conchoecinae. Dana-Rep. 84:1-224.
Reid, J. L. 1973. Transpacific hydrographic sections at Lats. 43°S and 28°S: the SCORPIO Expedition-III. Upper water and a note on southward flow at mid-depth. Deep-Sea Res. 20:39-49.
Sievers, H. 1960. Operation oceanografica 'Marchile I', mission cumplida. Revista Mar. 76:477-485.
Wooster, W. S. & Gilmartin, M. 1961. The Peru-Chile Undercurrent. J. Mar. Res. 19(3):97-122.
Wyrtki, K. 1963. The horizontal and vertical field of motion in the Peru Current. Bull. Scripps Inst. Ocean. 8(4):313-345.
Wyrtki, K. 1965. Surface currents of the eastern tropical Pacific Ocean. Inter-Am. Trop. Tuna Comm. Bull. 9(5):271-304.

Author's address:

University of Hamburg, Zoological Institute and Zoological Museum, Martin-Luther-King-Platz 3, D-2000 Hamburg 13, B.R.D.

Sixth Intern. Ostracod Symposium, Saalfelden

SOME OBSERVATIONS ON THE VERTICAL DISTRIBUTION AND STOMACH CONTENTS OF GIGANTOCYPRIS MUELLERI SKOGSBERG 1920 (OSTRACODA, MYODOCOPINA)

A. MOGUILEVSKY & A.J. GOODAY

Abstract

Gigantocypris muelleri Skogsberg 1920 was studied during Cruise 61 of RRS Discovery in the N.E. Atlantic. It was caught abundantly in hauls between 700 m and 1500 m. Males and gravid females predominated in 1250-1500 m hauls while juveniles and non-gravid females occurred mainly between 700 m and 1250 m. Thus there appears to be some slight downward migration during ontogeny. However, there is no evidence for diurnal vertical migration. The contents of 65 stomachs containing a total of 165 food items were examined, 10 (15.4 %) were empty and 5 (7.7%) contained only unidentifiable remains. Of the remaining 50 guts, 36 (72%) contained chaetognaths, 20 (40%) copepods, 14 (28%) probable coelenterate (medusa) remains, 8 (26%) mysids, 3 (6%) ostracods, 1 (2%) euphausiids, 1 (2%) fish remains while indeterminate crustacean fragments were found in 9 (18%) stomachs and totally unidentifiable items occurred in 10 (20%) stomachs. Feeding in the net occured but did not significantly affect the results. Many of the prey are highly active, suggesting that *G. muelleri* can also swim actively, and this is confirmed by observations on live animals. The diverse nature of the gut contents indicates that it belongs to a complex food web rather than a simple food chain.

Introduction

Gigantocypris muelleri is a large bathypelagic cypridinid ostracod, widely reported from the Atlantic, Indian and Antarctic Oceans (Müller, 1895 as *G. agassizii*, Fowler, 1909 as *G. pellucida,* Skogsberg, 1920, Cannon, 1940, Poulsen, 1962, Tibbs, 1965, Kornicker, 1976, Kornicker et al., 1976). Detailed descriptions of the limbs are given by Skogsberg (1920), Poulsen (1962), Kornicker et al. (1976) and of the soft part anatomy by Cannon (1940). There has been considerable recent interest in *G. muelleri* and related species as experimental animals. Thus MacDonald, Gilchrist & Teal (1972) and MacDonald (1972) studied the effect of high hydrostatic pressure on a variety of marine animals, including *G. muelleri.* The closely related *G. agassizii* was used by Childress (1971) in a study showing that the respiratory rates of marine animals decreased with depth. The chemical composition and buoyancy of the same species were analysed as a function of depth by Childress & Nygaard (1974) who found that in its high water content, *G. agassizii* was more akin to a jellyfish than a crustacean. MacDonald (1975) summarised laboratory studies on the buoyancy, oxygen consumption and effects of pressure on *G. muelleri.*

Despite this interest, there is little information on the depth distribution of *G. muelleri.* What data are available are either inaccurate because the depths were based on meters of trawl wire paid out (Skogsberg, 1920, p. 217, Poulsen, 1962, p. 83, MacDonald, 1972, p. 218) or inadequate because very few specimens were examined (Kornicker et al., 1976). In addition little is known about the diet of *Gigantocypris*. Cannon (1940, p. 193, 194) and Kornicker et al. (1976, p. 9) each examined the gut contents of three specimens, but only Cannon was able to identify remains. Here we report on the depth distribution and gut contents of a large collection of *G. muelleri* from a small area in the Northeast Atlantic, in the hope that this will add to our knowledge of the ecology of this important and interesting deep-sea animal.

Materials and methods

The material was collected in April and May 1974 during Cruise 61 of RRS 'Discovery' in the Northeast Atlantic in which both authors participated. All specimens were taken around 44° N, 13° W and were caught with an RMT 1+8 combination net (Baker et al., 1973). This has an acoustically controlled opening and closing mechanism and the depth of fishing, water temperature and net speed can be continuously monitored while the net is operating. Thus very exact data on depth distribution are obtained. At depths above 1000 m, the net was towed for 2 hours and filtered approximately 56,000 m³ of water; below 1000 m it was towed for 4 hours and filtered twice the volume.

All specimens of *G. muelleri* were picked out from the RMT 8 catches; most were immediately preserved in formalin but some were kept alive in an aquarium for observation. Measurements were made with calipers on board ship to avoid problems arising from the distortion of the flexible carapace which may arise during lengthy preservation. However, many specimens, particularly juveniles, were damaged while in the net. Examination of the gut contents was also done mainly on the ship.

Depth distribution

G. muelleri was caught in hauls taken between 700 m and 1500 m and 577 specimens were picked out from 21 of the 39 samples taken from this depth range. The remaining 18 samples contained no *Gigantocypris*. Fig. 1 is a plot of length against height for the 289 specimens that were not too distorted for accurate measurement; the depth ranges of all hauls yielding *G. muelleri* are summarised in Fig. 2.

Out of the 577 specimens 63 (11%) were adult males and these occured mainly in the 1250-1500 m hauls; very few were found at shallower depths. Their size ranged from 1.18 cm to 1.32 cm. Females were nearly twice as abundant, numbering 118 (20%). Gravid females also occurred mainly in the 1250-1500 m hauls whereas the more abundant females with only small eggs or ovaries were usually taken near the top of the range. Females were usually larger than males, ranging between 1.29 cm and 1.51 cm. Gravid specimens carried 27 to 48 large eggs or brooded first instars. The remaining 396 (69%) juvenile specimens occurred mostly in the 1000-1250 m hauls but some occurred at shallower depths up to 700 m.

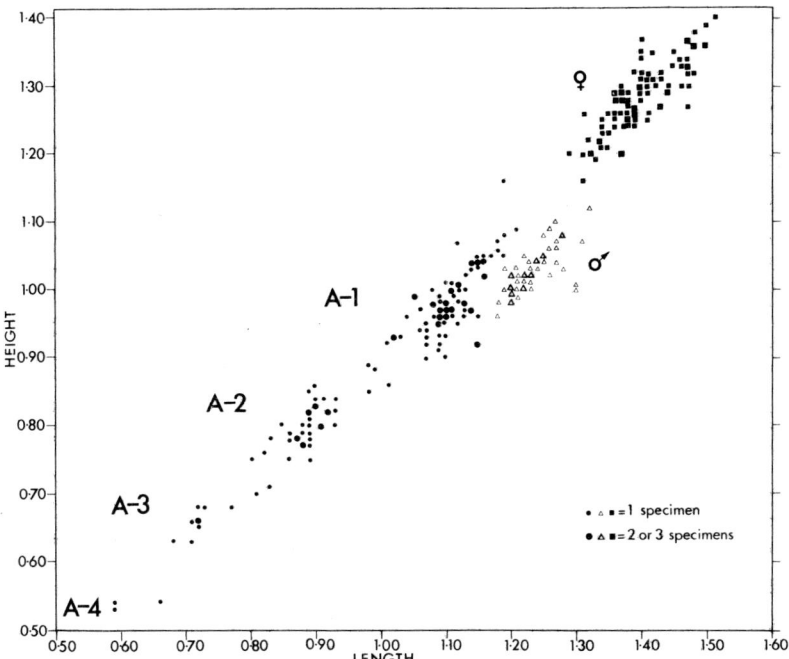

Fig. 1. Plot of length versus height for *Gigantocypris muelleri*. Juvenile instars are numbered A-1 to A-4. The height and length units are cm.

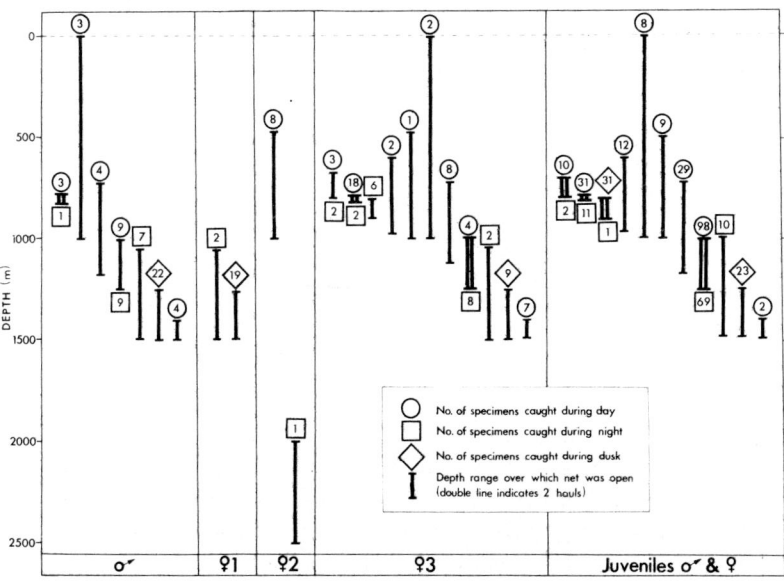

Fig. 2. Depth distribution of *Gigantocypris muelleri*. Female specimens are divided into three categories: 1, with large eggs of brooded first instar juveniles; 2, with intermediate sized eggs; 3, with small eggs or only ovaries.

Four juvenile instars, with an overall length range of 0.59–1.28 cm, were present. Females of the final juvenile stage overlapped in size with adult males but were easily distinguished by their more rounded outline in dorsal view.

It seems that the upper depth limit for *G. muelleri* in the area sampled is around 700 m since the species was totally absent in 23 hauls taken at 600 m over a 48 hr period to study vertical migration. During some hauls yielding *G. muelleri*, the net was open at depths above 700 m (Fig. 2), though in all these the net also fished below 700 m and it was presumably here that the specimens were caught. The lower depth limit for the species is probably around 1500m since of seven hauls taken below this depth only one (2000–2500 m) yielded a single *G. muelleri*.

The majority of juveniles were caught at rather shallower depths (1000–1250 m) than most adult males and gravid females (1250–1500 m) and thus there appears to be some slight downward migration during ontogeny. However, there is no evidence for diurnal vertical migration which is common amongst halocyprids living in the top 1000 m (Fig. 2).

Gut contents of G. muelleri

The stomach of *G. muelleri* is contained in the posterior part of the body and is relatively large, measuring 45–60% of the total carapace length. It is laterally compressed, even when distended with food. Of the 65 guts examined, 10 (15.4%) were empty and 5 (7.7%) contained only unidentifiable remains. The contents of the other 50 guts are summarised in Table 1. There were 165 food items belonging to 9 taxonomic groups; 12 genera and 10 species were identified, although some remains were determinable only above generic level.

The commonest group was the Chaetognatha (arrow worms) which occurred in 36 (72%) of the 50 guts and made up half (50.30%) of all food items. Their size ranged from 1.30 to 2.5 cm and in some cases the whole animal was present, in others they were chopped in half or beheaded and in varying states of digestion. At least two species were present although identification has not been attempted. The Copepoda were the next most abundant group with 23 specimens (13.94% of all food items) being found in 20 (40%) of the 50 guts. Nine species belonging to nine genera were identified. Up to three specimens were found per gut, often in varying stages of digestion and sometimes with only the head or some limbs remaining. Sixteen (9.75%) probable coelenterate fragments occurred in 14 (28%) guts. These were mainly bits of red gelatinous material which fluoresced in ultraviolet light after treatment with methanolic sulphuric acid. This was presumably due to prophyrin pigmentation within tissue probably derived from the deep-sea medusae *Atolla* and *Periphylla* (Herring 1972). Ten (6.01%) mysid shrimp remains were found in eight (16%) stomachs and two species have been identified. One or two specimens occurred in each gut in variable states of digestion; in some cases the original length must have approched 2 cm, longer than the largest *G. muelleri*. The other taxonomic groups were present in small numbers and included halocyprid ostracods, euphausiids, fish remains, radiolarians, amphipods and unidentifiable crustacean fragments (Table 1).

Table 1. Summary of stomach contents of 50 specimens of *Gigantocypris muelleri*

Taxonomic Group	Specimens in all stomachs.		Maximum No. of specimens per stomach.	Stomachs containing group.	
	No	%		No	%
Chaetognatha	83	50.30	5	36	72
Copepoda					
Indet. specimens	13	7.88	2	11	22
Disseta palumboi	4	2.42	2	3	6
Euchirella rostrata	1	0.61	1	1	2
Euchaeta gracilis	1	0.61	1	1	2
E. sp.	1	0.61	1	1	2
Clausocalanus furcatus	1	0.61	1	1	2
Metridia lucens	1	0.61	1	1	2
Pleuromammal robusta	1	0.61	1	1	2
Rhincalanus copepodite	1	0.61	1	1	2
Undeuchaeta plumosa	1	0.61	1	1	2
Gaidius cf. *tenuispinus*	1	0.61	1	1	2
Total	26	15.70	–	20	40
Coelenterata					
Porphyrin bearing tissue	14	8.48	2	13	26
Other remains	2	1.22	1	2	4
Total	16	9.75	–	14	28
Mysidacea					
Eucopia unguiculata	3	1.82	1	3	6
E. grimaldi	1	0.61	1	1	2
Indet. specimens	6	3.64	1	6	12
Total	10	6.06	–	10	20
Ostracoda					
Conchoecia spp.	3	1.82	2	2	4
Indet. appendages	1	0.61	1	1	2
Total	4	2.42	–	3	6
Euphausiacea	2	1.21	1	2	4
Amphipoda	1	0.61	1	1	2
Fish remains	2	1.21	1	2	4
Radiolaria	1	0.61	1	1	2
Indet. Crustacea	12	7.27	3	9	18
Indet. Items	11	6.67	2	10	20

Discussion

Cannon (1940, p. 194) identified the chaetognath *Sagitta*, parts of a Brachyuran zoea, ten copepods identified as *Pleuromamma robusta* Dahl, *P.* sp. and *Heterorhabdus* and the half-digested remains of a small fish, in the stomachs of three *G. muelleri* specimens. Kornicker et al. (1976, p. 9) also examined the guts of three specimens and found that one was practically empty, one had a small amount of unrecognisable material while the third had a full gut, the contents of which were not identified. The stomach contents reported by Cannon (1940) agree well with those found in our material.

Cannon (1940, p. 193) believed that 'from its shape and actual size it may be safely deduced that *Gigantocypris* is unable to move quickly through the water'. He goes on to quote from Kemp, who had seen living specimens and believed them incapable of rapid movement, noting that 'they rock and roll a lot, finding it difficult to keep on an even keel and swim feebly'. Because of this apparent sluggishness, Cannon was puzzled that *Gigantocypris* fed on active prey, and he proposed that it was a sedentary feeder, catching prey which was passing by. Cannon also believed that 'the valves of the shell cannot be opened, and the only passage to the mouth is through the small opening at the confluence of the antennal notches'. Since the mouth and mandibular palps may lie well inside the carapace there must be a method of pressing the mouthparts down onto the antennal notches so that the mandibular palps can be thrust out. Cannon suggests that this is brought about by a hydrostatic system controlled by the flexible criss-cross musculature in the flexible dorsal body wall of the region.

We believe that Cannon's ideas are untenable for two reasons. (1) Observations on live *G. muelleri* show it to be an active and strong swimmer, quite capable of catching the recorded prey. This was particulary true of specimens observed by Angel (pers. comm.) at 53° N, 20° W in April 1971 when the water column was almost isothermal to 800 m. These swam steadily with powerful strokes of the antennae slightly head up, with the ventral edge inclined at about 20–30° to the horizontal. Similar Hardy (1970, p. 230) mentions a *Gigantocypris* which was 'remarkably active and swam backward and foreward across the dish as fast as any shrimp' while MacDonald (1972) found that his animals 'maintained vigorous swimming activity for a few days at 5°C and 1 atm'. Although not as active as those of Angel, Hardy and MacDonald, our specimens swam quite well. However, there was a tendency for them to swim round in tight circles or spin around on a dorso-ventral axis. This was presumably not a natural behaviour. (2) We never observed our animals extruding their mouthpart through the rostral incisure. Instead the gape between the valves in live animals is quite sufficient (as shown by MacDonald's 1975, Fig. 1.11c) for the exopodites of the mandibles, which do extend through the gape, to pull into the carapace all the food items we have recorded. On preservation the valves are drawn tightly shut, giving the impression that the gape is much more restricted than in the living animal.

MacDonald (1972, p. 218) found that *G. muelleri* survives decompression from 200 atm (equivalent to a depth of 2000 m) with no change of locomotory activity. This presumably explains why the species remains active when brought to the surface in nets. In other experiments, MacDonald (1975, p. 311) found *G. muelleri* to have a fairly low rate of oxygen consumption which is not affected by the range of pressure that it normally experiences. He suggests (p. 312) that the food which it would require 'to sustain its oxygen consumption would be about one large copepod per month'. The results reported here suggest that *G. muelleri* consumes considerably more than one large copepod per month and hence MacDonalds animals were probably not behaving naturally.

The variety of animals in the guts of *G. muelleri* raises the possibility that it may feed in the nets where the density and diversity of food is greater. While in the net *G. muelleri* may potentially feed on two categories of animal: (1) those with a similar depth range, caught with *G. muelleri* during the 2–4 hrs that the net is open and being fished; (2) animals with shallower depth ranges than *G. muelleri* which enter through the 5 mm meshes of the RMT 8 while the net is being lowered or retrieved in a closed condition. Being sympatric with *G. muelleri* animals in the first category are obviously potential prey under natural conditions. However, those in the second category are allopatric and hence their presence in the gut content must indicate that net feeding has occured. The guts we examined yielded single fresh and undigested specimens of two copepod species whose normal depth range is shallower than *G. muelleri*; there must, therefore, have been some feeding in the net. However, because of the generally fragmentary and digested nature of the gut contents and the slow metabolic rate (Childress 1971) we believe feeding in the net to have been slight and a minor source of bias in these results. The fact that a significant minority (15.4%) of guts were entirely empty also suggests that feeding in the net is not significant (Kornicker pers. comm.).

It would seem from the diverse nature of its gut contents that *G. muelleri* is a predatory species belonging to a complex food web. This is to be expected since *Gigantocypris* lives at a depth where food is not abundant and hence a diverse diet is advantageous. It would be interesting to analyse the precise role *G. muelleri* plays in the bathypelagic food web, in the juvenile stages as well as adults, and to investigate any seasonal variations that might occur.

Acknowledgements

We are grateful to Dr. M.V. Angel for helpful discussions, critically reading the manuscript and suggesting a number of improvements. Dr. R.C. Whatley and Mr. P.M. David also criticised the manuscript. Mr. H. Roe kindly identified the copepods and Dr. P.J. Herring identified the probable medusoid remains. We are also grateful to Mrs C. Darter who prepared the figures.

References

Baker, A, de C., Clarke, M. R. & Harris, M. J. 1973. The N.I.O. combination net and further developments of rectangular midwater trawls. J. mar. biol. Ass. U.K., 53 (1): 167–184.
Cannon, G. 1940. On the anatomy of Gigantocypris mülleri. Discovery Rep. 19:185–244.

Childress, J. J. 1971. Respiratory rate and depth of occurrence of mid-water animals. Limnology and Oceanography, 16 (1):104–106.
Childress, J. J. & Nygaard, M. 1974. Chemical composition and buoyancy of midwater Crustacea as function of depth of occurrence off southern California. Marine Biol. 27 (3):225–228.
Fowler, G. H. 1909. Biscayan Plankton. Part XII – The Ostracoda. Trans. Linn. Soc. London (2) 10 219–313.
Hardy, A. 1956. The open sea, its natural history: the world of plankton. Collins, London, 335 pp.
Herring, P. J. 1972. Porphyrin pigmentation in deep-sea medusae. Nature, London 238 (5362):276–277.
Kornicker, L. 1976. Gigantocypris muelleri Skogsberg, 1920 (Ostracoda) in benthic samples collected in the vicinity of Heard Island and the Kerguelen Islands on Cruise MD 03 of the research vessel 'Marion-Dufresne' 1974. CNFRA, 39: 47–48.
Kornicker, L., Wirsing, S. & McManus, M. 1976. Biological studies of the Bermuda Ocean Acre: Planktonic Ostracoda. Smith. Contr. Zool. 223:1–34.
MacDonald, A. G. 1972. The role of high hydrostatic pressure in the physiology of marine animals. Symp. Soc. Exp. Biol. 26:209–232.
MacDonald, A. G. 1975. Physiological aspects of deep sea biology. Cambridge University Press, Cambridge. 450 pp.
MacDonald, A. G., Gilchrist, I. & Teal, J. M. 1972. Some observations on the tolerance of oceanic plankton to high hydrostatic pressure. J. mar. biol. Assn. U.K. 52 (1):213-223.
Müller, G. W. 1895. Reports on the dredging operations off the west coast of Central America to the Galapagos. No. 5, Die Ostracoden. Bull. Mus. Comp. Zool. 27 (5):155–159.
Poulsen, E. M. 1962. Ostracoda – Myodocopa, 1: Cypridiniformes – Cypridinidae. Dana Report 57:1-414.
Skogsberg, T. 1920. Studies on marine ostracods, 1: Cypridinids, Halocyprids and Polycopids. Zool. Bidr. Uppsala, Supplement 1:1-784.
Tibbs, J. F. 1965. Observations on Gigantocypris (Crustacea: Ostracoda) in the Antarctic Ocean. Limnology and Oceanography 10 (3):480–482.

Authors' addresses:

Alicia Moguilevsky, Dept of Geology, University College of Wales, Aberysthwyth, Wales, U.K.

Andrew J. Gooday, Institute of Oceanographic Sciences, Wormley, Godalming, Surrey, U.K.

Sixth Intern. Ostracod Symposium, Saalfelden

OSTRACODS FROM THE RICE-FIELDS OF SRI LANKA (CEYLON)

JOHN W. NEALE

Abstract

Study of ostracod faunas from the rice-fields of Sri Lanka (Ceylon) shows that the fauna in the central and southern parts of the island has close affinities with that of Indonesia, while there is apparently no such affinity in the faunas from the Jaffna Peninsula in the North. It is suggested that the most probable explanation of this affinity lies in the exchange of rice seed or rice plants between Indonesia and Sri Lanka in the past.

The ostracoda of Sri Lanka (Ceylon) are poorly known and have been neglected for half a century. Hitherto, we have been dependent on a few papers which examined isolated and occasional samples such as those of Brady (1886), Daday (1898), Apstein (1907) and Gurney (1916). During the last six years Professor C. H. Fernando has collected samples throughout the island in connection with a detailed limnological study. The systematic study of the ostracods from these samples has commenced. This contribution deals only with the faunas from rice-fields in localities stretching from Belihuloya (6°43'N) and Nugegoda (6°52'N) in the south to Karainagar and Vaddukoddai (9°44'N) in the Jaffna Peninsula in the north. Altogether 36 samples were examined of which 19 came from the Nugegoda area, 4 from the Belihuloya area, 8 from localities in the Jaffna Peninsula and 5 from ricefields elsewhere in the island (Fig. 1).

Taxonomy

The taxonomy of the fresh-water ostracods from SE Asia is at present the subject of a thoroughgoing revision and a number of taxa are here left under open nomenclature. The unidentified ostracods of the table (q.v.) are mainly juvenile, decalcified specimens. The commonest species include a number of well-known and easily recognisable taxa. *Cypris subglobosa* J de C Sow. 1840 (Pl. 1, Fig. 8) is a large species some 1400-1500 μm long with serrate postero-ventral margins and has recently been covered by Okubo (1972) and Neale (1976). The two common Cyprettas, both about 500-550 μm long, are easily differentiated. *Cypretta globosa* (Brady 1886) has short, strong bristles on the carapace in addition to the more general pilosity (Pl. 1, Fig. 5), whilst *C. globula* (Sars 1889) lacks these spines (Pl. 1, Fig. 3). The former tends to be pale brown to gold in colour; the latter pale yellowish-green. The small species *Pseudocypretta maculata* Klie, 1932 (Pl. 1, Fig. 4) about 300 μm long is most distinctive and is characterised by three circular, bright violet spots on each valve. The oval *Hemicypris pyxidata* (Moniez 1892) is beautifully marked with dark-brown patches, measures about 1000 μm in length and has the left valve considerably smaller than the right (Pl. 1, Fig. 6). *Hetero-*

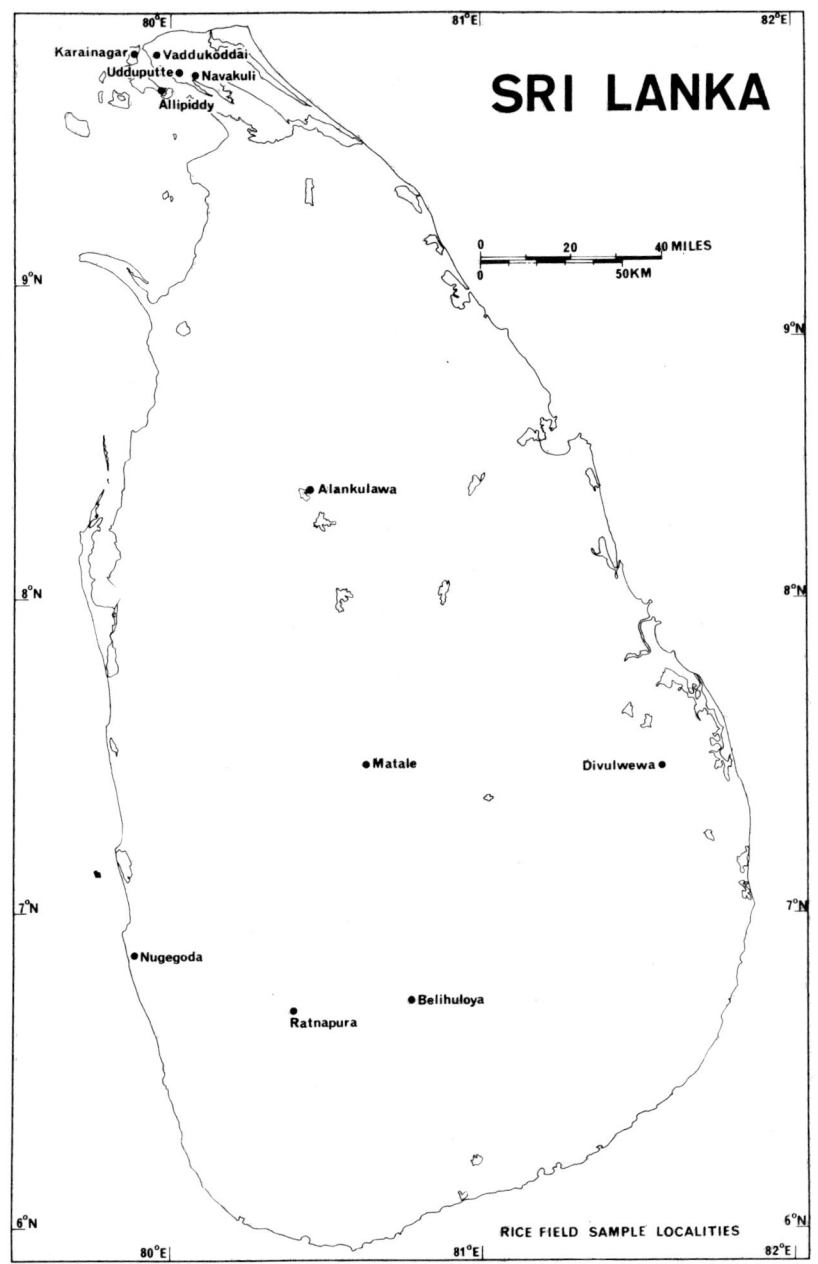

Fig. 1. Map of rice field localities.

Table 1. Distribution of Ostracoda.

Locality	Date	Sample Number	Centrocypris viridis (Neale 1976)	Cypretta globula (Sars 1889)	Cypretta subglobosa J. de C. Sowerby, 1840	Cypridopsis sp. C.44	Cypridopsis sp. C.45	Hemicypris pyxidata (Montez 1892)	Heterocypris dentatomarginatus (Baird 1859)	Ilyocypris luxata (Brady 1886)	Indiacypris maculata Klie 1932	Pseudocypretta spp. (juveniles)	Stenocypris major (Baird 1859)	Strandesia elongata Hartmann 1964 subsp. nov.	Strandesia marmorata (Brady 1886)	Strandesia purpurascens (Brady 1886)	Strandesia wiertejskii (Grochmalicki 1915) s.l.	Strandesia sp. nov.	UNIDENTIFIED	TOTAL
JAFFNA																				
UDDEPUTTE	16.XII.71	89	1			3										9			3	16
ALLIPIDDY	17.XII.71	76				3											7?			10
KARAINAGAR EAST (2)	15.XII.71	63(2)				16											7			23
KARAINAGAR EAST (1)	15.XII.71	63(1)	38																	38
VADDUKODDAI	14.XII.71	58		1?		12				2						2		2	9	28
KARAINAGAR (2)	15.XII.71	54(2)	1			13		1	12							2		3		32
KARAINAGAR (1)	15.XII.71	54(1)	2			5		1	3											11
NAVAKULI	6.XII.71	19																	3	3
ALANKULAMA	12.VIII.72	64					1				1?				10				4	16
DIVULWEWA	11.VIII.72	40					2													2
RATNAPURA	18.IX.72	42			8							3	1		2	12			1	27
NEAR RATNAPURA	18.VII.72	35		24												22			3	49
MATALE	2.I.65	34	1							8?						3				12
BELI-HULOYA																				
	16.XII.72	100									1	2								3
	16.VIII.72	22f		3								3				15?			3	24
	16.VIII.72	22b						1				8	1			18			6	34
	16.VIII.72	22a		3															2	5
NUGEGODA																				
GANGODAWILA	22.XII.70	78(2)			6											7?				13
GANGODAWILA	22.XII.70	78(1)		3								13				1			1	18
	11.VI.70	103					1		49							35	3		1	89
	11.VI.70	86											1							1
	7.VIII.70	70		4?								2				2	6		2	16
	19.X.71	65		1?					1								5		2	9
	30.X.71	62		3					2			1				3?	2			11
	22.XII.70	53		1												3				4
	22.VII.71	51	4														2		5	11
	31.X.71	46		1															1	2
	10.XI.71	32b		1					2											3
	10.XI.71	32		9					1									1		11
	22.VI.72	30	1									1	1?			1?			1	4
	15.VII.71	26														1				1
	2.IV.71	23	1													3?			3	7
	15.VII.71	20																	10	10
	19.III.71	17														1?			1	2
	22.VII.71	5										8							2	10
	19.X.71	2		2					2							1			1	6

cypris dentatomarginatus (Baird 1859) about 900 μm long, has serrate antero-ventral and postero-ventral margins in the left valve and is brown in colour (Pl. 1, Fig. 2). *Stenocypris major* (Baird 1859), about 1900-2000 μm long (Pl. 1, Fig. 1) has been the subject of a recent study by Ferguson (1969) and *Centrocypris viridis* Neale 1976, is a large dark-green species just over 1000 μm in length (Pl. 1, Fig. 7). Among the Strandesias the commonest are the purple marked *Strandesia purpurascens* (Brady 1886) whose lectotype has a left valve length of 972 μm, and the

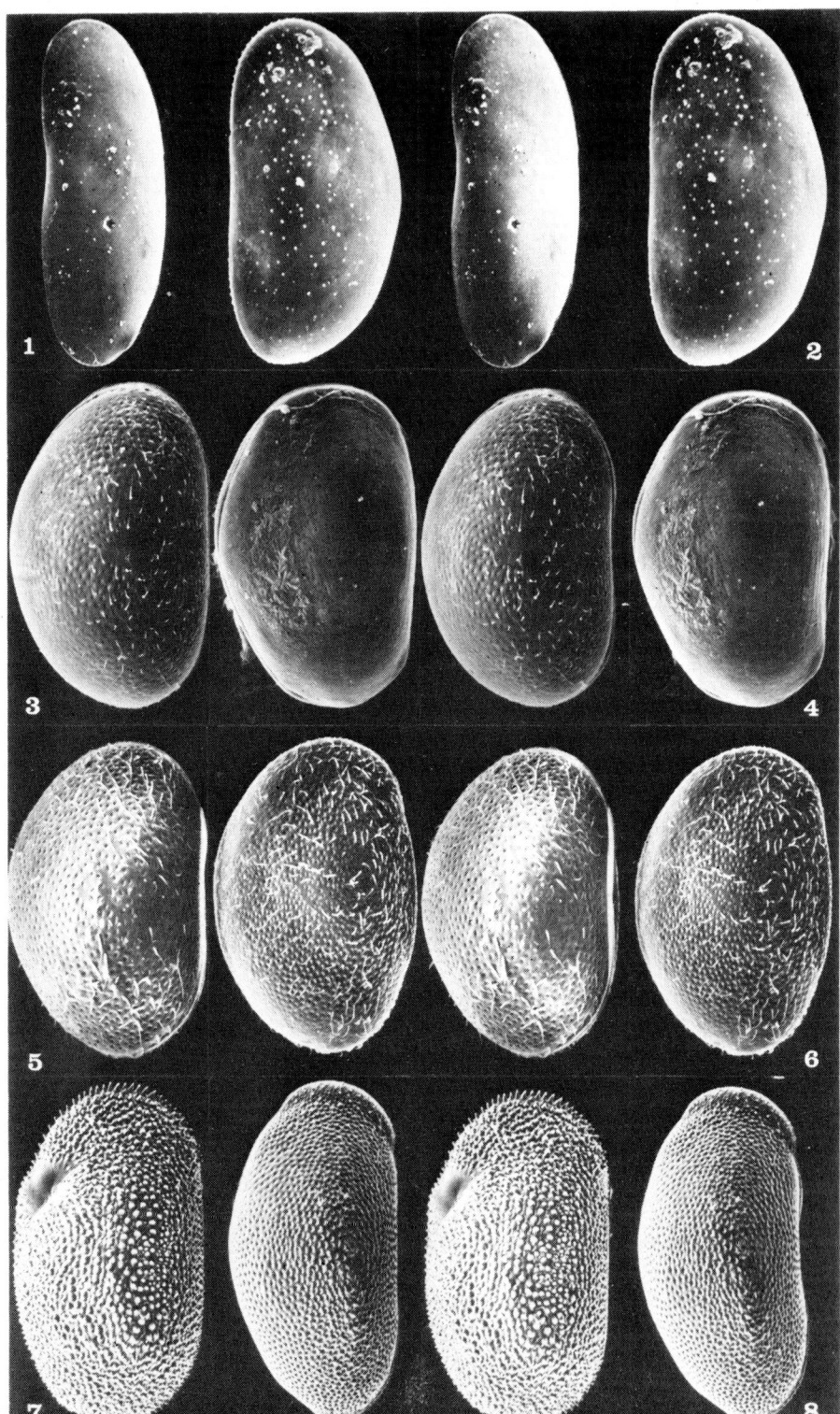

◄ Plate 1. Paired stereoscopic photographs of rice field species of Ostracoda from Sri Lanka.

1. *Stenocypris major* (Baird). ♀ LV. × 25. HU.239.R.14a. Pond, Moratuwa. 2. *Heterocypris dentatomarginatus* (Baird). ♀ LV. × 54. HU.247.R.2a. Paddy field, Vadukkodai. 3. *Cypretta globula* Sars. RV × 84. HU.254.R.13a. Pond, Marawila. 4. *Pseudocypretta maculata* Klie. Carapace from right. × 186. HU.254.R.16. Pond, Marawila. 5. *Cypretta globosa* (Brady). Carapace from right. × 91. HU.254.R.6. Talawila, Wilpattu. 6. *Hemicypris pyxidata* (Moniez). ♀ RV. × 46. HU.248.R.2a. Rice field, Nugegoda. 7. *Centrocypris viridis* Neale. ♀ RV. × 49. HU.250.R.13a. Rice field, Karainager East, Jaffna. 8. *Cypris subglobosa* Sowerby. RV. × 35. HU.241.R.7. Pond, Moratuwa.

white and black or dark blue-purple mottled *S. marmorata* (Brady 1886) whose left valve lectotype is 576 μm long. *Strandesia* is one of the more difficult genera to deal with at the species level, as becomes apparent in dealing with the *S. wierzejskii* – *S. striatoreticulata* group. In 1915, Grochmalicki established *Cypris wierzejskii* for an ostracod with characteristic fingerprint-like ornamentation of the shell found at Sitoe Sampora and Tjitajam in Java. In 1932, Klie described *S. striatoreticulata* from Java and Sumatra, his first recorded locality being the palm section of the Botanic Garden at Buitenzorg close to Grochmalicki's localities. The shell had the same type of ornamentation but Klie was able to examine some of Grochmalicki's material and defined the differences in the appendages. In *S. striatoreticulata* the terminal claws of the second antenna are longer than the anterior margin of the penultimate joint, the third endite of the maxilla does not carry toothed or feathered bristles (Dornen gefiedert) and the furca is straight and the anterior furcal seta is longer than the sub-terminal claw. In *Strandesia wierzejskii* the terminal claw of the second antenna is as long as the anterior margin of the penultimate joint, the third endite of the maxilla carries toothed bristles and the anterior furcal seta is shorter than the sub-terminal claw.

The second antennal claws in the present material appear to be fairly constant (although the matter is complicated by sexual dimorphism) and approximately the same length as the anterior margin of the penultimate joint. The third endite of the maxilla appears to be toothed in some specimens but in specimens from some localities it is difficult to be positive. The furca shows considerable variation in the various localities. Sometimes the anterior furcal seta is markedly shorter than the sub-terminal claw, sometimes shorter (typical *S. wierzejskii*, Fig. 2, 4), sometimes the two are almost equal and sometimes the anterior furcal seta is longer (typical *S. striatoreticulata*, Fig. 2, 5). The matter is further complicated by the fact that there is a tendency for the male furcal ramus to be more curved and slender than that of the female. For the present, the material covered here is referred to *Strandesia wierzejskii* (Grochmalicki) *sensu lato*. Since this species group has not been found in the Jaffna Peninsula, this interpretation does not affect the discussion on distribution given below.

The remaining taxa consist of rare species or new species which are left under open nomenclature.

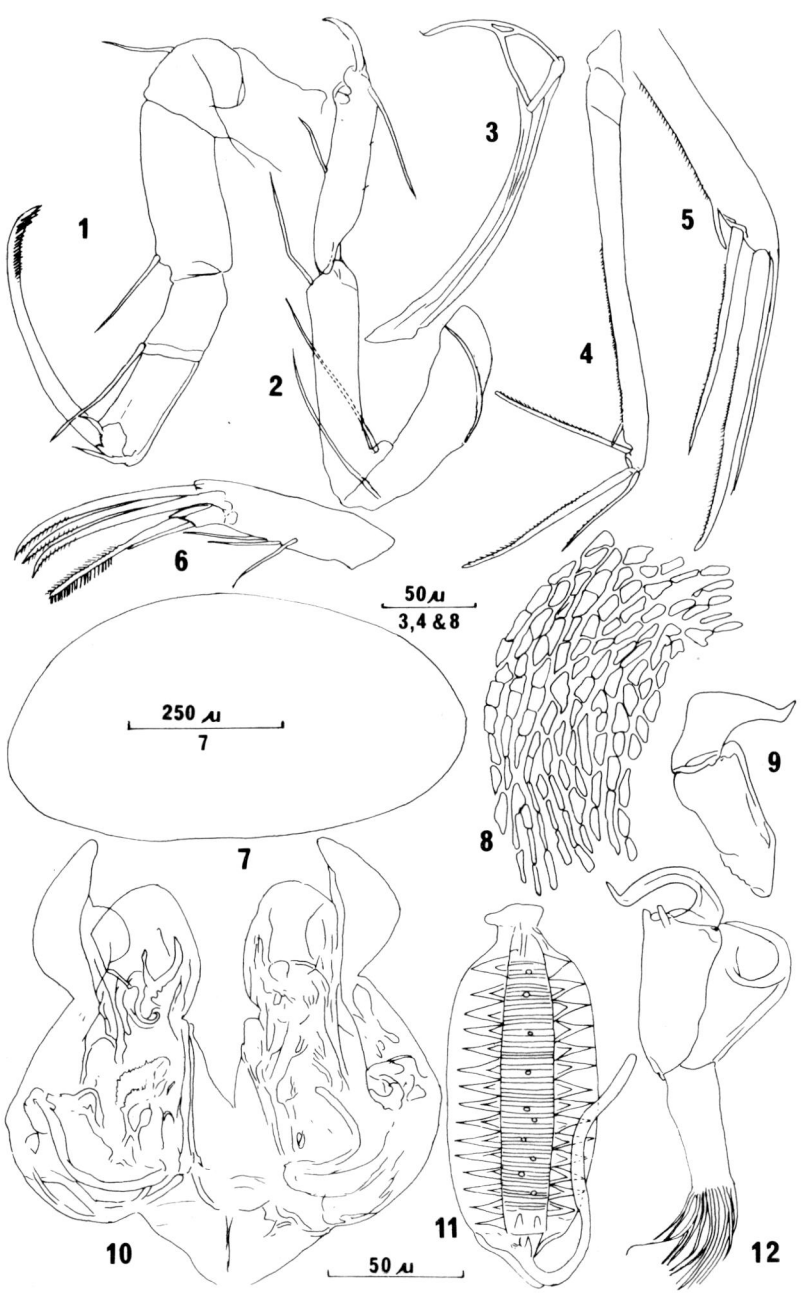

◀ Fig. 2. *Strandesia wierzejskii* (Grochmalicki 1915) sensu lato.
1. Limb 6 × 370, 2. Limb 7 × 370, 3. Furcal attachment × 260, 4. Furcal ramus × 260, 5 Detail of furcal ramus × 370, 6. Termination of left antenna x 370, 7. Outline of female left valve × 84, 8. Detail of shell reticulation, posterior part of left valve, arrow indicates anterior. × 260, 9. Male clasping organ, termination of right limb 5 × 370, 10. Male copulatory appendage × 370, 11. Zenker's Organ × 370, 12. Male left limb 5 × 370.
1,3. Female, HU.252.R.7., 2, 4, 6, 9, 10, 11, 12. Male, HU.252.R.8. Roadside pools, near Mundel, Sri Lanka, 4.1.65.
5,8. Female, HU.252.R.11. Tank, Drasastrawelliya, Sri Lanka, 3.1.65. 7. Female from spirit collection. Tank, Drasastrawelliya, Sri Lanka, 3.1.65.

Distribution

In view of the different times of collecting and the smallness of some of the faunas, one would be ill-advised to make too much of the differences between the various areas. Some specimens are obviously very rare. Only a single specimen of *Indiacypris luxata* (Brady 1886), a species recently covered by Neale & Victor (in press), was found and came from Belihuloya, although outside the ricefields it is known from two other localities. Similarly two small juveniles of *Ilyocypris* were found – one at Belihuloya and the other, referred only tentatively to the genus, came from Alankulama. Both were small and poorly preserved and it was impossible to say whether they were conspecific with *I. taprobanensis* Neale 1976 which is known from the single locality of Battuluoya. *Centrocypris viridis* Neale 1976 occurs in the ricefields at Karainagar East and Uddeputte in the Jaffna Peninsula, but whilst not known from the ricefields further South, it is known from a reservoir at Ma-Eliya. It appears to be an endemic species which occurs in some abundance at the rare localities in which it has been found (Fig. 5). Another apparently endemic species is the large new species of *Strandesia* which has been found at four localities in the Jaffna Peninsula ricefields and also in Nugegoda sample 32. *Cypris subglobosa* Sowerby 1840 only occurs in the Jaffna ricefields where it was found in six of the eight samples, and has not been recorded in the ricefields further south. This is misleading, however, since it is widely distributed elsewhere in central and southern Sri Lanka (Fig. 6), and its non-occurrence in the ricefields outside the northern peninsula is probably due to collection failure.

The species that occur in the other ricefields but not in those of the northern peninsula would appear to be more significant. *Pseudocypretta maculata* (Fig. 10) has been found in the central and southern ricefields examined, but has not been found north of Marawila. *Hemicypris pyxidata* (Fig. 4) has been found in a number of samples from the Nugegoda ricefields and in other water bodies in central and southern Sri Lanka but has not been found north of the latitude of Anuradhapura. *Strandesia marmorata, S. purpurascens* and *S. weirzejskii* show a similar sort of distribution. At this stage attention can be drawn to the fact that all five species occur in Indonesia but do not appear to be recorded from India, Malaysia or the rest of SE Asia, except for *S. purpurascens* which has been recorded from Bangkok, Thailand, by Vavra (1906).

Lastly three species, namely *Cypretta globula, Heterocypris dentatomarginatus* and *Stenocypris major,* have a wide geographical distribution.

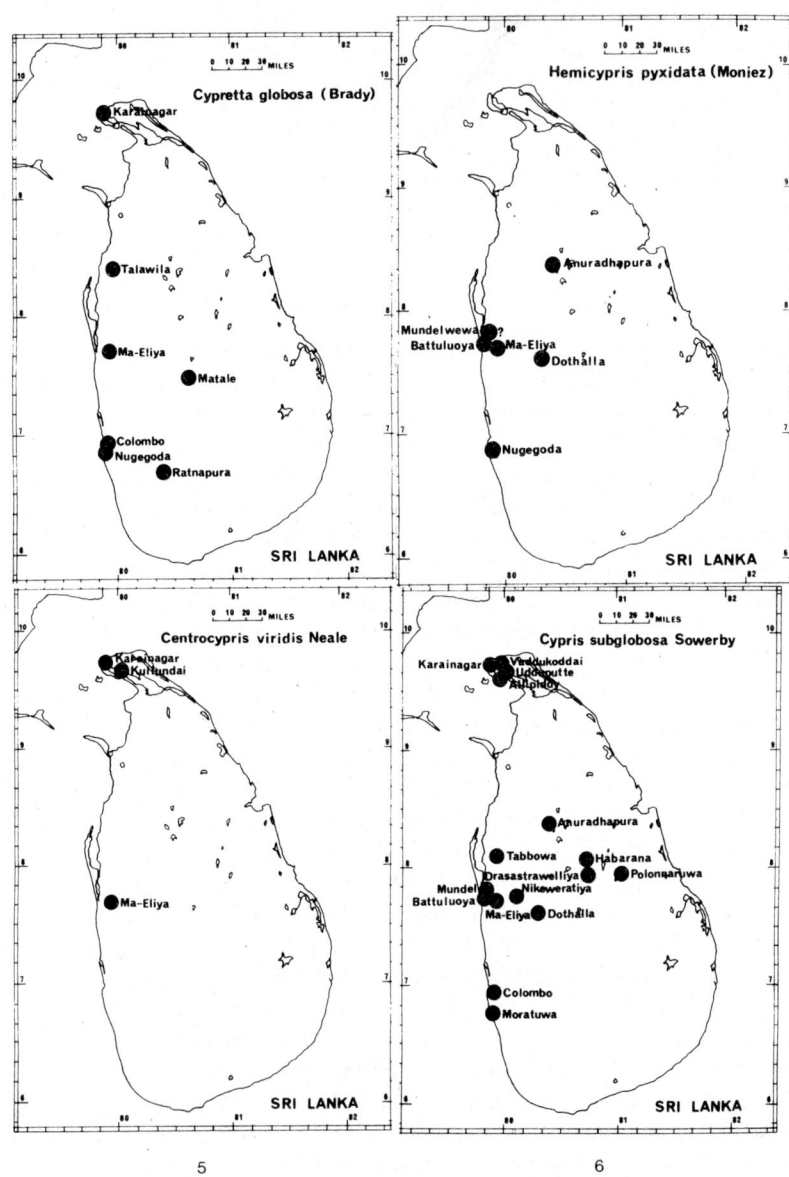

Fig. 3-10. Total known distribution in Sri Lanka of various species.

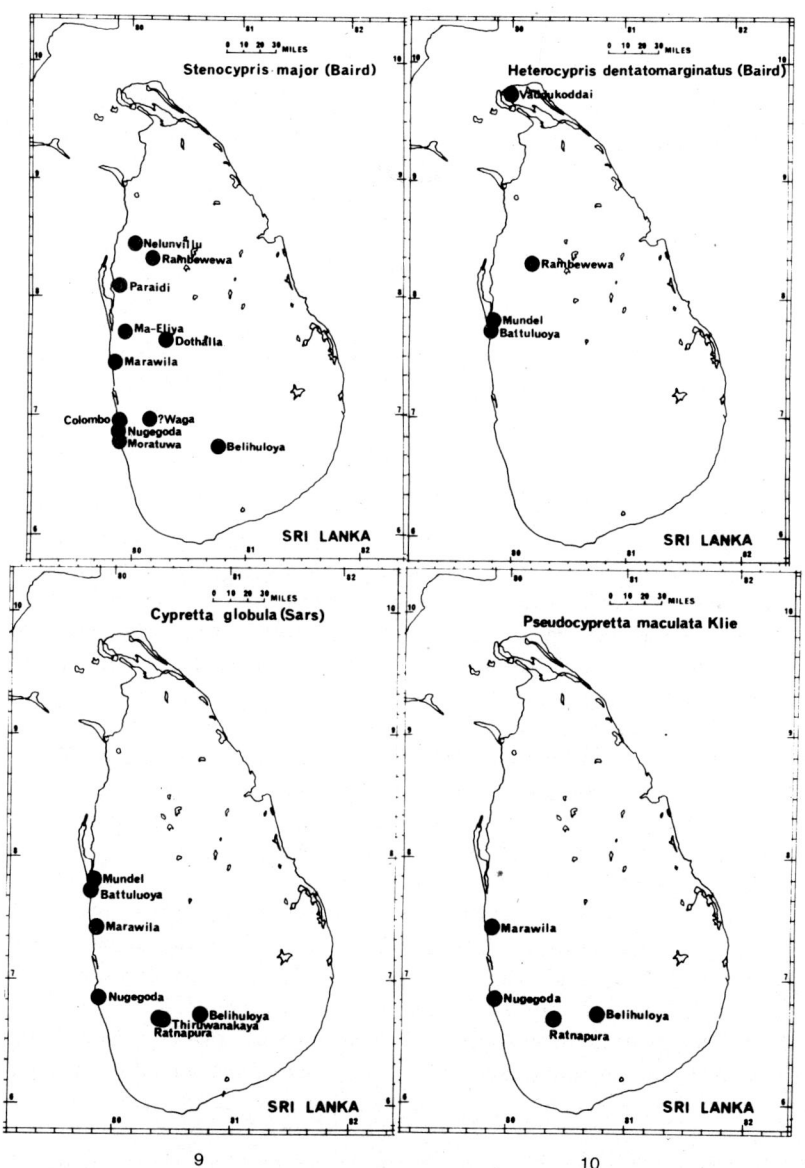

Faunal comparisons with other areas

1. Europe

Considerable work has been carried out during the last decade on the fauna of Europe and Asia Minor, especially on the Italian rice fields by Fox, Ghetti and Moroni. Moroni (1967) deals with the taxonomy and Ghetti (1973) gives relevant references up to that date. Apart from the geographically widespread *Cypretta globula* and *Stenocypris major* there are no other apparent links with the Sri Lanka fauna.

2. Africa

Some species from ponds and tanks in Sri Lanka have affinities with African species but this does not apply in the case of the ricefields species where there is no noticeable connection.

3. Indonesia

As noted above, a number of species are also found in Indonesia but not, apparently, in other parts of SE Asia. Clearly there is a very strong connection between the two areas which raises the problem of the distribution of freshwater species across the intervening ocean. Three ways suggest themselves. First, one of the most obvious ways is by the transfer of rice seed or rice plants between the two areas. Fox (1965) has suggested that dessication-resistant eggs among rice seed is the mechanism by which exotic species were introduced into the Italian ricefields and the same may well apply between Indonesia and Sri Lanka. No historical evidence has yet come to light but from the known pattern of rice culture and the distribution of the species it would appear more probable that the species were introduced into Sri Lanka from Indonesia than vice versa. It also suggests that such introduction was into the more southern ricefields and that the species generally have not yet extended northwards into the Jaffna Peninsula.

Secondly, the species may have been transferred in mud on the legs or plumage of migrating water birds such as duck or wagtails. This has been suggested in the past to explain the appearance of *Potamocypris humilis*, a South African species, on the south coast of Finland, localities which lie on the migratory route of the Arctic and Common Terns (Green 1961). There is little relevant information on bird migration in this region and this explanation seems much less likely than the above.

Third, there is a remote possibility that the species could have been introduced by some other mechanism such as from the ballast or freshwater tanks of visiting ships, but this seems very unlikely.

Conclusions

The ricefields of Sri Lanka contain an interesting ostracod fauna which consists

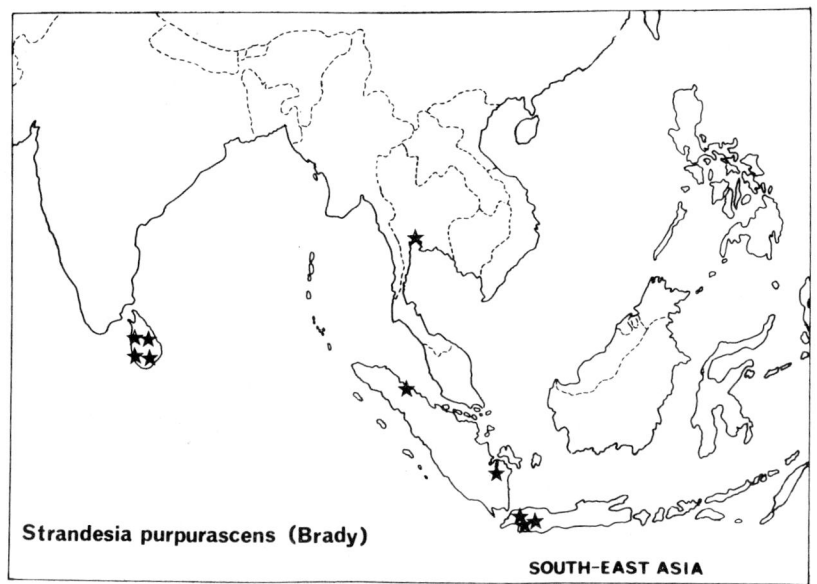

Fig. 11. Known distribution of *Strandesia purpurascens* (Brady 1886).

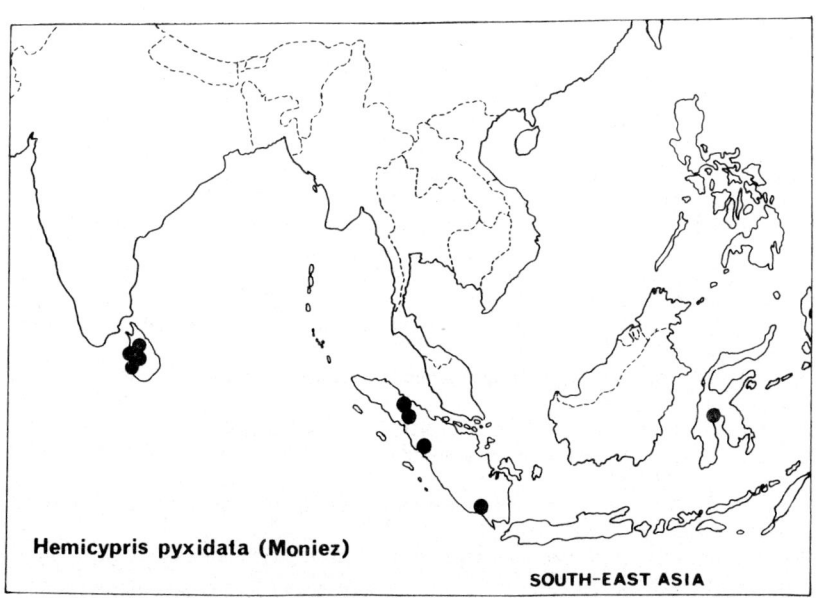

Fig. 12. Known distribution of *Hemicypris pyxidata* (Moniez 1892).

of rare or indigenous species, ubiquitous species, and species which have very strong links with Indonesia. The latter have not been found in the northern ricefields of the Jaffna Peninsula. Pending further work, it is tentatively suggested that the Indonesian links and present pattern of distribution are best explained by postulating the introduction of Indonesian species to central and southern Sri Lanka in rice seed or rice plants, and that, so far, these have not spread to the North.

Acknowledgements

I wish to thank Professor C. H. Fernando for providing the material and kindly making available laboratory facilities in the Department of Biology, University of Waterloo, Canada. I am also indebted to Dr. D. Goodwin of the British Museum (Natural History), Lord Medway of the British Ornithologists' Union and Dr. C. Mead of the British Trust for Ornithology who patiently answered my enquiries on bird migration.

References

Apstein, C. 1907. Das Plankton im Colombo-See auf Ceylon. Zool. Jahrb. 25; 201–244.
Baird, W. 1859. Description of some new recent Entomostraca from Nagpur collected by the Rev. S. Hislop. Proc. zool. Soc. Lond. 398; 231–234.
Brady, G. S. 1886. Notes on Entomostraca collected by Mr. A. Haly in Ceylon. J. Linn. Soc. Zoology, 19; 293–317.
Daday, E. von. 1898. Mikroskopische Süßwassertiere aus Ceylon. Termes. Füzetek. 21, Ostracoda 69–85.
Ferguson, E. 1969. The type species of the genus Stenocypris Sars 1889 with descriptions of two new species. In: Neale, J. W. (ed.), The Taxonomy, Morphology and Ecology of Recent Ostracoda. Oliver & Boyd, pp. 67–75.
Fox, H. M. 1965. Ostracod Crustacea from ricefields in Italy. Mem. Ist. Ital. Idrobiol. 18:205–214.
Ghetti, P. F. 1973. Dynamique des populations d'ostracodes de douze rizières italiennes. Notes d'écologie. Ann. Stat. biol. Besse-en-Chandesse. 7:273–294.
Green, J. 1961. A Biology of Crustacea. Aspects of Zoology. H. F. G. Witherby Ltd., London. 180 pp.
Grochmalicki, J. 1915. Beiträge zur Kenntnis der Süßwasserfauna Javas. Phyllopoda, Copepoda und Ostracoda. Bull. Int. Akad. Sci. Cracovie. Ser. B. Sci. Nat.:217–242.
Gurney, R. 1916. On some Fresh-water Entomostraca from Ceylon. Proc. zool. Soc. Lond.:333–343.
Klie, W. 1932. Die Ostracoden der Deutschen Limnologischen Sunda-Expedition. Archiv. f. Hydrobiol. 11:447–502.
Moroni, A. 1967. Ostracodi delle risaie italiane. Studium Parmense:1–79.
Neale, J. W. 1976. On Centrocypris viridis Neale sp. nov. Stereo-Atlas of Ostracod Shells, Leicester. 3:13–20.
Neale, J. W. 1976. On Ilyocypris taprobanensis Neale sp. nov. Stereo-Atlas of Ostracod Shells, Leicester 3:37–41.
Neale, J. W. 1976. On Cypris subglobosa J. de C. Sow. Stereo-Atlas of Ostracod Shells, Leicester 3:125–132.
Neale, J. W. & Victor, R. (in press). On Indiacypris luxata (Brady) a freshwater ostracod (Crustacea : Entomostraca) from Sri Lanka. J. Linn. Soc. Zool.

Okubo, I. 1972. Freshwater Ostracoda from Japan II. Cypris subglobosa Sowerby, 1840. Res. Bull. Shujitsu Jun. Coll., Okayama, No. 1:61–72.

Vavra, W. 1906. Ostracoden von Sumatra, Java, Siam, den Sandwich-Inseln und Japan. Zool. Jahrb. 23:413-438.

Author's address:

Dept of Geology, University of Hull, Hull, U.K.

Discussion

Kempf: I find Recent and Holocene specimens of *Cypris subglobosa* in the terminal lakes of the River Helmand on the Afghan-Iranian border. In my opinion, Recent and Quaternary specimens of *Cypris subglobosa* have been well identified in most cases, so that there is little taxonomic confusion. However, the type of *Cypris subglobosa* is a fossil from India described by Sowerby (1840) from sediments within a Tertiary volcanic series. At that time the sediments were not well dated but a Miocene or even older age has to be assumed, so that I wonder whether the Tertiary specimens of *Cypris subglobosa* are really identical with the Quaternary ones. Did you compare Recent specimens of *Cypris subglobosa* with Sowerby's type material?

Neale: So far it has not been possible to trace Sowerby's type, which came from the Sichel Hills, and it seems to me unlikely that it will now be found. On the other hand, the interpretation of the species is satisfactorily stabilised as far as the living species is concerned and this presents no taxonomic problem. Sowerby's figure is small and gives little information beyond general shape which is not inconsistent with the present interpretation so that there is much to be said for maintaining the status quo.

Sixth Intern. Ostracod Symposium, Saalfelden

THE DISTRIBUTION OF THE 'GIANT' OSTRACODS (FAMILY: CYPRIDIDAE BAIRD, 1845) ENDEMIC TO AUSTRALIA

PATRICK DE DECKKER

Abstract

Australocypris, Mytilocypris and *Trigonocypris* can tolerate waters of wide ranging salinities (some species tolerate up to 112.5‰ salinity). These 'giant' ostracods are recorded in most parts of Australia, except in the centre and northern part of the continent.

It is suggested that their dispersal could come about by some of the continuously migrating waterfowl (often seen inhabiting brackish waters) carrying ostracods and ostracod eggs.

Introduction

A number of species of 'giant' ostracods ($>$ 3 mm) of the tribe Mytilocypridini are widely, and often disjunctly, distributed in Australia. During field work, it was noticed that large flocks of waterfowl typically inhabit the same waters. Since it is possible that waterfowl are inadvertently responsible for the dispersal of ostracods, an analysis of the distribution of these ostracods, dispersal mechanisms of ostracods, and distribution of waterfowl and their possible role in dispersal of ostracods is presented.

Distribution of ostracods of the tribe Mytilocypridini

Most of the ostracods belonging to the tribe Mytilocypridini De Deckker, 1974 are grouped within three already known genera (*Mytilocypris* McKenzie, 1966, *Australocypris* De Deckker, 1974 and *Trigonocypris* De Deckker, 1976) (Table 1).

Table 1. List of occurrences of studied Mytilocypridinid Ostracods recorded in the literature and from my unpublished data.

Australocypris De Deckker, 1974.

A. *Australocypris hypersalina* De Deckker, 1974
– cut-off arm of Lake Eliza, S.A. (De Deckker, 1974)
– Lake Mitre, Vic.
– Lake near Nora Creina Bay, S.A. (De Deckker, 1974)
– Lake Robe, S.A. (De Deckker, 1974)
– Lake St. Clair, S.A. (De Deckker, 1974)
– Pond near Salt Works, 12 km E. of Langhorne Ck., S.A.
– Salt Lake, Beachport, S.A.
– shallow saline swamp near southern end of Lake Eliza, S.A. (De Deckker, 1974)
– small lake near Errington's Hole on way to the boundary, S.A. (De Deckker, 1974)
– small lake on opposite side of road to southern end of Lake Eliza, S.A. (De Deckker, 1974)
– south of The Coorong, S.A.
– St. Mary's Lake, Vic.

Table 1 ctd.

B. *Australocypris insularis* (Chapman, 1966)
- Birida, 8.5 km south of Salt Works, Shark Bay, W.A. (26°33'S 113°24'E) = Loc. 21
- Birida, 40 km west of Hamilton Homestead, Shark Bay, W.A. (26°33'S 113°55'E) = Loc. 23
- Birida, 12 km south of Salt Works, Shark Bay, W.A. (26°12'S 113°24'E) = Loc. 24
- Houtman's Abrolhos, West Wallabi Island, W.A. (Chapman, 1966)

C. *Australocypris robusta* De Deckker, 1974
- Lake Beeac, Vic. (De Deckker, 1974)
- Lake Gnarpurt, Vic. (De Deckker, 1974)
- Lake Gnotuk, Vic. (De Deckker, 1974)
- Lake Keilambete, Vic. (De Deckker, 1974)
- Salt Works, Turnbridge, Tas.
- Samphire Flats, Barrilla Bay, Tas. (= *A.* sp. De Deckker, 1974)
- south of The Coorong, S.A.

D. *Australocypris? rectangularis* n.sp.
- Centre Lake, Vic.

Trigonocypris De Deckker, 1976

A. *Trigonocypris globulosa* n. sp.
- Lake Buchanan, Qld.
- The Salt Lake, 80 km south of Tibooburra, N.S.W.

B. *Trigonocypris timmsi* De Deckker, 1976
- Pine tree Creek Lagoon, via Hughenden, Qld. (De Deckker, 1976)

C. *Trigonocypris* sp. De Deckker, 1976
- Fossil, Tertiary, Central Qld. (De Deckker, 1976)

Mytilocypris McKenzie, 1966.

A. *Mytilocypris henricae* (Chapman, 1966)
- Bulldozer Swamp, Rottnest Island, W.A.
- Currawong via Salt Creek, S.A.
- Lake Coradgill, Vic.
- Lake Linlithgow, near Hamilton, Vic. (Chapman, 1966)
- Lake Lomond, Vic. (F. Chapman's collection)
- Lake Purdigulac,Vic.
- Oldfield River, Rottnest Island, W.A.
- pond near Lake White, Vic.

B. *Mytilocypris mytiloides* (Brady, 1886)
- Blackmans Lagoon, NE Tas.
- Creek off Logans Lagoon, Flinders Island.
- Kangaroo Island (Brady, 1886)
- Lake Edward, S.A. (37°38'S 140°36'E)
- Lake Hopetoun, Vic. (Chapman, 1966)
- Lake Wallace, Vic. (Chapman, 1966)
- Sheepwash Lagoon,Vic. (38°08'S 141°11'E)
- White Lagoon, Tas. (42°04'S 147°28'E)

C. *Mytilocypris splendida* (Chapman, 1966)
- Collins' Lake, near Lake Yallaker, Vic.
- Lake Bolac, Vic. (Chapman, 1966)
- Lake Bookar, Vic. (Chapman, 1966)
- Lake Buchanan, Qld.
- Lake Colac, Vic.
- Lake Cooper, Vic. (Chapman, 1966)
- Lake Gnarpurt, Vic.
- Lake Kariah, Vic.
- Lake Learmouth, Vic. (Chapman, 1966)
- Muddah Lake, N.S.W.
- The Salt Lake, 80 km S. of Tibooburra, N.S.W. (30°08'S 142°07'E)
- Salt Works, Turnbridge, Tas.

D. *Mytilocypris tasmanica* McKenzie, 1966
- Calverts Lagoon, South Arm, near Hobart, Tas. (McKenzie, 1966)
- Boneo Swamp, Vic. Pleistocene

E. *Mytilocypris ambiguosa* n. sp.
- Barkers Swamp, Rottnest Island, W.A.
- Lighthouse Swamp, Rottnest Island, W.A.
- Pool near Lake Goldsmith, Vic.
- 5 km east of Ongerup, W.A. (M.A. Chapman's collection)

F. *Mytilocypris minuta* n. sp.
- Barkers Swamp, Rottnest Island, W.A.
- Currawong, via Salt Creek, S.A.
- Lake Preston, 50 km south of Perth, W.A.

G. *Mytilocypris praenuncia* (Chapman, 1936)
- Lake Bookar, Vic.
- Lake Coradgill, Vic.
- Lake Corangamite, Vic.
- Lake Kariah, Vic.
- Lake Keilambete, fossil
- small lake near Centre Lake, Vic.
- The Mallee, Vic. Pleistocene (Chapman, 1936)

H. *Mytilocypris* sp. Fossil
- Late Tertiary, Morwell Coal Seam, Geelong, Vic. (Waldman & Handby, 1968)

I. genus *Mytilocypris*
- Lake Modewarre, Vic. (Pollard, 1970)
- Murchinson River, W.A. (McKenzie, 1971)
- Swampy Lagoon, near Kilcunda, Vic. (McKenzie, 1971)

All are endemic to Australia. They are recorded from inland waters of wide ranging salinities (0 to 112.5‰); typically they live in at least slightly saline waters or in ephemeral lakes.

These ostracods are abundant in the western and south-eastern parts of Australia (including Tasmania, Flinders and Kangaroo Islands) but there are also records from Central Queensland (Fig. 1 & Table 1). There are only two records from New South Wales which could be due to lack of collecting in this state, but there are not many athalassic saline lakes that are accessible in the eastern parts, nor are there many accessible in South Australia. Similar comments apply to Queensland.

Outstanding examples of disjunct distributions are provided by *Mytilocypris henricae*, *M. mytiloides* and *M. splendida* (Table 1).

Dispersal mechanisms of ostracods

The sex ratio in all my samples of mytilocypridinid ostracods is unity, suggesting that this tribe lacks parthenogetic species. Hence for a species to populate a new water body successfully, either many specimens of both sexes or many eggs are needed. Eggs of mytilocypridinids (as for most cypridinids) can withstand desic-

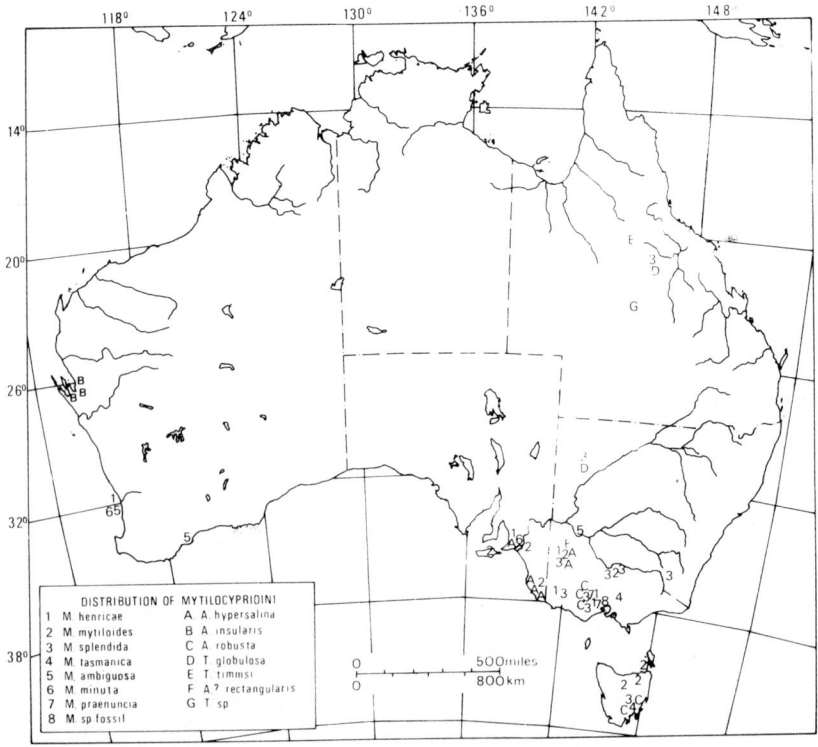

Fig. 1. Distribution of Mytilocypridinid Ostracods

cation. This is shown by their protective double-wall structure (McKenzie 1971) and proved by Sars who raised ostracods from dried mud.

Ostracods and ostracod eggs can be carried either by moist winds, or inside the alimentary canals of fishes (Kornicker & Sohn 1971) and birds (Proctor 1964 and Proctor & Malone 1965) or attached to the body of birds (Sandberg & Plusquellec 1974) visiting the waters in which the ostracods live.

In the Australian context, it is difficult to visualize that eggs or ostracods could be carried by moist winds for long distances (some localities of *Mytilocypris henricae* are separated by up to 2,400 km) across the dry centre, or from south to north or vice versa (in the case of *M. splendida*). Even if this was possible it would be an extreme coincidence for ostracods and their eggs to be dropped into a suitable environment i.e. a small saline lake.

Dispersal by fishes does not apply to species in the tribe Mytilocypridini, since the saline or ephemeral lakes they inhabit are endorheic.

The remaining agents of dispersal for mytilocypridinids are birds, especially migrating waterfowl. This has been suggested for other ostracods by many authors (Klie 1926, Sandberg 1964, Proctor 1964, Proctor & Malone 1965, Mckenzie 1971 and Sandberg & Plusquellec 1974).

Similarly, Baker (1945) reported that the distribution of freshwater snails in the West Indies corresponded to the migration pattern of birds which transported and distributed them. Further, Yen (1947) mentioned the modes of distribution of molluscs which are attached to the feet of wading birds and transported for long distances.

Distribution of waterfowl and their possible role in dispersal of ostracods

Among the 19 species of Australian waterfowl, 12 (Table 2) are found over most of the continent, particularly in northern parts. Most inhabit various types of water, ranging from fresh to saline, and are often seen migrating to new places when shortage of food occurs, or during drought or flood conditions. Norman (1971) studied the movement of the Black Duck, Mountain Duck and Grey Teal populations in South Australia. He concluded that the Grey Teal disperse widely in most directions, some up to 1000 km or more, and the Black Duck and Mountain Duck populations appear to contain a proportion which move long distances (even up to at least 500 to 1000 km).

It is known from stomach content analyses of these species (Frith 1967 and Frith et al. 1969) (Fig. 2) that ostracods are ingested by at least 7 of them. Two of these are plant feeders and hence must accidentally swallow ostracods while feeding; ingestion of ostracods (or at least of microinvertebrates) is probably more deliberate in the plant-animal feeders (see Table 2). With reference to the former situation, ostracods are known to attach their eggs to plant matter. The author has also observed *Australocypris robusta* laying eggs in empty gastropod shells as well as attaching them to algae in an aquarium.

These waterfowl could disperse mytilocypridinid ostracods in a number of ways:

(a) Ostracod eggs could be swallowed in one lake by the waterfowl and ejected

Table 2. Composition of food and presence of ostracods in gizzards of 12 species of Australian waterfowl (data after Frith, 1967 and Frith et al., 1969).

		Presence of ostracods in gizzards.
Plant feeder (<80% total volume	Black swan	
	Freckled duck	× (0.2%)
	Grey teal	
	Black duck	×
	Wood duck	
Plant-animal feeder (± 50–50% total volume	Hard head	×
	Blue-billed duck	×
	Pink-eared duck	×
	Shoveler	×
	Musk duck	×
	Chestnut teal	
	Mountain duck	

in the faeces of the bird in another lake if the animal has been travelling just after eating. As mentioned before, this could occur when the plant-feeder birds accidentally swallow ostracod eggs attached to plants. Proctor (1964) fed a wide variety of crustacean eggs, among which were 5 species of ostracods (4 of which belong to the family Cyprididae) and recovered them from the lower digestive tract of both domesticated and wild ducks. These have been hatched and raised to maturity. Similarly, Proctor & Malone (1965) fed the eggs of 2 cypridids to mallard ducks and recovered them from the faeces of the birds. Both species grew from these eggs but it was noticed that the eggs of one species seemed capable of withstanding avian digestion processes better than the others. It is understood that many eggs have to be dropped in a lake in order to populate a new environment successfully, but this could easily be achieved as many eggs can be carried in the faeces of only one bird.

(b) Ostracods could also be egested by the waterfowl, especially by the plant-animal feeders. Kornicker & Sohn (1971) said that none of the ostracods they fed to fishes in their experiment was defecated alive. This could well be the case for the mytilocypridinids, being larger in size and therefore easily crushed inside the birds or damaged by the digestive fluids. However, some of the carapaces enclosing and protecting many eggs could be defecated and eggs would be hatchable if dropped in slightly saline waters.

However, I have seen many specimens of *Mytilocypris splendida* (Chapman 1966) collected by Dr. E. F. Riek from the stomach of a trout in Lake Learmouth, Victoria. These were in a perfect state of preservation and contained eggs inside their carapaces. Sandberg & Plusquellec (1974) mention specimens of an ostracod taken from the stomach of a duck showing little evidence of valve dissolution.

(c) Ostracods and their eggs could be carried for long distances from one lake to another attached to feathers and the waterfowl's legs. (While collecting ostracods in some lakes where they were abundant, I often found that after putting my arm in the water, a few ostracods stayed attached to the hairs of my arm. Occasionally, the ostracods were surrounded by a bubble of water allowing them to stay moist as they closed their carapaces in order to avoid drying up). McKenzie & Hussainy (1968) tested some ostracods by leaving them for 9 days on moist filter paper protected by 2 petri dishes. After that period of time, the ostracods were seen returning to normal activity after immersion in lake water. This would support the idea that ostracods could easily be transported among the feathers of the waterfowl or

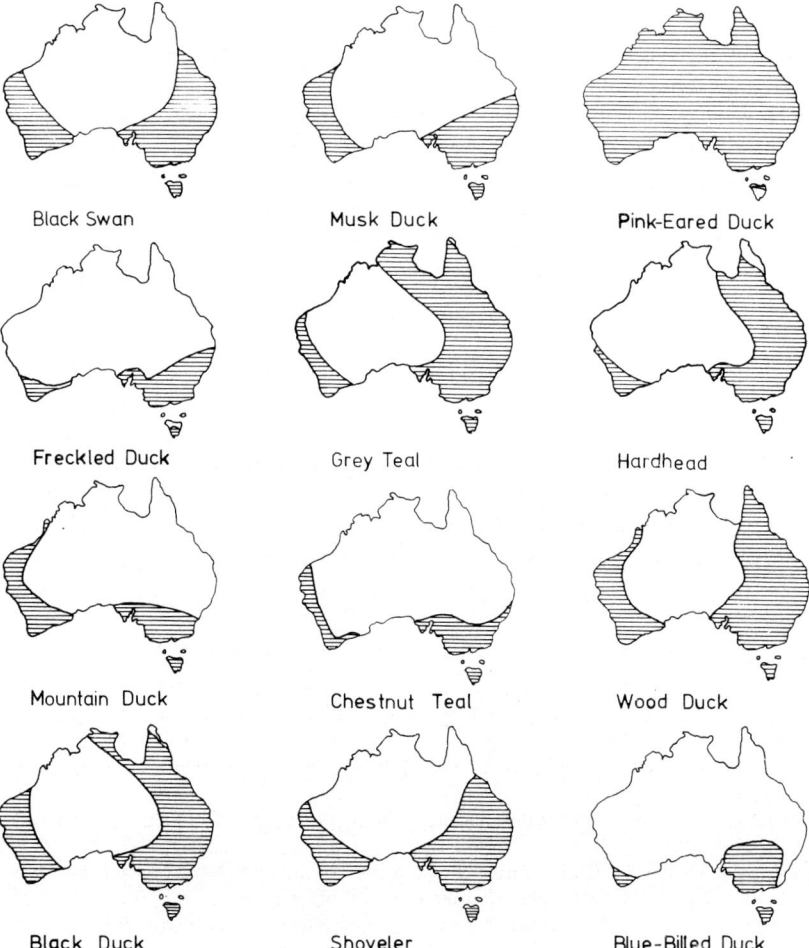

Fig. 2. Maps showing the distribution (in shaded) for 12 species of the Australian Waterfowl. (Data from Frith 1967 and MacDonald 1973.)

in a mud cake attached to the bird's legs without suffering desiccation and therefore being transported into a new body of water.

For example, to explain the presence of *Mytilocypris splendida* at an altitude of 990 m in Muddah Lake near Cooma, N.S.W., one could say that this ostracod would probably have been carried there by waterfowl (e.g. Mountain Duck).

Conclusions

It has been suggested for some time that ostracods and their eggs could be transported by means of moist winds, fishes and migrating waterfowl. The former two agents are considered rather unlikely in Australia for the mytilocypridinid ostracods which often inhabit saline waters. The migrating waterfowl, often visiting such waters and having the same geographical distribution as the mytilocypridinids, could be acting as dispersal agents for these ostracods, this being achieved in the following ways:
– ostracod eggs could be egested by the waterfowl and dropped in another lake within the faeces of the birds;
– ostracods and their eggs could be carried attached to the bird's legs and feathers prior to and during flight and deposited later in another body of water.

Finally, similar conclusions could be drawn for the distribution of *Diacypris*, a small ostracod from saline waters in Western Australia, South Australia, Victoria and Tasmania.

Acknowledgements

Dr. B. V. Timms read the manuscript and kindly suggested improvements. I also wish to record my thanks to Drs. I.A.E. Bayly, P.S. Lake, B.V. Timms who provided me with mytilocypridinid ostracods and to Mr. G.L. Dean-Jones who helped in the collection of samples.

References

Baker, F.C.1945. The molluscan family Planorbidae. Univ. of Illinois Press, Urbana, pp. 1-530.
Bayly, I.A.E. & Williams, W. D. 1973. Inland Waters and Their Ecology (Longman, Australia).
Chapman, F. 1936. Cypridiferous limestone from the Mallee. Rec. Geol. Surv. Vic. 5 (2):296-298.
Chapman, M. A. 1966. On Eucypris mytiloides (Brady) and three New species of Eucypris Vavra (Cyprididae, Ostracoda). Hydrobiologia 27:368-378.
De Deckker, P. 1974. Australocypris, a new ostracod genus from Australia. Aust. J. Zool. 22:91-104.
De Deckker, P. 1976. Trigonocypris, a new ostracod genus from Queensland. Aust. J. Zool. 24:145-157.
Frith, H. J. 1967. Waterfowl of Australia. Angus & Robertson, Sydney.

Frith, H. J., Braithwaite, L. W. & McLean, J. L. 1969. Waterfowl in inland swamp in New South Wales. II. Food. CSIRO Wild. Res. 14:17-64.
Klie, W. 1926. Ostracoda, Muschelkrebse. In: Schulze, P., Biologisches der Tiere Deutschlands 22 (16):1-56.
Kornicker, L. S. & Sohn, I. G. 1971. Viability of ostracode eggs egested by fish and effect of digestive fluids on ostracode shells-ecologic and paleontologic implications. Bull. Centre Rech. Pau SNPA 5 suppl.:125-135.
MacDonald, J. D. 1973. Birds of Australia. Reed, Sydney.
McKenzie, K. G. 1971. Palaeozoogeography of freshwater Ostracoda. Bull. Centre Rech. Pau SNPA 5 suppl.:207-237.
McKenzie, K. G. & Hussainy, S. U. 1968. Relevance of a freshwater cytherid (Crustacea, Ostracoda) to the Continental Drift Hypothesis. Nature 220 (5169):806-808.
Norman, F. I. 1971. Movement and Mortality of Black Duck, Mountain Duck and Grey Teal in South Australia, 1953-1963. Trans. R. Soc. S. Aust. 95 (1):1-7.
Proctor, V. W. 1964. Viability of Crustacean Eggs recovered from Ducks. Ecology, 45:656-658.
Proctor, V. W. & Malone, C. R. 1965. Further evidence of the passive dispersal of small aquatic organisms via the internal tract of birds. Ecology 46:728-729.
Sandberg, P. A. 1964. The ostracod genus Cyprideis in the Americas. Stockh. Contr. Geol. 12:1-178.
Sandberg, P. A. & Plusquellec, P. L. 1974. Notes on the anatomy and passive dispersal of Cyprideis (Cytheracea, Ostracoda). Geoscience and Man 6:1-26.
Yen, Teng-Chien. 1947. Distribution of fossil fresh-water mollusks. Geol. Soc. Amer. Bull. 58:293-298.

Author's address:

Institut Géologique, Université de Louvain, 1348 Louvain-la-Neuve, Belgium

Discussion

Löffler: The coeca of waterfowl provide storage of duration stages up to 10 days, that is, much longer than any digestive tract passage would provide for. (See: Löffler, H. & Leibetseder, J. 1965 Daten zur Dauer des Darmdurchganges bei Vögeln. Zool. Anz. 177 (5/6):334–340.)

Kornicker: The high diversity of species of this group in Australia suggest fragmentation of an ancestral population once widespread there. The present species evolved in geographically isolated ponds after fragmentation of the original ancestral species. Dispersal by ducks may be a late occurrence and could account for sympatric species, but not for the high diversity of species in Australia.

De Deckker: You just summarized what I tried to say: the occurrence of sympatric species today is caused by the ever migrating waterfowl which carry ostracods from one place to another. The high diversity of mytilocyprinidinid species is surely caused by fragmentation and isolation of an ancestral population which occurred on vast areas of the Australian continent like the Murray Basin, the Darling-Warrego Basin and Lake Eyre Basin which were intermittently wetter than today during the Late Tertiary and Early Quaternary.

Neale: One reason why I preferred the rice seed hypothesis to the transference of ostracods by bird hypothesis is that whilst there is a dearth of information on bird migration in S.E. Asia, my impression is that migration follows a general north-south direction rather than an east-west one. Does this accord with your data in the Australian area ?

De Deckker: I think that waterfowl in Australia has a rather complex movement pattern. These birds are mostly found in the wetter coastal areas of the continent and only disperse inland, in all directions, when rain and flooding conditions occur, thus creating new temporary habitats.

Whatley (to Neale): Herons and similar birds are quite as likely to transport ostracods as anatids. Herons, throughout much of the world, undergo fairly widespread longitutinal migrations, particularly in times of drought or flooding.

Sixth Intern. Ostracod Symposium, Saalfelden

ON THE ORIGIN AND DIVERSITY OF EUROPEAN FRESHWATER INTERSTITIAL OSTRACODS

DAN L. DANIELOPOL

Abstract

The representatives of the few ostracod groups which live in interstitial habitats derive mostly from freshwater epigean lines. Crawling species, lacking or with reduced antennal swimming bristles, are favoured to colonize interstitial systems.

The Candoninae, a predominantly amphigonic group, has had a high rate of diversification and plentiful opportunities for specialisation since the Tertiary. This partly explains the large number of candonine species in interstitial habitats. The Darwinulidae and Metacyprinae of the genus *Kovalevskiella*, with few interstitial species, have parthenogenetic reproduction and, therefore, a low rate of diversification.

Introduction

Information on freshwater subterranean (interstitial and cave dwelling) ostracods has increased during recent years. A comprehensive bibliography on this subject was presented by Danielopol (1976). Investigations on freshwater interstitial ostracods have concentrated mainly on European forms. Figure 1,A shows the groups of ostracods which have representatives known exclusively from freshwater interstitial and from interstitial and cave habitats. The subfamily Candoninae is represented by 6 genera and 36 species; *Pseudocandona* – 19, *Cryptocandona* – 4, *Fabaeformiscandona* – 2, *Candonopsis* – 3, *Mixtacandona* – 7, *Phreatocandona* – 1); the Metacyprinae has one genus *Kovalevskiella* (= syn. *Cordocythere*) with 5 hypogean species; the Pseudolimnocytherinae has only one representative (*Pseudolimnocythere hypogea*); the Kliellidae has 2 monotypic genera (*Kliella* and *Nannokliella*); the Darwinulidae has only one hypogean species which lives in an interstitial habitat (*Darwinula boteai*).

Such representatives are referred to as hypogean animals or troglobites (for a discussion, see Danielopol 1976). Figure 1, B presents the groups which have troglophillic species and which are frequently encountered in interstitial systems as well as in surface habitats (e.g. running waters). The subfamily Candoninae is represented by 4 genera (*Candona* – 3 sp., *Pseudocandona* – 3 sp., *Fabaeformiscandona* – 1 sp., *Cryptocandona* – 1 sp.); the subfamily Cpridinae by 4 genera (*Ilyodromus* – 1 sp., *Eucypris* – 1 sp., *Cypridopsis* – 1 sp., *Potamocypris* – 2 sp.); the Cyclocypridinae by one genus (*Cypria* – 1 sp.); the Limnocytherinae by one genus (*Limnocythere* – 1 sp.).

The number of ostracod groups which have troglobitic and troglophilic representatives living in freshwater interstitial environment is reduced as compared to the number of groups with epigean species. Figure 1,C presents information on 273 ostracod species which live in freshwater habitats of Europe (data from Löf-

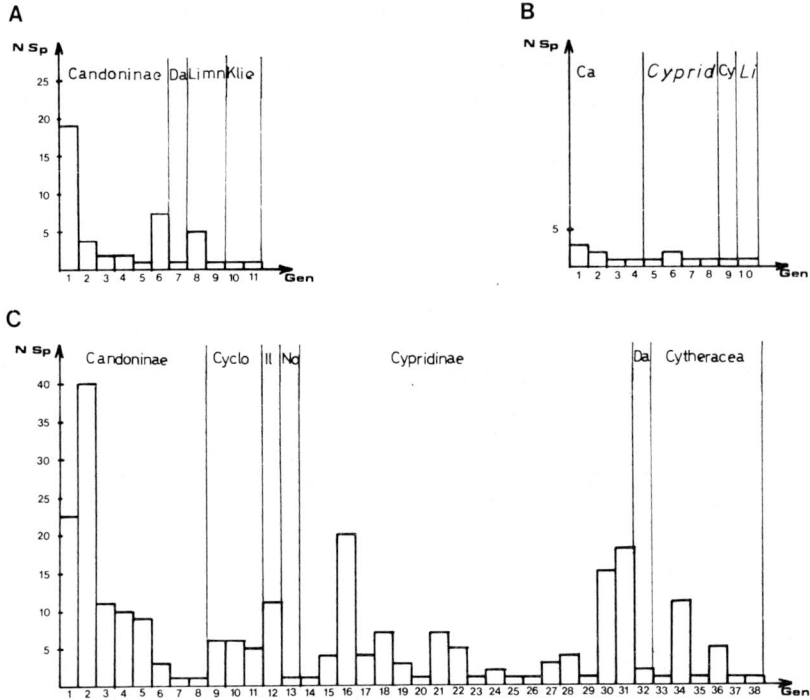

Fig. 1. A. The freshwater ostracod groups which have exclusively interstitial or interstitial and cave dwelling species (Da- Darwinulidae, Limno-Limnocytheridae, Klie-Kliellidae; 1-*Pseudocandona*, 2-*Cryptocandona*, 3-*Fabaeformiscandona*, 4-*Candonopsis*, 5-*Phreatocandona*, 6-*Mixtacandona*, 7-*Darwinula*, 8-*Kovalevskiella*, 9-*Pseudolimnocythere*, 10-*Kliella*, 11-*Nannokliella*). B. The freshwater ostracod groups which have troglophilic species (Ca-Candoninae, Cy-Cyclocypridinae, Li.Limnocipridinae; 1-*Candona*, 2-*Pseudocandona*, 3-*Fabaeformiscandona*, 4-*Cryptocandona*, 5-*Ilyodromus*, 6-*Eucypris*, 7-*Cypridopsis*, 8-*Potamocypris*, 9-*Cypria*, 10-*Limnocythere*). C. A general view of the freshwater ostracod fauna of Europe (data from Löffler 1967); 1-*Candona*, 2-*Pseudocandona*, 3-*Fabaeformiscandona*, 4-*Cryptocandona*, 5-*Mixtacandona*, 6-*Candonopsis*, 7-*Nannocandona*, 8-*Paracandona*, 9-*Cyclocypris*, 10-*Cypria*, 11-*Physocypria*, 12-*Ilyocypris*, 13-*Notodromas*, 14-*Cyprois*, 15-*Cypris*, 16-*Eucypris*, 17-*Cypricercus*, 18-*Heterocypris*, 19-*Dolerocypris*, 20-*Stenocypria*, 21-*Herpetocypris*, 22-*Ilyodromus*, 23-*Hungarocypris*, 24-*Strandesia*, 25-*Candonocypris*, 26-*Stenocypris*, 27-*Isocypris*, 28-*Cypretta*, 29-*Scottia*, 30-*Cypridopsis*, 31-*Potamocypris*, 32-*Darwinula*, 33-*Leucocythere*, 34-*Limnocythere*, 35-*Cytherissa*, 36-*Leptocythere*, 37-*Cythereis*, 38-*Metacypris*; Cyclo-Cyclocypridinae, Il-Ilyocypridinae, No-Notodromadinae, Da-Darwinulidae. The following genera are not included in the figure: *Syphlocandona* (a junior synonym of *Candona*), *Cytherois*, *Cyprideis*, *Cytheromorpha* and *Loxoconcha* (their representatives are not living in freshwater habitats in Europe). Not included also data on the genera *Sphaeromicola* and *Hemicythere* and on some doubtful species belonging to Candoninae and Cypridinae. The names of the Candoninae groups, according to Danielopol, 1976.

fler 1967). the Candoninae and the Cypridinae appear to be the most abundant groups. The former not only has the largest number of species encountered in European groundwaters (see also Danielopol 1971), but also many epigean forms.

The present contribution discusses some aspects of the origin of the main freshwater interstitial ostracod groups and attempts to explain their diversity.*

Marine and freshwater origins

Delamare-Deboutteville (1960) showed that many freshwater interstitial groups have a marine origin. Noodt (1974) mentioned: 'Das Limnopsammal wurde zum geringen Teil von epibenthischer Limnofauna besiedelt. Häufiger ist der Weg über die Stationen marines Mesopsammal – Sandstrand limnisches Mesopsammon verlaufen...' pp. 448-449. A similar statement on the cavernicoles was also made by Vandel (1965): 'Strangely enough the majority of aquatic cavernicoles have not been derived from freshwater epigeous forms but from marine forms'. p. 272. In the case of ostracods, most of the marine groups which colonized coastal groundwaters could not overcome the salinity barrier and hence did not penetrate into the freshwater realm (Fig. 2,A). This is the case of the Polycopids, of most of the cytheraceans (Hartmann, 1973), and of the Bairdiids of the subfamily Pussellinae (Danielopol 1973). Most of the freshwater interstitial ostracods, such as the Metacyprinae belonging to the genus *Kovalevskiella*, the Darwinullidae, the Candoninae, the Cyclocypridinae and the Cypridinae, are derived from epigean freshwater forms. The Kliellidae and the Pseudolimnocytherinae are the only groups for which a direct marine origin can be inferred. Schäfer (1951) suggested that this freshwater group is related to the Psammocytheridae inhabiting the marine interstitial habitats of Europe (Klie 1938). The author postulated (1971) that the two groups are related because they have thoracopods with four endopodial segments in common, a characteristic which is very peculiar in the Cytheracea. The difference in carapace structure of the Kliellidae and Psammocytheridae suggests that the former group did not originate directly from the latter.**

As the sediments of running waters allow a direct contact between the epigean and the hypogean habitats for long periods of time (in sense of evolutionary time) in many cases, it is most probable that many of the actual interstitial species of Europe originated from species which originally lived in surface running waters (Fig. 2B). An example of such a species is *Cypridopsis subterranea* which occurs in epigean running waters on gravel and sand bottoms (Klie 1938), in the hyporheal (Löffler 1963) and in deep groundwater (Wolf 1919). *Darwinula boteai* has

---

* The author expresses his gratitude to Miss Susan Powell (Wien) and Dr. R. Whatley (Aberystwyth) who improved the English form of the text.
** Schäfer (1945) noted that the Kliellinae have been found in 'Grundwasser' or interstitial water. Some years later (1951) he changed views mentioning that the animals came from 'Spaltwasser' which means water in large crevices. However, taking into account that the faunistic association is a typical interstitial one (isopods belonging to the genus *Microcharon*, harpacticoids of the genus *Parastenocaris*) it is very likely that the typical habitat of the Kliellidae is an interstitial one. Hartmann (1973) arrived at the same conclusion using morphological arguments.

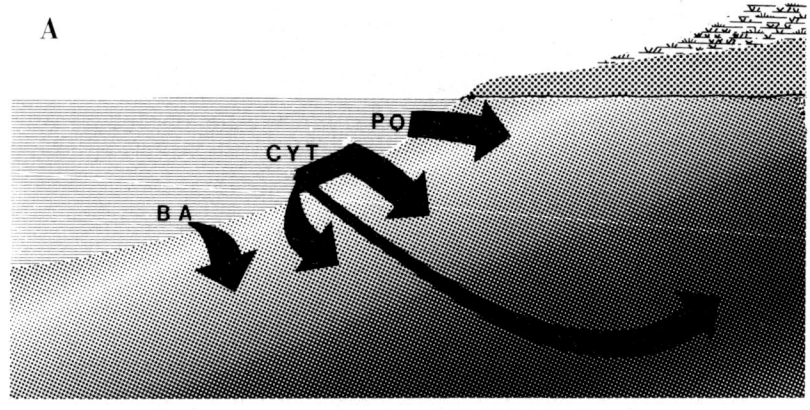

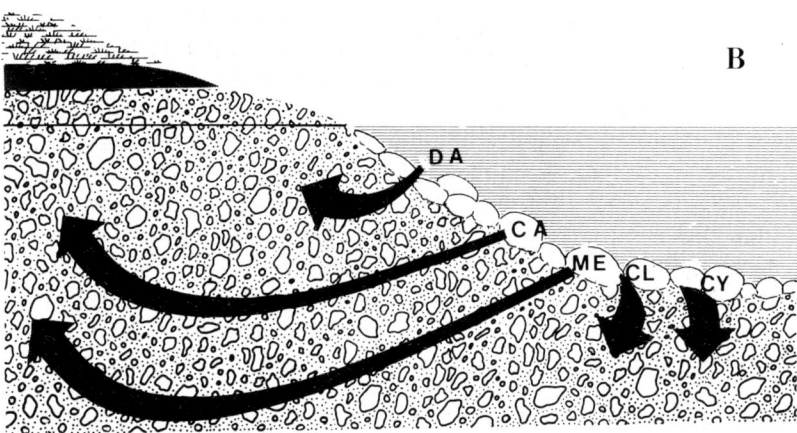

fig. 2. The main ways of colonization of the coastal and inland interstitial habitats. A. The marine origin, B. The freshwater origin from the epigean running waters. Po- Polycopidae, Cyt-Cytheracea, Ba-Bairdiacea, Da-Darwinulacea, Ca-Candoninae, Cl-Cyclocypridinae, Cy-Cypridinae, Me-Metacypridinae.

only been found in the hyporheal of the Mraconia stream (Danielopol, 1970). This species has many affinities with a *Darwinula* which I found in the outflow of a limnocrene of a thermal spring (26°C) in the south of France. Absolon (1974) showed that *Pseudocandona spelaea, Ps. bilobatoides* and *Ps. bilobata*, which are hypogean species, have close affinites with *Pseudocandona brevicornis*, known from epigean running waters. The species* are common in surface running wa-

---

* Forms with reduced or devoid of 'swimming' antennal bristles (fig. 4,C). Freshwater swimming ostracods have antennal bristles which attain or exceed the top of the endopodite of the 2nd antenna (fig. 4,E).

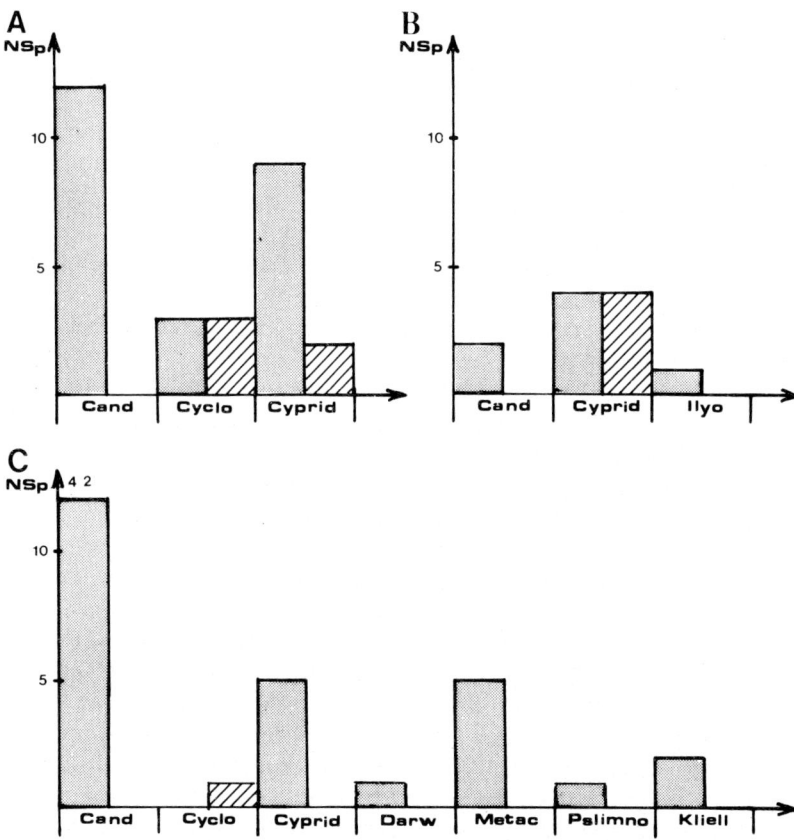

Fig. 3. The groups of ostracods and the number of species with long (▨) and short (or devoid of) swimming bristles (▩) in surface running water and interstitial habitats. A. Ostracod fauna in 24 springs in Thuringia (after Klie 1925). B. Ostracod fauna in 12 mountain tributaries of the Danube in the Banat, Romania (after Brezeanu et al. 1970). C. The ostracods living in freshwater interstitial habitats of Europe (both troglobites and troglophiles are recorded). Cand- Candoninae, Cyclo-Cyclocypridinae, Cyprid-Cypridinae, Darw-Darwinula, Ilyo-Ilyocypridinae, Metac-Metacypridinae, Pslimno- Pseudolimnocytherinae, Kliell-Kliellinae.

ters (Fig. 3 A,B). In the interstitial habitats, crawling species are dominant(Fig, 3,C). There is only one troglophilic species, *Cypria ophthalmica,* which has long 'swimming' bristles and which has been recorded in interstitial groundwater (Klie 1950). It is obvious that a crawling type of movement well fits the ostracods for life not only in running waters but also in interstitial systems. At the present time no examples of freshwater interstitial troglobites which could have directly originated from lake forms or troglophilic species which live in lakes and in interstitial waters have been certainly recorded. The crawling movement is characteristic of the following groups: Candoninae, Cytheracea, Darwinulidae as well as of some small Cypridinae groups (e.g. *Ilyodromus*).

From these data, it is possible to suggest that the freshwater interstitial habitats have mainly been colonized, at least in Europe, by a restricted group of ostracods already possessing the ability to crawl on and in the substrate. Most of these hypogean ostracods derived from epigean freshwater forms which could live in surface running waters.

The diversity of the freshwater interstitial Ostracods

Figures 1, A and B show that, as a group, the Candoninae has the highest number of interstitial species (both troglobites and troglophiles), as compared to the Cyclocypridinae, the Cypridinae, the Darwinulidae, the Kliellinae, the Metacyprinae and the Pseudolimnocytherinae.

Jeannel (1928) and Vandel (1965) discussed the problem of the selective co lonization of the subterranean realm. For cavernicolous Coleoptera, Jeannel (1928) quoted by Vandel (1965, pp. 485, 486) considered that: (species) 'entered caves because their degree of evolution made it impossible for them to exist among the epigeous fauna under the prevailing climatic conditions. It thus remains to establish the orthogenetic evolution which resulted in cavernicoles. It has been viewed as an accident during different areas among ancient lines only. Perhaps it is very simply a result of senility of lines . . .'. In discussing the distribution of the cavernicoulous Oniscoida (terrestrial isopods) Vandel (1965) states: 'The distribution of the cavernicolous species between the 21 families is most unequal. The cavernicolous state is related to the following conditions, one an ecological and the other a phylogenetic one' p. 130 . . . . . . 'While the cavernicoles represent relicts of an ancient fauna, the recently evolved species which are well adapted to life in the world as it is today possess few hypogeous representatives' p. 130. Vandel gives the following example: 'The family Armadillidiidae may be devided in two phyletic series, the elumean series and the armadillidian series. The former group is characteristic by a primitive cephalic structure whilst in the latter group the cephalic morphology is complex. But, of the thirteen cavernicolous Armadillidiidae only two belong to the armadillidian series...'.

In a recent contribution, Danielopol (1976) compared the degree of evolution of the Candoninae with that of other Cypridacea groups. From a detailed morphological analysis it appeared that within the superfamily Cypridacea, the Macrocypridiidae and the Pontocypridiidae have many plesiomorphic characters and could be considered as more primitive than the other Cypridiidae groups. The Candoninae are neither primitive nor specialized compared to the other Cypridiidae groups (Cypridinae, Ilyocypridinae, Notodromadinae and Cyclocypridinae). From such comparison it was concluded that the degree of evolution of the Candoninae can not be the cause of the dominance of this group in the subterranean realm.

If one compares the morphology of the Darwinulidae, the Metacyprinae, the Candoninae and the troglophilic Cypridinae it appears that the first group has the best morphological characteristics for habitation of the interstitial environment i.e. the carapace is fusiform and elongated (Fig. 4 A,B), the first antenna is very powerful with stout setae, the second antenna (Fig. 4,C) has a short endopodite

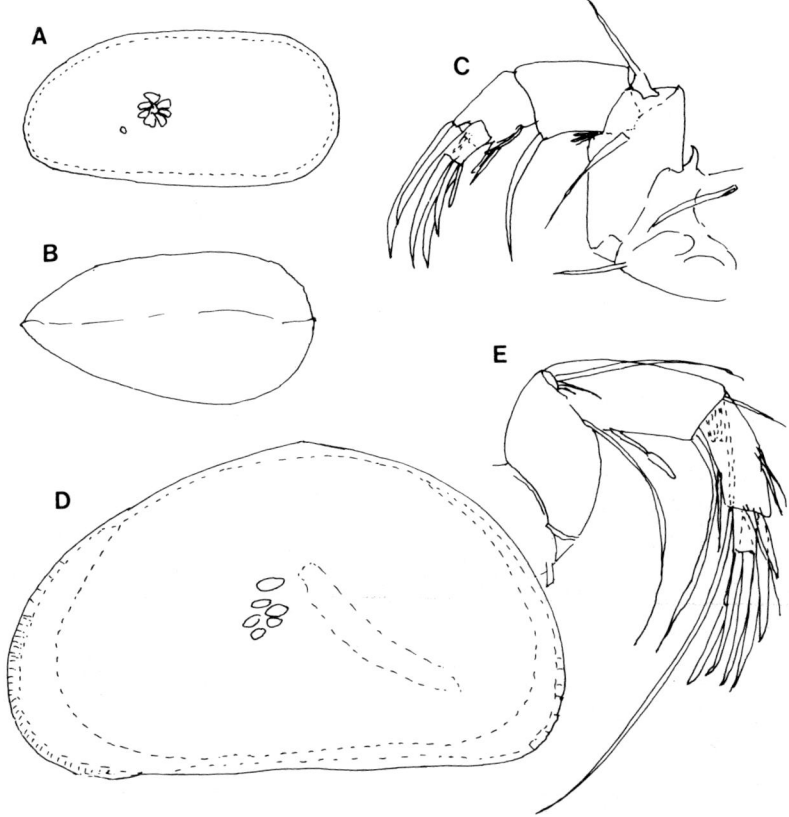

Fig. 4. A-C. *Darwinula boteai* Dan., A. left valve, B. carapace, dorsal view, C. 2nd antenna, D. *Pseudocandona* n.sp. gr. *eremita*, ♀, left valve, E. *Cypria* sp., 2nd antenna.

lacking swimming bristles, but with a distal segment joined subdistally (thus producing a large area for substrate contact). In spite of these characters, the Darwinulidae are both low in number of species and of reduced distribution in interstitial habitats (see Danielopol 1968, 1970, 1971). The Candoninae usually have triangular and trapezoidal carapaces (Fig. 4,D), which are generally shorter than those of the surface Candoninae (Danielopol, 1976). The Metacyprinae and the Cypridinae which occur in the interstitial habitats are also crawling forms, with short to medium length carapaces. From Figure 5,A which summarizes information data from 'Limnofauna Europaea' (Löffler 1967) it appears that the Candoninae and the Cypridinae have the highest diversity in Europe as compared to the other ostracod groups (Cytheracea, Darwinulidae, Cyclocypridinae, Ilyocypridinae, Notodromadinae). One can infer that the Candoninae has a high potential for diversification, much higher than that of the other freshwater groups with crawling representatives i.e. the Cytheracea, the Darwinulidae and the crawling Cypridinae. This can be partly explained by the fact that the Candoninae is a predominantly amphigonic group. The Darwinulidae, the Metacyprinae

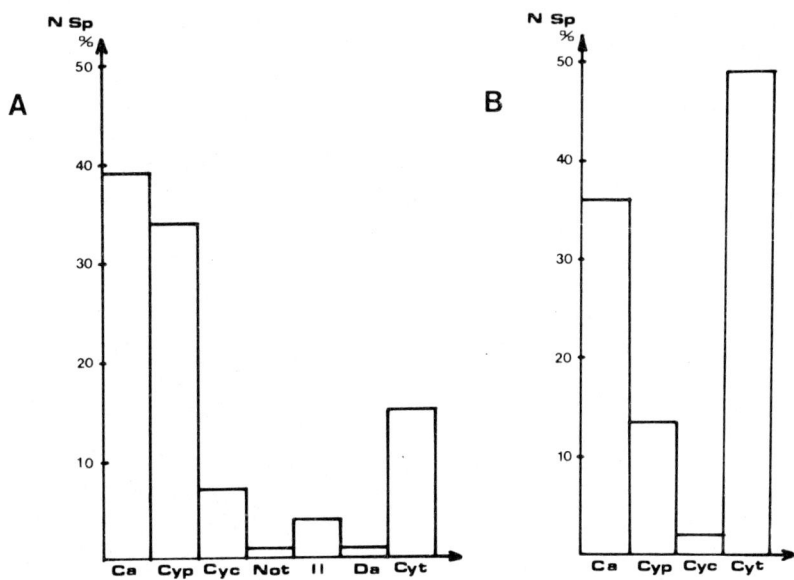

Fig. 5. A. The main ostracod groups and the number of species living in the freshwater habitats of Europe (data from Löffler 1967). B. The Tertiary ostracod fauna (Pannonian-Pontian) of Medvenica, Yugoslavia (data from Sokac, 1972). Ca-Candoninae, Cyp-Cypridinae, Cyc-Cyclo-cypridinae, Not-Notocypridinae, Il-Ilyocypridinae, Da-Darwinulidae, Cyt-Cytheracea.

of the genus *Kovalevskiella* and the troglophilic Cypridinae are parthenogenetic. Of the Darwinulidae, Metacyprinae and the Candoninae, the latter group shows the greatest adaptation to different environments. The Candoninae live in standing and running waters, in fresh and mesohaline habitats and in permanent and astatic water bodies. The Darwinulidae and the Metacyprinae have a more restricted distribution. For instance, species of these groups do not live in astatic water bodies.

The main freshwater groups found in Europe at the present time are already known as fossils from Miocene and Pliocene deposits. In the Neogene only 5 or 6 *Darwinula* species are known from Europe (Danielopol 1968, and Malz & Moyedpur 1972). the Metacyprinae of the genus *Kovalevskiella* are represented by 3 fossil species. Figure 5,B shows the ostracod groups and species numbers recorded by Sokac (1972) in the Pannonian-Pontian deposits from a Paratethys basin (brackish water environment) in Yugoslavia. Two groups show a high diversity: the Cytheracea and the Candoninae. The Cypridinae has lower diversity and the Cyclocypridinae are poorly represented. A similar situation is indicated by the data of Mandelstamm & Schneider (1963) and Krstic (1972). This last author pointed out: 'In the conditions of the isolated Pannonian basin, the ostracodan fauna evolves quickly generating a range of new forms..... Among the Cyprididae, the *Candona* change most rapidly. The diversity of the *Candona* represents a very significant factor since they constitute one half of all ostracods from the *Congeria* beds...' p. 139 (The concept of *Candona* used by Krstic is equivalent to the Candoninae belonging to the genera *Candona, Pseudocandona, Cryptocandona* and *Mixtacandona* in the system of Petkovski (1969). These data show that, during the Tertiary, the Candoninae were already highly diversified.

One can conclude that as compared to the other freshwater ostracod groups with crawling representatives, the Candoninae has a higher evolutionary plasticity (as defined by Dobzhansky 1968) allowing a higher rate of diversification and greater possibilities of specialisation.

References

Absolon, A. 1974. Ostracoda der Qellbiotope von Sulovske Skaly. In: Martin, Monographie Sulovske Skaly. Vlastisved. Sbor. Povazia 1:285-295.
Brezeanu, Gh. et al. 1970. Componenta si structura biocenozelor din afluenti, zoocenozele bentonice. In: Busnita et al., Monografia zonei Portilor de Fier.:122-136, Edit. Acad. R.S.R., Bucuresti.
Danielopol, D. L. 1968. Microdarwinula N.G. et quelques remarques sur la repartition de la famille Darwinulidae Br. et Norm. (Crustacea, Ostracoda). Ann Limnol. 4,2:153-174.
Danielopol, D. L. 1970. Sur une nouvelle espèce du genre Darwinula Br. et Rob. des eaux souterraines de Roumanie et quelques remarques sur la morphologie des Darwinulidae (Ostracoda, Podocopida). Trv. Inst. Speol. 'E. G. Racovitza' 10:189-207.
Danielopol, D. L. 1971. Quelques remarques sur le peuplement ostracodologique des eaux souterraines d'Europe. In: Oertli (ed.), Paleoecologie des Ostracodes. Bull. Centre Rech. Pau-SNPA. 5-Suppl.:179-190.
Danielopol, D. L. 1973. Preliminary report on Pussella botosaneanui n.g.n.sp., type of the new family Pussellidae. In: Résultats Scientifiques des Expeditions Biospéologiques Coubano-roumaines à Cuba, pp. 145-153, Edit. Acad. R.S.R., Bucuresti.
Danielopol, D. L. 1976. Beiträge zur Kenntnis der hypogäischen Ostracoden-Daten zur Morphologie der Candoninae. Diss. Univ. Wien.
Delamare-Deboutteville, Cl. 1960. Biologie des eaux souterraines litorales et continentales. Hermann:1-740, Paris.
Dobzhansky, Th. 1968. On some fundamental concepts of Darwinian biology. In: Dobzhansky, Hecht & Steeres, Evolutionary Biology 2, 1-34, North-Holland Publ. Comp., Amsterdam.
Hartmann, G. 1973. Zur gegenwärtigen Stand der Erforschung der Ostracoden interstitieller Systeme. Ann. Speleol. 28,3:417-426.
Klie, W. 1925. Entomostraken aus Quellen. Arch. Hydrobiol. 16,2:243-301.
Klie, W. 1938. Krebstiere oder Crustacea. III Ostracoda, Muschelkrebse. In: Dahl, Die Tierwelt Deutschlands 37:1-230.
Klie, W. 1950. Entomostraken aus Unterfranken. Mitt. Nat. Mus. Aschaffenburg. 4:15-28.
Krstic, N. 1972. Genus Candona (Ostracoda) from Congeria beds of Southern Pannonian basin. Serbian Acad. Sc. and Arts Monographs 140, 39:1-145.
Löffler, H. 1961. Zur Ostracodenfauna des obersten Donau Einzugsgebietes Arch. Hydrobiol. Suppl. 25: 332-340.
Löffler, H. 1963. Die Ostracodenfauna Österreichs in: Beiträge zur Fauna Austriaca. Sitz. Ber. Österr. Akad. Wiss. Math. Nat. Kl. 1,3-5: 199-211.
Löffler, H. 1967. Ostracoda, In: Limnofauna Europea, pp. 162-172, Stuttgart. Malz, H. & Moayedpur, E. 1973. Miozäne Susswasser Ostracoden aus der Rhön. Senckenbergiana Lethaea 54,2/4:281-309.
Mandelstamm, M. I. & Schneider, G. F. 1963. Petrified Ostracods of the USSR, Family Cyprididae. VNIGRI Publ. 203:1-242.
Noodt, W. 1974. Anpassung an interstitielle Bedingungen: Ein Factor in der Evolution Höherer Taxa der Crustacea? Faun. Ökol. Mitt. Kiel 4:445-452.

Petkovski, T. 1969. Über die Notwendigkeit einer Revision der Susswasserostracoden Europas, In: Neale (ed.), Taxonomy, Morphology and Ecology of Recent Ostracoda. Oliver & Boyd, pp. 76-84, Edinburgh.
Schäfer, H. W. 1945. Grundwasser Ostracoden aus Griechenland. Arch. Hydrobiol. 40,4:857-866.
Schäfer, H. W. 1951. Über die Besiedlung des Grundwassers. Verhl. Int. Verein. Theor. Angewand. Limnol. 11:324-330.
Vandel, A. 1965. Biospeology, the biology of cavernicolous animals. Pergamon Press, pp. 1-524.
Wolf, J. P. 1919. Die Ostracoden der Umgebung von Basel. Arch. für Naturgesch. 85:1-101.

Note added in proof.

During August 1976 I had the opportunity of collecting a new species of *Pseudolimnocythere*, living in fresh water in a well (interstitial habitat) on the Eubea Island coast, in Greece (Danielopol, in prep.). The ostracod shows similarities with *Loxoconchidea minima* Bondaduce, a sub-littoral marine species (with the same carapace and sieve shapes and the same dimensions) and the marine interstitial species *Tuberoloxoconcha tuberosa* Hartmann and *T. nanna* Marinov (similarities in the carapace shape and size and in the general form of the antennula, the maxilla respiratory plate and the male copulatory organ). Dr. G. Hartmann informed me (letter of 11.7.1977) of a *Tuberoloxoconcha* species from a marine interstitial habitat on the American Atlantic coast. The carapace sculpture and the hinge (smooth-henodont) closely resemble those of my new *Pseudolimnocythere* species. These data suggest that Pseudolimnocytherinae is a group which contains the genera *Tuberoloxoconcha, Pseudolimnocythere* and probably *Loxoconchidea* (in this last case living material must be found and description of the limbs also made). In my opinion the Pseudolimnocytherines belong to the Loxoconchinae (the peculiar 'aberrant' seta on the maxilla plate and the fused sensorial and normal bristle on the distal segment of the antennula also occur in the Loxoconchinae e.g. *Loxoconcha* aff. *granulata*). The Pseudolimnocytherinae Hartmann and Puri could be considered to be a tribe of the subfamily Loxoconchinae. Finally it should be pointed out that the Pseudolimnocytherini are the second known example of a marine ostracod group which has colonized the freshwater interstitial habitats via marine costal interstitial habitats.

(I acknowledge all the people with whom I discussed this problem: Dr. G. Hartmann, Dr. H. J. Oertli, Dr. R. Benson, Dr. G. Bonaduce, Dr. G. Carbonnel, Dr. J. W. Neale and Dr. W. Sissingh.)

Author's address:

Limnologisches Institut, Oesterreichische Akademie der Wissenschaften, Berggasse 18, 1090 Vienna, Austria.

Discussion

Whatley: Is there any evidence to suggest that in certain particular environments a species which, in terms of its total ecological range, is an inhabitant of both surface waters and interstitial environments, would select the latter because of such environmental factors as high turbulence levels?

Danielopol: In mountain streams, the interstitial habitat (the hyporheal) supports an abundant fauna represented mainly by surface running water animals. This is well documented in Williams & Hynes, 1974 (The occurrence of benthos deep in the substratum of a stream, Freshwat. Biol. 4:233-256). For marine ostracods Hartmann & Hartmann-Schröder (1975) showed that on 'open beaches where the action of the waves can operate without hindrance' the surface ostracods are lacking but a rich and 'typical' ostracod fauna exists in the interstitial habitats.*

Oertli: Do you think that subterranean/interstitial ostracods occurred at any geological time i.e. als in the Pre-Tertiary? If so, shouldn't we find many 'archaic' forms, presuming that, due to little environmental stress, the evolution had been slow, or at least different.

Danielopol: It is commonly accepted that the interstitial fauna could have existed in any geological time. Evidence for Pre-tertiary interstitial Crustacea has been documented by Schminke (1973) in his study on Parabathynellids.** Such an assumption is based primarily on the present geographical distribution of this group. To date, there is no evidence of any Pre-Tertiary ostracod group (at the generic level) which could continue to exist nowadays due exclusively to interstitial representatives. Considering the freshwater interstitial ostracods there are no 'archaic' groups at least at suprageneric level; this was thoroughly documented by Danielopol (1976) in the case of Candoninae.

One should notice that, in the case of the Candoninae, most of the hypogeans (interstitial species) belong to groups with poorly or moderately developed male copulatory organs i.e. *Mixtacandona* (which has exclusively recent subterranean species), *Candonopsis, Cryprocandona* and *Pseudocandona*. The genus *Fabaeformiscandona* which has species with well developed copulatory organs has only two hypogean representatives. Finally, the genus *Candona* (gr. candidoïda and neglectoïda) has no exclusively subterranean species. The representatives of this genus have the most developed (specialized) male copulatory organ. Based on these morphologic assumptions one could speculate that *Mixtacandona* is a primitive group as compared with *Pseudocandona, Candona* and *Fabaeformiscandona*.

---

* Cf. Zoogeography and biology of littoral ostracoda from South Africa, Angola and Mozambique. Bull-Am. Paleont. 65:353–368.
** Evolution, System und Verbreitungsgeschichte der Familie Parabathynellidae. Akad. Wiss. Lit. Mainz. 24:1–192.

IV. PALEOECOLOGY AND PALEOZOOGEOGRAPHY

Sixth Intern. Ostracod Symposium, Saalfelden

PALEOECOLOGICAL IMPLICATIONS OF HOLOCENE AND LATE PLEISTOCENE OSTRACODA, LAKE LAHONTON BASIN, NEVADA

FREDERICK M. SWAIN

Abstract

Sediments of the shores and recently exposed bottom of evaporative Pyramid Lake, a remnant of Pleistocene Lake Lahonton, contain abundant *Cyprideis* sp., *Limnocythere* spp., and other ostracodes and gastropods, although the present lake is too alkaline (4700 ppm dissolved solids) to support living populations of ostracodes. Former higher beach levels of Pyramid Lake up to 20 m or more above present lake level and more than 3,700 years old lack brackish-water *Cyprideis*, but have a variety of freshwater ostracodes. These generally decrease in variety and abundance upward. The history of the lake apparently involved gradual increase in ostracode populations and in alkalinity as the lake level dropped through cessation of pluviation until the alkalinity became too high to support the *Cyprideis-Limnocythere* population. ^{14}C dates of the youngest shells range from about 300 to about 1800 years old based on work by Broecker et al. (1958, 1965). Pleistocene peaty clays south of Pyramid Lake and about 250,000 years old contain only *Candona* and *Limnocythere* spp. and probably represent conditions of relatively low alkalinity. Comparisons are made with other lake deposits in the Great Basin.

The succession of ostracode populations in other Quaternary pluvial lakes should be useful in interpreting the history of such lakes.

Introduction

The area discussed in this paper lies in the basin of Pleistocene Lake Lahonton, a former multi-basin glacial lake in northwestern Nevada, northeastern California and south-central Oregon (Russell 1885). Most of the ostracodes studied here came from Pyramid Lake, Nevada, the largest remnant of Lake Lahonton. Additional samples were studied from dry Carson Sink, east of Pyramid Lake and from Pleistocene deposits along Truckee River south of Pyramid Lake.

Pyramid Lake (Fig. 1) occupies about 2070 km^2 and is up to 100 m. deep. The composition of the lake water in 1951 is given in Table 1. The water is brackish, but concentration of salts is low enough to permit several species of fresh-water fish to occur and to allow consumption by cattle (Swain & Meader 1958). The lake lies in a complex of Tertiary volcanic rocks of the Esmeralda Formation. Faulted and metamorphosed Mesozoic limestones are exposed near the southern end the lake and underlie part of the lake basin (Turner 1900). The basin is bounded by one or more faults along the eastern side (Hutchinson 1957) from which hot springs and gas vents issue forth. Calcareous tufa (thinolite and other forms) has accumulated in large quantities along the beaches and cliffs of the lake. The tufa has been derived from the lake water as a result of drainage into the lake of carbonate-charged surface water or from sub-lacustral hot springs. The processes involved in precipitating the CaCO$_3$ may have been inorganic by evaporation of

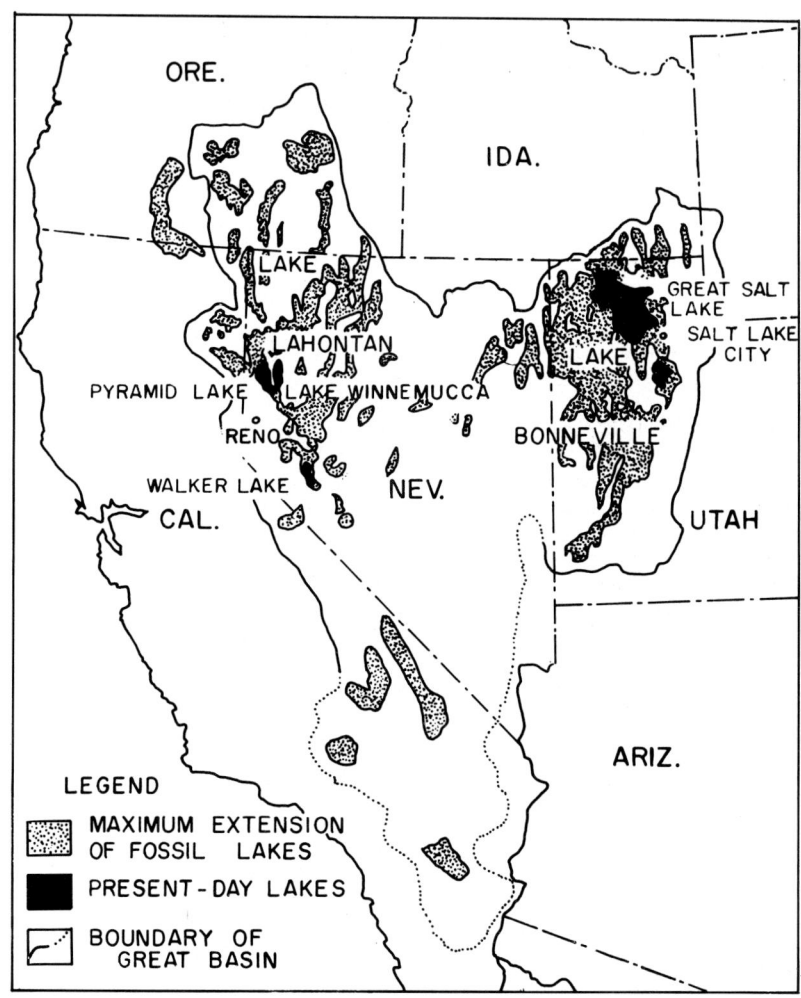

Fig. 1. Pleistocene Great Lakes of the Western United States (after Broecker & Orr 1958)

wave spray (Russell 1885), organic by blue-green algae metabolism (Jones 1925) or a combination of both processes.

Formal study of the past and present biota of Pyramid Lake has not been made but included, in addition to fossil ostracodes, are fossil and living? melosiroid, campylodiscoid, and naviculoid diatoms in large numbers, and living *Chara* sp., *Cladophora* sp., *Potamogeton pectinatus*, myxophytic algae, *Grantia* sp., *Osphranticum* sp., *Daphnia pulex parapulex*, *Ceriodaphnia laticaudata*, diaptomid larvae, midge larvae, and 8 species of fish including *Chasmites cujus*, an endemic

Lahonton species not found living elsewhere in the world (Hutchinson 1937). Anaho Island in Pyramid Lake is a famous pelican rookery. Numerous gastropod shells, including *Pyrgulopsis nevadensis* and *Paraphalyx effusa* occur in Pyramid Lake sediments (Hutchinson 1937) but are not found living in the lake.

A series of terraces mark former levels of Lake Lahonton around Pyramid Lake (Russell 1885) fig. 2, after Broecker & Orr 1958); these lie at about 173 m, 164 m, 97 m, and 45 m above present lake level. The lake level fluctuates several meters in response to prolonged wet and dry periods.

Carbon 14 dating of the terraces (Broecker & Orr 1958, Broecker & Kaufman 1965, Broecker 1965) suggests that tufa formation in Pyramid began within the last 30,000 years and that the lake has mainly been shrinking during the last 11,700 years following a minimum about 12,500 years ago.

In the present study ostracode bearing samples were obtained from the present sediments and beach sands of Pyramid Lake and at intervals of 4.5 m and 20 m above the lake level; intervening samples below 20 m and additional samples above 20 m were devoid of ostracodes. The age of the uppermost ostracode-bearing samples is interpreted to be approximately 3,700 years old based on the data of Broecker & Orr (1958) whereas that of the 4.5 m sample is about 860 years.

Additional ostracode-bearing samples were obtained near Carson Sink, about 40 miles east of Pyramid Lake, a dry Lahonton remnant. The age of these samples is uncertain. Radiocarbon pages obtained by Broecker & Kaufman for Carson Sink deposits range from less than 200 to 1400 ($\pm$ 200) years B.P.

Table 1. Composition of Pyramid Lake water, in pm (after Swain & Meader 1958)

	A	B
Silica	2.6	2.1
Iron	0.03	0.04
Calcium	7.5	8.0
Magnesium	111.0	111.0
Sodium	1570.0	1540.0
Potassium	128.0	160.0
Carbonate	285.0	280.0
Bicarbonate	830.0	830.0
Sulfate	263.0	265.0
Chloride	1920.0	1920.0
Fluoride	2.9	2.9
Nitrate	1.8	1.8
Boron	12.0	12.0
Total dissolved solids	4700.0	4700.0
$CaCO_3$ hardness	478.0	476.0
Specific conductance microohms at 25°C	7610.0	7660.0
pH	9.4	9.2

A. 3 m from surface, 13 February 1951, U.S. Geol. Survey, Salt Lake City Laboratory, analysis No. 6139
B. 30 m from surface

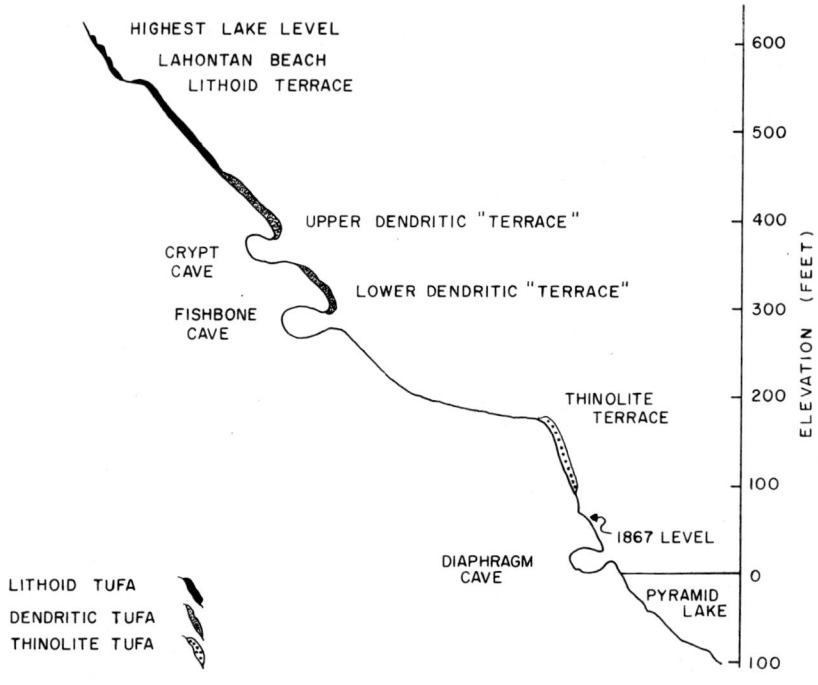

Fig. 2. Terrace Levels of Lake Lahontan at Pyramid Lake (after Broecker & Orr 1958)

Drill cores of presumable Pleistocene lake sediments were obtained through the courtesy of the U.S. Bureau of Reclamation from test holes near Nixon, Nevada, about 2 miles south of Pyramid Lake. These deposits lie just above the volcanic Esmeralda Formation and may be about 250,000 years old based on data of Broecker & Kaufman (1965).

A few species of fresh- and brackish-water ostracodes from a former beach level of Lake Coahuila, ancestral to Salton Sea, Imperial Valley are also included here for comparison.

Lists of species

Many of the recorded species are identified tentatively; some of them may be new species.

1. Ostracoda from modern sediments and beach deposits of Pyramid Lake.
Candona bretzi Staplin, 1963 (Pl. 1, Figs. 1a, b)
Candona protzi Hartwig, 1898 (Pl. 1, Figs. 2a, b)
Candona cf. *C. crogmaniana* Turner, 1894 (Pl. 1, Fig. 3)
Candona sp. aff. *C. patzcuaro* Tressler, 1954 (Pl. 1, Fig. 4, 7)

Candona acuminata (Fischer, 1854) (Pl. 1, Figs. 5, 6, 9)
Candona cf. *C. ohioensis* Furtos, 1933 (Pl. 1, Fig. 8)
Limnocythere cf. *L. posterolimbata* Delorme, 1970 (Pl. 1, Fig. 14–16)
Limnocythere cf. *L. ceriotuberosa* Delorme
Cyprideis cf. *C. beaconensis* (LeRoy, 1943) (Pl. 2, Fig. 12–17)

2. Ostracoda from raised beach sediment 4.5 m above lake level, Pyramid Lake.
Prionocypris? sp. (Pl. 1, Fig. 17)
Heterocypris? sp. (Pl. 1, Figs. 18–20)
Cypridopsis sp. (Pl. 1, Fig. 19)
Cyprinotus cf. *C. glaucus* Furtos, 1933 (Pl. 1, Fig. 21)
Limnocythere sp. aff. *L. varia* Staplin 1963 (Pl. 2, Fig. 4, 5, 19)
Limnocythere sp. aff. *L. verrucosa* Hoff 1942 (Pl. 2, Fig. 7)
Elkocythereis spp. (Pl. 2, Figs. 9–11)

3. Ostracoda from raised beach sediment 20 m above lake level, Pyramid Lake.
Limnocythere sp. aff. *L. varia* Staplin (pl. 1, Fig. 29, Pl. 2, Fig. 6, 8)

4. Ostracoda from sediments of dry Carson Sink Nevada.
Cypricercus? sp. (Pl. 1, Figs. 10–12)
Herpetocypris? sp. (Pl. 1, Figs. 13)
Limnocythere sp.cf. L. varia Staplin (Pl.1, Figs. 23–27; Pl. 2, Figs. 2)
Limnocythere cf. *L. ceriotuberosa* Delorme (Pl. 1, Fig. 28, Pl. 2, Figs. 1, 3)

5. Ostracodes from cored samples, U.S. Bureau of Reclamation, DH-1, Washoe Project, Marble Bluff Dam, Upper Site, 12.1–22.7m, gray silt with sand laminae. Ostracoda shells are very thin and disintegrate on drying.
Candona sp. indet.
Limnocythere sp. indet.

6. Ostracodes from former beach level, Salton Sea, Truckhaven, Imperial Valley, California.
Cyprideis torosa (Jones, 1850) (Pl. 2, Figs. 20-25)
Cypridopsis vidua (O.F. Müller, 1776) (Pl. 2, Figs. 26, 27))

Discussion

The present salinity of Pyramid Lake water (about 3.5‰) is far lower than that tolerated by living *Cyprideis* elsewhere (Vesper 1975, Sandberg 1964). This has been stated to range up to 6‰, but reaches optimum condition between 2 and 16.5‰ (Wagner 1957). Therefore some other factors such as high K, SO_4, B, etc., would appear to prevent *Cyprideis* from living in the present lake waters. There should be ample opportunity for re-stocking of the lake with this and other ostracodes by means of pelicans which roost here and by other migrating birds, but ostracodes brought to the lake by such means have apparently not survived.

The apparent absence of ostracodes from the higher levels of former lake beaches in the Lahonton basin may be due to inadequate sampling, poor preservation, or unfavorable environment for their existence. It appears that environmental conditions for ostracodes in the main body of Lake Lahonton around present-day Pyramid Lake were unfavorable until the lake level had dropped from the high of 173 m to below 45 m above present lake level. In some areas, perhaps re-

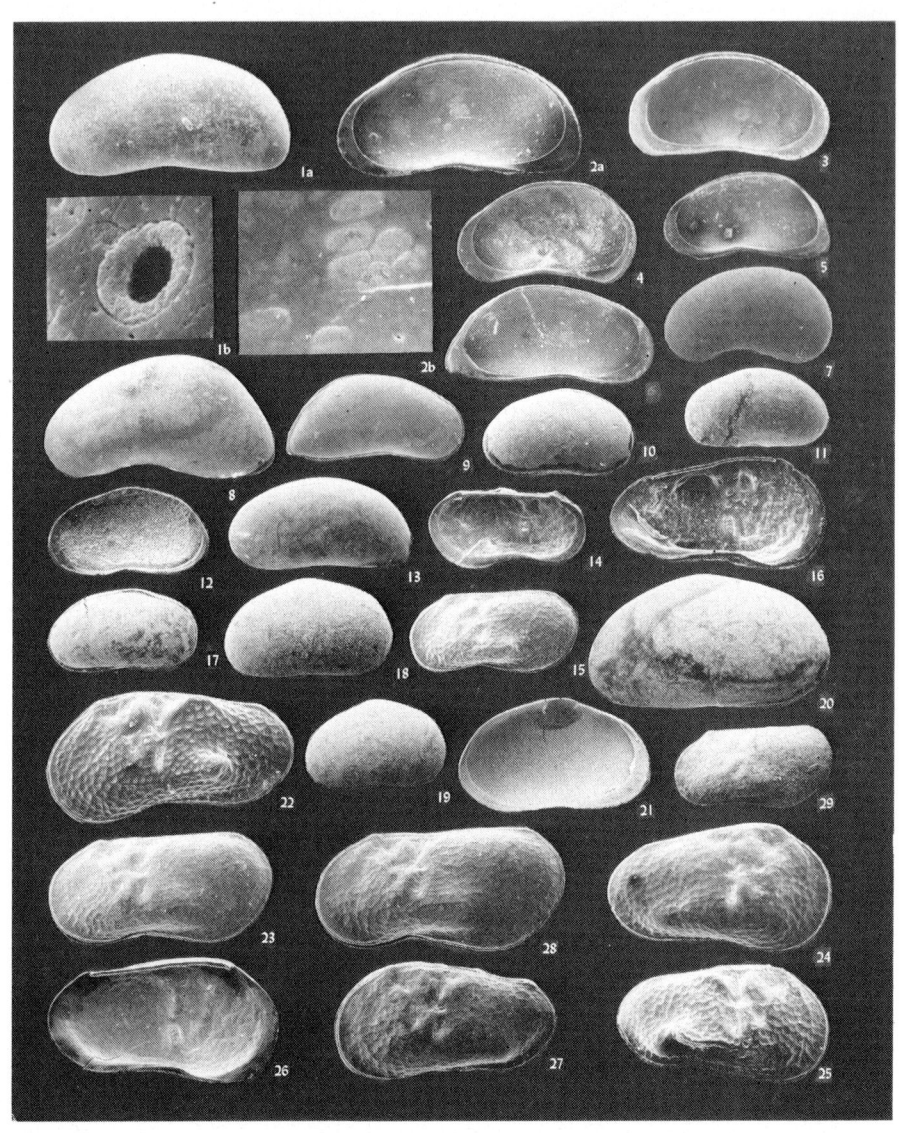

◄ Plate 1.

Fig. 1a, b. *Candona bretzi* Staplin. 1a, Exterior of male left valve, x58; 1b, Enlargement of normal pore with crateriform rim, x11,000; sub-Recent, Pyramid Lake.

Fig. 2a, b. *Candona protzi* Hartwig. a, Interior of female right valve, x58; b, Enlargement of adductor muscle scar area, x315; sub-Recent, Pyramid Lake.

Fig. 3. *Candona* cf. *C. crogmaniana* Turner. Interior of male left valve, x30; sub-Recent, Pyramid Lake.

Fig. 4. *Candona* aff. *C. patzcuaro* Tressler. Interior of female right valve, x32; sub-Recent, Pyramid Lake.

Fig. 5. *Candona acuminata* (Fischer). Interior of female left valve, x30; sub-Recent, Pyramid Lake

Fig. 6. *Candona acuminata* (Fischer). Interior of female right valve, x30; sub-Recent, Pyramid Lake.

Fig. 7. *Candona* aff. *C. patzcuaro* Tressler. Exterior of female left valve, x28; sub-Recent, Pyramid Lake.

Fig. 8. *Candona* cf. *C. ohioensis* Furtos. Left side of female shell, x34; sub-Recent, Pyramid Lake, Nevada.

Fig. 9. *Candona acuminata* (Fischer). Right side of shell, x30; sub-Recent, Pyramid Lake, Nevada.

Fig. 10. *Cypricercus*? sp. Right side of shell, x32; sub-Recent, Carson Sink, Nevada.

Fig. 11. *Cypricercus*? sp. Exterior of left valve, x32; sub-Recent, Carson Sink, Nevada.

Fig. 12. *Cypricercus*? sp. Interior of left valve, x35; sub-Recent, Carson Sink, Nevada.

Fig. 13. *Herpetocypris*? sp. Exterior of left valve, x62; sub-Recent, Carson Sink, Nevada.

Fig. 14. *Limnocythere* cf. *L. posterolimbata* Delorme. Interior of male left valve, x37; sub-Recent, Pyramid Lake, Nevada.

Fig. 15. *Limnocythere* cf. *L. posterolimbata* Delorme. Exterior of male left valve, x37; sub-Recent, Pyramide Lake, Nevada.

Fig. 16. *Limnocythere* cf. *L. posterolimbata* Delorme. Interior of female left valve, x58; sub-Recent, Pyramid Lake, Nevada.

Fig. 17. *Prionocypris*? sp. Right side of shell, x33, late Pleistocene or sub-Recent, Pyramid Lake, Nevada. (elevated beach sample 3).

Fig. 18. *Heterocypris*? sp. Exterior of left valve, x35; late Pleistocene or sub-Recent, Pyramid Lake, Nevada (elevated beach sample 3).

Fig. 19. *Cypridopsis* sp. Exterior of right valve, x32; late Pleistocene or sub-Recent, Pyramid Lake, Nevada. (elevated beach sample 3).

Fig. 20. *Heterocypris*? sp. Exterior of left valve, x56; late Pleistocene or sub-Recent, Pyramid Lake, Nevada. (elevated beach sample 3).

Fig. 21. *Cyprinotus* cf. *C. glaucus* Furtos. Interior of left valve, x33; late Pleistocene or sub-Recent, Pyramid Lake, Nevada. (elevated beach sample 3).

Fig. 22. *Limnocythere* cf. *L. ceriotuberosa* Delorme. Exterior of female left valve, x61; sub-Recent, Pyramid Lake, Nevada.

Fig. 23. *Limnocythere* cf. *L. varia* Staplin. Left side of male shell, x58; sub-Recent, Carson Sink, Nevada.

Fig. 24. *Limnocythere* cf. *L. varia* Staplin. Exterior of female right valve, x63; sub-Recent, Carson Sink, Nevada.

Fig. 25. *Limnocythere* cf. *L. varia* Staplin. Left side of male shell, x50; sub-Recent, Carson Sink, Nevada.

Fig. 26. *Limnocythere* cf. *L. varia* Staplin. Interior of female left valve, x63; sub-Recent, Carson Sink, Nevada.

Fig. 27. *Limnocythere* cf. *L. varia* Staplin. Exterior of female left valve, x60; sub-Recent, Carson Sink, Nevada.

Fig. 28. *Limnocythere* cf. *L. ceriotuberosa* Delorme. Exterior of immature female right valve, x62; sub-Recent, Carson Sink, Nevada.

Fig. 29. *Limnocythere* aff. *L. varia* Staplin. Exterior of left valve, x35, late Pleistocene, former lake level, sample 13, Pyramid Lake, Nevada.

◀ Plate 2.

Fig. 1. *Limnocythere* cf. *L. ceriotuberosa* Delorme or subspecies, x63; sub-Recent, Carson Sink, Nevada.

Fig. 2. *Limnocythere* cf. *L. varia* Staplin, x69; sub-Recent, Carson Sink, Nevada.

Fig. 3. *Limnocythere* cf. *L. ceriotuberosa* Delorme, x62; sub-Recent, Carson Sink, Nevada.

Fig. 4. *Limnocythere* aff. *L. varia* Staplin. Exterior of distorted female left valve, x58; late Pleistocene or early Holocene, former lake level, sample 3, Pyramid Lake, Nevada.

Fig. 5. *Limnocythere* aff. *L. varia* Staplin. Exterior of male right valve, x60; late Pleistocene or early Holocene, former lake level, sample 3, Pyramid Lake, Nevada.

Fig. 6. *Limnocythere* aff. *L. varia* Staplin. Interior of male left valve, x55; late Pleistocene, former lake level, sample 13, Pyramid Lake, Nevada.

Fig. 7. *Limnocythere* aff. *L. verrucosa* Hoff. Exterior of male left valve, x60; late Pleistocene or early Holocene, former lake level, sample 3, Pyramid Lake, Nevada.

Fig. 8. *Limnocythere* aff. *L. varia* Staplin. Interior of right valve, x57; late Pleistocene, former lake level, sample 13, Pyramid Lake, Nevada.

Fig. 9. *Elkocythereis* sp. Exterior of left valve, x60; late Pleistocene or early Holocene, former lake level; sample 3, Pyramid Lake, Navada.

Fig. 10. *Elkocythereis* sp. Interior of right valve, x60; late Pleistocene or early Holocene, former lake level, sample 3, Pyramid Lake, Nevada.

Fig. 11. *Elkocythereis* sp. Exterior of right valve, x60; late Pleistocene or early Holocene, former lake level, sample 3, Pyramid Lake, Nevada.

Fig. 12. *Cyprideis* cf. *C. beaconensis* LeRoy. Right side of female shell, x39; sub-Recent, Pyramid Lake, Nevada.

Fig. 13. *Cyprideis* cf. *C. beaconensis* LeRoy. Exterior of female left valve, x33; sub-Recent, Pyramid Lake, Nevada.

Fig. 14. *Cyprideis* cf. *C. beaconensis* LeRoy. Interior of female left valve, x37; sub-Recent, Pyramid Lake, Nevada.

Fig. 15. *Cyprideis* cf. *C. beaconensis* LeRoy. Right side of male shell, x32; sub-Recent, Pyramid Lake, Nevada.

Fig. 16. *Cyprideis* cf. *C. beaconensis* LeRoy. Interior of male right valve, x35; sub-Recent, Pyramid Lake, Nevada.

Fig. 17. *Cyprideis* cf. *C. beaconensis* LeRoy. Interior of female left valve, x35; sub-Recent, Pyramid Lake, Nevada.

Fig. 18. *Cyprideis* cf. *C. beaconensis* LeRoy. Enlargement of normal pore and sieve plate, x2992; sub-Recent, Pyramid Lake, Nevada.

Fig. 19. *Limnocythere* sp. aff. *L. varia* Staplin. Interior of distorted female left valve, x62; late PLeistocene or early Holocene, former lake level, sample 3.

Fig. 20. *Cyprideis torosa* (Jones). Exterior of female left valve, x40; sub-Recent, former beach level, Salton Sea, Imperial Valley, California.

Fig. 21. *Cyprideis torosa* (Jones). Exterior of male right valve, x43; sub-Recevot; former beach level, Salton Sea, Imperial valley, California.

Fig. 22. *Cyprideis torosa* (Jones). Dorsal view of shell, x39; sub-Recent, Salton Sea, Imperial Valley, California.

Fig. 23. *Cyprideis torosa* (Jones). Ventral view of male left valve, x41; sub-Recent, former beach level, Salton Sea, Imperial Valley, California.

Fig. 24. *Cyprideis torosa* (Jones). Interior of male left valve, x37; sub-Recent, former beach level, Salton Sea, Imperial Valley, California

Fig. 25. *Cyprideis torosa* (Jones). Ventral view of nodose immature right valve, x62; sub-Recent, former beach level, Salton Sea, Imperial Valley, California.

Fig. 26. *Cypridopsis vidua* (O. F. Müller). Right side of shell, x72; sub-Recent, former beach level, Salton Sea, Imperial Valley, California.

Fig. 27. *Cypridopsis vidua* (O. F. Müller). Ventral view of shell, x72; sub-Recent, former beach level, Salton Sea, Imperial Valley, California.

presented by shallow peaty bays, bordering the main lake, ostracodes existed well back into the Pleistocene, as represented in the Bureau of Reclamation drill holes near Nixon, Nevada. As the lake continued to decline *Limnocythere* sp. aff. *L. varia* populated the littoral areas, but was discontinuous in its vertical distribution. By the time the lake had dropped to 4.5 m above present level, several other fresh-water ostracodes had appeared, along with *Elkocythereis,* a cytheracean genus previously recorded in this region from Plio-Pleistocene deposits (Dickinson & Swain 1967). This genus is thought to have lived in a high-calcium lacustrine environment, of mestrophic to early eutrophic nature (Swain, Becker & Dickinson 1971). By the time the lake had dropped to somewhat above its present level, *Cyprideis* cf. *C. beaconensis* and several *Candonas* were established. Radiocarbon dates on the snail shells suggest that this took place from about 1800 to about 300 years ago. Subsequently the waters became too alkaline for the ostracode population to survive.

The sediments of dry Carson Sink, a Lahonton remnant 50–60 km east of Pyramid Lake contain an ostracode population resembling one of the intermediate levels of Pyramid Lake. That area may have dried up before the *Cyprideis* population developed.

Former beach levels of Lake Coahuila, the ancestral Salton Sea, Imperial Valley, yielded *Cypridopsis vidua* and *Cyprideis torosa*. These forms, the first freshwater and the second brackish water, perhaps represent stages of overflow of Colorado River into the Salton Sea depression in relatively modern times. Whether the two forms lived in the lake at the same time or represent intervals of varying salinity cannot be determined at present.

Conclusion

The succession of ostracode populations in pluvial lakes undergoing progressive shrinking through decrease in pluviation is useful in interpreting lacustrine history.

An example of a reversed condition in this region is that of the *Cyprideis*-bearing late Miocene Middle Esmeralda Formation of western Nevada (Swain, Becker & Dickinson 1971). The *Cyprideis* beds probably represent brackish-water lake deposits that are succeeded by fresh-water Pliocene deposits with *Elkocytherieis* and other species. Increased pluviation during the Pliocene in the area is a plausible explanation for the changes in the ostracode populations.

The history of late Pleistocene and Holocene ostracode populations in Lake Bonneville northern Utah and southern Idaho is different from that of Lake Lahonton (Swain, Becker & Dickinson 1971). The Lake Bonneville Group contains fresh-water ostracodes at the older, higher Alpine level and the younger, lower Provo level. One species in both assemblages *Limnocythere inopinata* (Baird) occurs in brackish water estuaries elsewhere, and possibly suggests slightly saline conditions for those intervals in Lake Bonneville. Subsequent levels of Lake Bonneville are devoid by ostracodes supposedly owing to the increasingly high salinity of the lake. The reason for absence of *Cyprideis* populations in the post-Provo Bonneville sediments is not clear.

Acknowledgements

Appreciation is expressed to R.W. Meader who assisted in collecting the Lake Lahonton samples and to L.W. LeRoy who supplied the Salton Sea specimens. Dr N.P. Prokopovich of the U.S. Bureau of Reclamation supplied the drill core samples from Marble Bluff dam site. SEM work was done at the University of Delaware with the assistance of Takako Nagase.

References

Broecker, W.S. 1965. Isotope geochemistry and the Pleistocene climatic record. In: The Quaternary of the U.S. (Wright & Frey, eds.), Princeton Univ. Press, pp. 737–753.
Broecker, W.S. & Kaufman, A. 1965. Radiocarbon chronology of Lake Lahonton and Lake Bonneville. II. Great Basin. Geol. Soc. Am. Bull. 76:537–566.
Broecker, W.S. & Orr, P.C., 1958, Radiocarbon chronology of Lake Lahonton and Lake Bonneville. Geol.Soc. Am. Bull. 69:1009–1032.
Dickinson, K.A. & Swain, F.M. 1967. Late Cenozoic freshwater Ostracoda and Cladocera from northeastern Nevada. Jour. Paleontology 41:335–350.
Hutchinson, G.E., 1937. A contribution to the limnology of arid regions. Conn. Acad. Arts and Sci., Trans. 33:47–132.
Russell, I.C. 1885. Geological history of Lake Lahonton U.S. Geol. Survey Monograph 2:1–288.
Sandberg, P.A. 1964. The ostracod genus Cyprideis in the Americas. Acta Uni. Stockholm., Stockholm Contr. in Geol. 12, 128 pp.
Swain, F.M., Becker, J. & Dickinson, K.A. 1971. Paleoecology of Tertiary and fossil Quaternary non-marine Ostracoda from the western interior. U.S. In Paleoecologie Ostracodes (Oertli, ed.), Bull. Centre Rech. Pau-SNPA, 5 (suppl.):461–487.
Swain, F.M. & Meader, R.W. 1958, Bottom sediments of southern part of Pyramid Lake, Nevada. Jour. Sed. Petrology 28:286–297.
Turner, H.W. 1900. The Esmeralda Formation. Am. Geologist 25:168–170.
Vesper, B. 1975. To the problem of noding on Cyprideis torosa (Jones 1850), In: Biology and Paleobiology of Ostracodes (Swain, Kornicker, Lundin, eds.) Paleont. Res. Inst., Bull. 65 (282):205–216.
Wagner, C.W. 1957. Sur les ostracodes du quaternaire Récent des Pays-Bas et leur utilization dans l'étude géologique des déposits Holocènes. Mouton & Co., 's-Gravenhage, 259 pp.

Author's address:

Department of Geology & Geophysics, University of Minnesota, Minneapolis, Minn. 55455, USA

Discussion

Whatley: Is it not possible to invoke temperature changes to explain some of the disappearance phenomena which you have observed?

Swain: All of the ostracode genera cited in the paper live now under a variety of temperature conditions that include those likely to have occurred during the Pleistocene and Holocene times represented in Lake Lahonton. Some of the species are apparently extinct so it is not possible to make a definitive statement as to limiting temperatures of the species themselves. At the generic level, however, I would not expect temperature changes alone to account for the observed disappearances, but temperature may have been one of the controlling factors.

Sixth Intern. Ostracod Symposium, Saalfelden

'FOSSIL' MEROMIXIS IN KLEINSEE (CARINTHIA) INDICATED BY OSTRACODES

H. LÖFFLER

Abstract

Kleinsee, west of Klopeinersee and with respect to its origin considered a kettle, at present behaves holomictic. A core taken at its deepest part (9.3 m) shows that in the late Pleistocene (Bölling) the lake entered a meromictic stage indicated by a sudden decrease of ostracods (especially *Cytherissa lacustris*) as well as of chironomids. With increasing sedimentation, however, the lake became holomictic again, probably during the Subatlanticum. This again is suggested by a dramatic increase of benthic ostracods and chironomids.

Kleinsee, at present a holomictic lake of 0.125 km² and with a maximum depth of 9.3 m (Hartl & Sampl 1973, Löffler 1972), is considered a kettle. Together with its twin lake Klopeinersee it formed part of the former Kühnsdorf Lake which was fed by the Drau River. Since the collapse of Kühnsdorf Lake the Kleinsee basin is fed only by a small brook draining the Littermoos. Thus Kleinsee at the beginning of its sedimentation process must have had the ideal conditions for entering a meromictic state, which means that the lake no longer had its full circulations during springtime and fall.

Cores taken in the littoral zone of Kleinsee and Klopeiner See and at a depth of 33 m in Klopeiner See (z_{max} : 46 m) strongly indicate an abrupt collapse of the Kühnsdorf Lake: a sudden change in the ostracod fauna (Lower infralittoral – Upper infralittoral) due to a lowered lake level as well as the end of any mica sedimentation due to the Drau River both demonstrate a shift from a big late Pleistocene lake into a small kettle basin (Löffler 1972, 1975). Most likely this shift took place in Ic (Younger Dryas) or even earlier, which is not exactly known yet.

In 1975 a core was taken from the lake's deepest part in order: 1. to learn whether Kleinsee had undergone a meromictic period and 2. to get more information about the collapse of Kühnsdorfer Lake. The core taken by a Kullenberg-Livingstone modification was used for the analysis of both subfossil organisms and pollen. The results of the latter with respect to the identification of late and postpleistocene periods are indicated on Fig. 2. Their more detailed description will be given elsewhere.

The section 660-390 cm Ia (Elder Dryas) obviously starts with an already lacustrine state: *Bosmina* cf. *coregoni*, never very abundant, is present in its deepest portion. The ostracod fauna, which in the Kleinsee includes at least 13 species, starts with *Cypria ophthalmica* from then on found throughout the core though with striking change in its abundance. From 540 cm onward *Candona* (*Candida* group) appears and becomes abundant in the last section of this Late Pleistocene period (470-390 cm). At the same time more species like *Erpetocypris* cf. *reptans*, *Eucypris* sp., *Potamocypris* sp., *Cypridopsis* sp., *Ilyocypris* sp. appear. It is also char-

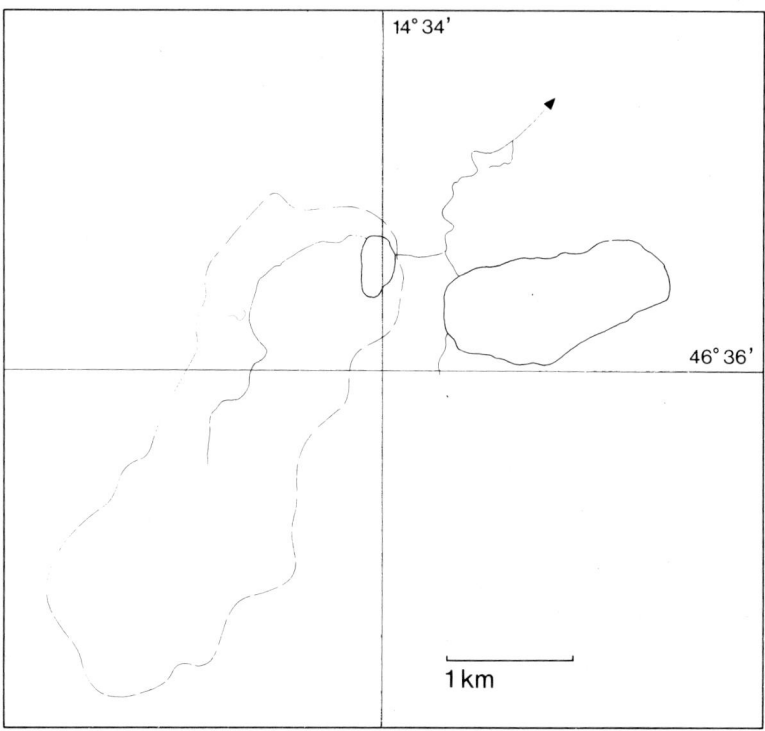

Fig. 1. Geographic position of Klein- and Klopeiner See. Broken line: catchment area of Kleinsee.

acterized by the presence of *Cytherissa lacustris*, a valuable indicator species with respect to the structure of sediments (Löffler 1969, 1975, Powell 1976). Throughout the section mentioned Ephippia of *Daphnia* (pulex group) could be observed for the first time and head capsulae of chironomids become abundant here.

The next section (most likely Ib and Ic, 390-365 cm) is lacking all ostracods some of which, however, may be present in very small numbers. *Erpetocypris, Eucypris, Ilyocypris,* and above all *Cytherissa*, never reoccur again. At the same time *Bosmina* reaches a maximum. Decrease of chironomids and the end of mica sedimentation are further characteristics of this period. The latter could be explained by the collapse of Kühnsdorfer See mentioned above. It is of interest that in a littoral core taken in 1971 the end of mica sedimentation as well as the presence of *Cytherissa, Ilyocypris* and *Erpetocypris* coincide, although at a depth of sediment of approximately 9 m. Along with a much lower water level (according to Stiny 1934:15-20 m) in modern Kleinsee the phytal belt must have changed its position. However, in contrast to the littoral core mentioned, an increase in upper littoral species cannot be observed. At the beginning of Alleröd (365-350) *Chaoborus* (Diptera) starts which during VIII (Subboreal) reaches its maximum. Sofar in most of the Austrian meromictic lakes a decrease in ostracod species (mainly *Cytherissa*) and abundance in chironomids together with the onset of *Chaoborus*

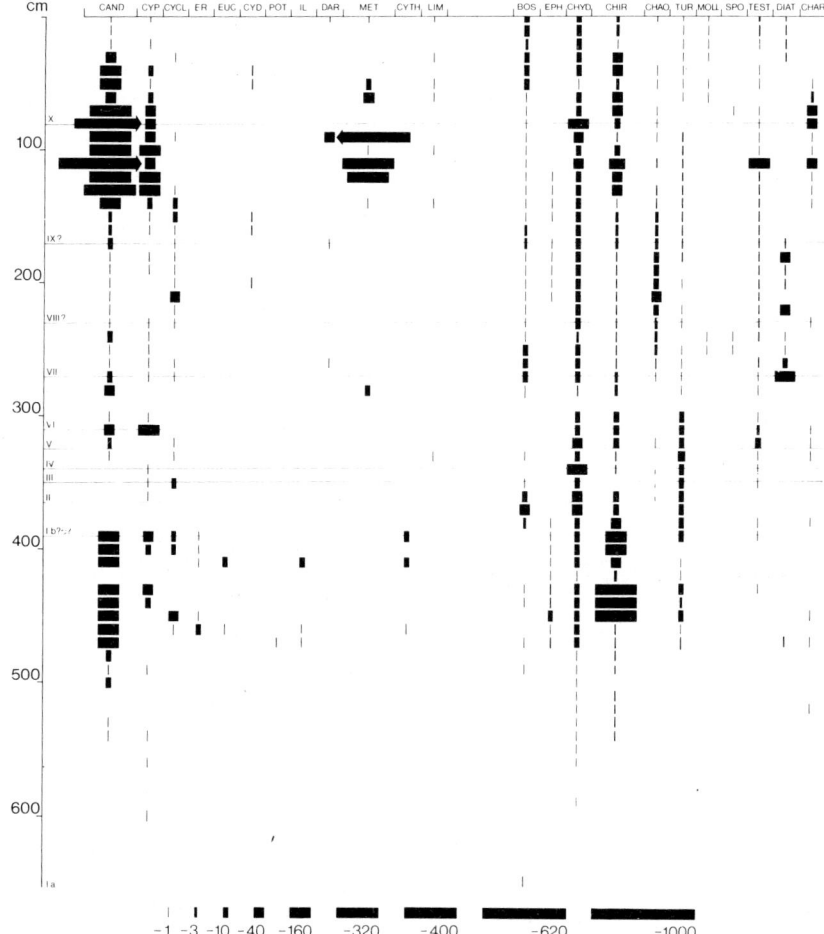

Fig. 2. Distribution of organisms in the Kleinsee-Core. Ostracods, Daphnia-Ephippia and Chironomids are quantitatively indicated; the rest semiquantitatively.
Cand: *Candona;* Cyp: *Cypria ophthalmica;* Cycl: *Cyclocypris* sp.; Er: *Erpetocypris reptans;* Euc: *Eucypris* sp.; Cyd: *Cypridopsis* sp.; Il: Ilyocypris cf. *lacustris;* Dar: *Darwinula stevensoni;* Met: *Metacypris cordata;* Cyth: *Cytherissa lacustris;* Lim: *Limnocythere inopinata;* Bos: *Bosmina;* Eph: *Daphnia*-Ephippia; Chyd: Chydoridae; Chir: Chironomidae; Chao: *Chaoborus;* Tur: Egg-cases *Turbellaria;* Moll: Molluscs; Spo: Spongilla; Test: Testacea; Diat: Diatomeae (mainly *Campylodiscus*); Char: *Chara.*
Ib – X according to FIRBAS

has been interpreted as the beginning of meromictic conditions and it seems to be true also in Kleinsee: thus from about 350 cm until 150 cm (elder Subatlanticum) chironomids and ostracods are scarce and mainly restricted to swimming and littoral species (*Cypria, Cyclocypris,* less *Cypridopsis*). Shells of *Limnocythere inopinata* (for the first time during IV, Praeboreal), *Metacypris cordata (for the first time during VI, Atlanticum) and Darwinula stevensoni* which appears a little

323

later than *Metacypris* could have either drifted from the littoral zone to the lake's deepest area or are indicating casual full circulation periods as has been suggested also for Goggausee (Löffler 1975). Such reoxygenation periods would also explain the abundance of Candona in some of the sections (320, 310, 280, 270, 240, 170-150 cm). Reoxygenation has recently also been reported from the meromictic Längsee (Berger 1973), which also belongs to the group of meromictic lakes with a maximum depth of less than 25 m.

The upper part of the core (140-30 cm) is characterized by a most striking increase of *Candona* (besides a *Candida* group species also *C. rostrata*) *Cypria*, chironomids and a recent onset and increase of *Metacypris cordata*. *Limnocythere inopinata* also becomes more abundant; however, being mainly a littoral species, its shells may result from drifting. Compared with the species just mentioned their numbers are rather insignificant. At the same time *Chaoborus* decreases but never disappears totally from the lake. This upper core section is in good accordance with that one analyzed from another core taken in 1973 which in its deeper sections, however, was incomplete and therefore could not be described fully (Löffler 1975).* Especially the distribution of *Candona* and *Metacypris* in both cores concurs excellently. Because of very loose sediment (very high water content) the top section (20-0 cm) shows a decrease in almost all of the organisms present in the sectrion just discussed.

The increase of ostracods from 140 cm onward doubtlessly indicates the end of deoxygenation or even meromictic period. In this context it is of interest that this happened when Kleinsee had a maximum depth of approximately 11 m. Goggausee with about the same area and a maximum depth of 12.8 m still behaves meromictic (Findenegg 1963, Löffler 1975) although casual reoxygenations of the monimolimnion do occur. In Kleinsee the critical depth for meromictic conditions has obviously been surpassed with increasing sedimentation and thus recolonization of the profundal zone by different organisms, mainly ostracods and chironomids, has been made possible. It is worthwhile to mention that such a recolonization, according to observations made in Wallersee (Jäger 1975), does not occur within one season (reoxygenation during fall-spring) but only if the reoxygenation lasts for more than one year. Thus, in Kleinsee the meromictic period lasting from Bölling until Subatlanticum must have had several of such long lasting events.

In both of the smallest meromictic lakes – Goggausee and at that time Kleinsee – the onset of the deoxygenation period is approximately at the same time, whereas in the bigger lakes Längsee and Klopeiner See (0.76 and 1.3 km²) this happened much later – somewhere between Alleröd and Preboreal. It is of interest then to learn whether in the biggest of the Carinthian meromictic lakes – Wörthersee, Millstädter See and Weissensee – according to their size the beginning of their present state was even later. Cores from Wörthersee, obtained in 1976 from both the western and eastern bassin, will certainly give more general information about this.

---

* Erroneously in Fig. 2 the depths of 2.5 m and 5 m are labelled 5 and 10 m, respectively!

References

Berger, F. 1973. Einige physikalische und hydrochemische Beobachtungen am Längesee. In: Arbeitsbericht über die Limnologische Exkursion 1972 zum Längsee. Carinthia II, 163/83: 331–377.
Findenegg, I. 1963. Ein meromiktischer Kleinsee, der Goggausee in Kärtnern. Anz. math. nat. Kl. Österr. Akad. Wiss. 7: 1–11
Hartl, H. & Sampl, H. 1973. Beschreibung der Schutzgebiete. In: Die Unterkärnter Seen und Berge. Amt der Kärntner Landesregierung, Verfassungsdienst, Klagenfurt, pp. 9-28.
Jäger, P. 1974. Limnologische Untersuchungen im Wallersee unter besonder Berücksichtigung der Ostracodenpopulation. Diss. Univ. Wien.
Löffler, H. 1972. Arbeitsbericht der limnologischen Exkursion Klopeiner See 1971. Carinthia II, 162 (82): 235-274.
Löffler, H. 1975. The onset of meromictic conditions in Goggausee, Carinthia. Verh. Int. Ver. Limnol. 19: 2284-2289.
Löffler, H. 1975. The evolution of ostracod faunus in alpine and prealpine lakes and their value as indicators. Ostraxod. Symp. Newark, Del.
Löffler, H. 1975. The onset of meromictic conditions in alpine lakes. Royal Soc. New Zealand, Wellington 1975: 211-214.
Powell, S. 1976. Einige Aspekte der Beziehung zwischen Sedimenteigenschaften und der Fortbewegung benthischer Süßwasser-Ostracoden, mit spezieller Berücksichtigung der Cytherissa lacustris (SARS). Diss. Univ. Wien.
Stiny, J. 1934. Zur Kenntnis der Hochfläche von Rückersdorf (Kärnten). Jb. geol. Bundesanst. Wien 84: 1-12.

Author's address:

Limnologisches Institut der Österreichischen Akademie der Wissenschaften, –Berggasse 18, 1090 Vienna, Austria

Discussion

Whatley: This type of study is based upon the assumption that the species concerned have not, during the Holocene, changed their ecology. I am constantly faced with this problem in my studies involving environmental restrictions in the Quaternary, based upon the known ecology of Recent mostly cytheraean species, certain of which may have somewhat different ecological requirements in the past. Can you have complete confidence that none of your species have changed their ecology during the Holocene?
Löffler: It is very unlikely though not impossible that any major change in the ecology of ostracods did occur during the last 15,000–20,000 years.

Sixth Intern. Ostracod Symposium, Saalfelden

PALEOECOLOGY OF MARINE AND FRESHWATER HOLOCENE OSTRACODES FROM LAKE CHAMPLAIN, NEARCTIC NORTH AMERICA

FRED J. GUNTHER & ALLEN S. HUNT

Abstract

Bottom-muds of Lake Champlain have been sampled by piston cores taken during 1971–75. Samples from 17 cores have yielded more than 30 ostracode species representing two freshwater assemblages and one marine assemblage. Freshwater species are common to both North American and European Pleistocene, Holocene and modern Holarctic, Nearctic and Alpine lake deposits; marine species are circumpolar in distribution, mostly from North Atlantic and nearby Arctic shallow-water, marine environments. The ostracode assemblages represent changing postglacial environments: periglacial Lake Vermont; marine-estuarine Champlain Sea; and possibly the estuarine-freshwater transition from the Champlain Sea to Lake Champlain. Modern sediments in Lake Champlain have not yielded ostracodes.

Introduction

This report is part of a general study of Lake Champlain which has been undertaken at the University of Vermont (by ASH). Several reports have been published or have been given as papers (Chase & Hunt 1972, Fillon & Hunt 1974a, 1974b, Freed, Hunt & Fillon 1975, and Gunther & Hunt 1976) and more are planned.

Here we wish to consider the distribution of more than 30 species of ostracodes found in about 1000 samples taken from 17 piston cores. The results will be discussed in terms of the ecology and distribution of the individual species and in terms of the paleoecology of the ostracode assemblages.

Biogeography

The species of ostracodes found in the sediments of Lake Champlain have also been found in numerous areas. Both marine and freshwater species, except for three that appear to be new, have northern, nearly circumpolar distributions.

The freshwater species have been found across northern North America, and one species is well-known from Europe (Table 1). *Candona rawsoni* is considered to be Nearctic (Lister 1975) or Holarctic (Delorme 1970) in distribution, while *C. subtriangulata* appears to be restricted to the Canadian Shield (Delorme 1970) and to the Great Lakes (Benson & MacDonald 1963). *Cytherissa lacustris* has been found in the freshwater facies of the Alaskan North Slope on the one hand and in deep-water lakes in Canada (Delorme 1970) and Europe (Sars 1928) on the other. *Limnocythere friabilis* is known from Pleistocene and Holocene lake sediments in central and western North America (Benson & MacDonald 1963, Lister 1975) and as living in modern Arctic lakes (Lister 1975).

Table 1. Biogeography of freshwater Ostracodes from Lake Champlain cores.

Key: R – Recent H – Holocene Pl – Pleistocene	*Candona rawsoni*	*Candona subtriangulata*	*Cytherissa lacustris*	*Limnocythere friabilis*
Alaska			Pl	
Western Canada	Pl-R	Pl-R	Pl-R	
Western U.S.	Pl-R		Pl	Pl
Kansas	Pl		Pl	
L. Michigan		Pl		Pl
Illinois	Pl			Pl
L. Erie	H	H	H	H
Eastern Canada	R	R		R
Great Britain			Pl-R	
Scandinavia			Pl-R	
Alps			R	

The marine species have been found over a much wider range. Pertinent studies cover most of the Arctic and North Atlantic (Table 2) and have been published from 1785 to the present day; very helpful modern surveys are those of Neale (1964), Hazel (1967, 1970) and Neale & Howe (1975). Individual studies are too numerous to mention here.

Most of the marine species from Lake Champlain cores are found in the North Atlantic. In comparison with species-lists for previously described shallow-water provinces (see Neale & Howe 1975), we find that of the 24 species to which names can be assigned, 18 are found in the Celtic Province. In addition, 16 species are found in the Norwegian Province and 15 in the Western Arctic Province (which here includes the poorly-studied Gulf of Alaska fauna). The Eastern Arctic Province and the Nova Scotian Province are represented by 13 species each. Only five species are found in the Virginian Province. Most species are found in more than one province.

From the wide geographic range in circumpolar directions, and from the small contribution of the Virginian Province, we conclude that the marine fauna represents a geneal Arctic and North Atlantic ostracode fauna.

Ecology

Ecologic data is not abundant for ostracode species. Depth occurence and geographic distribution have often been noted, but not temperature, salinity nor oxygen content of the water. What data exists has been collected or compiled by Neale (1964), Delorme (1969) and Hazel (1970). Examination of this data for the species found in Lake Champlain cores (Figure 1) indicates that most species are tolerant of wide ranges in both temperature and salinity. Only a few species (*Argilloecia cylindrica*, *Bythocythere bilobatus* and perhaps *Cytheropteron vespertilio*) appear to be limited by the $30^0/\text{oo}$ boundary noted by Neale (1964);

other species are either euryhaline or freshwater. An interesting feature of the data is the apparent tolerance of colder water temperature by marine species rather than by freshwater species; one would expect the opposite because of the greater proximity of glacial ice during Lake Vermont time.

Paleoecology

Examination of the cores collected prior to 1975 indicated the existence of three distinct assemblages (Table 3) of ostracodes. In addition it was found (Gunther & Hunt 1976) that the cores had sampled two layers of sediment that did not con-

Table 2. Biogeography of marine Ostracodes from Lake Champlain cores. Areas within each province are (E.A.): 1. Franz Joseph Land, 2. Novaya Zemlya, 3. Russian Arctic, 4. Spitzbergen Islands; (N.): 1. Finland, 2. Norway, 3. Iceland; (C.): 1. North Sea, 2. England, 3. Scotland, 4. Ireland, 5. Shetland Islands; (W.A.): 1. Greenland, Davis Strait & Canadian Arctic Islands, 2. North Slope, Alaska, 3. Gulf of Alaska; and (N.S.): 1. Gulf of Maine, 2. Gulf of St. Lawrence, 3. Montreal Clay. The Virginian Province (V.) has been treated as a single unit. Data compiled from numerous authors.

Key: R – Recent F – Fossil X – both	Eastern Arctic 1 2 3 4	Nor- wegian 1 2 3	Celtic 1 2 3 4 5	Western Arctic 1 2 3	V. 1	Nova Scotian 1 2 3
Argilloecia cylindrica		R	X			
Bythocythere bilobatus						R
Cythere lutea		X X	R X X X	R	F	R F
Cytherois fischeri		R	R X X			R
Cytherois vitrea		R	R			
Cytheromorpha macchesneyi	F			F		F F
Cytheropteron arcuatum			F F F	R		
Cytheropteron inflatum	R	R	F	R F		
Cytheropteron montrosiense	F R	F	F F F	R F		
Cytheropteron paralatissimum	R R R			R F		R
Cytheropteron vespertilio			F	R		
Cytherura striata		X	X X X			F
Eucytheridea declivis		R	F		R	R R
Eucytheridea macrolaminata	R R	R		R F		
Eucytheridea punctillata	R R R R	R	R R F	R	R	R R
Haplocytheridea setipunctata			R			
Heterocyprideis sorbyana	R R R	R X	X R X	R F		R F
Hulingsina americana				R	R	
Palmenella limicola	R R	R	R R X R	R F	F	R F
Roundstonia globulifera	R R	R	R F F X	F F		F
Schlerochilus contortus	R R R		X R R	R	F	R
Semicytherura affinis	R R R	R				
Xenocythere cuneiformis			F			
Xestoleberis depressa	R R R R	X	X X F	R R		X

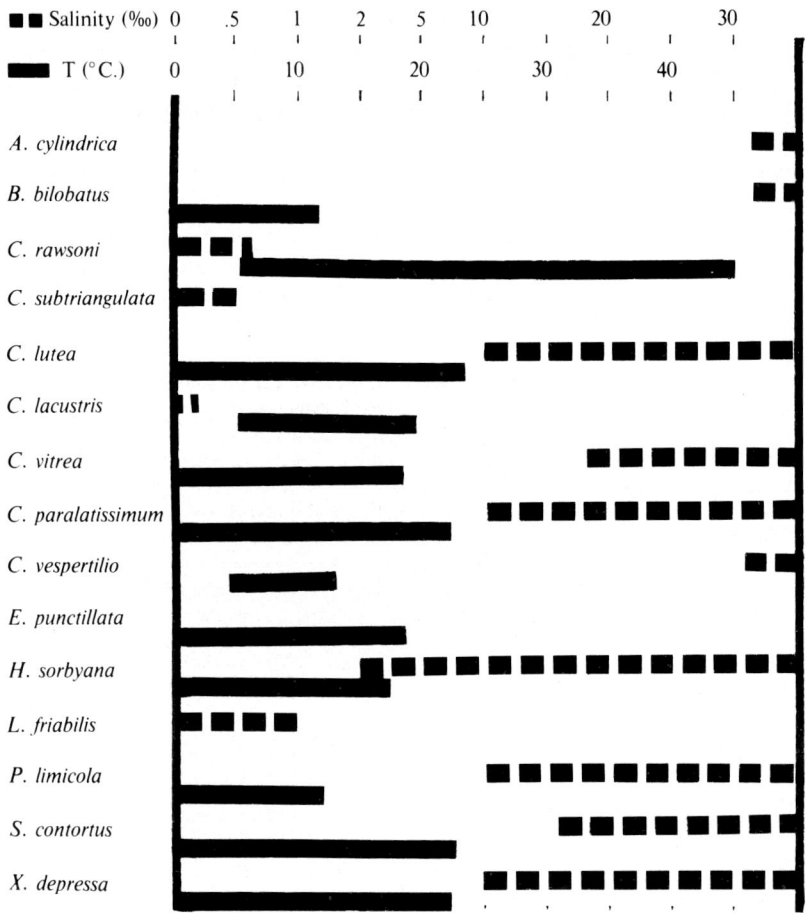

Fig. 1. Distribution of Ostracoda Species from Lake Champlain Cores over Gradients of Temperature and Salinity. The data for *Bythocythere bilobatus* is taken from that of the similar *B. constricta* and for *Cytheropteron vespertilio* from that for *C. alatum*. Data compiled from several authors.

tain ostracodes. Cores collected during the 1975 season confirm the existence of five zones. For this reason, the discussion that follows will be concerned with the paleoecology of each assemblage zone.

In each core, the uppermost layer lacks ostracodes. This barren zone varies in thickness from 10 cm to at least 4.5 m and is found in all parts of the lake. Seismic profiling indicates that this layer may be much thicker in central basins (Chase & Hunt 1972). The layer obviously represents sedimentation in the modern lake; we do not know why this layer has not yielded ostracode valves.

A distinct assemblage (Table 3, A) has been encountered in the northern part of the lake. One core contains mostly freshwater species, but with two euryhaline-marine species. Lack of core penetration into another assemblage-zone and lack of nearby cores does not permit correlation with other assemblage-zones at this time. The presence of *Cytherissa lacustris* indicates very low salinity but a wide range in temperature. It is tempting to consider that we have the transition from the marine Champlain Sea to the freshwater Lake Champlain.

In 16 cores taken from central Lake Champlain, the first ostracode assemblage (Table 3, B) encountered in sampling down the core is dominated by the genus *Cytheropteron*. Other species are present, but usually in low numbers; it is a rare sample where the non-*Cytheropteron* species predominate.

Table 3. Assemblages of Ostracode Species from Lake Champlain Cores. The assemblages are arranged in assumed order of age, youngest (A) to oldest (C).

Assemblage A	B	C	Species
	P		*Argilloecia cylindrica*
	P		*Bythocythere bilobatus*
S		C	*Candona rawsoni*
	P	C	*Candona subtriangulata*
	P		*Cythere lutea*
C	S		*Cytherissa lacustris*
	S		*Cytherois fischeri*
	P		*Cytherois vitrea*
	C	P	*Cytheromorpha macchesneyi*
	C		*Cytheropteron arcuatum*
	P		*Cytheropteron inflatum*
	C		*Cytheropteron montrosiense*
	C		*Cytheropteron paralatissimum*
	C		*Cytheropteron vespertilio*
	P		*Cytheropteron* sp. 1.
	S		*Cytheropteron* sp. 2.
	P		*Cytheropteron* sp. 3.
	S		*Cytherura* cf. *C. striata*
	P		*Eucytheridea declivis*
	P		*Eucytheridea macrolaminata*
	S		*Eucytheridea punctillata*
	S		*Haplocytheridea setipunctata*
	C		*Heterocyprideis sorbyana*
	P		*Hulingsina americana*
C			*Limnocythere friabilis*
	C		*Palmenella limicola*
	S		*Roundstonia globulifera*
S	S		*Sclerochilus contortus*
	S		*Semicytherura affinis*
	S		*Xenocythere cuneiformis*
	P		*Xestoleberis depressa*

Key: S – single occurrence
P – present: 2-several specimens
C – characteristic of the assemblage

The several species characteristic of the *Cytheropteron* Assemblage Zone are not found uniformly nor randomly throughout the zone. The local range of *C. montrosiense* usually covers only part of the assemblage-zone and can be used to divide it. When divided into subzones on this basis, we find that the upper subzone has a lower species diversity than the lower subzone. This change in diversity is similar to that seen by Whatly & Kaye (1971) in marine and estuarine Pleistocene deposits of southern England. The upper subzone is missing in some cores; since these cores are from steeper slopes, it is likely that erosion has removed material.

We consider that the *Cytheropteron* Assemblage Zone represents the marine-estuarine Champlain Sea of postglacial times. Marine species dominate; the occasional occurrence of freshwater ostracodes can be caused by displacement in a density-layered estuary (Kilenyi 1971). The observed change in diversity is considered to represent a rapid influx of salt water followed by gradual freshening, resulting in the gradual reduction in the marine fauna.

In seven cores, species of *Candona* are found below the marine assemblage-zone. The *Candona* Assemblage (Table 3, C) includes the species *C. subtriangulata* and *C. rawsoni* (*C. nyensis* of Gunther & Hunt 1976); many of the specimens are juveniles or fragmentary, so other species could be present. As noted above, these species are typical cold-, freshwater ostracodes; we consider that this layer represents periglacial Lake Vermont.

Four cores completely penetrate ostracode-bearing sediments. Lower layers in these cores lack ostracodes. Maximum penetration is 1.92m. However, the sediments give no sign that glacial deposits are being sampled. We do not know why this layer lacks ostracodes.

Conclusions

The ostracode assemblages of Lake Champlain bottom-muds represent both marine and freshwater environments. The marine assemblage represents the marine-estuarine environments of the Champlain Sea. A freshwater assemblage dominated by specimens of *Candona* represents periglacial Lake Vermont; a second freshwater assemblage may represent the transition between the Champlain Sea and the modern Lake Champlain. All three assemblages represents cold, Nearctic living conditions.

Notes on systematics

Most species are well known and various publications may be consulted for descriptions and illustrations. The problem cases are discussed below.

Cytheropteron vespertilio is used in the sense of Brady (1868), not that of Reuss (1850). This form was combined with another under the name *C. arcuatum* (Brady, Crosskey & Robertson 1874); we are keeping the forms separate in this study because we have found instars of each form that are the same length but that differ greatly in height and in lateral-view outline.

Cytheropteron montrosiense is the form illustrated by Brady et al. (1874) on plate 8; not that illustrated on plate 14.

Acknowledgements

This study has been aided by the cooperaton of many people. Thanks are freely given to all. To begin with, the core samples were processed under the direction of Mrs. Shelley Snyder of the University of Vermont. Considerable assistance with library materials and working space was given by Mr. Norman Hillman of the American Museum of Natural History. Mr. Bob Schulster of Kean College of New Jersey assisted with the debugging and running of computer programs.Dr. I.G. Sohn of the U.S. Geological Survey provided assistance with reference material concerning the freshwater ostracodes.

References

Benson, R.H. & MacDonald, H.C. 1963. Postglacial (Holocene) Ostracodes from Lake Erie. Kansas Univ. Paleont. Contr. Arthropoda 4, 26 pp.
Brady, G.S. 1868. Contribution to the study of the Entomostraca. 1. Ostracoda from the Arctic and Scandinavian Seas. Ann. Mag. Nat. Hist., s4 2(7):30–35.
Brady, G.S., Crosskey, H.W. & Robertson, D. 1874. A Monograph of the Post-Tertiary Entomostraca of Scotland, including Species from England and Ireland. Palaeontological Society, London. 232 pp.
Chase, J.S. & Hunt, A.S. 1972. Sub-bottom Profiling in central Lake Champlain — A Reconnaissance Study. Great Lakes Research, 15th Conf., Procd., pp. 317–329.
Delorme, L.D. 1969. Ostracodes as Quaternary Paleoecological Indicators. Canadian J. Earth Sci. 6(6):1471–1476.
Delorme, L.D. 1970. Freshwater Ostracoda of Canada. Pt. 4. Families Ilyocyprididae, Notodromadidae, Darwinulidae, Cytherideidae, and Entocytheridae. Canadian J. Zool. 48(6):1251–1259.
Fillon, R.H. & Hunt, S.A. 1974a. Foraminiferal Paleoecology of the Champlain Sea in the Champlain Valley. Geol. Society America, Abstr. Programs 5(1):27.
Fillon, R.H. & Hunt, A.S. 1974b. Late Pleistocene benthic Foraminifera of the southern Champlain Sea: Paleotemperature and Paleosalinity Indications. Maritime Sed. 10(1):13–17.
Freed, W.K., Hunt, A.S. & Fillon, R.H. 1975. The Use of Magnetic Properties as Stratigraphic Indicators in Sediment Cores from Lake Champlain, Vermont. Geol. Society America, Abstr. Programs 7(1):60.
Gunther, F.J. & Hunt, A.S. 1976. Ostracodal Biostratigraphy of Champlain Sea Sediments. Geol. Society America, Abstr. Programs 8(2):186–187.
Hazel, J.E. 1967. Classification and Distribution of the Recent Hemicytheridae and Trachyleberididae (Ostracoda) off Northeastern North America.U.S. Geol. Survey, Prof. Paper 564. 49 pp.
Hazel, J.E. 1970. Atlantic Continental Shelf and Slope of the United States. Ostracode Zoogeography in the Southern Nova Scotian and Northern Virginian Faunal Provinces. U.S. Geol. Survey, Prof. Paper 529–E. V + 21 pp.
Kilenyi, T.I. 1971. Some Basic Questions in the Paleoecology of Ostracods. Pau – SNPA, Bull. Centre Rech., 5 suppl.: 31–44.

Lister, K.H. 1975. Quaternary freshwater Ostracoda from the Great Salt Lake Basin, Utah. Kansas Univ. Paleont. Contr., Paper 78.

Neale, J.W. 1964. Some Factors influencing the Distribution of Recent British Ostracoda. Napoli staz. zooi., Pubbl. 33 suppl.: 247–296.

Neale, J.W. & Howe, H.V. 1975. The marine Ostracoda of Russian Harbour, Novaya Zemlya, and other high latitude Faunas. Bulls Amer. Paleont. 65(282):381–431.

Reuss, A.E. 1850. Die fossilen Entomostraceen des Österreichischen Tertiärbechens. Naturw. abh. (Vienna) 3, 81 pp.

Sars, G.O. 1928. An Account of the Crustacea of Norway. 9. Ostracoda. Bergen Museum. 277 pp.

Whatley, R.C. & Kaye, P. 1971. The Paleoecology of Eemian (Last Interglacial) Ostracoda from Selsey, Sussex. Pau – SNPA, Bull. Centre Rech., 5 suppl.: 311–330.

Authors' addresses:

F.J. Gunther, Consultant, 9464 Wandering Way, Columbia, Maryland 21045, USA

A.S. Hunt, University of Vermont, Burlington, Vermont 05401, USA

Sixth Intern. Ostracod Symposium, Saalfelden

ENVIRONMENTAL INTERPRETATION OF THE OSTRACODE SUCCESSION IN LATE QUATERNARY SEDIMENTS OF PLUVIAL LAKE COCHISE, SOUTHEASTERN ARIZONA

SUZANNE P. CAMERON & ROBERT F. LUNDIN

Abstract

Analysis of the ostracodes from a 42 m core drilled in pluvial Lake Cochise sediments has allowed definition of three 'zones' representing change from interpluvial climate through a transitional climate into full pluvial conditions during the late Pleistocene. These 'zones' are defined on species distribution, species proportions, abundance of ostracodes and percent of barren section. The 'zones' indicate that Lake Cochise was a temporary, relatively small, oligotrophic alkaline lake during pre-Wisconsinan times and evolved into a large, permanent, mesotrophic, alkaline and moderately saline lake during the Wiconsinan.

Introduction

General statement

Numerous playas mark the locations of Pleistocene lakes in the western United States. Some of these lake beds have been cored and, largely on the basis of pollen analysis, paleoclimatic interpretations have been made (e.g. Martin 1963). In 1961, the Geochronology Laboratories in Tucson, Arzizona drilled a 42 m core in lake beds beneath Willcox Playa (Fig. 1). The core was drilled near the center of the lake basin to obtain as complete a record as possible and to avoid sediments deposited near the shoreline which fluctuated with climate changes. This paper represents an analysis of the ostracodes of this core and surface (or near-surface) samples from the west wide of the playa.

Locality and geologic setting

Willcox Playa is the surface of a sequence of lake muds which are at least 100 m thick (Schreiber et. al. 1972). These sediments were deposited in Pleistocene Lake Cochise which existed in the Sulphur Springs Valley, a typical structural trough of the Great Basin of southwestern United States. The playa is located south of the town of Willcox, Cochise County, Arizona (Fig. 1). Surrounding the playa are the Dos Cabezas, Winchester, Little Dragoon, Dragoon and Chiricahua Mountains. All major rock types, except ultramafics, are exposed in these mountains and these rocks range in age from Precambrian to late Cenozoic.

Sulphur Springs Valley, which formed during mid-Tertiary block faulting, has been the site of deposition of alluvial gravels, sands, silts and muds of Tertiary-Quaternary age and Quaternary lake muds and associated sediments of the playa (Schreiber et al. 1972). Schreiber et al. (1972) have presented a complete summary of the geologic setting of Willcox Playa.

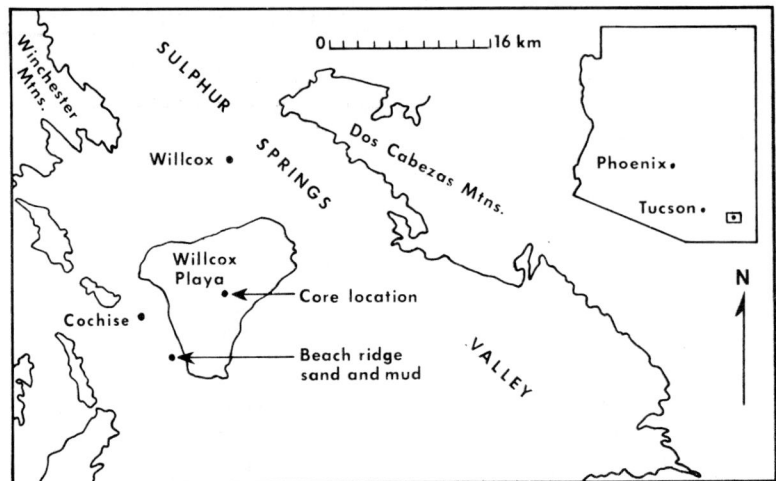

Fig. 1. Index map showing the location of Willcox Playa, the core site and sample locations of the beach ridge sand, and mud below the beach ridge.

Previous investigations

The first major study of the geology of Sulphur Springs Valley was that of Meinzer & Kelton (1913). Since then numerous investigations (Jones & Cushman 1974, Coates 1952, Cooper 1960, Brown et al. 1963, Brown & Shumann 1969) of the geology and water resources of the area have been completed. More specific studies of the playa sediments and their source area and the mineralogy of the lake muds have been made by Pine (1963) and Pipkin (1964). Robinson (1965) analysed the sediments of the beach ridges on the east and west sides of the playa and Wilt (1965) studied eolian features of the playa.

Previous studies of the lake deposits which are of special interest to us are the pollen analysis of the core sediments (Martin 1963, Martin & Mosimann 1965) and shore deposits (Hevly & Martin 1961). The pollen analysis of the core sediments is especially significant because Martin (1963) divided the 42 m core into five pollen divisions ('zones') on which he based his interpretation of the recent climatic history of the area. Finally, Long (1966) established a chronology for Lake Cochise sediments on the basis of radiocarbon dates taken from organic carbon and inorganic carbonates. Although enough dates were not obtained to establish a detailed radiometric chronology, the dates which were obtained demonstrate that the core sediments used in this study are late Pleistocene. Long (1966) also made Eh and pH measurements of the core sediments.

Although important contributions to our knowledge of Pleistocene and Holocene fresh water ostracodes have been made by Delorme (1967, 1968, 1969, 1970a, 1970b, 1970c, 1970d, 1971a, 1971b, 1971c), Hoff (1942), Staplin (1963a, 1963b), Furtos (1933), Benson (1967, 1969), Benson & MacDonald (1963), Gutentag & Benson (1962), Teeter (1970) and others, relatively little has been published on these organisms from the Great Basin. The studies of Swain (1947),

Swain, Becker & Dickinson (1971) and Dickinson & Swain (1967) represent the primary investigations of Tertiary-Quaternary freshwater ostracodes. Delorme (1971b, 1971c) and Löffler (1975) and others have demonstrated the potential value of freshwater ostracodes for paleoenvironmental analysis.

Purposes

Critical to the studies mentioned above which involved paleoenvironmental interpretations are data on the environmental requirements or tolerances of various freshwater ostracode species. Generally these data are not known for the species which existed in pluvial Lake Cochise. Nevertheless, the ostracode faunas of pluvial Lake Cochise changed during the late Pleistocene and some paleoenvironmental inferences can be made. Accordingly, the purpose of this paper is to 1) describe the succession of ostracodes in the 42 m core of Lake Cochise sediments, 2) compare those faunas with the faunas of the beach ridge sand and lake muds below the beach ridge, 3) interpret the ostracode faunas in terms of the late Pleistocene history of the lake and 4) to compare this interpretation to the pollen-based interpretation of climatic history for southeastern Arizona.

Materials and methods

The core samples available for study after the pollen analysis and mineralogical studies were completed included the interval from 6 m to 42 m. The upper 6 m of the core had been destroyed prior to our study. Core diameter was 10 cm to a depth of 22 m and 6 cm below that. The core was initially cut into samples which varied in length from 2.5 cm to 90 cm but most of them were approximately 20 cm long. By the time the individual core samples were transferred to Arizona State University, the sediments had dried making further refinement of sample lengths impossible. Because of this and because of the varying core diameter it was not possible to standardize our sample size. Accordingly, the data on faunal content has been reduced to percentages. More than 8800 ostracode specimens were obtained from the core samples. Preservation was excellent, but many specimens were broken.

In addition to the core samples, one sample of sand from the beach ridge west of the playa (Fig. 1) and one sample of lake mud directly below the beach ridge were analysed. More than 8000 specimens were identified from the beach ridge sand and 352 specimens were identified from the sample of mud below the beach ridge.

Table 1 summarizes the faunal content of the core samples, the beach ridge sand and the lake mud below the beach ridge. The number of specimens per sample for each species present was approximated by dividing the total number of isolated valves by two and adding the number of carapaces. Because some specimens were broken, absolute abundances could not be not be precisely determined. Nevertheless, the data presented below on numbers of specimens gives a good indication of abundance trends through the cored interval. A sample by sample analysis of the ostracode content and abundance of ostracodes in the core is presented by Cameron (1971).

Table 1. Abundance of ostracode species of Lake Cochise core and beach ridge samples

Species	Core		Beach Ridge Mud		Beach Ridge Sand	
	No. of Specimens	% Fauna	No. of Specimens	% Fauna	No. of Specimens	% Fauna
Cyprinotus glaucus	8	0.0	–	–	32	0.4
Cypridopsis vidua	5	0.0	–	–	997	11.3
Potamocypris granulosa	5	0.0	–	–	349	4.4
Physocypria pustulosa	1	0.0	–	–	170	2.1
Candona devexoidea	–	–	–	–	4	0.1
Candona pearlensis	–	–	–	–	48	0.6
Candona patzcuaro	1	0.0	–	–	290	3.6
Candona simpsoni	–	–	–	–	1702	21.4
Candona truncata	–	–	–	–	28	0.4
Candona wantlesi	8	0.0	–	–	4213	52.8
Limnocythere robusta	4868	55.0	98	27.1	3	0.0
Limnocythere herricki	–	–	–	–	6	0.1
Limnocythere ceriotuberosa	300	3.4	2	0.5	221	2.8
Limnocythere pterygoventrata	3575	40.0	252	72.4	7	0.1
Limnocythere staplini	75	0.8	–	–	–	–
TOTALS	8846	99.2	352	100.0	8070	99.7

Lake Cochise sediments

General statement

The drainage within Sulphur Springs Valley is nonintegrated. Accordingly, sediments transported by streams from the surrounding mountains were trapped in Lake Cochise. The primary streams entered Lake Cochise at its northern and southern ends. As would be expected, a facies change occurs toward the center of lake basins. Detailed sediment analyses are presented by Schreiber et al. (1972).

Nearshore deposits

The nearshore deposits are generally sands, muddy sands and sandy muds (Robinson 1965) but some samples even contain gravel. Although the width of this relatively coarse grained facies is variable, at the surface it extends about 0.8 km basinward from the beach ridge. Schreiber et al. (1972) show the beach ridge sediments to be predominantly sands and muddy sands. The sand fraction of the nearshore sediments contains quartz, feldspar, rock fragments and accessory minerals. Calcium carbonate is present in the form of rounded grains, cement and ostracode shells, the latter of which forms the majority of the sand fraction in some samples.

Central lake deposits

Data on the central lake deposits are primarily from analyses of the sediments of the 42 m core which was drilled in what is thought to have been the deepest part of the lake. These sediments, except for the upper 3–4 m, which are oxidized, are black to greenish gray clays. Grain size is rather uniform averaging 1 percent sand, 26 percent silt and 73 percent clay (Schreiber et al. 1972). Organic matter content for the unoxidized part of the core ranged from 0.44 to 2.20 percent. Although an iron sulfide mineral could not be identified, evidence for presence of iron sulfide in the sediment is presented by Schreiber et al. (1972). Eh and pH values were obtained for the oxidized as well as unoxidized parts of the core. Fore the unoxidized sediments Eh ranged from -106 to -306 mv. The pH values ranged from 8.7 to 9.6 and averaged 9.2. The clay-sized fraction of the core sediments is dominated by illite (Schreiber et al. 1972). Calcium cabonate was present in all size grades of the core sediments in the form of small caliche-like lumps, gastropod shell fragments an ostracode valves. Gypsum or other primary evaporites were not present. Details of the sand and silt fraction mineralogy are presented by Schreiber et al. (1972).

Pollen analysis

Martin (1963) divided the sediments of the 42 m core into five divisions on the basis of their pollen content. Division 1, from the surface to 1.8 m lacked pollen except for accumulations of recent years. Martin concluded that the pollen of this interval had been destroyed by oxidation in postglacial time. Between 1.8 and 3.6 m, division 2, pine pollen comprises 95–99 percent of the total pollen present. Division 3 extends from 3.6 to 14 m and is characterized by more Gramineae, Compositae, *Artemisia* and *Quercus* than in division 2, but pine pollen still dominates. A further decrease in pine pollen and increase in *Artemisia* and Gramineae occur in division 4, from 14 to 23 m. Pine pollen frequency fluctuates noticeably in this interval but still generally comprises 40 percent or more of the total pollen. In division 5, from 23 to 42 m, grass pollen increased substantially and *Ephedra* and Compositae other than *Artemisia* were more abundant than in the upper divisions. Also pollen was generally less abundant. In fact, between 23 and 29 m pollen was virtually absent from the core.

Martin (1963) interpreted the high pine pollen frequencies as an indicator of a cool wet (pluvial) climate and relatively low pine pollen frequencies and poor pollen preservation as an indicator of interpluvial climate. On this basis he concluded that division 1 represents the post-Wisconsinan, divisions 2–4 and upper 2 m of division 5 the Wisconsinan pluvial, that part of division 5 between 23 and 29 m the Sangamon interpluvial, and the remainder of division 5 the Illinoian pluvial. Martin (1963) was cautious, however, about the interpretation of an Illinoian pluvial interval. An alternative interpretation which Martin suggested was that the entire core interval below 23 m represents the Sangamon.

Radiometric chronology

Long (1966) established a chronology for Lake Cochise based on radiocarbon dates (Damon, Long & Sigalove 1963) derived from organic material and inorganic carbonates. Radiocarbon dates of 22,000 ± 500 and 23,200 ± 500 years on inorganic carbonate from core depths between 1.9 and 2.1 m confirm that the upper part of Martin's (1963) division 2 is Wisconsinan.

Long (1966) further suggested that Lake Cochise was higher than 1280 m above sea level from before 30,000 B.P. until 13,000 B.P. and then receded due to arid climate. During the last short pluvial phase from 11,500 B.P. to 10,000 B.P. the lake level was close to that of the beach ridge (1265 m above sea level). Mollusks from the mud below the beach ridge yielded a date of 10,110 ± 400 B.P. (Long 1966) suggesting that this mud was deposited in a swamp or lagoon associated with the beach ridge during the last short pluvial interval. Other radiocarbon dates from Lake Cochise sediments are listed by Long (1966).

Lake Cochise Ostracodes

Ostracode content of the core sediments

Table 1 and Figure 2 summarize the ostracode content of the core sediments. *Limnocythere robusta* and *L. pterygoventrata* strongly dominate this fauna. Only two other species, *L. ceriotuberosa* and *L. staplini*, occur in any abundance. All other species collectively constitute less than 1 percent of the entire fauna of the core sediments.

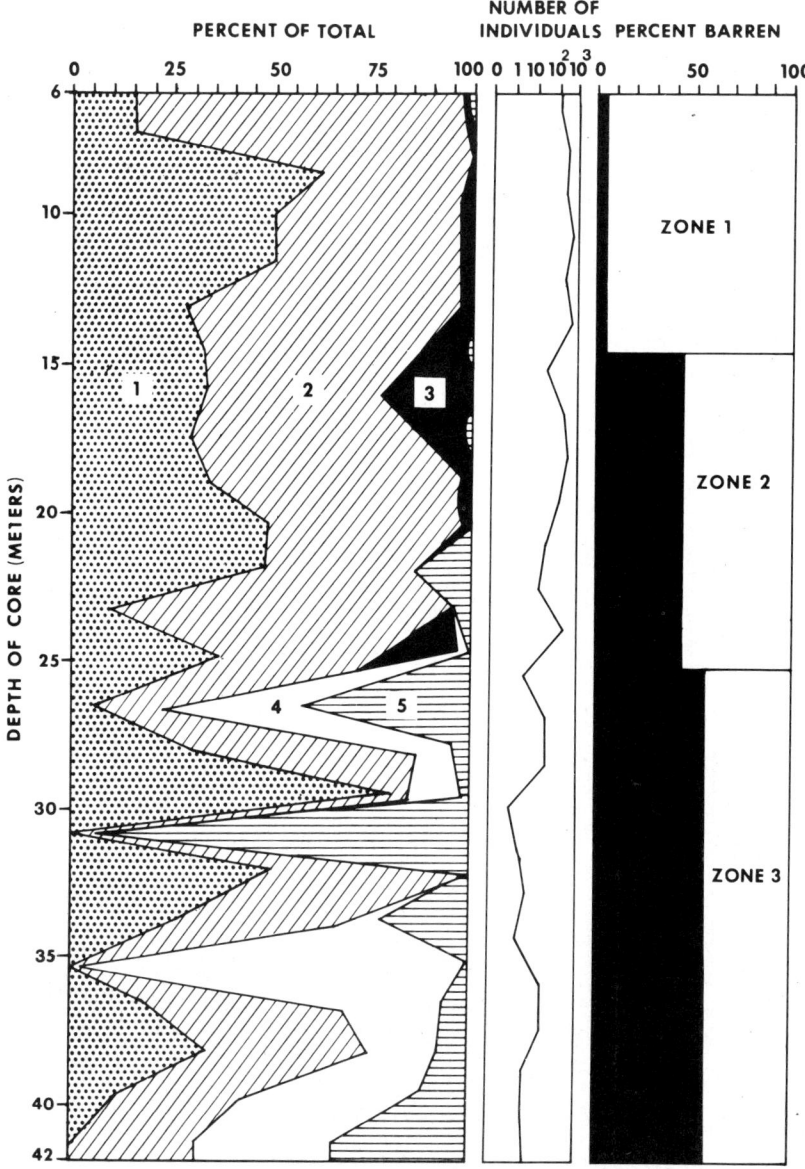

Fig. 2. Diagram showing content of the ostracode fauna (left), abundance of ostracodes, and percent of barren section in the 'zones' established for the 42 m core. Note that number of individuals is plotted on logarithmic scale. 1, *Limnocythere pterygoventrata*; 2, *L. robusta*; 3, *L. ceriotuberosa*; 4, *L. staplini*; 5, all other species (see Figs. 3 and 4).

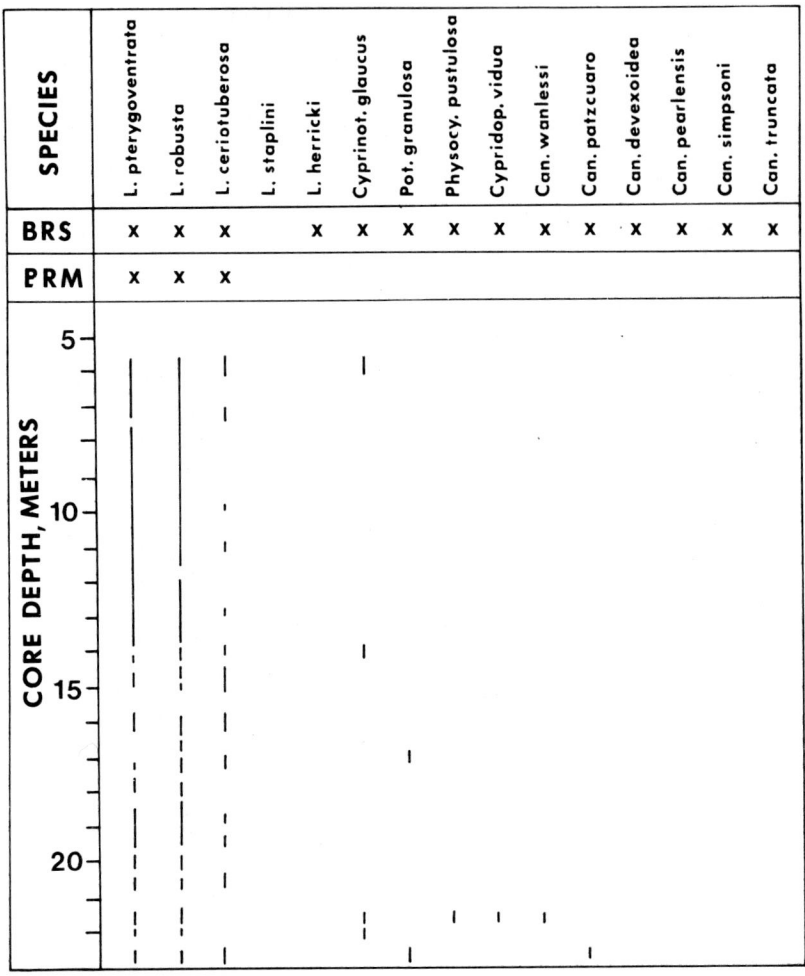

Fig. 3. Distribution of ostracode species in the 6 to 23 m interval of the core and the occurrence of them in the beach ridge sand (BRS) and mud below the beach ridge (BRM).

Ostracode content of the mud below the beach ridge

Table 1 and Figures 3 and 4 summarize the ostracode content of the lake mud below the beach ridge (Fig. 1). This fauna is virtually the same as that of the core except that *L. staplini* is absent and *L. pterygoventrata* constitues a larger proportion of the fauna than *L. robusta*. *L. ceriotuberosa* comprises less than 1 percent of this fauna.

Ostracode content of the beach ridge sand

Table 1 and Figures 3 and 4 summarize the ostracode content of the beach ridge

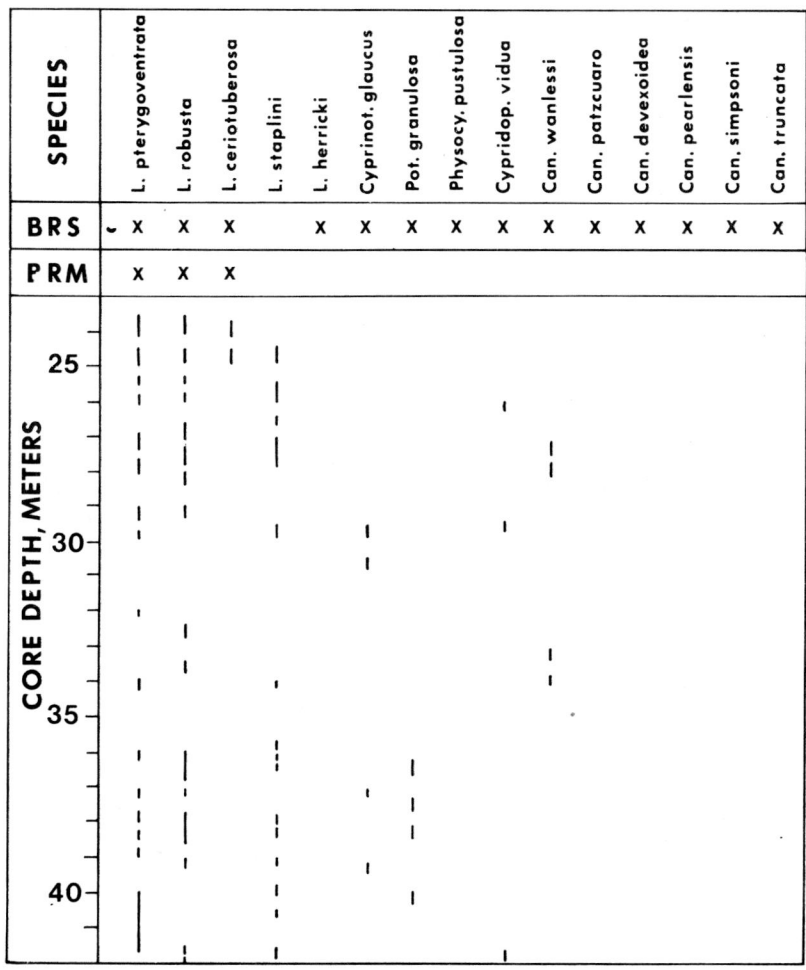

Fig. 4. Distribution of ostracode species in the lower 19 m of the core and occurrence of them in the beach ridge sand (BRS) and mud below the beach ridge (BRM).

sand (Fig. 1). This fauna is grossly different from those described above. The two most abundant species in the core sediments and lake mud below the beach ridge occur in only trace amounts in the beach ridge sand and these occurrences probably are the result of reworking of the lake muds. The beach ridge sand fauna is strongly dominated by species of *Candona*, especially *C. simpsoni* and *C. wanlessi*. *Cypridopsis vidua*, *Potamocypris granulosa*, *Physocypria pustulosa* and *Candona patzcuaro* are the only other species which are reasonably abundant in this fauna.

Stratigraphic distribution

Figures 3 and 4 show the stratigraphic distribution of ostracodes in the core sediments and the occurrence of the various species in the beach ridge sand, and mud below the beach ridge. The most significant observation about the stratigraphic distribution of ostracodes in the core is that *L. staplini* is restricted to the lower 17m of the core and *L. ceriotuberosa* is restricted to the upper 25m of the core. *L. pterygoventrata* and *L. robusta* occur throughout the cored interval where ostracodes are present.

Abundance of Ostracodes

Figure 2 (center column) shows the abundance of ostracodes through the cored interval. These data have been derived by summing the total number of specimens (see section on Materials and methods) for 1.5m intervals. The record shows a general increase in abundance of ostracodes toward the top of the core. A noticeable increase in abundance occurs between 25 and 15m and a dramatic increase in abundance occurs in the interval above 15m.

The right column of Figure 2 shows the proportion of the section which is barren of ostracodes for the three 'zones' which have been defined on the basis of species distribution and abundance of ostracodes. As would be expected, decreases in the percent of barren section correspond closely with increases in the total number of individuals. Although many individual samples, especially in the lower part of the core, are barren, this does not show on the center column of Figure 2 due to the effects of summing the number of individuals through 1.5m intervals.

Zonation

General statement

On the basis of abundance of ostracodes, percent of barren section and/or stratigraphic distribution and proportions of ostracode species, the core sediments can be divided into three 'zones'. These 'zones' are not considered formal biostratigraphic zones because each is based upon different and/or multiple criteria. Nevertheless, we believe that the definition of these 'zones' forms a basis for interpreting the environmental history of Lake Cochise.

Zone 1

Zone 1 occurs from 6 m (uppermost core sample available to use) to 14.8 m. It is characterized by a strong dominance of *L. robusta* and *L. pterygoventrata*, generally more than 90 percent of the fauna, and a small proportion of *L. ceriotuberosa*. This interval contains very few barren samples. The base of this zone is readily recognized by a dramatic increase in abundance of ostracodes and the entire zone is characterized by large numbers of individuals.

Zone 2

This zone occupies the interval from 14.8 m to 24.8 m. The base of it is defined by the highest occurrence of *L. staplini* and the lowest occurrence of *L. ceriotuberosa*. The assemblage of this zone is dominated by *L. robusta* and *L. pterygovertrata*. There is a decrease in the number of individuals compared to zone 1 and a corresponding increase in the amount of sediment which is barren of ostracodes. Samples in this interval characteristically contain tens of individuals as opposed to the hundreds of individuals commonly found in samples of zone 1.

Zone 3

Zone 3, from 24.8 m to the bottom of the core (42 m), is defined by the stratigraphic distribution of *L. staplini* in the core. The proportions of the various ostracode species are exceedingly variable partly because total abundance is low and therefore a few specimens of a particular species might represent a significant percentage of the total population. Figures 2-4 show that some of the species present in the beach ridge sand occur more commonly in zone 3 than they do in zones 1 and 2. This zone is further characterized by many barren intervals.

Interpretation of paleoenvironments

Zone 1

We believe that this interval represents full pluvial conditions of the Wisconsinan. The virtually continuous record of ostracode-bearing sediments indicates that Lake Cochise was permanent during the time represented by this stratigraphic interval. Although specific environmental tolerances and requirements for the dominant species of this zone are not known, Swain, Becker & Dickinson (1971) considered abundance of *Limnocythere pterygoventrata* as an indicator of an alkaline mesotrophic lake. The alkaline nature of the lake is supported by the abundance of *Ruppia,* a monocotyledon which inhabits saline and alkaline water, from 10 m to 23 m (Martin 1963). Further evidence of full pluvial conditions and a permanent lake is the great abundance of ostracodes in this interval. The lake must have developed to a trophic level capable of supporting substantial populations of ostracodes. Nevertheless, the waters were probably moderately saline as indicated by the presence of *L. ceriotuberosa* (see Delorme 1967).

Zone 2

This interval represents a time of transition into full pluvial conditions (zone 1) from the interpluvial conditions indicated by zone 3. Although the same species dominate this interval as in zone 1, ostracodes are less abundant and barren intervals are more common. This suggests certainly a lower nutrient level for the lake than that indicated for zone 1. Additionally this supports our contention that the lake was smaller and suffered periodic dessication, especially during the time represented by the lower part of zone 2. Figures 2 and 3 show an increase in the occurrence of species common in the beach ridge sand. We believe these represent a nearshore assemblage and this increased occurrence indicates that the shoreline was closer to the core site during the early part of the interval represented by zone 2 than it was during the time represented by zone 1. Again, the presence of *L. ceriotuberosa* indicates moderately saline waters.

Zone 3

This interval, which is the lower 17.2 m of the core, represents interpluvial climatic conditions which preceded the Wisconsinan. The occurrence of *L. staplini* in zone 3 indicates a possible increase in salinity. Delorme (1969) reported finding this species in waters with total dissolved salts ranging from 149-199000 ppm. It appears that *L. ceriotuberosa* replaced *L. staplini* in the upper parts (zones 1 and 2) of the core. Although we do not know the water depth tolerances for *L. ceriotuberosa*, Delorme (1969) indicates a relatively restricted water depth range (0.06-1.2 m) for *L. staplini*. We conclude from this that the water at the core site during deposition of zone 3 sediments was very shallow. Increased occurrence of the nearshore assemblage supports the conclusion that the shoreline commonly was near the core site during this interval. Finally, the numerous barren intervals suggest poor nutrient conditions and periodic dessication. If *L. ceriotuberosa* prefers deeper water this could explain its replacement of *L. staplini* in zones 1 and 2. Indeed during the time represented by zone 3 Lake Cochise may well have been represented by a series of temporary ponds or small lakes. Nevertheless, the water chemistry was not sufficiently changed to drastically alter the content of the fauna. This is attested to by the occurrence of *L. pterygoventrata* and *L. robusta*, which dominate zones 1 and 2.

Mud below the beach ridge

The sample of lake mud from directly beneath the beach ridge, no doubt, is younger than any of the core samples available to us. Nevertheless, the ostracode fauna of this sample (Table 1) closely resembles that of the upper part of zone 1, especially the fauna which occurs at a core depth of about 8 m. Accordingly, we conclude that the mud below the beach ridge was deposited under conditions similar to those indicated by zone 1, namely full pluvial climate and mesotrophic, alkaline and moderately saline waters. If Long's (1966) interpretation is correct, this sample represents part of the sediments deposited during a high stand of the lake during the last short pluvial interval (11,500-10,000 B.P.).

Beach ridge sand

The beach ridge deposit does not, of course, represent the environment in which the ostracodes contained therein lived. We believe that the assemblage of ostracodes from the beach ridge sand sample was derived from shallow water muddy sands and sands adjacent to the beach ridge. Ostracodes were thrown on to the beach ridge along with the sand by storm waves which swept the lake. Lagoons behind the beach ridge, no doubt, also served as a source of ostracode-bearing sediment for the beach ridge.

The species common in the beach ridge sand (Table 1) are compatible with the interpretation of a shallow water nearshore environment. Though some of these species have wide environmental tolerances, the most common, *Candona wanlessi* and *C. simpsoni,* have been found in shallow slackwater or lake deposits and in shallow weedy ponds, marshes or sluggish streams (Staplin 1963a).

Comparison with the pollen analysis

General statement

Various kinds of data are available for comparison with the ostracode data. Long (1966) questioned the value of the Eh measurements of the sediments in interpreting bottom conditions during deposition. He further doubted the significance of the occurrence of large calcite euhedra in the lower 18 m of the core, a criterion which Pipkin (1964) used to infer a warm dry climate. The presence of hydrotroilite in the lake sediments supports the contention that the environment of deposition was alkaline. An increase in the number of silty layers between 23 and 35 m in the core might suggest proximity to the shoreline which would be compatible with our interpretation of zone 3. But these silty layers might also indicate stronger streams or increased seasonal runoff. The best data with which to compare the ostracode data is that of the pollen study of Martin (1963).

Pollen Divisions 1, 2 and 3

Pollen divisions 1 and 2 are not represented by the core samples available to us. Our uppermost core sample (6 m) falls within pollen division 3 and represents the top of ostracode zone 1. Martin interpreted pollen division 3 as representing pluvial climate but noted a decrease in pine pollen frequency and an increase in grass pollen as compared to pollen division 2. This interpretation fits well with that derived from the ostracode data. In fact the base of pollen division 3 and ostracode zone 1 are placed at virtually identical levels in the core.

Pollen Division 4

Martin placed the base of this division at about 21 m and placed the base of the Wisconsinan at about 23.5 m. Accordingly, this division is still interpreted as pluvial but Martin noted still further decrease pine pollen and increase in grass and *Artemisia*. Of significance here is Martin's (1963) interpretation of two pine pollen peaks which occur in pollen division 4. By continuous sampling at close intervals (6 mm) Martin (1963, p. 443) was able to demonstrate that the erratic behavior (peaks) of the pine pollen profile, especially noticeable in division 4, represented "authentic stratigraphic events and not artifacts of extraction or sampling". Martin concluded that sudden decreases in pine pollen frequency were the result of

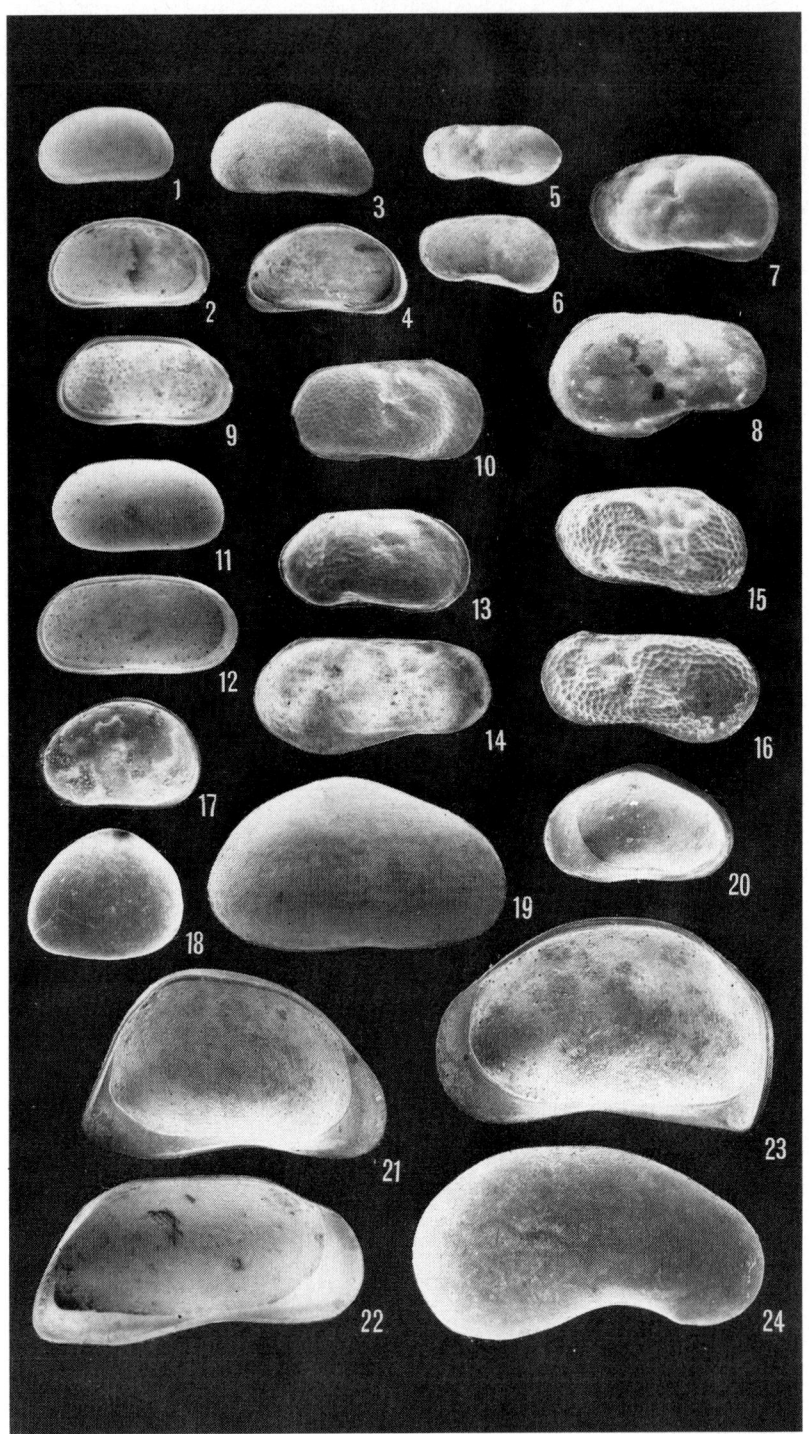

◀ Plate 1. All figures ×40

Fig. 1. *Candona simpsoni* Sharpe, 1897. Lateral view, right valve.
Fig. 2. *Candona simpsoni* Sharpe, 1897. Interior view, left valve.
Fig. 3. *Potamocypris granulosa* Daday, 1902. Lateral view, right valve.
Fig. 4. *Potamocypris granulosa* Daday, 1902. Interior view, left valve.
Fig. 5. *Limnocythere staplini* Gutentag & Benson, 1962. Lateral view, male, left valve.
Fig. 6. *Limnocythere staplini* Gutentag & Benson, 1962. Interior view, female, left valve.
Fig. 7. *Limnocythere pterygoventrata* Dickinson & Swain, 1967. Lateral view, female, left valve.
Fig. 8. *Limnocythere pterygoventrata* Dickinson & Swain, 1967. Lateral view, male, right valve.
Fig. 9. *Candona pearlensis* Staplin, 1963. Interior view, left valve.
Fig. 10. *Limnocythere herricki* Staplin, 1963. Lateral view, right valve.
Fig. 11. *Candona wanlessi* Staplin, 1963. Lateral view, right valve.
Fig. 12. *Candona wanlessi* Staplin, 1963. Interior view, left valve.
Fig. 13. *Limnocythere robusta* Delorme, 1967. Lateral view, female, right valve.
Fig. 14. *Limnocythere robusta* Delorme, 1967. Lateral view, male, right valve.
Fig. 15. *Limnocythere ceriotuberosa* Delorme, 1967. Lateral view, female, right valve.
Fig. 16. *Limnocythere ceriotuberosa* Delorme, 1967. Lateral view, male, left valve.
Fig. 17. *Physocypria pustulosa* (Sharpe, 1897). Interior view, right valve.
Fig. 18. *Physocypria pustulosa* (Sharpe, 1897). Lateral view, left valve.
Fig. 19. *Cyprinotus glaucus* Furtos, 1933. Lateral view, right valve.
Fig. 20. *Cypridopsis vidua* (O. F. Mueller, 1776). Interior view, right valve.
Fig. 21. *Candona truncata* Furtos, 1933. Interior view, left valve.
Fig. 22. *Candona devexoidea* Dickinson & Swain, 1967. Interior view, left valve.
Fig. 23. *Candona patzcuaro* Tressler, 1954. Interior view, female, right valve.
Fig. 24. *Candona patzcuaro* Tressler, 1954. Lateral view, male, right valve.

dessication of Lake Cochise. This is compatible with our interpretation that ostracode zone 2, which nearly coincides with pollen division 4, represents conditions transitional between interpluvial and pluvial. The base of ostracode zone 2 corresponds closely with Martin's placement of the base of the Wisconsinan, even though he placed the base of pollen division 4 somewhat higher in the core.

Pollen Division 5

Martin (1963) interpreted the interval between 23 and 29 m as an interglacial ("perhaps the Sangamon"), and on the basis of an increase in grass pollen frequency and 'scattered records of a summer rain flora' the interval between 29 and 42 m was considered to represent a climate similar to that of the modern Mexican Plateau. This he suggested to be a wetter climate than the climate represented by the 23-29 m interval but not a full pluvial climate as indicated by pollen division 2.

The ostracode data agree well with Martin's interpretation for the 23-29 m interval. We believe that ostracode zone 3 represents an interglacial (interpluvial) climate. We see no evidence, however, of a 'semipluvial' climate or 'Illinoian glacial' for the interval between 29 and 42 m. If Martin's interpretation of the lower 13 m of the pollen record is correct, the climatic change must have been sufficiently subtle as to not have been impressed on the ostracode record. We conclude that the entire ostracode zone 3 represents a pre-Wisconsinan interpluvial.

Summary

1) The ostracode succession in the lower 36 m of the 42 m core in pluvial Lake Cochise sediments defines three 'zones' based on species distribution, species proportions and abundance of ostracodes.
2) Zone 1, 6-14.8 m, represents full pluvial conditions and an alkaline mesotrophic permanent lake with moderately saline water.
3) Zone 2, 14.8-24.8 m, represents conditions transitional between full pluvial and interpluvial climates. Nutrient levels of the lake were lower than for zone 1, the lake was smaller, and periodic dessication occurred.
4) Zone 3, 24.8-42 m, represents interpluvial climate. Dessication was common, salinity was probably higher than for zones 1 and 2 and nutrient levels were low.
5) The mud below the beach ridge was deposited under conditions equivalent to those indicated by ostracode zone 1.
6) The beach ridge sand is dominated by species of *Cadona* as opposed to *Limnocythere* which dominates the core sediments and lake muds below the beach ridge. The *Candona*-dominated assemblage is a nearshore assemblage and the *Limnocythere*-dominated assemblage is an offshore (deeper water) assemblage. Interpretations based on ostracodes agree closely with those based on pollen analysis except in the lower part of the core where the ostracode record shows no evidence of a wet climate as indicated by the pollen record.

Acknowledgements

We wish to express our thanks to Paul Martin of the Geochronology Laboratories in Tucson, Arizona for arranging for us to have the core samples, to Joseph Schreiber, Jr, of the University of Arizona for supplying the beach ridge samples and to Dick Watkins and Jim Swafford of Arizona State University for their aid in preparation of the scanning electron micrographs. We further thank L. D. Delorme and F. L. Staplin for their helpful comments regarding the taxonomic position of the Lake Cochise ostracodes. Finally, funding for this research from the Arizona State University Grants Committee is gratefully acknowledged.

References

Benson, R. H. 1967. Muscle-scar patterns of Pleistocene (Kansan) ostracodes. Kansas Univ. Geol. Dept. Special Publ. 2, pp. 211-241.
Benson, R. H. 1969. Ostracodes of the Rita Blanca Lake Deposits. Geol. Soc. America, Mem. 113:107-115.
Benson, R. H. & MacDonald, H. C. 1963. Postglacial (Holocene) ostracodes from Lake Erie. Kansas Univ. Paleont. Contr., Arthropoda, Art. 4:1-26.

Brown, S. G., et al. 1963. Basic ground-water data of the Willcox Basin, Graham and Cochise Counties, Arizona. Arizona State Land Dept. Water-Resources Rept. 14:1-93.
Brown, S. G. & H. H. Schumann. 1969. Geohydrology and water utilization in the Willcox Basin, Graham and Cochise Counties, Arizona. U.S. Geol. Survey Water-Supply Paper 1959-F, Washington, 32 pp.
Cameron, S. P. 1971. Ostracodes of Pluvial Lake Cochise, Cochise County, Southeastern Arizona. Unpubl. M. S. Thesis. Arizona State Univ., Tempe, 65 pp.
Coates, D. R. 1952. Willcox Basin, Cochise and Graham Counties. In: Halpenny, L. C., et al., Groundwater in the Gila River Basin and adjacent areas – A summary. U.S. Geol. Survey open-file report, pp. 177-186.
Cooper, J. R. 1960. Reconnaissance map of the Willcox, Fisher Hills, Cochise, and Dos Cabesas Quadrangles, Cochise and Graham Counties, Arizona. U.S. Geol. Survey Mineral Inv. Field Studies Map MF-231.
Damon, P. E., Long, A. & Sigalove, J. J. 1963. Arizona Radiocarbon dates IV. Radiocarbon 5:283-301.
Delorme, L. D 1967. New freshwater Ostracoda from Saskatchewan, Canada. Can. J. Zool. 45:357-363.
Delorme, L. D. 1968. Pleistocene freshwater Ostracoda from Yukon, Canada. Can. J. Zool. 46:859-876.
Delorme, L. D. 1969. Ostracodes as Quaternary paleoecological indicators. Can. J. Earth Sci. 6:1471-1476.
Delorme, L. D. 1970a. Freshwater ostracodes of Canada. Part I. Subfamily Cypridinae. Can. J. Zool. 48:153-168.
Delorme, L. D. 1970b. Freshwater ostracodes of Canada. Part II. Subfamily Cypridopsinae and Herpetocypridinae, and family Cyclocyprididae. Can. J. Zool. 48:253-266.
Delorme, L. D. 1970c. Freshwater ostracodes of Canada. Part III. Family Candonidae. Can. J. Zool. 48:1099-1127.
Delorme, L. D. 1970d. Freshwater ostracodes of Canada. Part IV. Families Ilyocyprididae, Notodromadidae, Darwinulidae, and Enterocytheridae. Can. J. Zool. 48:1251-1259.
Delorme, L. D. 1971a. Freshwater ostracodes of Canada. Part V. Families Limnocytheridae, Loxoconchidae. Can. J. Zool. 49:43-64.
Delorme, L. D. 1971b. Paleoecology of Holocene sediments from Manitoba using freshwater ostracodes. Geol. Ass. Canada, Special Paper 9, pp. 301-304.
Delorme, L. D. 1971c. Paleoecological determinations using Pleistocene fershwater ostracodes. Colloquium on the paleoecology of ostracodes. Soc. Nat. des Pétroles d'Aquitaine, Centre des Recherches 5:341-347.
Dickinson, K. A. & Swain, F. M. 1967. Late Cenozoic freshwater Ostracoda and Cladocera from northeastern Nevada. J. Paleont. 41:335-350.
Furtos, N. C. 1933. The Ostracods of Ohio. Ohio Biol. Survey Bull. 29 (5):411-524.
Gutentag, E. D. & Benson, R. H. 1962. Neogene (Plio-Pleistocene) freshwater ostracodes from the central high plains. State Geol. Survey Kansas, Bull. 157, pt 4, pp. 1-60.
Hevley, R. & Martin, P. S. 1961. Geochronology of pluvial Lake Cochise, southern Arizona I. Analysis of shore deposits. J. Arizona Acad. Sci. 2:24-31.
Hoff, C. 1942. The ostracodes of Illinois. Ill. Biol. Mon. 19, Nos. 1-2, pp. 1-196.
Jones, R. S. & Cushman, R. L. 1947. Geology and groundwater resources of the Willcox basin, Cochise and Graham Counties, Arizona. U.S. Geol. Survey. mimeogr. report, 35 pp.
Löffler, H. 1975. The evolution of ostracode faunas in alpine and pre-alpine lakes and their value as indicators. p. 433, in: F. M. Swain (ed.), Biology and Paleobiology of Ostracoda. Paleontological Res. Institution, Ithaca.
Long, A. 1963. Late Pleistocene and Recent chronologies of playa lakes in Arizona and New Mexico. Unpubl. Ph.D. diss., Univ. of Arizona, Tucson, 141 pp.
Martin, P. S. 1963. Geochronology of pluvial Lake Cochise, southern Arizona II. Pollen analysis of a 42 meter core. Ecology 44:436-444.
Martin, P. S. & Mosiman, J. E. 1963. Geochronology of pluvial Lake Cochise, southern Arizona III. Pollen statistics and Pleistocene metastability. Amer. J. Sci. 263:313-358.

Meinzer, O. E. & Kelton, F. C. 1913. Geology and water resources of the Sulfur Springs Valley, Arizona. U.S. Geol. Survey Water-Supply Paper 320, pp. 9-231.
Pine, G. L. 1963. Sedimentation studies in the vicinity of Willcox Playa, Cochise County, Arizona. Unpubl. M. S. Thesis, Univ. of Arizona, Tucson, 38 pp.
Pipkin, B. W. 1964. Clay mineralogy of the Willcox Playa and drainage basin, Cochise County, Arizona. Unpubl. Ph.D. diss., Univ. of Arizona, Tucson, 160 pp.
Robinson, R. C. 1965. Sedimentology of beach ridge and nearshore deposits, pluvial Lake Cochise, southeastern Arizona. Unpubl. M. S. Thesis, Univ. of Arizona, Tucson, 111 pp.
Shreiber, J. F., et al. 1972. Sedimentologic studies in the Wilcox Playa area, Cochise County, Arizona. In: Playa Lake Symposium (C. C. Reeves, ed.), ICASALS Publ. no. 4, pp. 133-184.
Staplin, F. L. 1963a. Pleistocene Ostracoda of Illinois. Part I. Subfamilies Candoninae, Cyprinae, general ecology, morphology. J. Paleont. 37:1164-1203.
Swain, F. M. 1947. Tertiary non-marine Ostracoda from the Salt Lake Formation, northern Utah. J. Paleont. 21:518-528.
Swain, F. M., Becker, J. & Dickinson, D. A. 1970. Paleontology of Tertiary and fossil Quaternary non-marine Ostracoda from the western interior United States. Colloquium on the Paleoecology of Ostracodes, Soc. Nat. des Pétroles d'Aquitaine, Centre des Recherches 5:461-487.
Teeter, J. W. 1970. Paleoecology of a Pleistocene microfossil assemblage at the Fairlawn, Ohio, mastodon site. Amer. Mid. Nat. 83:583-594.
Wilt, J. C. 1965. Some features of wind deposits near Willcox Playa, Cochise County, Arizona. Unpubl. Honors Program Senior Thesis, Univ. of Arizona, Tucson, 41 pp.

Authors' addresses:

S. P. Cameron, 125 W. Rose Lane, Phoenix, Arizona 85013, USA
R. F. Lundin, Dept. of Geology, Arizona State University, Tempe, Arizona 85281, USA

Sixth Intern. Ostracod Symposium, Saalfelden

SOME OSTRACODES FROM THE UPPER AMAZON BASIN, BRAZIL. ENVIRONMENT AND AGE.

IVONE PURPER

Abstract

Brackish and fresh water ostracods with several new genera and species are registered from sediments of the Upper Amazon Basin. The sediments are of a reworked material presenting Mesozoic genera associated with a great quantity of well-preserved Tertiary ostracods of a probable Pliocene or more Recent age.

Introduction

The material under study was found in samples collected by Prof. Dr. Eurico Rômulo Machado from the Instituto de Geociências – UFRGS – who looked into the economic use of the lignite of that area. The material proceeds from three drill cores – CPCAN-I (Tamanduá); CPCAN-II (Poreré) and CPCAN-III (São Paulo de Olivença) with a depth range from 19.50 m to 215 m and from one outcrop (Benjamin Constant) (see map). All the samples contain ostracods, molluscs and fish teeth and spines. The ostracods are represented by several new species belonging to nine known genera and four new genera which are being studied in a complete systematic way by the present author in her doctoral Thesis. For this reason new species and genera are registered by letters and or numbers. Despite this work's representing a preliminary study about ostracods of a restricted area of this vast and not very well-known region, the author tries to reach some conclusions about paleogeography, environment and age.

The author expresses her gratitude to Prof. Dr. Irajá Damiani Pinto for his orientation and suggestions during the course of this work.

History

The first discovery of fossils in the Upper Amazon beds was done by James Orton at Pebas (Peru) in 1867 on the Ambyacu River two miles above its confluence with the Marañon River. This discovery of fossil molluscs enabled Orton to refute Agassiz's theory that the Amazonian beds were entirely composed of glacial drift. The first description of this material was done by Gabb (1868) describing 6 species of molluscs.

In subsequent years, other authors made other identifications of many new genera and species of Pelecypoda and Gastropoda collected in Peru and Brazil. These authors are: Conrad (1871), Woodward (1871), Conrad (1874), Boettger (1878), Etheridge (1879), Roxo (1924, 1937), Maury (1937), Greve (1938). The latter made a complete revision of all studies from the region.

Table 1.

	CPCAN-I 131,19–135,27 m	CPCAN-I 211,86–214,96 m	CPCAN-II 154,05–156,36 m	CPCAN-III 19,50–20,78 m	CPCAN-III 31,52–32,62	outcrop
New Genus A	2c	8v 3f		9v 3f	13v 1c	
New Genus B		4v			2v 1c 5f	
New Genus C		4v	1v	2v	1c	
New Genus D			1v	2v 1f		
Cytheridea sp.nov. A	1v				2v 1c 5f	
Cytheridea sp.nov. B					8v	
Cytheridea sp.nov. C		10v		137v 21f		9v 12c 1f
Cytheridea sp.nov. D			2v 4f	9v 10f		
Cytheridea sp.nov. E						1v 5c
Cyprideis sp.nov. A	4v 1c					
Cyprideis sp.nov. B	1v			18v 8f	1v	241v 17c 100f
Cytheridella sp.nov. A		1v 1f		4v 1f	6v 1f	
Darwinula sp.					3v	26v 10c
? Cypria sp.						
? Paracytheridea sp.1		2v				
? Paracytheridea sp. 2			1v	3v 1f		
Bisulcocypris sp.						2v 12c
Perissocytheridea sp.nov. A		7v 2c		10v 2c 3f		
Cypridea sp.						1c

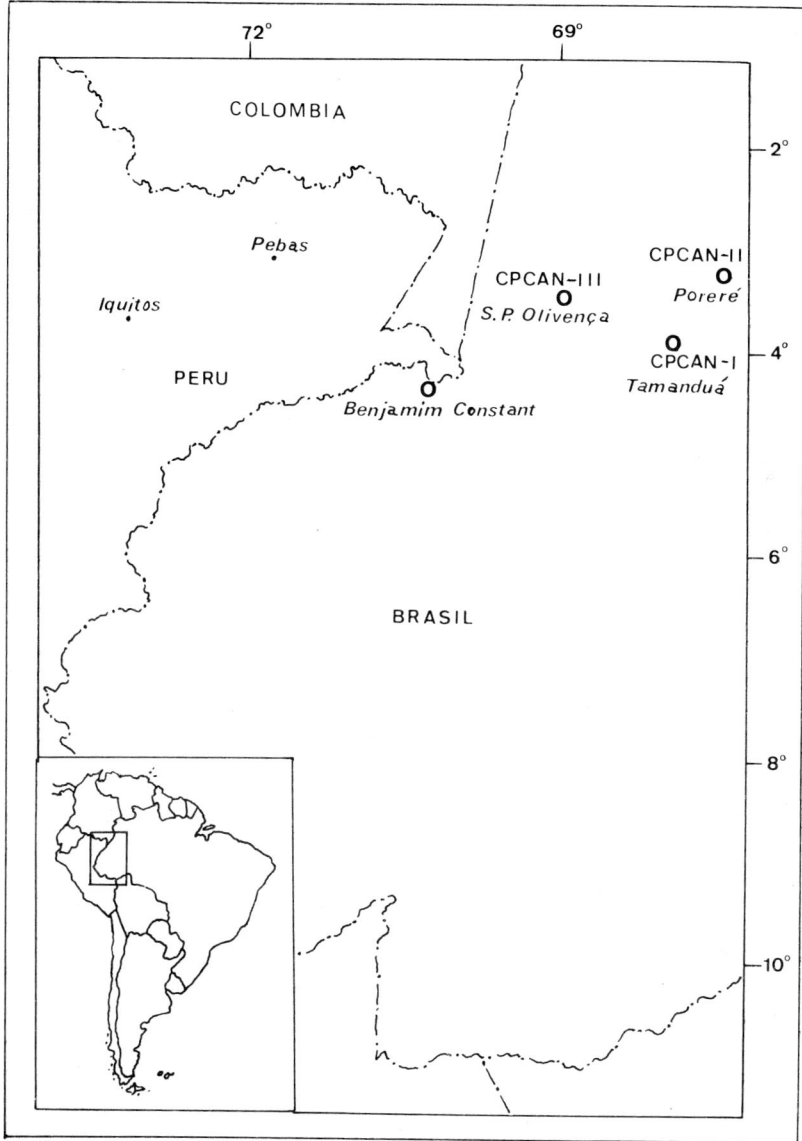

The first reference to ostracods was made by Gardner (1927) who mentioned *Cypris* sp., a fresh-water form. Another reference to the presence of Ostracoda was made by Maury (1937) but without taxanomic determination.

Environment and age

The fauna of the Upper Amazon Basin has been studied principally on its molluscs and, as practically all the fauna is represented by new genera and species,

Table 2

Author	Environment	Age
1868, Agassiz, L. (in Greve 1938)	fresh-water	–
1868, Gabb, W. M.	'marine or perhaps rather brackish water'	'very recent era'
1869, Agassiz, L. (in Greve 1938)	–	Lower Tertiary
1870, Orton, J.	marine	Tertiary
1871, Conrad, T. A.	'fresh or brackish-water but it is certainly not of marine origin'	'it cannot be later than Tertiary'
1871, Woodward, H.	'quite lake or estuary'*	Tertiary – may be Miocene*
1872, Hartt, Ch. Fr.	estuary	Tertiary
1872, Dall, W. H.	marine water to estuary	'Pliocene or perhaps later'
1874, Conrad, T. A.	fresh-water – brackish	Tertiary – older than Pliocene
1874, Conrad, T. A.	'fresh-water to which brackish water had access at times'	
1878, Boettger, O.	brackish water (estuary)	Lower Tertiary (Oligocene or may be Eocene)
1879, Barrington Brown	delta	Tertiary
1903, Katzer, Fr.	brackish-water	Lower Miocene
1924, Roxo, M. G. de Oliveira	brackish water	Upper Tertiary – Pliocene
1924, Maury, C. J.	'puzzling mixture of freshwater, brackish and even marine molluscan genera'	Tertiary
1924, Oliveira, A. I. & Carvalho, P. F.	fresh or brackish (Branner)	Tertiary – may be Miocene
1927, Gardner, J.	fresh-brackish water	Upper Pliocene
1928, Marshall, W. B.	fresh-water	–
1929, Steimann, G.	brackish water	Upper Tertiary, may be Pliocene
1932, Marshall, W. B. & Bowles, E.	estuary	'the age can not be later than Pliocene and it may be earlier'
1935, Roxo, M. G. de Oliveira	brackish water	Pliocene
1937, Maury, C. J.	rather freshwater than brackish	Pliocene
1937, Berry, E. W.	fresh-water	Upper Pliocene
1937, Roxo, M. G. de Oliveira	fluvio-lacustrine or terrestrial	Pliocene
1938, Greve, L.	predominantly fresh water	Upper Tertiary or more recent age
1961, Simpson, G. G.	–	Recent or Pleistocene

* Woodward reported the words of Orton of which he wrote to the Geological Magazine. Probably they did not have the same stratigraphical bed.

some of them without recent representation, it was difficult to make a comparison with other material. So, the previous authors tried to assess the approximate age of those sediments based on geological interpretation and tried also to determine the environment. The interpretation of these subjects – age and environment – by each author is registered in table 2, showing a great variety of opinion which could be due to several aspects, besides the fact that the fauna is almost entirely new.

First of all, the material studied was not provided by one place only but it covered a vast region. In this case there is the possibility of finding a variety of facies in the same bed which permitted the material to be differently influenced by salinity. Another thing to be pointed out is the fact that there is no agreement among the authors on the systematics, which is fundamental for a good interpretation of environment.

Based on the study of the ostracods, we have great difficulties in reaching some conclusion about paleoecology and stratigraphy because a great part of the fauna is represented by new genera and/or new species, giving no chance to compare and correlate them with some other material already known. Nevertheless, many species of *Cytheridea* (Pl. 2, Pl. 3. Fig. 1–7) and *Cyprideis* (Pl. 3, Fig. 8–16) which are eurihaline genera suggesting a brackish water environment are present and predominant. Analysing some predominant aspects of structural details influenced by a chance of salinity (Morkhoven 1962) we can observe that almost all material has smooth surfaces or, if ornamented, it is with punctae or fossetae. The hinges of the majority are very characteristic being the antimerodont-entomodont type, both being the type frequent in brackish-water forms. The presence of fresh-water genera as *Darwinula* (Pl. 4, Fig. 5–8), *Cytheridella* (Pl. 4, Fig. 1–4) and ?*Cypria* (Pl. 4, Fig. 9) suggests the proximity of freshwater. This seems to be a normal mixture at the boundaries of brakish zones. The fresh-water species may be brought into these conditions by rivers or streams. Part of this fauna is certainly allochthonous. With respect to the age of these sediments, it is evident that at least the outcrop is composed of a reworked material because in it cretaceous genera – *Bisulcocypris* and *Cypridea* were found together with those of younger age. Concerning the younger material Pliocene or more recent age is suggested provisionally. At first the presence of *Cyprideis* restricts the age of these beds to an age at least no earlier than Miocene (Sandberg 1964). Otherwise, the species of *Cyprideis* and *Cytheridea* recall those forms from Talparo Formation (Trinidad) dating back to the Pliocene age and La Cruz Formation of the same age, from Cuba.

The preservation and the exceptional fragility of some specimens suggest that this material could also be younger than Pliocene or could be a result of a very slow movement of water.

Paleogeography

To justify the presence of such a mixohaline fauna in a place so far from the Atlantic Ocean at the present date, with a geographic barrier to the Pacific Ocean, it may be supposed that in the past these beds have had an approximity to the sea.

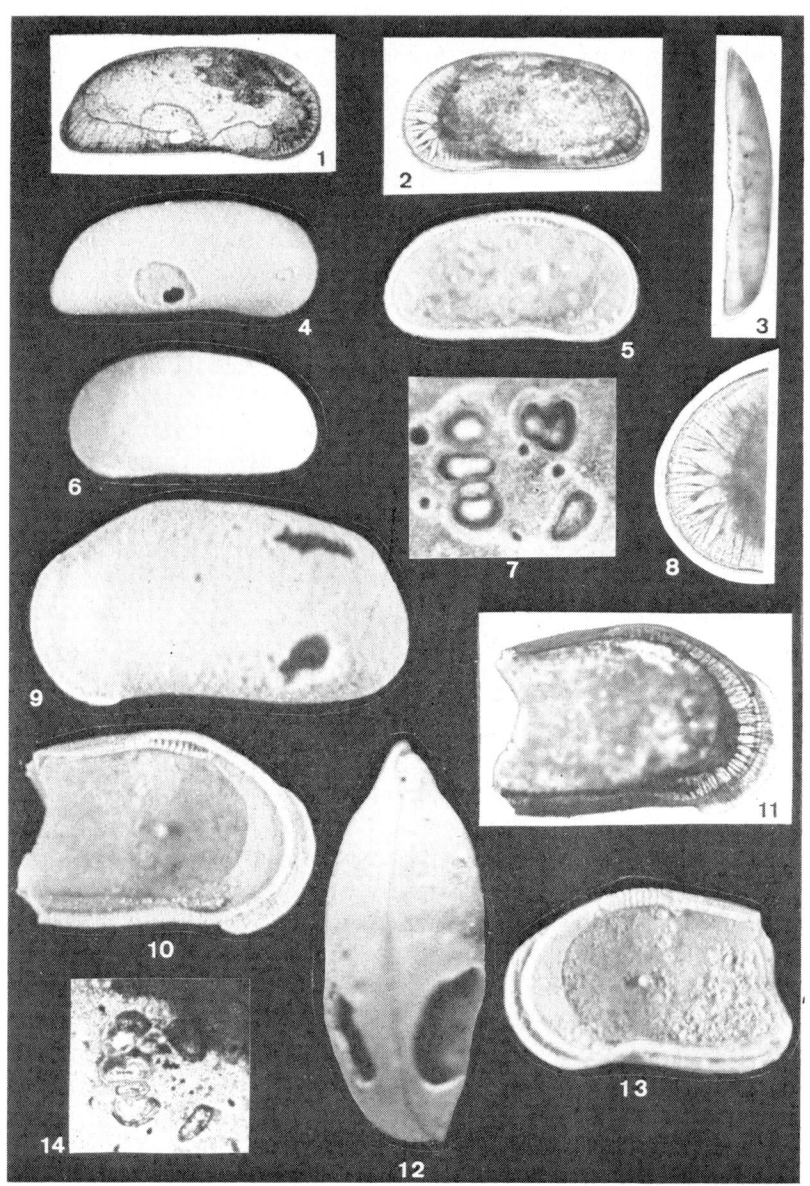

◄ Plate 1
Fig. 1–8. Ostracoda A n.g.,n.sp.
1. Male right valve in lateral view no. MP-O-501   x42
2. Female left valve in lateral view no. MP-O-502   x42
3. Male right valve in dorsal view no. MP-O-503   x42
4. Male right valve in lateral view no. MP-O-501   x42
5. Female left valve in internal view no. MP-O-502   x42
6. Female left valve in lateral view no. MP-O-502   x42
7. Male right valve. Muscle scar detail no. MP-O-504   x160
8. Female left valve. Anterior marginal pores no. MP-O-502   x74.

Fig. 9–14. Ostracoda B n.g.,n.sp.
9. Left valve external view no. MP-O-505   x33
10. Broken left valve showing part of the hinge no. MP-O-506   x33
11. Broken left valve showing anterior marginal pores no. MP-O-506   x33
12. Carapace in dorsal view. no. MP-O-505   x33
13. Broken right valve in internal view no. MP-O-507   x33
14. Right valve. Muscle scar detail. no. MP-O-507   x84.

Up to now research have been very limited, not permitting accurate conclusions on the source of this brackish environment, though it is thought there are hypotheses on the existence of a preterite sea which would explain the existence of such an environment there.

According to Katzer (1903) the brackish influence of these sediments could be associated with the Pacific Ocean before the Andean uplift. So, at the beginning of the Mesozoic time, with the linking of the Guianas lands to those of Brazil the Mediterranean Paleozoic Sea would be withdrawn to the western instead of to eastern side, draining into the Pacific Ocean at the Guayaquil Gulf. With the Andean uplift this situation began to reverse: at first the water bodies were prisoners in lakes on the edges of which clays and lignites with a fresh and brackish fauna were deposited. But with the continuation of the Andean uplift movements these lakes were all linked together forming a sea lake covering the entire depression. Only later this water was forced to the other side having the movement that is found today.

This hypothesis is very much discussed and not accepted by many Brazilian geologists (e.g. Mendes 1971, Loczy 1966) due to the lack of geological and paleontological studies in that part of Brazil. Another hypothesis believes that there was a link with the Caribbean Sea during the Tertiary. This suggestion was not only accepted by some authors but extended in such a way that some believe that a sea arm entered and crossed the whole continent. Ihering (1827, in Boltovskoy 1958) was the first to propose the existence of such an arm of the Tethys. He based his conclusion on a comparison of the Patagonian fauna and flora with those of Australia and New Zealand, and the Brazilian fauna and flora with those of Africa. Camp (1952) independently reached a conclusion similar to that of Ihering, dividing the South American continent into three parts, during the Tertiary, on botanical and geological grounds. Later, Szidat (1955, in Boltovskoy 1958) studyng fish and their parasites drew the same conclusion concerning the existence of a Miocene sea separating Archiplata from Archibrazil. Boltovskoy (1958),

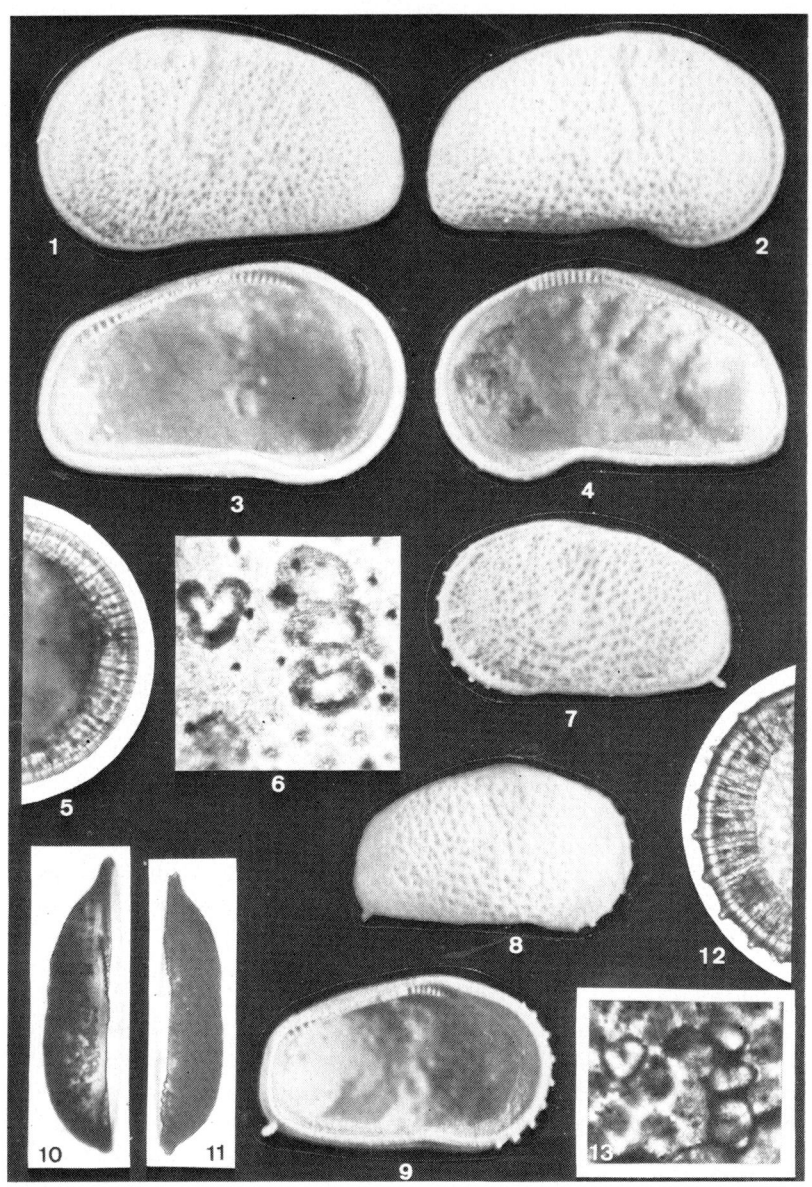

◄ Plate 2
Fig. 1–6 *Cytheridea* sp.nov. A
1. Female left valve in lateral view no. MP-O-508   x42
2. Male right valve in lateral view no. MP-O-509   x42
3. Female left valve in internal view no. MP-O-508   x42
4. Male right valve in internal view no. MP-O-509   x42.
5. Female left valve. Anterior marginal pores. no. MP-O-508   x76
6. Female left valve. Muscle scar detail. no. MP-O-508   x169.

Fig. 7–13. *Cytheridea* sp.nov. B
7. Female left valve in lateral view no. MP-O-510   x42
8. Male right valve in lateral view no. MP-O-511   x42
9. Female left valve in internal view no. MP-O-510   x42
10. Female left valve in dorsal view no. MP-O-510   x42
11. Male right valve in dorsal view no. MP-O-512   x42
12. Female left valve. Anterior marginal pores no. MP-O-510   x84
13. Male right valve. Muscle scar detail no. MP-O-511   x169.

studying foraminifera from Rio de La Plata, found benthonic antillean forms which, he supposed, had passed through the Tethys arm since these forms are not found in Brazilian waters.

Actually, as was accepted before, it is necessary to have much material from various places to follow the past path of this sea that probably once existed. At this moment, it is an open question.

Distribution of the ostracods

The distribution of the ostracods in the drill cores and outcrop is shown in Table 1 where the following abbreviations are used: v = vale; c = carapace; f = fragment.

Resumo

Novos gêneros e espécies de ostracodes de ambiente mixohalino e de água doce são registrados para os sedimentos da Bacia do Alto Amazonas. Os sedimentos são retrabalhados, registrando-se gêneros mesozóicos associados com uma grande quantidade de formas terciárias bem conservadas, de provável idade pliocênica ou mais recente.

References

Berry, E. W. 1937. Late tertiary plants from the Territory of Acre, Brazil. Stud. in Geology 12:81-90.

Boettger, O. 1878. Die Tertiärfauna von Pebas am oberen Marañon. Jahrbuch der K.K. Geol. Reichsanst., Wien 28 (3):485-504.

Bold, W. A. van den. 1963. Upper Miocene and Pliocene Ostracoda of Trinidad. Micropaleontology 9 (4):361-424.

◀ Plate 3
Fig. 1-4 *Cytheridea* sp.nov. C
1. Female left valve in lateral view no. MP-O-513   x42
2. Female right valve in lateral view no. MP-O-514   x42
3. Male left valve in lateral view no. MP-O-515   x42
4. Male right valve in lateral view no. MP-O-516   x42

Fig. 5-6. *Cytheridea* sp.nov. D
5. Female right valve in lateral view no. MP-O-517   x42
6. Male left valve in lateral view no. MP-O-518   x42

Fig. 7. *Cytheridea* sp.nov. E
7. Female left valve in lateral view no. MP-O-519   x42

Fig. 8-10. *Cyprideis* sp.nov. A
8. Female right valve in lateral view no. MP-O-520   x42
9. Female in dorsal view no. MP-O-521   x42
10. Female left valve in lateral view no. MP-O-522   x42

Fig. 11-16. *Cyprideis* sp.nov. B
11. Female in dorsal view no. MP-O-523   x42
12. Female left valve in lateral view no. MP-O-524   x42
13. Female right valve in lateral view no. MP-O-525   x42
14. Male left valve in lateral view no. MP-O-526   x42
15. Male right valve in lateral view no. MP-O-527   x42
16. Male in dorsal view no. MP-O-528   x42

Bold, W. A. van den. 1975. Neogene Biostratigraphy (Ostracoda) of Southern Hispanola. Bull. Amer. Paleont. Soc. 66 (286):549-639.
Bold, W. A. van den. 1975. Ostracodes from the Late Neogene of Cuba. Bull. Amer. Paleont. Soc. 68 (289) 121-167.
Boltovskoy, E. 1958. The Foraminifera Fauna of the Rio de la Plata and its relation to the Caribbean area. Contr. Cushman Found. Foraminiferal Res. 9 (1) 17-21.
Brown, B. 1879. On the Tertiary deposits on the Solimoes and Javary rivers in Brazil. Quart. J. Geol. Soc. Lonon 35:76-81 (with app. by Etheridge, R. 1879, pp 82-88).
Camp, W. H. 1952. Phytophyletic patterns on land bordering the South Atlantic basin. Bull. Amer. Mus. Nat. History 99 (3):205-212.
Conrad, T. A. 1871. Descriptions of a new fossil shell of the upper Amazon. Amer. J. Conchology 6 (3):192-198.
Conrad, T. A. 1874. Remarks on the Tertiary clay of the Upper Amazon, with descriptions of new shells. Proc. Acad. Nat. Sci. Philadelphia 26:25-32.Conrad, T. A. 1874. Descriptions of two new fossil shells of the Upper Amazon. Proc. Acad. Nat. Sci. Philadelphia 26:82-83.
Dall, W. H. 1872. Note on the genus Anisothyris Conrad, with a description of new species. Amer. J. Conchology 7:89-92.
Etheridge, R. 1879. Notes on the Mollusca collected by Barrington Brown from the Tertiary deposits of Solimoes and javary Rivers, Brazil. Quart. J. Geol. Soc. Lonon 35:82-88.
Gabb, W. 1868. Descriptions of fossils from the clay deposits of the Upper Amazon. Amer. J. Conchology 4:197-200.
Gardner, J. 1927. A recent collection of late Pliocene invertebrates from the head-waters of the Amazon. J. Wash. Acad. Sci. 17 (20):505-509.
Greve, Leonard de. 1938. Ein Molluskenfauna aus dem Neogen von Iquitos am Oberen Amazonas in Peru. Abh. Schweiz. Ges. 61:1-133.

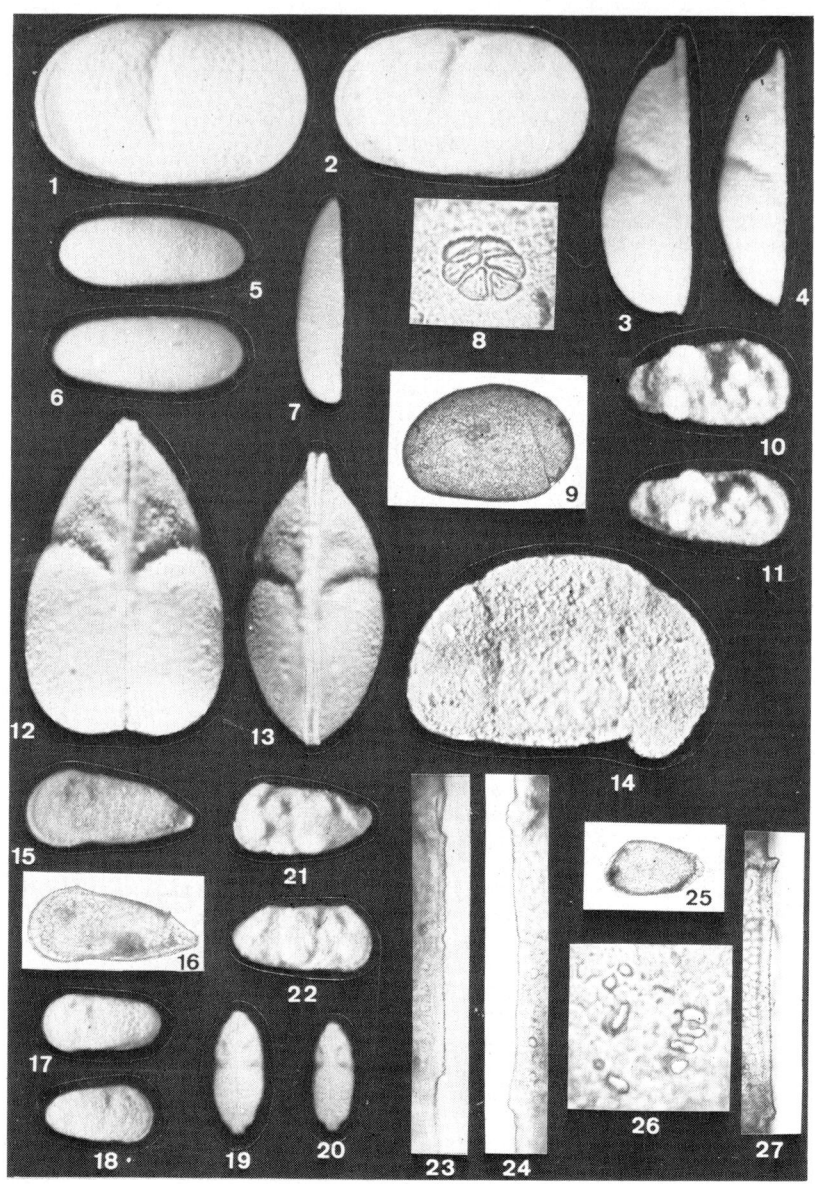

◄ Plate 4

Fig. 1–4. *Cytheridella* sp.nov. A
1. Female left valve in lateral view no. MP-O-529   x42
2. Male left valve in lateral view no. MP-O-530   x42
3. Female in dorsal view no. MP-O-529   x42
4. Male in dorsal view no. MP-O-530   x42

Fig. 5–8. *Darwinula* sp.
5. Male right valve in lateral view no. MP-O-531   x42
6. Female left valve in lateral view no. MP-O-532   x42
7. Female in dorsal view no. MP-O-532   x42
8. Female left valve. Muscle scar detail no. MP-O-532   x253

Fig. 9. ?*Cypria* sp.
9. Lateral view no. MP-O-533   x50

Fig. 10–11. ?*Paracytheridea* sp.1
10. Female right valve in lateral view no. MP-O-534   x84
11. Male right valve in lateral view no. MP-O-535   x84

Fig. 12–13. *Bisulcocypris* sp.
12. Female in dorsal view no.MP-O-536   x42
13. Male in dorsal view no. MP-O-537   x42

Fig. 14. *Cypridea* sp.
14. Lateral view no. MP-O-538   x42

Fig. 15–16. Ostracoda C n.g.,n.sp.
15. Female left valve in lateral view no. MP-O-539   x50
16. Female left valve in lateral view no. MP-O-529   x50

Fig. 17–20. *Perissocytheridea* sp.nov. A
17. Female left valve in lateral view no.MP-O-540   x42
18. Male right valve in lateral view no. MP-O-541   x42
19. Male in dorsal view no. MP-O-542   x42
20. Female in dorsal view no. MP-O-543   x42

Fig. 21–24. ?*Paracytheridea* sp.2
21. Male left valve in lateral view no. MP-O-544   x42
22. Female right valve in lateral view no. MP-O-545   x42
23. Male left valve. Hinge detail no. MP-O-544   x169
24. Female right valve. Hinge detail no. MP-O-545   x169

Fig. 25–27. Ostracode D n.g.,n.sp.
25. Lateral view no. MP-O-546   x42
26. Muscle scar detail no. MP-O-546   x253
27. Hinge detail no. MP-O-546   x169

Katzer, K. 1933. Geologia do Estado de Pará (Brasil). Bol. Mus. Paraense E. Goeldi de Hist. Nat. e Etnografia 9:1-269. (Trad. 1903).

Loczy, Louis de. 1966. Contribuições à Paleogeografia e Historia do Desenvolvimento geológico da Bacia do Amazonas. Bol. Div. Geol. e Mineral. DNPM 223:1-96.

Marshall, W. B. 1928. New fossil pearly fresh-water mussels from deposits on the Upper Amazon of Peru. Proc. U.S.Nat. Mus. 74 (3):1-6.

Marshall, W. B. & Bowles, E. O. 1932. New fossil fresh-water mollusks from Ecuador. Proc. U.S. Nat. Mus. 82 (5):1-7.

Maury, Carlotta J. 1924. Fosseis terciários do Brazil com descrição de nuovas formas cretaceas. Monogr. Serv. Geol. e Mineral. do Brasil DNPM 4:1-705.

Maury, Carlotta J. 1937. Argillas fossiliferas do Pliocenio do Territorio do Acre. Brasil. Bol. Serv. Geol. e Mineral. do Brail DNPM 77:1-29.

Mendes, Josué Camargo. 1971. Geologia do Brasil. Rio de Janeiro, Instituto Nacional do Livro, 207 pp. (Enc. brasileira, Bibl. Universitária. Geosciências. Geologia, 9).

Moore, Raymond C. 1961. Treatise on Invertebrate Paleontology. Part. Q. Arthropoda 3 Crustacea-Ostracoda. Geol. America & Univ. Kansas Press, New York/Lawrence, 442 pp.

Morkhoven, F. P. C. M. van. 1962. Post-Palaeozoic Ostracoda. Their morphology, taxonomy and economic use: General. Elsevier Publ. Comp., Amsterdam, Vol. 1, pp. 1-79.

Morkhoven, F. P. C. M. van. 1963. Generic descriptions. Elsevier Publ. Comp., Amsterdam, Vol. 2, Fig. 1-763.

Oliveira, A. I. & Leonardos, O. H. 1943. Geologia do Brasil. 2nd ed. Serv. Inf. Agricola, Min. Agricultura, Rio de Janeiro, 813 pp. (Ser. Didática No. 2).

Oliveira, A. I. & Carvalho, P. F. de. 1924. Estudos geológicos na fronteira com o Perú Linhito no Alto Solimoes). Bol. do Serv. Geol. e Mineral. do Brasil DNPM 8:55-76.

Orton, James. 1870. The Andes and the Amazon. Harper Bros, New York, 356 pp.

Pinto, I. D. & Sanguinetti, Y. T. 1962. A complete revision of the Genera Bisulcocypris and Theriosynoecum (Ostracoda) with the world geographical and stratigraphical distribution (including Metacypris, Elpidium, Gomphocythere and Cytheridella). Publ. Esp. Escola de Geol. 4:1-165.

Roxo, G. de Oliveira. 1924. Breve notícia sôbre os fósseis terciários do Alto Amazonas. Bol. Serv. Geol. e Mineral. do Brasil DNMP 11:41-52.

Roxo, G. de Oliveira. 1935. Considerações sobre a geologia a paleontologia do Alto Amazonas. Ann. Acad. Brasil. Scienc. 7 (1):63-67.

Roxo, G. de Oliveira. 1937. Fosseis Pliocenios do Rio Juruá, Estado do Amazonas, Brazil. Notas Prelim. Est. Div. de Geol. e Mineral. DNPM 9:4-14.

Sandberg, Phillip A. 1964. The Ostracod genus Cyprideis in the Americas. Stockholm Contr. Geol. 12:1-178.

Simpson, G. G. 1961. The supposed Pliocene Pebas Beds of the Upper Juruá River, Brazil, J. Paleont. 35 (3):620-624.

Steinmann, G. 1929. Geologie von Perú. C. Winters Universitätsbuch., Heidelberg, 448 pp.

Woodward, H. 1871. The Tertiary shells of the Amazon Valley. Ann. & Mag. Nat. Hist., 4th Ser. 7:59-64, 101-109.

Author's address:

Universidade Federal do Rio Grande do Sul, Departamento de Paleontologia e Estratigrafica, 90.000 Porto Alegre, RS, Brasil.

Discussion

Oertli: The species on the lower ⅔ of Plate 4 are most probably of lowermost Cretaceous origin.

Purper: The species under numbers 12, 13 and 14 certainly have their range from Upper Jurassic to Lower Cretaceous. Thank you for your suggestion about the species Fig. 15-27 also. For me they are new species and by its preservation they seem to be Tertiary, but as this is not a safe characteristic I must check again to see if they can be compared to any Cretaceous species.

Sixth Intern. Ostracod Symposium, Saalfelden

LES OSTRACODES, INDICATEURS PALEOCLIMATIQUES ET PALEOGEOGRAPHIQUES DU QUATERNAIRE TERMINAL (HOLOCENE) SUR LE PLATEAU CONTINENTAL SENEGALAIS

J. P. PEYPOUQUET

Abstract

On the continental shelf of Senegal (South Dakar 17°10 - 17°15 N and 14°30' - 14°40'S) between isobaths - 15 - 25 m, an oceanographic cruise allowed us to take 44 core samples.
The analysis of the Ostracodes microfauna shows several faunal assemblages which mean that different ecological environments follow each others.
The evolution, in space and time of these faunal assemblages is the result of nouakchottian transgression on the continental shelf of Senegal, and involves paleoclimatic and paleogeographic changes.

1. Introduction

Au cours de la Campagne 'ROSILDA' financée par le C.N.E.X.O. et dont les études ont été menées par l'I.G.B.A. et le B.R.G.M., une cinquantaine de carottages ont été exécutés sur le plateau continental sénégalais au Sud de Dakar.

Les prélèvements ont été réalisés au vibrocarottier Zenkowitch dans une zone limitée par les isobathes -15 et -25 mètres et par les parallèles 14°32' et 14°39' N.

Le secteur défini couvre environ une serface de 65 km² (Fig. 1). La campagne de caottage a été réalisée selon des profils perpendiculaires à la côte, d'une longueur moyenne de 5 km, l'espacement entre les sites successifs est de 400 à 600 m, excepté dans la zone médiane (radiale 36 S) où il n'est que de 100 à 260 m. La longueur cumulée des carottes est environ de 180 m; elle correspond à une pénétration moyenne d'environ 3 m par sondage.

A partir de ces prélèvements et par la technique des ostracodes, un essai de reconstitution paléogéographique a été tenté. Il a pour but de montrer comment s'est effectuée la transgression nouakchottienne dans ce secteur du plateau continental ouest-africain.

2. Les associations fauniques (Fig. 2)

En accord avec des principes établis précédemment (Peypouquet 1971, Carbonel et al. 1975) nous avons regroupé les faunes d'ostracodes en associations, en distinguant la biocenose s. l. de la thanatocenose.

2.1. *Association littorale marine (zone phytale)*

a) *Biocenose*
   Aurila sp. 1
   Semicytherura div. sp.
   Cytherura sp. 1

   Callistocythere
   Cytheropteron div. sp.
   Cytherella div. sp.
   Puriana div. sp.

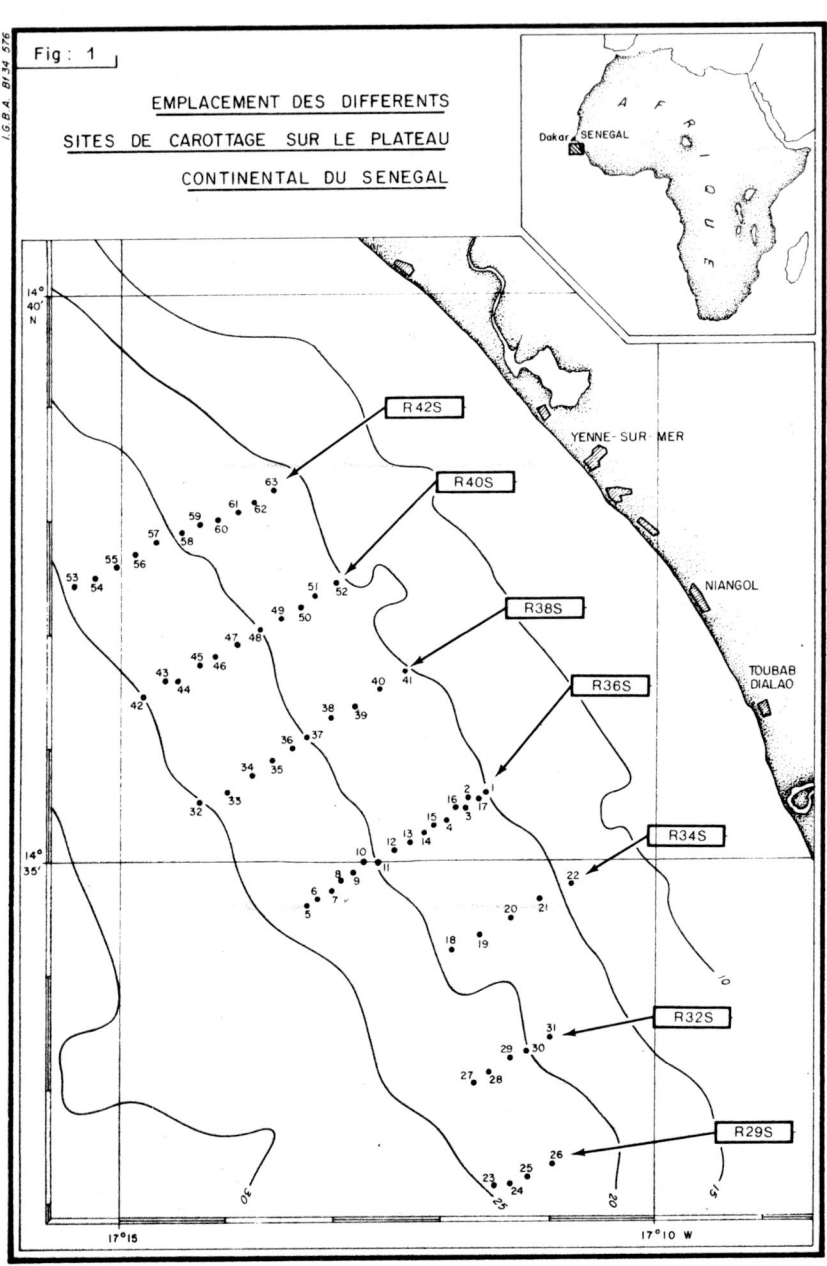

Fig: 1
EMPLACEMENT DES DIFFERENTS SITES DE CAROTTAGE SUR LE PLATEAU CONTINENTAL DU SENEGAL

Ruggieria div. sp.
Falunia
Loxoconcha div. sp.
Chrysocythere
Microcythere
Bythocypris

Paracytheridea
Paijenborchellina
Neocyprideis
Neocytherideis
Campilocythereis

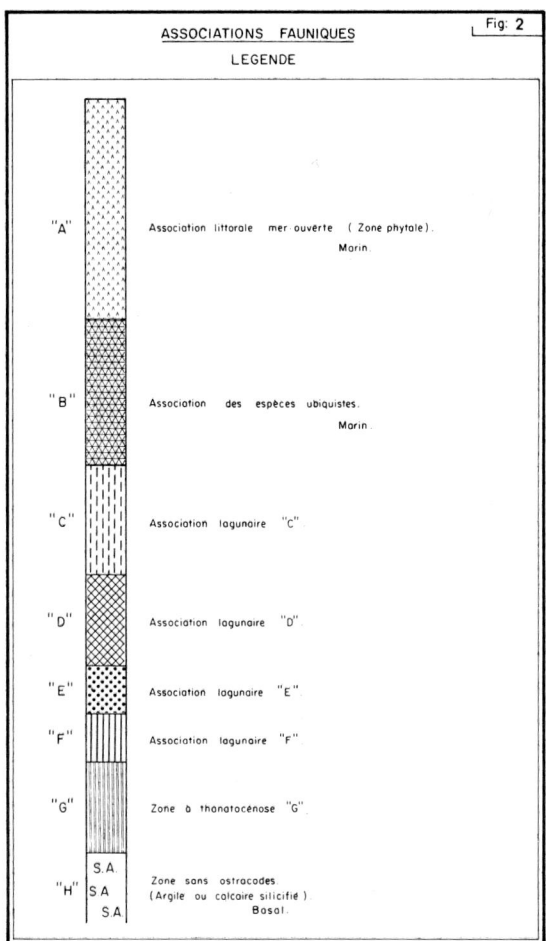

 b) *Thanatocenose*
    Cytheropteron
    Xestoleberis
    Neomonoceratina
    Candona
    Loxoconcha sp. 3

 c) *Paléothanatocenose*
    Carinocythereis carinata
    Loculicytheretta sp.
    Aurila sp. 2

2.2. *Association des espèces ubiquistes marines 'B'*

a) *Biocenose*
   Aurila
   Soudanella
   Bythocypris
   Cytherella
   Loxoconcha sp. 1
   Falunia

b) *Thanatocenose*
   Neomonoceratina

2.3. *Association lagunaire 1 'C'*

a) *Biocenose*
Xestoleberis
Paracypris
Occultocythereis
Loxoconcha sp. 3
Neomonoceratina
Aurila sp. 2

Propontocypris

b) *Thanatocenose*
Soudanella
Bythocypris
Puriana
Cytherella

2.4. *Association lagunaire 2 'D'*

a) *Biocenose*
Identique a (1 'C') mais les Neomonoceratina disparaissent complètement et sont remplacées par Occultocythereis.

2.5. *Asssociation lagunaire 3 'E'*

a) *Biocenose monospécifique*
Neomonoceratina
+ quelques Xestoleberis, Loxoconcha sp. 3

2.6. *Association lagunaire 4 'F'*

a) *Biocenose monospécifique*
Cyprideis

b) *Thanatocenose*

Soudanella
Bythocypris
Aurila
etc.

2.7. *Zone sans ostracodes en biocenose*

a) Zone de la tourbe 'G'

b) *Thanatocenose*
Neomonoceratina
Xestoleberis

Soudanella
Cyprideis
Cytherella
Cytheropteron
Aurila

2.8. *Zone du calcaire silicifié basal 'H'*

Pas de microfaune

3. Valeur écologique des associations

L'association 'A' représente une faune marine franche soit de mer ouverte, soit de baie largement en communication avec la mer. De nombreuses espèces liées au domaine des algues et des herbiers témoignent d'une eau bien oxygénée et d'une salinité jamais < à 34‰; de plus la diversité des espèces et le nombre des individus récoltés indiquent que la zone est très propice au développement de ces microfaunes.

L'association 'B' dite des 'espèces ubiquistes' montre des carapaces très souvent usées, ayant été légèrement remaniées. Ces formes sont de bons indicateurs des zone marines aux conditons de vie difficile pour le benthos (milieu de haute énergie, zone de bas-estran, zone d'action des vagues). Elles indiquent souvent des milieux où les conditions marines ne sont pas encore bien établies (début de transgression marine).

L'association 'C', de par sa composition faunique restreinte du point de vue générique et spécifique, représente un milieu abrité, une zone lagunaire où les influences marines sont encore fortes, et où les plantes sont bien représentées, (Xestoleberis dominant), la salinité doit être toujours supérieure à 20‰ (eau polyhaline). De nombreuses valves d'huîtres rencontrées dans les sédiments confirment cette hypothèse.

L'association 'D' témoigne également d'un biotope lagunaire sensiblement identique au précédent. L'absense de Neomonoceratina et la présence de Occultocythereis sont des arguments en faveur d'une salinité qui, en moyenne, est légèrement supérieure au milieu précédent.

L'association 'E' est composée presque exclusivement de Neomonoceratina et représente un environnement mésohalin à oligohalin. C'est un biotope de fond de lagune, ou de zone, dans laquelle les influences continentales sont notables.

L'association 'F' est composée exclusivement de Cyprideis possédant une réticulation très forte. Cette microfaune d'ostracodes, monospécifique, est probablement le reflet d'une lagune ou d'une zone estuarienne, en climat aride à semi-aride. Ce type de Cyprideis n'exclut pas des salinités $> $ à 35‰.

L'association 'H' est caractérisée par : soit une thanatocenose d'ostracodes composée essentiellement de Neomonoceratina, soit par une absence totale d'individu. Le faciès sédimentaire est composé par des tourbes dérivant d'une mangrove à paletuviers.

Le calcaire silicifié basal 'beach-rock' et les argiles sus-jacentes sont dépourvus d'ostracodes.

4. Analyse des diverses carottes

4. 1 *La radiale 42 S* (Fig. 3)

De la région Ouest en se dirigeant vers l'Est, nous avons:

La carotte 42 S 53 (longueur 2,40 m)
- à la base (2,40 M) le 'beach-rock' calcaire coquillier silicifié.
- 2,20 m à 1,90 m, l'association 'F' composée de Cyprideis dont la richesse décroit en remontant vers le sommet.
- 1,80 m à 1,30 m, l'association 'B' au sein des sables fins vaseux gris noirs.
- 1,30 m au sommet, l'association 'A' contenant de nombreuses formes phytales.

N.B.: Il faut signaler la présence de formes d'eau douce au niveau 0,80 m. De plus le sommet de la carotte contient une grande quantité de spicules d'éponges.

La carotte 42 S 54 (longueur 1,50 m)
- à la base le 'beach-rock' calcaire coquillier silicifié.
- 1,30 m à 1 m, l'association 'B' au sein des sables fins vaseux gris noirs.
- 1 m au sommet, l'association 'A' relativement pauvre en espèces phytales.

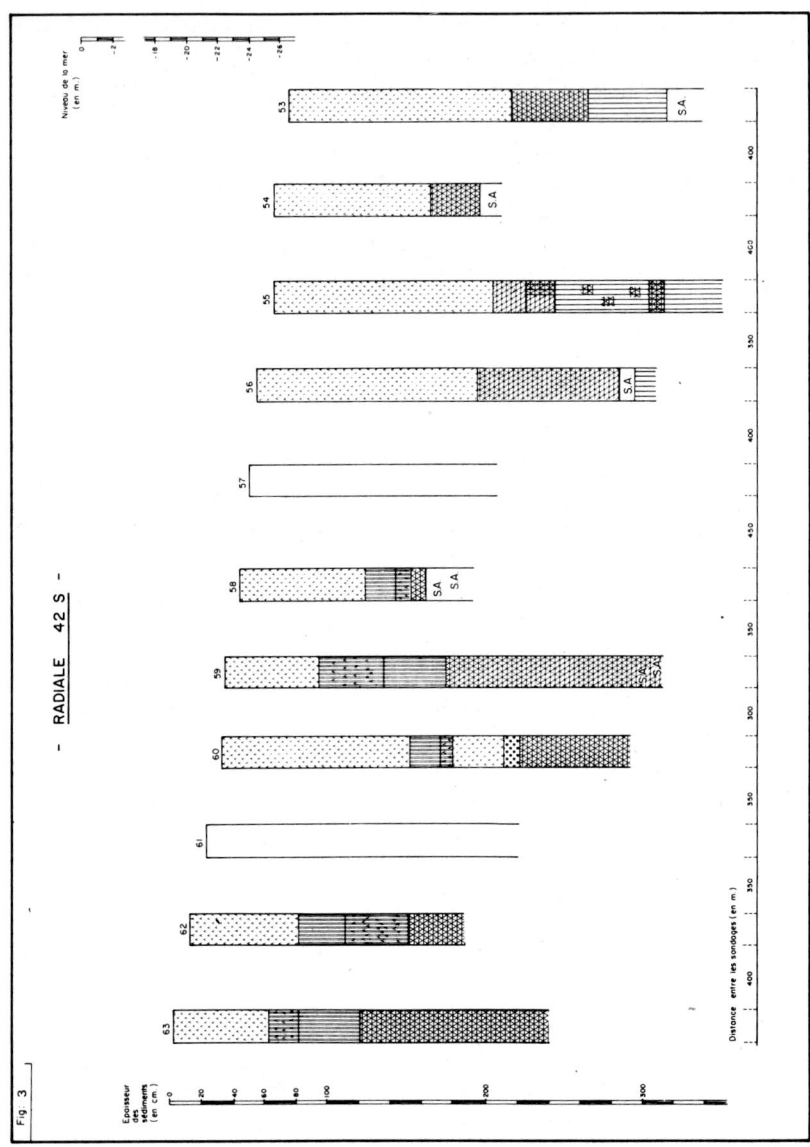

La carotte 42 S 55 (longueur 2,80 m)
- à la base (2,80 m) le 'beach-rock' calcaire coquillier silicifié.
- 2,80 m à 1,80 m, l'association 'F'. Les Cyprideis sont extrêmement abondants et souvent mélangés avec les espèces ubiquistes (Aurila, Soudanella, Bythocypris) de l'association 'B'.
- 1,80 m à 1,60 m. Seule subsiste l'association 'B'.
- 1,60 m au sommet, la faune devient marine. 'Association 'A'.

La carotte 42 S 56 (longueur 2,60) m
- à la base (2,60 m), le 'beach-rock' calcaire coquillier silicifié.
- 2,60 m à 2,40 m, l'association 'F' est bien individualisée.
- 2,40 m à 2,20 m, une zone azoïque sans ostracodes dans des sables fins gris noir.
- 2,20 m à 1,40 m, l'association 'B' est faiblement représentée.
- 1,40 m au sommet, l'association 'A' est très mal représentée par la faiblesse des espèces phytales et une pauvreté constante sur le plan quantitatif des individus.

La carotte 42 S 58 (longueur 1,50 m)
- à la base une argile plastique, azoïque.
- 1,20 m à 1,10 m, l'association 'B'.
- 1,10 m à 0,80 m, l'association 'G', niveaux de tourbe complètement dépourvus d'ostracodes.
- 0,80 m au sommet, association 'A', faune marine très riche, très développée.

La carotte 42 S 59 (longueur 2,80 m)
- à la base: argile beige azoïque comprenant des fragments de calcaire silicifié.
- 2,70 m à 1,30 m, l'association 'B' dans un sable fin beige à brun.
- 1,30 m à 0,80 m, l'association 'G' à thanatocenose faiblement représentée.
- 0,80 m au sommet, l'association 'A' peu développée sur le plan quantitatif et qualitatif.

La carotte 42 S 60 (longueur 2,60 m)
- de la base à 1,90 m, l'association 'B' dans un sable fin vaseux gris noir comprenant de nombreuses formes ubiquistes.
- 1,90 m à 1,80 m; l'association 'E' forme lagunaire monospécifique à Neomonoceratina.
- 1,70 m à 1,40 m, association 'A' + association 'G'. Les sédiments tourbeux alternent avec des sédiments marins contenant de nombreux ostracodes caractéristiques.
- 1,40 m au sommet, association 'A' caractéristique de l'infra-littoral marin très riche et diversifié.

La carotte 42 s 62 (longueur 1,70 m)
- à la base (1,70 m) une argile beige à blanche, azoïque.
- 1,70 m à 1,50 m, l'association 'B'.
- 1,50 m à 0,90 m, association 'G' contient une thanatocenose développée.
- 0,80 m au sommet, association 'A' très riche et très diversifiée.

La carotte 42 S 63 (longueur 2 m)
- de la base à 1,20 m, association 'B' au sein d'une argile gris bleu plus ou moins sableuse.
- 1,20 m à 0,90 m, association 'G' (tourbe sans aucun ostracode).
- 0,90 m à 0,70 m, association 'G' et association 'A' mélangées. Alternances de tourbes et de sédiments marins.
- 0,70 m au sommet, association 'A' contenant une microfaune relativement pauvre.

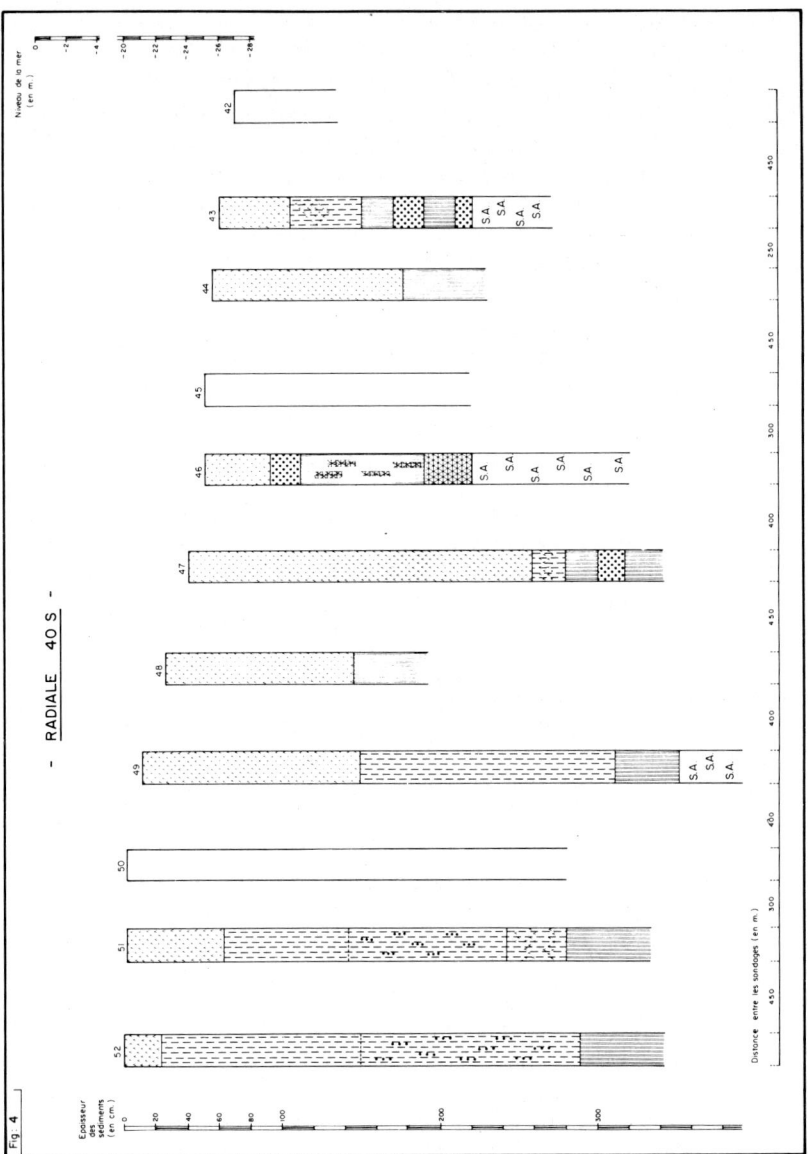

Interprétation de la radiale 42 S

En intégrant les données fournies par l'étude des différentes carottes constituant la radiale, deux zones différentes se dessinent de par leur évolution dans le temps.

a) *A l'Ouest* (carottes 53, 54, 55, 56). Sur un substratum ancien constitué de calcaires silicifiés, formés en domaine marin proche du rivage, il s'est établi une faune à Cyprideis à forte réticulation, 'F', qui suggère soit un milieu estuarien sous un climat aride à semi-aride, analogue à celui mis en évidence par Benson (1959) dans la région de Baja California (Mexique) ou bien une lagune du type 'laguna Madre' au Texas du Sud. Puis une transgression marine semble s'établir difficilement dans un premier temps (faune 'B') plus franchement par la suite (faune 'A'). L'environnement est riche en plantes et constitue un biotope favorable au large développement des Semicytherura et Callistocythere.

b) *A l'Est* (carottes 58, 59, 60, 63). On ne trouve plus le substratum calcaire, ni la faune 'F'. La base des carottes correspond au début de la transgression marine, puis il y a probablement établissement d'une lagune en climat humide et chaud, dans laquelle la mangrove se développe ainsi que les grèves sablo-vaseuses à Neomonoceratina.

Enfin comme dans la partie occidentale, une transgression marine franche caractérisée par l'association 'A' termine la série.

4.2. *La radiale 40 S* (Fig. 4)

De la région Ouest en se dirigeant vers l'Est, nous avons:

La carotte 40 S 43 (longueur 1,90 m)
– de la base (1,90) à 1,60 m, 'H' zone sans ostracodes dans une argile blanchâtre contenant des débris de calcaire coquilliers silicifiés.
– 1,60 m à 1,50 m, association 'E' zone Neomonoceratina abondants.
– 1,50 m à 1 m, association 'F' alternant avec des bancs tourbeux 'G'.
– 1 m à 0,50 m, association 'G' lagunaire riche en Neomonoceratina.
– 0,50 m au sommet, association 'A' peu développée.

La carotte 40 S 44 (longueur 1,70 m)
– de la base à 1,10 m, association 'H' composée exclusivement d'une thanatocenose peu importante au sein des bancs tourbeux.
– 1,10 m au sommet, association 'A' faiblement développée.

La carotte 40 S 46 (longueur 2,30 m)
– de la base à 1,90 m, le 'beach-rock' calcaire coquillier silicifié et sans ostracodes.
– 1,90 m à 1,60 m, association 'B' dans un sable vaseux gris noir peu importante.
– 1,60 à 0,70 m, association 'H' contenant une thanatocénose relativement riche avec des lits tourbeux.
– 0,60 m à 0,50 m, association 'E' momospécifique à Neomonoceratina abondantes.
– 0,50 m au sommet, association 'A', toujours relativement peu développée sur les plans qualitatifs et quantitatifs.

La carotte 40 S 47 (longueur 2,90 m)
– de la base à 2,80 m, association 'H' constituée essentiellement par la tourbe.
– 2,80 m à 2,70 m, association 'E' à Neomonoceratina.
– 2,70 m à 2,60 m, 'H' constituée essentiellement par la tourbe.

- 2,60 m à 2,40 m, association lagunaire 'C' faiblement représentée.
- 2,40 m au sommet, association 'A' toujours relativement peu développée.

La carotte 40 S 48 (longueur 1,80 m)
- de la base à 1 m, association 'H' sans aucune thanatocenose.
- de 1 m au sommet, association 'A' ne contenant pratiquement aucune forme phytale.

La carotte 40 S 49 (longueur 3,70 m)
- de la base à 3,40 m, 'G' argile dépourvue d'ostracodes.
- 3,40 m à 3,10 m, association 'H' contenant quelques valves de Neomonoceratina.
- 3,10 m à 1,50 m, association 'C' faune lagunaire extrêmement bien développée avec forte prédominance de Xestoleberis, puis on rencontre Paracypris, Neomonoceratina et Loxoconcha sp. 3.
- 1,50 m au sommet, association 'A' bien développée et riche en espèces phytales.

La carotte 40 S 51 (longueur 3,30 m)
- 3,30 m à 2,90 m, association 'H' sans thanatocenose.
- 2,90 m à 0,70 m, association 'C' très riche sur le plan quantitatif. Xestoleberis est toujours largement dominant. Neomonoceratina est bien développée entre 2,60 m et 1,40 m, puis Paracypris prend une position prépondérante entre 1,40 m et 0,70 m.
- 0,60 m au sommet, 'Á' faune marine bien développée très riche en formes phytales.

La carotte 40 S 52 (longueur 3,40 m)
- de la base à 3 m, 'H' zone tourbeuse sans ostracodes.
- 2,90 m à 0,30 m, 'C' faune lagunaire en tout point identique à la carotte précédente. Ici également on observe la succession dans le temps de Neomonoceratina (2,90 m à 1,60 m) remplacée par Paracypris (1,50 m – 0,60 m).
- 0,30 m au sommet, association 'A' très riche en formes phytales.

Des datations au 14 C réalisées par le Laboratoire de faibles Radio-activités de Gif-sur-Yvette ont donné pour la tourbe (3,30–3,40 m) un âge de 8400 & 180 ans; pour les huîtres (2,90–3 m) un âge de 8180 & 180 ans; pour le niveau supérieur de la carotte (0,30–0,40 m) un âge de 8150 ans & 180 ans.

La phase lagunaire correspond donc à une sédimentation très forte (2,60 m environ) qui pourrait résulter sur le continent de l'installation d'un climat humide donnant des apports sédimentaires importants. Depuis l'installation d'un régime marin franc (à partir de 0,30 m), vers 8000 ans, la sédimentation de cette carotte a été très faible.

Interprétation écologique de la Radiale 40 S

En intégrant l'ensemble des données fournies par les carottes de cette radiale, certains traits se dégagent sur le plan paléogéographique. On peut séparer la partie orientale de la partie occidentale.

a) *A l'Ouest* (carottes 43, 44, 46, 47). Sur des substrats ne possédant pas de microfaunes d'ostracodes se développent des niveaux tourbeux à thanatocenose souvent constitués de Neomonoceratina.

Ces niveaux mal circonscrits sont surmontés par l'association 'E' monospécifique à Neomonoceratina. Quelquefois on observe l'alternance des bancs tourbeux avec les zones à Neomonoceratina. Ce fait évoque des environnements de mangrove peu développés sur des grèves lagunaires proches des apports continentaux. Au-dessus la faune marine 'A' est toujours très pauvre et relativement peu diversifiée.

b) *A l'Est* (carottes 52, 51, 49, 47). Ici les tourbes forment le substratum des carottes. Au-dessus s'installe une faune lagunaire très riche en individus et en espèces. Xestoleberis est dominant et implique un système phytal abondant.

Les successions dans le temps entre Neomonoceratina et Paracypris semblent indiquer tantôt une influence continentale tantôt une influence plus marine dans la lagune.

Au sommet, nous rencontrons une faune marine très riche et très diversifiée.

c) La comparaison entre la radiale 42 S et la radiale 40 S nous amène à formuler quelques remarques:

– Dans l'une comme dans l'autre, deux domaines sont séparés, un secteur oriental et un secteur occidental.

– Dans l'une comme dans l'autre la faune marine 'A' termine la série.

– Par contre on ne retrouve plus dans la radiale 40 S l'association 'F' monospécifique à Cyprideis.

De plus, on constate dans les carottes de la radiale 40 S la présence de deux associations nouvelles:

– Association 'E' monospécifique à Neomonoceratina.
– Association 'C' lagunaire à Xestoleberis dominant.

Enfin les tourbes, 'H', qui, dans la radiale 42 S étaient surmontées par des faciès marins 'A', sont dans la radiale 40 S surmontées par les faciès lagunaires 'E' ou 'C'. Ceci implique une disposition géographique de la lagune N-S.

4.3. *La radiale 38 S* (Fig. 5)

De la région Ouest en se dirigeant vers l'Est, nous avons:

La carotte 38 S 32 (longueur 2,90 m)
– De la base au sommet, cette carotte est marine, association 'A'. Mais beaucoup de niveaux sont dépourvus d'ostracodes.

La carotte 38 S 33 (longueur 2,30 m)
– De la base au sommet, l'association 'A' est très riche et très diversifiée.

La carotte 38 S 35 (longueur 3,60 m)
– De la base au sommet, l'association 'A' est très riche et très diversifiée. On doit noter tout de même quelques apports lagunaires aux niveaux 2,80 m et 1,20 m.

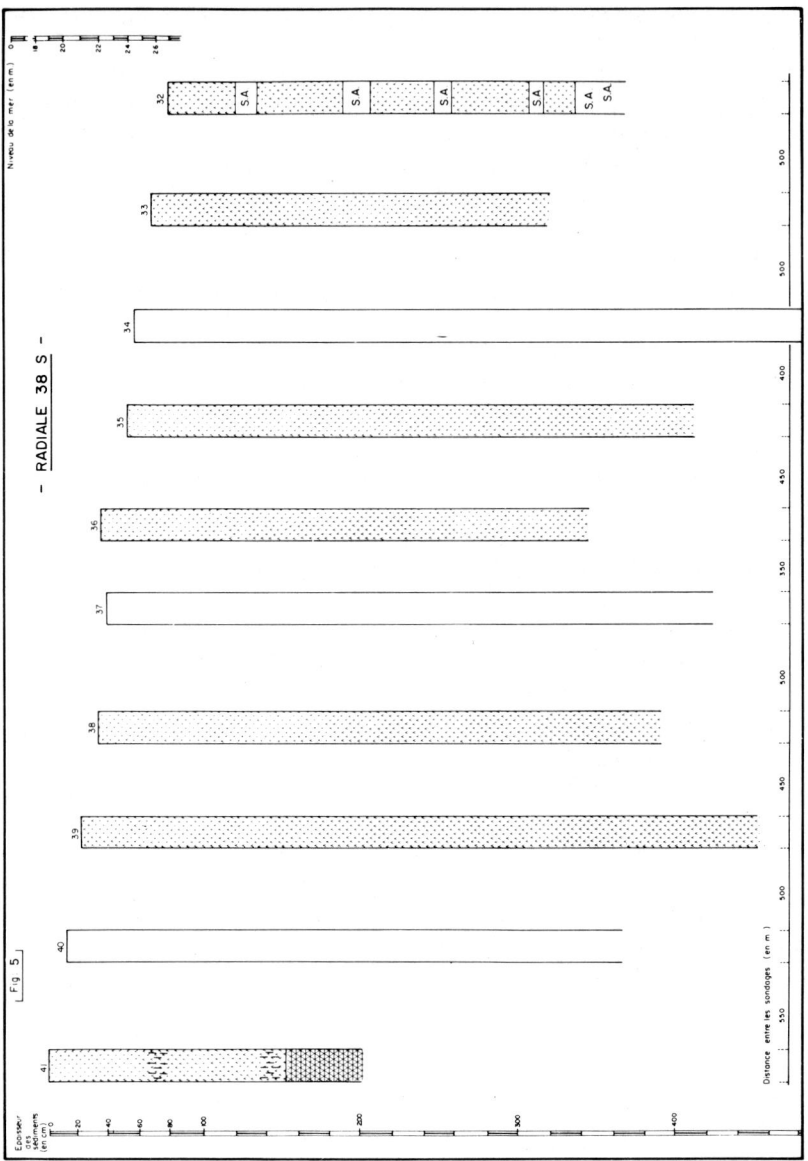

La carotte 38 S 36 (longueur 2,90 m)
- De la base au sommet, l'association 'A' est en tout point identique à la précédente. On constate également quelques apports de microfaunes lagunaires au niveaux. 2,90 m et 1,60 m.

La carotte 38 S 38 (longueur 3,60 m)
- De la base au sommet, l'association 'A' est très riche et très diversifiée. On note quelques apports de microfaunes lagunaires aux niveaux 2,90 m et 0,60 m.

La carotte 38 S 41 (longueur 2 m)
- De la base à 1,50 m, l'association 'B' est bien développée dans les sables vaseux gris noir.
- De 1,50 m au sommet, l'association 'A' se développe faiblement. On constate des apports lagunaires relativement importants aux niveaux 1,40 m et 0,80 m.

Interprétation écologique de la radiale 38 S

a) L'examen des microfaunes d'ostracodes de cette radiale montre une homogénéité remarquable de la base au sommet. Partout se développe une microfaune infralittorale marine, ou de baie marine ouverte sur l'ocean, riche en plantes, en herbiers nécessaires au développement des espèces phytales.

Quelques apports de microfaunes d'origine lagunaire sont perçus à plusieurs moments dans l'ensemble des carottes.

La carotte 38 S 41 semble avoir été davantage influencée par ces apports lagunaires.

b) La comparaison de cette radiale avec les précédentes montre évidemment un changement profond des environnements. Nous ne trouvons plus aucune trace des milieu lagunaires 'C', 'E', 'F', ou 'H'.

4.4. *La radiale 36 S* (Fig. 6)

De la région Ouest en se dirigeant vers l'Est, nous avons:

La carotte 36 S 6 (longueur 2,10 m)
- de la base à 2 m, le 'beach-rock' calcaire coquillier silicifié.
- 2 m au sommet, l'association 'A' très pauvre et contenant de nombreux niveaux azoïques.

La carotte 36 S 10 (longueur 3,90 m)
- de la base à 3,10 m, 'H' niveau tourbeux sans ostracodes.
- 3 m au sommet, l'association 'A' relativement pauvre alterne avec quelques niveaux azoïques.

La carotte 36 S 11 (longueur 4,20 m)

La carotte 36 S 12 (longueur 2,40 m)

La carotte 36 S 13 (longueur 2,50 m)
- de la base au sommet, on ne rencontre que l'association 'A' très riche et très diversifiée.

La carotte 36 S 4 (longueur 3,60 m)
- de la base à 3 m, l'association lagunaire 'D' se développe avec Xestoleberis comme forme dominante. La présence de Paracypris et de Occultocythereis, l'absence de Neomonoceratina, induit des conditions lagunaires très proches

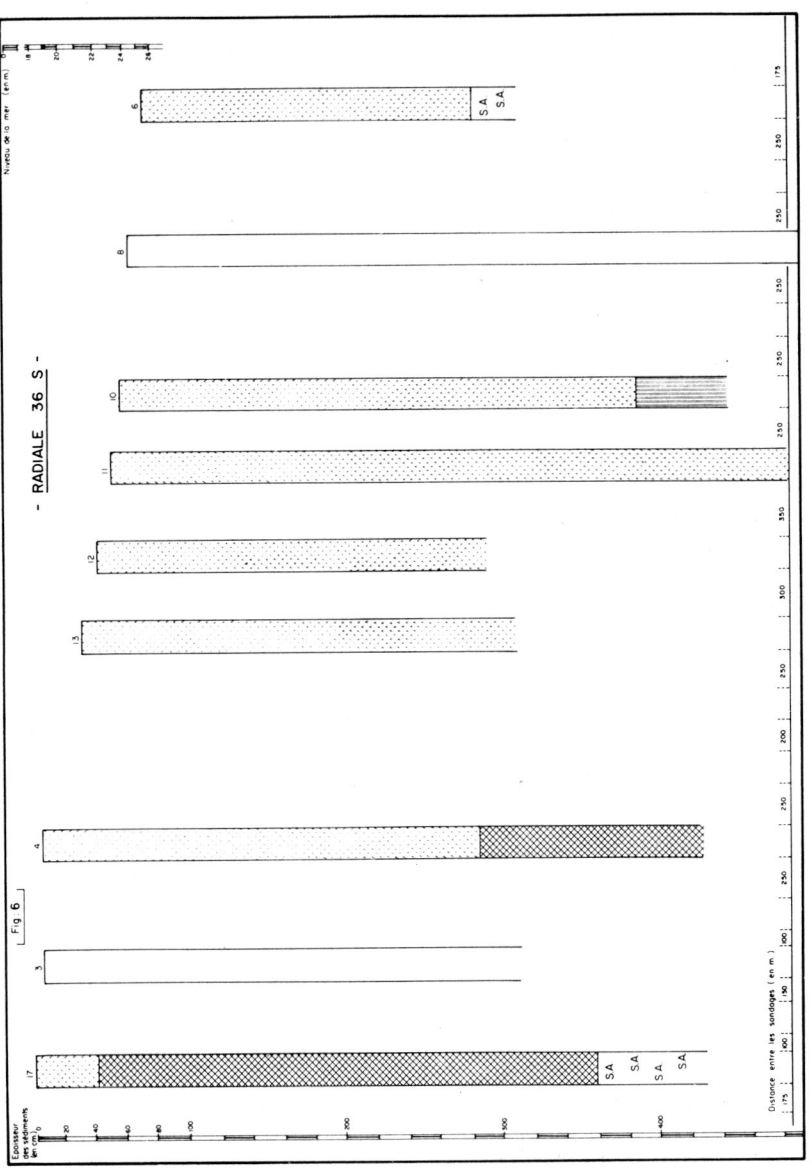

d'une salinité normale. Ce fait est confirmé par la presence d'espèces ubiquistes telles que Soudanella et Aurila.
- 2,90 m au sommet, l'association 'A' relativement pauvre se développe avec un nombre restreint de formes phytales.

La carotte 36 S 17 (longueur 4,20 m)
- de la base à 3,70 m, 'G' une zone d'argile gris bleu est totalement dépourvue d'ostracodes.

- 3,60 m à 0,60 m, l'association 'D' se retrouve à nouveau identique à la carotte précédente.
- 0,60 m au sommet, l'association 'A' devient riche et deversifiée.

Interprétation écologique de la radiale 36 S

Si nous intégrons les résultats de l'analyse des microfaunes d'ostracodes de ces différentes carottes, nous constatons que 3 secteurs s'individualisent.
a) *A l'Ouest* (carotte 6). L'association 'A' marine repose directement sur le 'beach-rock'. La faune marine est relativement pauvre.
b) *Au centre* (carotte 10). Le faciès marin transgressif repose directement sur les tourbes. Ici la faune marine est très riche et très diversifiée.
c) *A l'Est* (carottes 4 et 17). Les faciès marins 'A' reposent directement sur des faciès lagunaires 'D' très riches et très diversifiés, où on note l'absence total de Neomonoceratina.
d) La comparaison de cette radiale avec les précédentes nous permet de formuler les remarques suivantes:

Ainsi après la radiale 38 S qui était exclusivement composée de faciès 'A' marins homogènes, nous retrouvons de nouveau des successions d'associations comparables aux radiales 40 S et 42 S. Par exemple, 36 S 6 présente un enchaînement comparable à la carotte 42 S 54. La carotte 36 S 10 est à rapprocher de la carotte 42 S 59. Ceci implique, que, par rapport à la radiale 38 S essentiellement marine, il existe au Sud et au Nord de cet axe marin ainsi défini, deux entités lagunaires parfaitement individualisées.

4.5. *La radiale 34 S* (Fig. 7)

De la région Ouest en se dirigeant vers l'Est, nous avons:

La carotte 34 S 18 (longueur 3,40 m)
- de la base au sommet, l'association 'A' est relativement mal représentée, et les espèces phytales sont très rares.

La caratte 34 S 20 (longueur 4,80 m)
- de la base à 4,30 m, les sables fins vaseux gris noir sont dépourvus d'ostracodes.
- 4,20 m à 2,90 m, l'association 'B' est extrêmement mal représentée, pauvre en individus et en espèces.
- 2,90 m à 2,30 m, l'association 'H' est composée d'une thanatocenose relativement bien développée.
- 2,30 m à 2,20 m, association 'E' monospécifique à Neomonoceratina.
- 2,20 m au sommet, l'association 'A' est mal définie à la base puis mieux représentée vers le sommet de la carotte.

La carotte 34 S 21 (longueur 3,80 m)
- de la base au sommet, on rencontre l'association 'A' très riche sur le plan quantitatif et qualitatif.

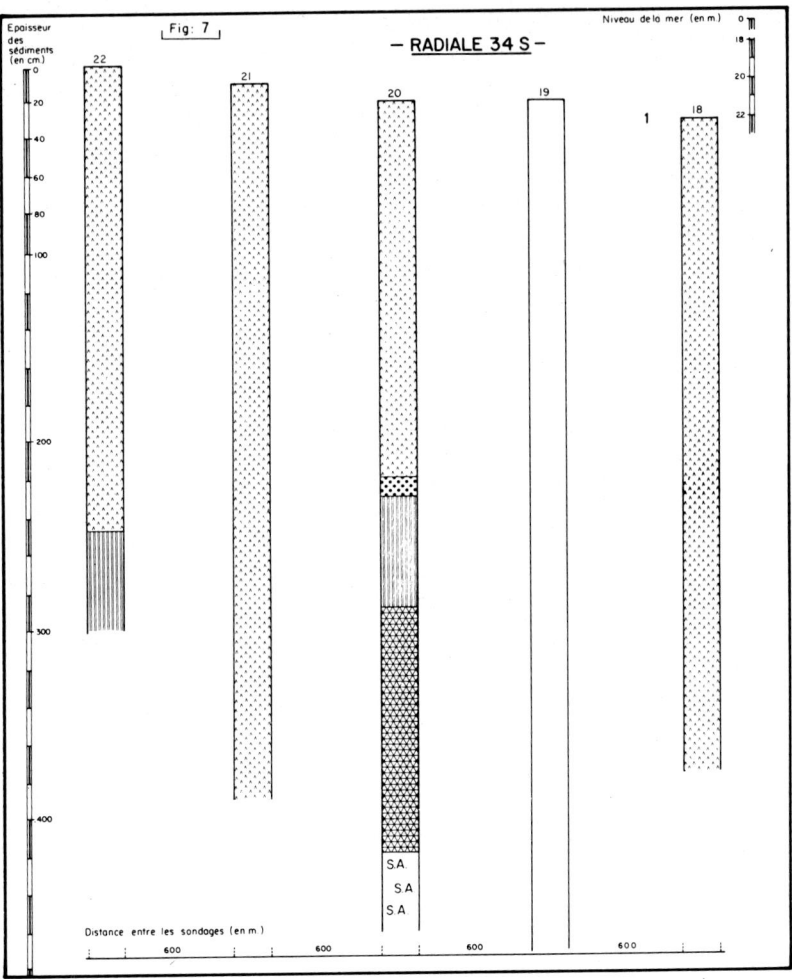

La carotte 34 S 22 (longueur 3 m)
- de la base à 2,40 m, l'association 'H' domine dans les niveaux tourbeux.
- 2,40 m au sommet, l'association 'A' est très pauvre et mal représentée en espèces phytales.

Interprétation écologique de la radiale 34 S

Les carottes 22 et 20 montrent la présence d'une microfaune lagunaire avant la transgression marine représentée par l'association 'A'. Il semble donc que la lagune déjà perçue dans la radiale 36 S semble se poursuivre vers le sud.

4.6. *La radiale 32 S* (Fig. 8)

De la région Ouest en se dirigeant vers l'Est, nous avons:

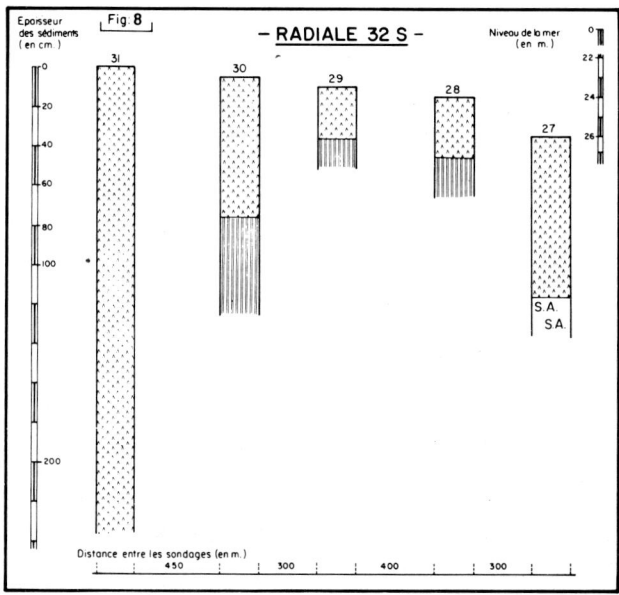

La carotte 32 S 27 (longueur 1 m)
- de la base à 0,80 m, l'association 'F' monospécifique à Cyprideis apparaît au sein d'un sédiment argileux jaunâtre.
- 0,80 m au sommet, l'association 'A' est pauvrement représentée.

La carotte 32 S 28 (longueur 0,50 m)
- de la base à 0,30 m, 'H' zone tourbeuse avec thanatocenose.
- 0,30 m au sommet, l'association 'A' mal représentée.

La carotte 32 S 29 (longueur 0,30 m)
- de la base à 0,20 m, 'H' zone tourbeuse sans ostracodes.
- 0,20 m au sommet, l'association 'A' mal représentée.

La carotte 32 S 30 (longueur 1,30 m)
- de la base à 0,70 m, 'H' zone tourbeuse sans ostracodes.
- 0,70 m au sommet, l'association 'A' est mal représentée.

La carotte 32 S 31 (longueur 2,40 m)
- de la base au sommet, on ne distingue que l'association 'A'.

Interprétation écologique de la radiale 32 S

En intégrant les résultats de l'analyse des ostracodes de ces 5 carottes il se dégage 3 zones.

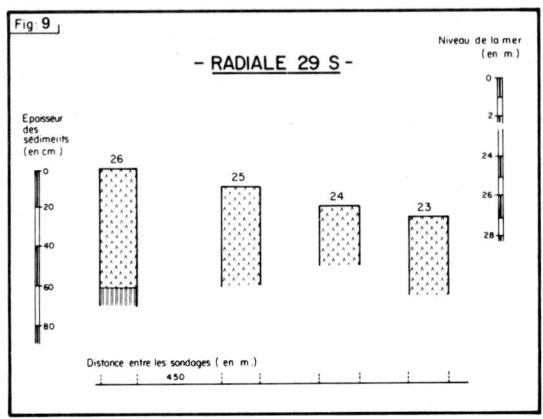

a) *A l'Ouest*. La faune marine 'A' transgresse directement la faune lagunaire à Cyprideis 'E'.
b) *Au Centre*. La faune marine 'A' transgresse directement les zones tourbeuses 'H'.
c) *A l'Est*. On retrouve plus que l'association 'A'. Il faut noter la présence de la faune lagunaire à Cyprideis 'E' qui était localisée précédemment dans la radiale 42 S.

4.7. *La radiale 29 S* (Fig. 9)

De la région Ouest en se dirigeant vers l'Est, nous avons:

La carotte 29 S 23 (longueur 0,40 m)

La carotte 29 S 24 (longueur 0,30 m)

La carotte 29 S 25 (longueur 0,50 m)
– de la base au sommet, ces carottes très courtes possèdent l'association 'A' mal représentée.

La carotte 29 S 26 (longueur 0,70 m)
– de la base à 0,60 m, 'H' zone tourbeuse sans ostracodes.
– 0,60 m au sommet, l'association 'A' est toujours mal représentée.

Interprétation écologique préliminaire de la radiale 29 S

Malgré leur faible longueur, la carotte 26 montre que les tourbes 'H' sont encore présentes en se dirigeant vers le Sud. La faune marine est toujours très faiblement réprésentée qualitativement et quantitativement.

5. Evolution paléogeographique et paléoclimatologique du plateau continental sénégalais pendant la fin du Quaternaire

Si on intègre les résultats des interprétations écologiques relatives à chaque radiale, il se dégage le schéma paléogéographique suivant–:

Avant 8 500 ans B.P. (Fig. 10) dans une zone proche du rivage se développe un système lagunaire sous un climat aride à semi-aride. En effet, la microfaune à Cyprideis 'F' (base des carottes 53-55-56, Radiale 42 S) et C. 27, radiale 32 S) implique des conditions arides à semi-arides sur le plan climatique. La salinité déduite des tests très fortement réticulés des Cyprideis pourrait être > à 35 ‰. L'âge de cette microfaune est probablement pré-nouakchottien au sens de Faure & Elouard (1967). Elle appartient peut-être à une époque légèrement antérieure à 8 500 B.P., car Faure (1969) a noté que plusieurs cycles climatiques différents étaient possibles entre 12 000 et 3 200 B.P. dans ces régions.

Cette lagune était abritée derrière un haut-fond (beach-rock) et sa communication avec la mer devait être relativement faible.

Toujours avant 8 500 ans (Fig. 11) la transgression nouakchottienne se développe et envahit toute la région étudiée. Elle atteint probablement l'isobathe -17-18m et tant que les conditions du milieu, soit marin, soit lagunaire ne sont pas complètement établies, seule s'installe la faune marine 'B' dite des espèces ubiquistes (Radiale 42 S, C. 46, C. 41, C. 20).

Vers le centre de la zone étudiée (Radiale 38 S et 36 S) les conditions marines s'installent rapidement (biotopes phytaux).

Peu avant 8 500 ans (Fig. 12) il semble que la transgression marque un temps d'arrêt; des faciès 'régressifs', lagunaires, s'installent. On peut raisonnablement envisager la formation d'un cordon littoral Nord-Sud situé très probablement aux alentours de l'isobathe -25-26 m, se fixant sur le haut-fond ancien (beach-rock).

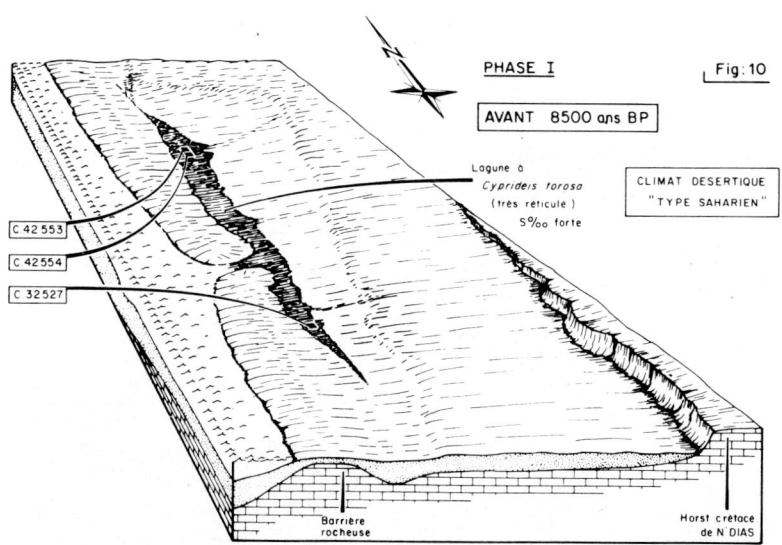

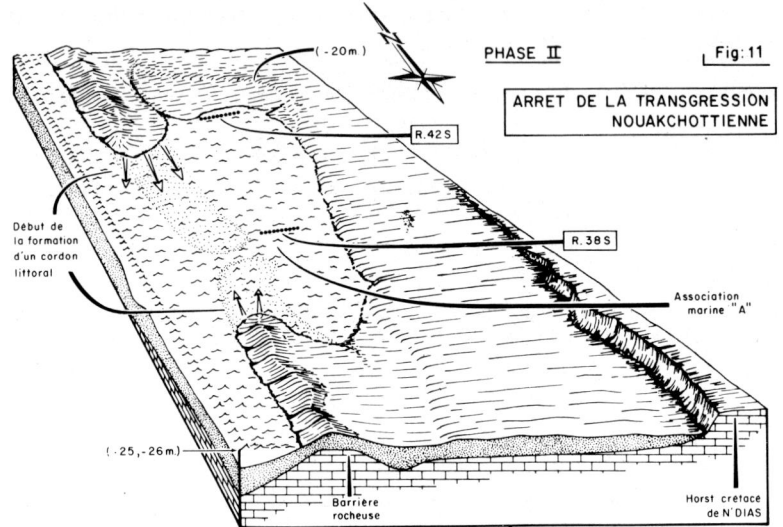

La mise en place de cette barrière naturelle permettra l'individualisation d'une lagune qui restera en communication avec la mer au niveau de la radiale 38 S.

Vers 8 500 ans (Fig. 13) et simultanément avec le phénomène paléogéographique décrit ci-dessus, un changement climatique provoque l'établissement d'une mangrove, aux conditions écologiques peu viables pour la microfaune, on note quelques thanatocénoses d'ostracodes.

Très rapidement entre 8 400 et 8 150 ± 180 ans (Fig. 14) les faciès à mangrove des radiales 40 S et 36 S sont rapidement recouverts par la mer qui poursuit sa route vers l'Est. Il semble qu'à ce moment là, la lagune atteint d'une part une ex-

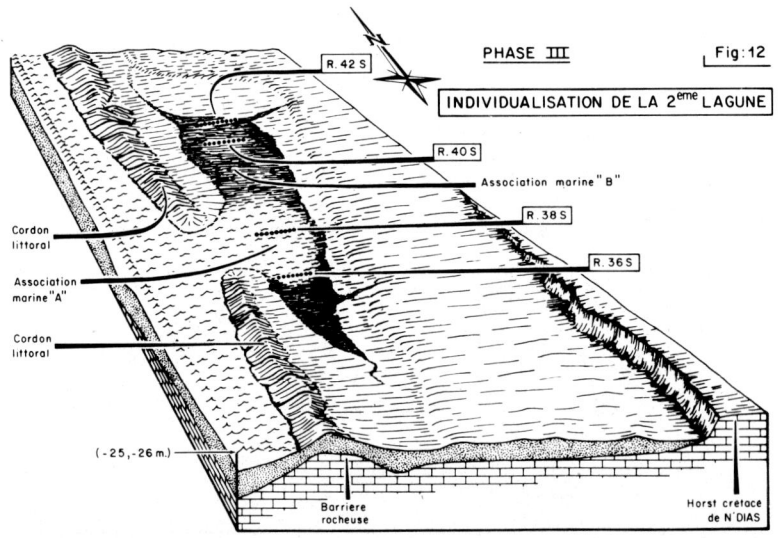

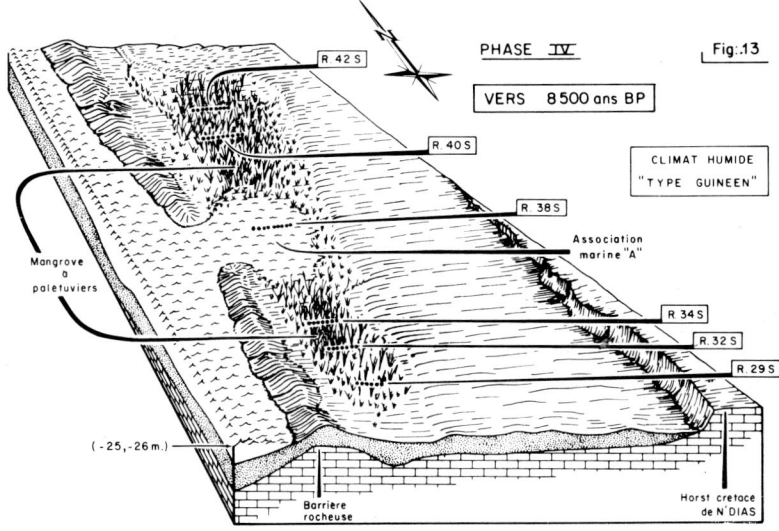

tension maximum, et d'autre part une diversité de faciès étonnante. Il est possible d'établir un panorama détaillé de la région datant de cette époque.
—Au Nord (Radiale 42), la mangrove préalablement bien développée aux alentours de la radiale 40 S va, sous l'effet de la transgression reculer vers le Nord et se développer essentiellement dans la zone nord-orientale. Sur les grèves sablo-vaseuses s'installent des biotopes à Neomonoceratina (42 S 60) attestant de la proximité des apports oligohalins au fond de la lagune, zone où les salinités peuvent être basses (0-17‰).

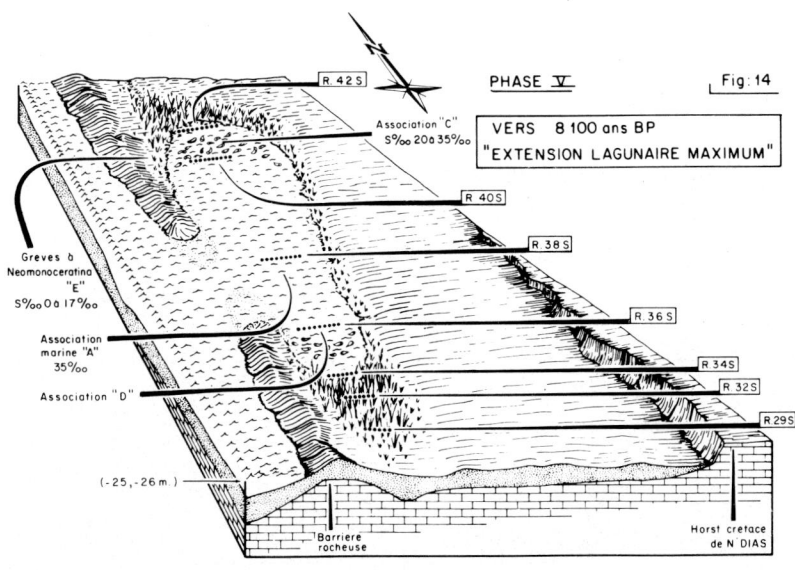

— Plus méridionalement (Radiale 40 S), deux biotopes lagunaires sensiblement différents s'établissent:

Dans la partie orientale, des faciès à huîtres, d'une part, microfaunes d'herbiers favorisant le développement de Xestoleberis, d'autre part, impliquent que les salinités peuvent varier entre 20 et 35‰. L'alternance entre les Neomonoceratina et les Paracypris pourrait indiquer des conditions mésohalines à polyhalines.

Dans la partie occidentale, seules les Neomonoceratina ont pu s'établir probablement au cours des apports continentaux plus importants. Ici les grèves sablovaseuses ont des salinités plus faibles 0 ‰-17‰ et les mangroves sont très proches.

— Au centre de la lagune (Radiale 38 S). Nous sommes dans la zone de permanence marine. C'est la baie largement ouverte en communication constante avec la mer. La faune est riche et très diversifiée.

Quelquefois des apports continentaux dûs probablement à de fortes précipitations entrainent quelques thanatocenoses provenant des faciès mesohalins. Ces phénomènes sont perçus dans les associations recueillies dans les carottes de cette radiale.

Dans la région Est (Carotte 41) les faunes allochtones sont plus importantes.
— Au Sud de cet 'axe marin' représenté par la radiale 38 S. nous retrouvons à nouveau des faciès lagunaires (radiale 36 S).

Dans la partie orientale, des biotopes lagunaires identiques à ceux rencontrés dans la radiale 40 S s'établissent. Mais l'absence totale de Neomonoceratina, la forte proportion de Paracypris et la présence d'Occultocythereis, nous amènent à penser que les influences marines sont plus fortes que dans la zone Nord pour des faciès similaires.

Dans la partie occidentale, il reste une petite mangrove qui se maintient derrière le cordon littoral.

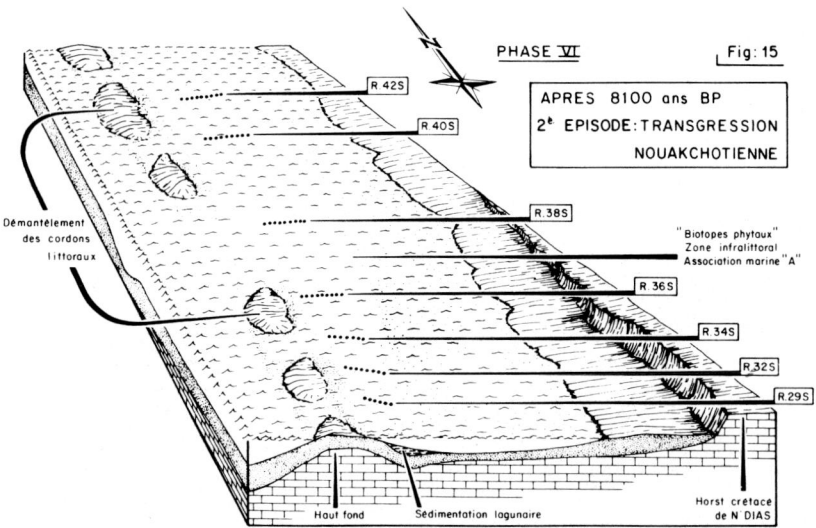

– Tout à fait au Sud, dans le fond de la lagune (Radiales 34-32-29 S) quelques biotopes de mangroves associés à des grèves à Neomonoceratina témoignent du caractère continental de ce secteur géographique.

Ce long 'travelling' Nord-Sud montre qu'il existe une certaine symétrie des faciès par rapport à la radiale 38 S. En effet de part et d'autre de cet axe central marin, nous trouvons deux entités lagunaires, de superficies inégales mais dont les biotopes sont sensiblement équivalents par leur composition qualitative.

Après 8 100 ans (Fig. 15) la transgression marine nouakchottienne poursuit sa route vers l'Est, elle démantèle le faible cordon dunaire situé sur le haut-fond.

Dans ce secteur s'installe la microfaune 'B' relativement pauvre et peu diversifiée. Puis les biotopes phytaux (Fig. 16) se développent largement sur toute la zone d'étude. Nous sommes dans l'infra-littoral interne du plateau continental sénégalais.

6. Conclusions

L'étude des ostracodes de l'ensemble des carottes du plateau sénégalais met en relief deux pôles d'intérêt différent.

Sur le plan paléogéographique, les Ostracodes permettent de montrer l'évolution spatiale de deux lagunes superposées dans le temps, finalement transgressées par la mer à la fin du Quaternaire. De plus ils suggèrent une ligne de rivage aux alentours de l'isobathe de –25 –26 m vers 8 500 ans B. P. environ. Ces résultats sont tout à fait en accord avec les données générales concernant les remontées des niveaux marins des régions de l'Afrique de l'Ouest (Faure & Elouard 1967, Delibrias 1973).

Sur le plan paléoclimatique, la succession dans le temps de la lagune à Cyp-

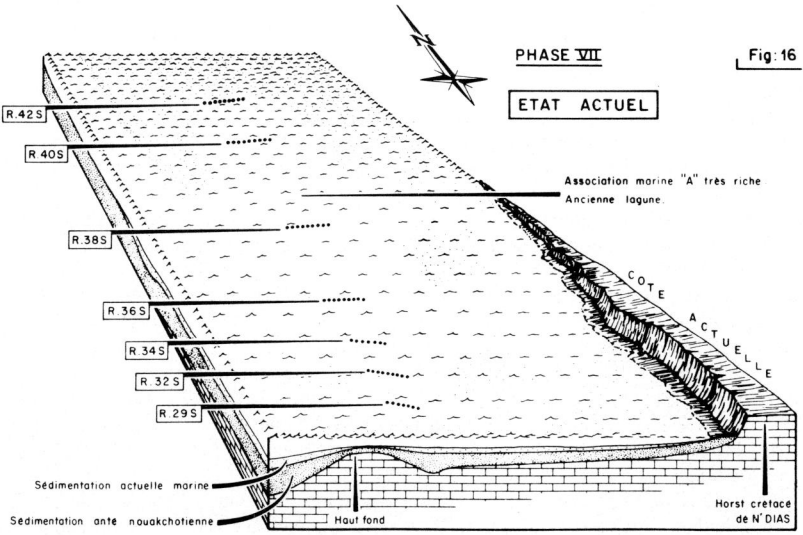

rideis 'F' de climat aride à semi-aride (type saharien) suivie de la lagune à faune d'ostracodes diversifiée ('C', 'E', 'H') de climat très humide (type Guinéen) datant de 8 400 ans environ, montre une étroite correspondance avec les variations paléoclimatiques enregistrées aux mêmes latitudes sur le continent africain.

En effet, Servant (1973), au lac Tchad et Butzer et al. (1972) dans les lacs d'Afrique orientale ont parfaitement montré la présence de climats très 'pluvieux' aux alentours de 8 500 ans.

Résumé

Sur la plateforme continentale du Sénégal, au Sud de Dakar (17°10'-17°15'N et 14°30'-14°40'W) entre les isobathes -15-25 m; une campagne océanographique nous a permis de récolter 44 carottes.

L'analyse de la microfaune d'ostracodes a livré plusieurs associations fauniques traduisant des environnements écologiques variés.

L'évolution dans le temps et dans l'espace de ces ensembles fauniques est la conséquence de la progression de la transgression nouakchottienne sur le littoral sénégalais. Cet épisode de la fin du Quaternaire se traduit par des variations dans les environnements inventoriés, impliquant des changements paléogéographiques et paléoclimatiques.

Ainsi, avant 8 500 B.P. on constate, dans la zone étudiée la présence d'une lagune à forte salinité de climat aride à semi-aride très prononcé, puis entre 8 500 et 8 100 ans B.P., un deuxième système lagunaire plus grand que le précédent s'installe sous un climat beaucoup plus humide; enfin des biotopes infra-littoraux marins terminent la série.

Bibliographie

Benson, R. M. 1969. Ecology of recent Ostracodes of the todos Santos Bay Region California Univ. Kansas Paleont. Contr. Arthropoda, art. 1, pp. 1-180.

Butzer, K. W., Isaac, G. L., Richardson, J. L. & Washbourn-Kamau, C. 1972. Radiocarbon dating of East African lake levels. Science 175 (4027):1069-1076.

Carbonel, P., Moyes, J. & Peypouquet, J. P. 1975. Utilisation des Ostracodes pour la mise en évidence et l'évolution d'une lagune Holocène à l'Ouest de la Gironde, Golfe de Biscaye. Bull. of American Paleontology 65:445-462.

Delibrias, G. 1973. Variation du Niveau de la Mer sur la côte Ouest Africaine depuis 26 000 ans. In: Colloques Internationaux du C.N.R.S. n° 219, 'Les Méthodes quantitatives d'étude des Variations du climat au cours du Pleistocène', pp. 127-134.

Faure, H. 1969. Recherches sur le Quaternaire littoral du Sénégal et de la Mauritanie, In: Paleoecology of Africa covering the years 1966, 1968. Vol. IV, 274 pp, (Ed. E. M. Van Zinderen, Bakker, Sr.) A. A. Balkema, Cape Town.

Faure, H. & Elouard, P. 1967. Schema des variations du niveau de l'ocean Atlantique sur la côte de l'Ouest d'Afrique depuis 40 000 ans C. R. Acad. Sc. Paris 265, série D:784-787.

Peypouquet, J. P. 1971. La distribution des biocenoses, thanatocenoses, paleothanatocenoses: Problème fondamental sur une plateforme continentale. Bull. Inst. Géol. Bassin d'Aquitaine 11 (1):191-208.

Servant, M. 1973. Séquences continentales et variations climatiques: Evolution du Bassin du Tchad au Cenozoïque supérieur. Thèse doct. Etat Science Nat., Paris VI, 348 pp.

Adresse de l'auteur:
Laboratoire de Géologie et d'Océanographie, Ave des Facultés, 33405 Talence, France.

Discussion

Carbonnel: Etes-vous certain de la détermination du genre *Neomonoceratina* dans vos profils? En effet, ce genre est proche de *Sulcostocythere* Maddocks. Ce dernier est plutot tropical alors que *Neomonoceratina* est un genre de préférence plus nordique: Tethys et Paratethys.

Peypouquet: Les genres *Neomonoceratina* Kingma 1948 et *Sulcostocythere* Benson & Maddocks 1962 sont effectivement très proches l'un de l'autre. Cependant, la charnière de *Neomonoderatina* est composée de 4 éléments: Type Amphidonte/schizodonte dans la classification élaborée par Van Morkhoven (1962); celle de *Sulcostocythere* est composée de 3 elements: Type schizodonte selon Maddocks (1966), mais en fait de type merodonte/antomadonte dans la classification mise au point par Van Morkhoven (1962).

De plus, le genre *Neomonoceratina* possède un 'processus caudal' très remarquable, cet élément morphologique semble être absent ou peu marqué chez *Sulcostocythere*.

Après vérification et pour les raisons indiquées ci-dessus il s'avère que c'est bien *Neomonoceratina* qui colonise la lagune Holocène du plateau continental sénégalais.

Carbonel: Comment déterminez-vous l'écologie du genre *'Neomonoceratina'* dans vos profils? La détermination correcte du genre conditionne en effet les implications paléoécologiques.

Peypouquet: L'écologie ou plutôt la paléoécologie de *Neomonoceratina* est déterminée en fonction de:
– sa présence au sein d'associations à caractère lagunaire très marqué.
– son absence dans les associations à caractère marin franc. (Associations 'A' ou 'B').
– l'information à caractère écologique donnée par l'etude de Omatsola (1970) concernant les Ostracodes de la lagune de Lagos (Nigeria).

Cet auteur a mis en évidence la présence de 'vrais *Neomonoceratina* dans les milieux déssalés de cette région.

La remarque concernant la détermination correcte du genre et les implications paléoécologiques qui en résultent est indiscutable. Mais dans le cas qui nous préoccupe, il s'avere que *Sulcostocythere* possède un habitat très proche de celui qui est attribué à *Neomonoceratina:* 'intertidal mud among mangroves on the estuarine banks' (Maddocks 1966, P, 68);

Il en résulte que l'erreur qui aurait pu être commise au niveau générique n'en ut point entrainé de modifications paléoécologiques importantes.

Keen: I have found *Cyprideis* in samples from the outer shelf off Sierra Leone. Perhaps similar conditions existed along the other parts of the west African coast prior to the Flandrian transgression.

Peypouquet: La valeur écologique des *Cyprideis* que vous avez trouvé sur le plateau continental de la Sierra Leone peut être une fonction des paramètres morphologiques des tests (Taille, degré de réticulation, présence ou absence de nodes . . .). Cependant, nous savons, grâce à d'autres techniques géologiques (Pollen, sédimentologie) que des conditions climatiques plus arides ont existé dans ces régions de l'Afrique de l'Ouest, avant, ou, au début de la transgression Flandrienne. (Assemien et al. (1971), Tastet (1975), Fredoux & Tastet (1976)

Bibliographie supplémentaire

Assemien, P., Filleron, J. C., Martin, L. & Tastet J. P. 1971. Le Quaternaire de la zone littorale de la Côte d'Ivoire. Bull. Ass. Sénégal et Quatern. Ouest Afr. Dakar, n° 25, Mars. Quaternaire Roma XV, 1971, pp. 305-316.

Fredoux, A. & Tastet J. P. 1976. Apport de la palynologie à la connaissance paléogéographie du littoral ivoirien entre 8 000 et 12 000 ans B.P. VIIè African Micropaleontological Colloquium, Ile – Ife Nigeria.

Maddocks, R. F. 1966. Distribution patterns of living and subfossil podocopid Ostracodes in the Nosy Bé Area. Northern Madagascar. The University of Kansas. Paleontological contributions paper 12, 72 pp.

Omatsola, M. E. 1970. Podocopid Ostracoda from the Lagos lagoon Nigeria. Micropaleontology. 16 (4) 407–445.

Tastet, J. P. 1975. Les formations sédimentaires quaternaires à actuelles du littoral du Dahomey. Notice explicative de la carte géologique. Bull. Ass. Sénégal et Quatern. Ouest Afr. 46, déc.,21-44.

Van Morkhoven, F. P. C. M. 1962. Post palaeozoïc Ostracoda. Elsevier. Publ. Company, Amsterdam, 204 pp.

Sixth Intern. Ostracod Symposium, Saalfelden

THE OSTRACOD GENUS TYRRHENOCYTHERE

NADEŽDA KRSTIĆ

Abstract

The basic genus character is represented by the marginal pore canals grouped in bundles. Ten species have been described to date. These species may be divided into three groups: semicircular, rectangular and trapezoidal, but they also have some other common features. They were and are still living in the Paratethys and east and central Tethys.

Species of the genus *Tyrrhenocythere* Ruggieri 1955 occur locally, usually only few specimens at a time, and therefore this communication is based on collections of numerous ostracodologists, and on the available literature. In different papers species of *Tyrrhenocythere* have often included within various other genera, due to the fact that the nomenclature has not yet been entirely refined. This has introduced additional difficulties, particularly as in some papers (ecological and similar works) the studied taxa are not illustrated. Owing to all these reasons, this communication does not claim any definitive precision, it does, however, represent the present state of knowledge on the genus *Tyrrhenocythere*.

Known species

The following species of the genus *Tyrrhenocythere* have been described to date:
Cythere amnicola Sars, 1888
Cythere sicula Brady, 1902
Cythereis donetziensis Dubowsky, 1926
Cythere pseudoconvexa Livental, 1938
Cythereis truncata Schneider in Agalarova et al., 1940
Cythereis azerbaidjanica Livental in Agalarova et al., 1940
Tyrrhenocythere pignattii Ruggieri, 1955
Cythereis bailovi Livental in Suzin, 1956 /?/
Trachyleberis immediata Markova, 1959
Trachyleberis gavrilovi Markova, 1959
Cythereis praeazerbaidjanica Agalarova in Agalarova et al., 1961
Cythereis pontica Livental in Agalarova et al., 1961
Hemicytheria filipescui Hanganu, 1962
Trachyleberis cibaria Sharapova in Mandelstam et al., 1962
Trachyleberis sp. Mandelstam et al., 1962 /pl. XXI, fig. 4/
Nereina? sp. Grekoff et Molinari, 1963
Tyrrhenocythere triebeli Krstić, 1963
Trachyleberis imnadzeae Vekua, 1965 /?/
Trachyleberis pontica var. Agalarova, 1967
Tyrrhencythere ruggierii Devoto, 1967
Tyrrhenocythere ballesioi Carbonnel, 1969

Some of these names are in fact synonyms of already known species. *T. sicula* (Brady), *T. donetziensis* (Dubowsky), *T. pignattii* Ruggieri and *T.* sp. (Mandelstam et al. 1962) are synonyms of *T. amnicola* (Sars), as probably are also *T. dendropora* (Suzin), *T. bailovi* (Livental in Suzin), *T. immediata* (Markova), *T. gavrilovi* (Markova) and *T. cibaria* (Sharapova in Mandelstam et al.). It may be that *T. pseudoconvexa* (Livental) is a distinct species being shorter, higher and more convex than *T. sicula* and without ornamentation; this can easily be checked on the recent material from the Caspian Sea (compare Naidina 1970). The name *T. rugierrii* Devoto is probably a synonym of *T. truncata* (Schneider, in Agalarova et al.) and *T. ballesioi* Carbonnel a synonym or subspecies of *T. pontica* (Livental, in Agalarova et al.). A correct list would be:

Upper Pliocene – Quaternary:
Tyrrhenocythere amnicola (Sars)
Tyrrhenocythere azerbaidjanica (Livental, in Agalarova et al.)

Middle Pliocene:
Tyrrhenocythere imnadzeae (Vekua)
Tyrrhenocythere praeazerbaidjanica (Agalarova)

Lower Pliocene (Pontian/Upper Messinian):
Tyrrhenocythere truncata (Schneider, in Agalarova et al.)
Tyrrhenocythere pontica (Livental, in Agalarova et al.)
Tyrrhenocythere filipescui (Hanganu)
Tyrrhenocythere sp. (Grekoff & Molinari 1963)
Tyrrhenocythere triebeli Krstić
Tyrrhenocythere pontica var. (Agalarova)
Tyrrhenocythere pontica ballesioi Carbonnel

Along with a number of the above indicated species, there are some other species of the genus *Tyrrhenocythere* as yet undescribed and some of which are illustrated here. They are: *T.* aff. *praeazerbaidjanica* (Agalarova) and *T. taurica* Sinegub, nomen nudum, while in Sinegub's collection I also saw *T. notabilis* (Imnadze, nomen nudum). For the species *Tyrrhenocythere cavernosa* Sharapova I have no data.

Description and relationships

Semicircular to trapezoidal in shape, smooth or ornamented with faint to distinct reticulation combined with pits and, occasionally, ribs. Normal pores simple (only?), with or without a narrow lip (Pl. 1, Fig. 6). Marginal pore canals grouped in bundles of 6-11, starting from a common base and radiating fan-like; this is one of the basic generic characteristics (Fig. 1). The right valve hinge consists of one strong tooth at the anterior end, bearing against the inside valve face and lying beneath the hinge margin; behind the tooth there is a socket, then a shallow sulcus open towards the interior and slightly wider and deeper at its posterior end. Agalarova and co-authors (1961, Pl. 82, Fig. 1 b) drew the crenulated ridge above the sulcus; at the posterodorsal angle there is another tooth, lower and only slightly or not at all elongate, which may or may not be crenulated. Muscle scars are

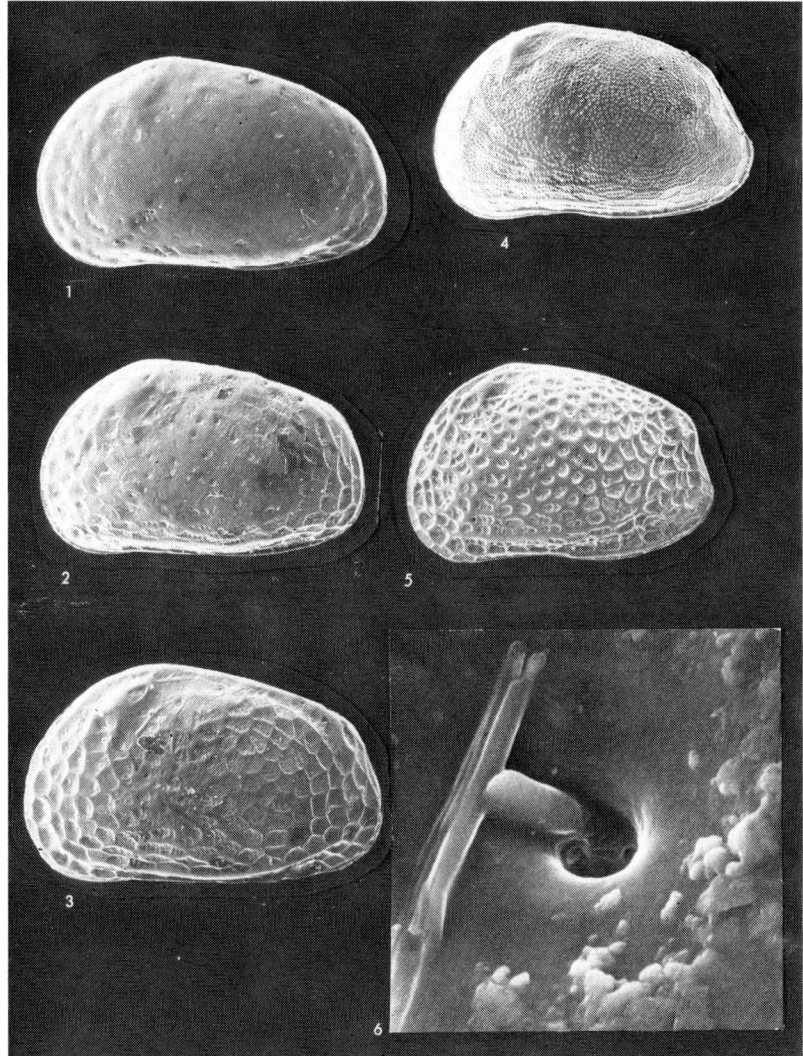

Plate 1. Photo: Dr. Ch. Samtleben, Kiel

1–3. *Tyrrhenocythere triebeli* Kr., L ♀;
Upper Pontian: 2. paratype, x 67, east Serbia, Kladovo, scan-photo 13151*; 1, 3. Greek Eastern Macedonia – sa-ple 98 M (coll. Niedersächsisches Landes-Amt für Bodenforschung), 1. No. 13148 (det. as *T. ruggierii* by Gramann, 1969), 3. No. 13149 (det. as *T.* cf. *pignatii* by Gramann, 1969), x 67.
4, 6. *Tyrrhenocythere amnicola* (Sars), L ♀;
Recent: north Caspian Sea (coll, N. N. Naidina): 4. x 62, No. 13270, 6. the same, a normal pore, x 6204, No. 13271.
5. *Tyrrhenocythere* cf. *praeazerbaidjanica* (Ag. in Ag. et al.), L ♀;
Kuyalnik beds: Odessa region (coll. V. V. Sinegub); scan-photo No. 13147, x 66.

* Numbers in the Photothece in Kiel.

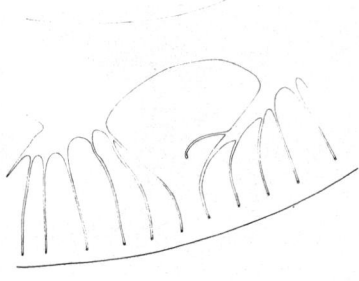

Fig. 1. The zone of concrescence at the anterior margin of *Tyrrhenocythere filipescui* (Hang), specimen from Pl. 2 Fig. 3.

the same as in other Hemicytherinae: the two middle adductor muscle scars are elongated and one or both of them divided into two parts; in front of the adductor scars there is an oblique row of three circular scars left by transverse and mandibular muscles.

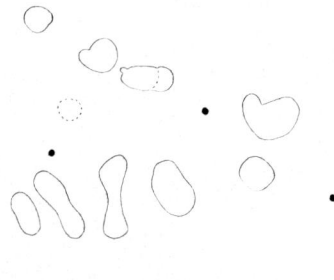

Fig. 2. Muscle scars of *Tyrrhenocythere taurica* (Sgb. nom. nud.), R ♀; Kuyalnik beds: vicinity of Scadovsk in south Ukraine (coll. V. V. Sinegub).

The carapace shapes and types of ornamentation are almost identical with those in the subgenus *Aurila (Hemicytheria)* Pokorny 1952. Certain species of these two genera converge so much that they can only be distinguished on the basis of the marginal pore canals of adults. *Aurila (Hemicytheria) josephinae* (Zalanyi 1929) from the Pannonian basin (Yugoslavia) and *Tyrrhenocythere notabilis* (Imnadze, nom. nud.) from the Pontian basin (Georgia, USSR) are almost identical – apart from the marginal pore canals they differ only in size.

Division

It is not my intention to describe new subgenera, although at least one of the groups mentioned below constitutes a new subgenus, since the material is not in my possession.

The species of *Tyrrhenocythere* can be divided into three groups according to shape: semicircular, rectangular and trapezoidal. Each of these groups features a different type of ornamentation.

Semicircular forms are: *T. truncata, T. amnicola, T. pontica* var. and *T. praeazerbaidjanica,* listed in order from smooth or punctate to fossal species.

Rectangular are: *T. triebeli, T. azerbaidjanica* and *T. pontica* ornamented with less to more distinct reticulation, in this order.

Trapezoidal are: *T. notabilis, T.* sp. (Grekoff & Molinari, 1963), *T. filipescui, T. taurica* and *T.? imnadzeae,* reticulate and ribed with a tendency to form an alar protuberance.

Ecology

The recent species *Tyrrhenocythere amnicola* (Sars) lives both in brackish and fresh water. In the Caspian Sea according to N. Naidina (1968, 1970, 1972) it is most numerous at a salinity of 9-13‰, but there are some specimens as low as 1‰. As substrate this species prefers shell debris. It usually lives at depths up to 30 m.

All fossil species of the genus *Tyrrhenocythere* lived in an environment of reduced salinity: on the basis of molluscs and the presence or absence of other fossils, such as foraminifers, in sediments with *Tyrrhenocythere,* the salinity can be estimated as 5-15‰. The substrate was sandy: *T. triebeli* was found in fine sands with shell debris, *T. filipescui* in sandy clay and falun, *T. cavernosa* and *T. azerbaidjanica* in clayey sand etc. The genus seems to be restricted to shallow water.

Zoogeography

The genus *Tyrrhenocythere* is known from Paratethys and Tethys from the Pontian or the Upper Messinian to the recent[1]. As it had a large number of species in the Pontian (the paleogeographic map shows the extent of the sea in Pontian), it is probable that some species can also be found in sediments older than the Pontian/Upper Messinian.

Due to insufficient infomation on Neogene-Recent ostracods, it is not possible to give any exact spacial distribution for the species, particularly for those of Pontian/Messinian age.

Tyrrhenocythere amnicola is the only species whose zoogeography is known. In the area of Paratethys, this species is known from the Recent[2] of Caspian and Pontian basins: lakes Issyk Kul and Aral, the Caspian Sea, Azov Sea, Black Sea[3], the rivers Donets and Dnester. In the Upper Pliocene-Pleistocene and Subrecent[4], it is known from the Caspian basin: west Turkmenia, the Volga region,

---

[1] The Upper Messinian and the Pontian are synchronous since they have identical mollusc (Stevanović 1969) and ostracod faunas (Ascoli, in Krstić 1971, Krstić 1975).

[2] Bronstein 1947, Shornikov 1969, Naidina 1970, 1972.

[3] I had the opportunity to compare directly the material from the Caspian Sea, which I obtained from N. Naidina, with the topotypes of the species *Cythere amnicola* Sars, 1888 (coll. Brady 1902, determined as *Cythere sicula* Brady), which were at that time (1969) with F. Gramann (Hannover). The specimens from the Caspian Sea and these from Sicily are identical.

[4] Suzin 1956, Mandelstam et al. 1962.

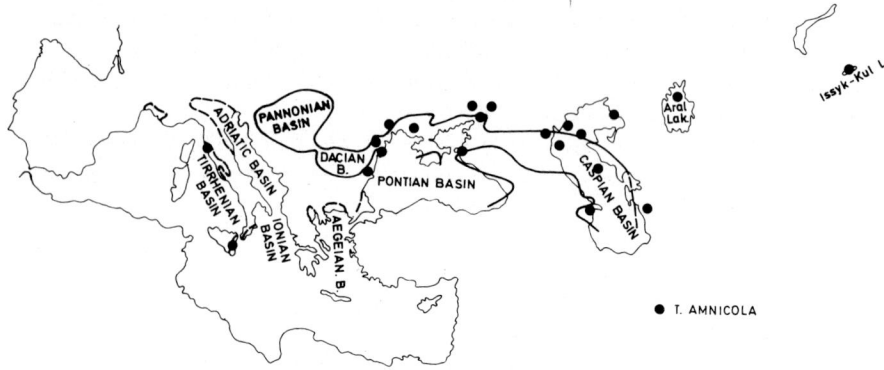

● T. AMNICOLA

north Caucasus and Azerbaydzhan. In the area of the Tethys this species is known, as Recent[5], only from the Ionian basin – Syracuse in Sicily. In the Tyrrhenian basin, in Tuscany (Italy), it occurs as a subfossil[6]. The finds outside the Mediterranean do not belong to this species or to this genus.

Tyrrhenocythere pontica is described from the Pontian of the Caspian basin[7], from Azerbaydzhan: the Apsheron Peninsula and Kobistan. In the Pontian basin[8] it is present in south Ukraine, on the Kerch Peninsula. The occurrence in southwest Moldavia[9] belongs to the same basin but at its boundary with the Dacian basin. Specimens found in northwest Bulgaria[10], near Lom, do not belong to this species. In the collection of P. Ascoli from the Adriatic basin, Emilia in Italy, this species is also present (Pl. 2, Fig. 2). The records from the west Mediterranean[11], in the Rhône valley (France) are of this species or its subspecies. Therefore, *T. pontica* also lived in the Upper Messinian of the Thetys.

---

[5] Sars 1888, Brady 1902.
[6] Ruggieri 1955.
[7] Agalarova 1967.
[8] Agalarova et al. 1961.
[9] Sinegub, in: Regional Stratigraphy of Moldavia, 1968.
[10] Stancheva 1965.
[11] Carbonnel 1969.

● T. PONTICA
▲ T. IMNADZEAE

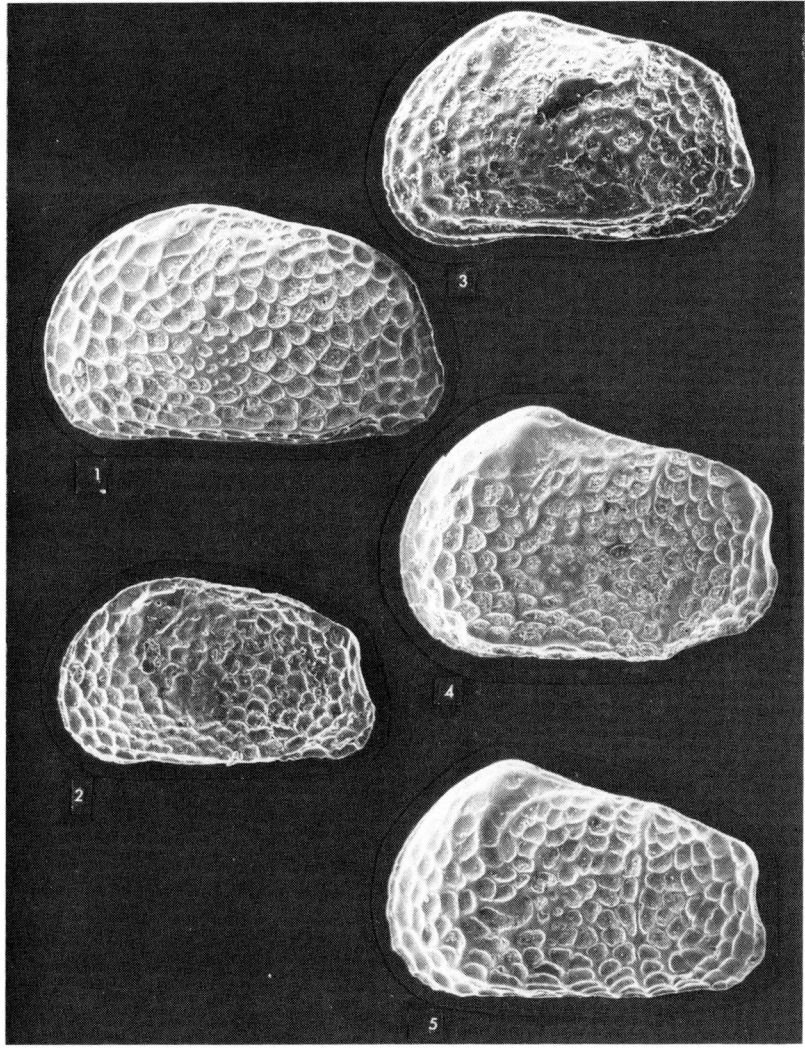

Photo: Dr. Ch. Samtleben, Kiel

Plate 2.

1, 2. *Tyrrhenocythere pontica* (Liv. in Ag. et al), L ♀;
1. Pontian II of N. Andrusov: Arshintsevo section on the Kerch Peninsula (coll. V. V. Sinegub), x 67, scan-photo No. 13150; 2. Uppermost Messinian: casa Giovani Bello in Emilia – Italy (coll. P. Ascoli), scan-photo No. 13269, x 62.
3. *Tyrrhenocythere filipescui* (Hang.), L ♀, paratype;
rhomboidea-beds of the Pontian: Valea Ursoaia Mare in Romania (coll. E. Hanganu), scan-photo No. 13188, x 73.
4, 5. *Tyrrhenocythere taurica* (Sgb., nom. nud.), L ♀;
4. *Phylocardium planum*-beds of the Pontian: Ploesti district in Romania (coll. E. Hanganu), scan-photo No. 13152, x 65; 5. Kuyalnik beds: vicinity of Skadovsk in south Ukraine (coll. V. V. Sinegub), scan-photo No. 13153, x 65.

● T. TRIEBELI
▲ T. TRUNCATA

Tyrrhenocythere triebeli comes from the western parts of the Dacian basin: in northwest Bulgaria[12] near Lom (where it is determined as *T. papilosa* by Stancheva 1965) and in east Serbia[13] near Kladovo. In the Pannonian basin[14] it was found on the mountain Frushka Gora, also in Serbia. In the area of the Tethys it has been found in the Aegean basin[15], in the Strimon valley of eastern Greek Macedonia (determined as *T. ruggierii* by Gramann 1969).

Tyrrhenocythere truncata is another species found in both the Paratethys and Tethys. It is described from the Caspian basin[16]: Turkmenia, north Caucasus and Azerbaydzhan. The Tethyan record was in the Tyrrhenian basin[17] at Lake Fucino (determined as *T. ruggierii* by Devoto 1967).

Tyrrhenocythere filipescui is known only from Paratethys. In the Pontian basin[18]

---

[12] Stancheva 1965.
[13] Krstić 1963, 1975.
[14] Krstić 1969, 1975.
[15] Gramann 1969.
[16] Agalarova et al. 1940, Agalarova 1967.
[17] Devoto 1967.
[18] Sinegub, in: Regional Stratigraphy of Moldavia, 1968.

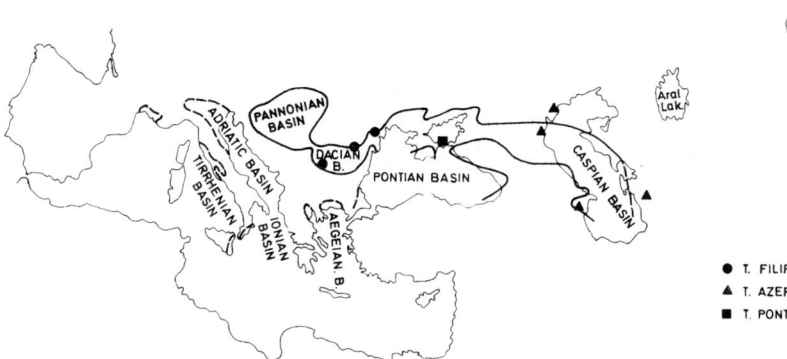

● T. FILIPESCUI
▲ T. AZERBAIDJANICA
■ T. PONTICA VAR. (Ag., 1967)

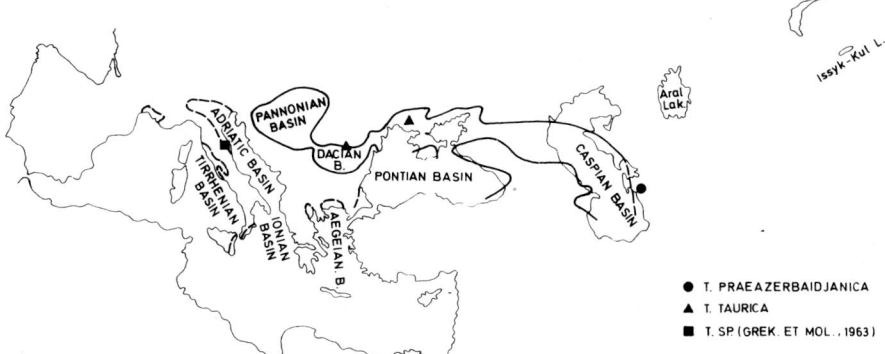

- ● T. PRAEAZERBAIDJANICA
- ▲ T. TAURICA
- ■ T. SP (GREK. ET MOL., 1963)

it was found in southwest Moldavia, in the Dacian basin[19] in the regon of Ploesti in Romania and near Lom in Bulgaria.

Tyrrhenocythere azerbaidjanica belongs to the Caspian basin[20]: Turkmenia, Volga region, north Caucasus and Azerbaydzhan, but it has also been found in the Pontian basin[21], in southwest Moldavia.

Tyrrhenocythere taurica has been found in the Pontian basin[22], in south Ukraine (Skadovsk), and in the Dacian basin[23] near Ploesti in Romania.

Tyrrhenocythere praeazerbaidjanica is known only from the Caspian basin[24]: Turkmenia and Azerbaydzhan.

Tyrrhenocythere imnadzeae has been found only in one locality in the Pontian basin[25]: in Abkhazia, Georgia.

The same applies to two species with open nomenclature: *Tyrrhenocythere pontica* var. (Agalarova 1967) found in the Pontian basin[26] in south Ukraine on the Kerch Peninsula, and *Tyrrhenocythere* sp. (Grekoff & Molinari 1963), from Tethys[27], found in the Adriatic basin in Italy (Emilia).

The three groups differentiated here are not restricted to a particular part of the Mediterranean area, but occur in both Paratethys and Tethys.

Outside the Mediterranean area, the genus *Tyrrhenocythere* has not been identified. This genus appeared after the big salinity crisis in the Mediterranean. It seems that its development was engendered by the decreasing salinity, in Paratethys, where it is more numerous and where the salinity decrease started earlier than in the Tethys.

---

[19] Hanganu 1962, Stancheva 1965.
[20] Schweyer 1949, Suzin 1956, Mandelstam et al. 1962, Klein 1975.
[21] Sinegub, in: Regional Stratigraphy of Moldavia, 1968.
[22] Pl. 2, Fig. 5.
[23] Pl. 2, Fig. 4.
[24] Agalarova et al. 1961.
[25] Vekua 1965.
[26] Agalarova 1967.
[27] Grekoff & Molinari 1963.

Acknowledgments

I would like to acknowledge the cooperation of several colleagues who helped me by passing on material for comparison: P. Ascoli, Agip Mineraria in Mailand; F. Gramann, Niedersächsisches Landes-Amt für Bodenforschung in Hannover; Elisabeta Hanganu, Facultet Geologie-Geografie in Bucuresti; Nina N. Naĭdina, Biologo-Počvenniĭ fakultet Moskovskogo universiteta; V.V. Sinegub, Kompleksnaya geologo-tematičeskaya partiya in Kišinev; Maria Stančeva, Geoložki institut BAN in Sofia. This research was supported by the Alexander von Humboldt Stiftung of the German Government, and a greater part of it was carried out at the Geologisch-Paläontologisches Institut der Universität Kiel, in the department of Prof. K. Krömmelbein, and in the Zoologisches Institut der Universität Hamburg, in the department of Prof. G. Hartmann.

References

Agalarova, D. A., Djafarov, D. I. & Halilov, D. M. 1940. Handboock of Tertiary microfauna from the Apsheron peninsula Baku.
Agalarova, D. A., Kadyrova, Z. K. & Kulieva, H. M. 1961. Ostracoda of Pliocene and Postpliocene sediments of Azerbaydzhan. Az.gos.izdat., Baku.
Agalarova, D. A. 1967. Microfauna of Pontian sediments from Azerbaydzhan and neighbour regions. AzNII po dobyche nefti, 'Nedra', Leningrad.
Ascoli, P. 1971. Discussion, in – Krstić N.: Ostracoda biofacies in the Pannone. Bull.Centre rech.SNPA, 5 suppl., Pau.
Brady, G. 1902. On new or imperfectly-known Ostracoda, chiefly from a collection in the Zoological Museum, Copenhagen. Trans. Zool. Soc. 16:4.
Bronstein, Z. S. 1947. Ostracodes des eaux douces. Fauna SSSR, Crustacea, II, 1, Izd.AN SSSR, Moskva-Leningrad.
Carbonnel, G. 1969. Les Ostracodes du Miocene Rhôdanien. Docum. labor. geol. fac. sci. Lyon 32.
Devoto, G. 1967. Studio delle ostracofaune. In: Colacicchi, R. Devoto, G. & Praturlon, A., Depositi messiniani oligoalini al bordo orientale del Fucino e descrizione di Tyrrhenocythere ruggierii Devoto, nuovo specie di ostracode. Boll. Soc. Geol. Italia, 86.
Dubowsky, N. V. 1939. Zur Kenntnis der Ostracodenfauna des Schwarzen Meer. Trudy Karadagskoi biol.stan. 5, Krym.
Elofson, O. 1945. On Cythereis amnicola (G. O. Sars) and Loxoconcha umbonata G. O. Sars, two ostracodes from the Caspian Sea. Arkiv Zool. B 36:2.
Gramann, F. 1969. Das Neogen im Strimon-Becken (griechisch-Ostmazedonien), Teil II. Ostracoden und Foraminiferen aus dem Neogen des Strimon-Beckens. Geol. Jahrb. 87, Hannover.
Grekoff, N. & Molinari, V. 1963. Sur une faune d'ostracodes saumatres du Neogen de Castell'arquato (Emilia). Geol.romana 2, Roma.
Hanganu, E. 1962. Specii noi de ostracode în ponțianul din Subcarpați. Comun.acad.Romîne 12:5.
Karmishina, G. I. & Skornikova, E. G. 1975. On the paleozoogeographical value of Pliocene ostracods of the region Black Sea – Caspian Sea; Mode of existence and regularities of settling on recent and fossil microfauna, 'Nauka', Moscow.
Klein, L. N. 1975. Palaeogeographical conditions of the distributions of ostracods in the Azerbaydzhan basin of the Apsheron age. Mode of existence... 'Nauka', Moscow.
Krstić, N. 1965 Pontian ostracodes of some Serbian localities with special references to the family Cytheridae; Bull.Inst. geol. geophys. A 21:1963.

Krstić, N. 1975. Pontian ostracodes in Paratethys and Tethys. Proc. VI congres R.C.Mediterranean Neogen Stratigraphy, Bratislava.

Livental, V. E. 1938. Sediments of Baku stage and their microfauna. Trudy Az.nauch.-issled.Instituta 1.

Mandelstam, M. I., Markova, L. P., Rozyeva, T. R. & Stephanaitys, N. E. 1962. Ostracoda of Pliocene and Pleistocene sediments from Turkmenistan. Izd. AN TurkSSR, Ashkhabad.

Markova, L. P. 1959. New ostracode species of family Cytheridae from Apsheronian sediments of west Turkmenia. Trudy Inst.geol. AN TurkSSR 2.

Naídina, N. N. 1970. Composition and distribution of ostracoda in North Caspian Sea. Kompleks.issled.Kasp.m. 1, Moskva.

Naídina, N. N. 1972. On a mediterranean genus more, living in the Caspian Sea. Kompleks.issled.Kasp.m. 3, Moskva.

Pokorný, V. 1955. Contribution to the morphology and taxonomy of the subfamily Hemi cytherinae Puri Acta Univ.Car.,Geol. 3, Praha.

Ruggieri, G. 1955. Tyrrhenocythere, a new recent ostracode genus from the Mediterranean; Journ.pal. 29, Menasha.

Sars, G. O. 1888. Nye Bidrag til Kundskaben om Middelhavets Invertebratefauna. Arch.Math.Naturvid., Kristiania.

Sinegub, V. V. Ostracode determination. In: Negadaev-Nikonov K. N. & Polev P. V. 1968. Regional stratigraphy of Moldavian SSR. Izd. AN MoldSSR, otd.pal.strat., Kishinev.

Stancheva M. 1965. Ostracoda from the Neogene in North-Western Bulgaria, IV. Pontian ostracodes. Trav.geol.Bulg.,ser.pal. 7, Sofia.

Stevanović, P. 1969. Impressions from the IV congress of Commite on Mediterranean Neogene Stratigraphy in Bologna. Compt.rend. seanc.soc.Serbe geol. 1967, Beograd.

Suzin, A. V. 1956. Ostracoda of the Tertiary sediments of northern For-Caucasus. Grozn. neft.inst., 'Gostoptehizdat', Moskva.

Schweyer, A. V. 1949. Fundamentals on the morphology and systematic of Pliocene and Pleistocene ostracoda. Trudy VNIGRI, n.ser., 10, Moskva – Leningrad.

Shornikov, E. I. 1969. Subclassis Ostracoda, in: Determinating-book for the fauna of Black and Azov Seas, II. Free living invertebrata – Crustacea. 'Naukova dumka', Kiev.

Vekua, M. L. 1965. New ostracode species from Kimmerian sediments of Abkhazia. Soonsch. AN GruzSSR.

Author's address:

Geoinstitute, Rovinjska 12, P.O.Box 158, Belgrade, Yugoslavia.

Sixth Intern. Ostracod Symposium, Saalfelden

LA CONQUETE DES MILIEUX DE PLATEFORME CONTINENTALE PAR L'ENSEMBLE CARINOCYTHEREIS ANTIQUATA/CARINATA DEPUIS LE MIOCENE MOYEN

P. CARBONEL

Abstract

At the present time, the *C. carinata* and *C. antiquata* group are living on the continental shelf of the Mediterranean sea and Atlantic Ocean. We find the first in the infralittoral zone and the second in the circalittoral (offshore) zone of the continental shelf.
  During the late Miocene, *C. carinata* appears for the first time in the mediterranean region, and arrives in the Atlantic ocean at the beginning of the Pliocene time: it is limivore form very near of the littoral shore.
  Then *C. antiquata*, a species well adapted to deeper environment, appears during middle Pliocene in the mediterranean basin, and only during Quaternary in the east Atlantic ocean.

L'ensemble qui fait l'objet de cette communication présente pour nous un double intérêt: sa très large distribution actuelle dans les bassins méditerranéen et atlantique oriental permet une étude éthologique à grande échelle; par ailleurs, sa répartition stratigraphique couvrant presque tout le Néogène amène à considérer sa progression depuis son foyer originel méditerranéen. La subdivision du groupe en deux types constants nous conduit enfin à examiner l'évolution possible de celui-ci.
  Nous articulons notre étude en trois parties:
- Distribution géographique actuelle
- Répartition stratigraphique
- Modalités de la migration et de l'évolution du groupe

Pour plus de commodité dans la suite de l'exposé, nous gardons la subdivision adoptée en 1971 (Carbonel & Moyes), à savoir: les formes à côte ventrale se prolongeant dans la partie antérieure sont groupées sous la dénomination *'carinata'*; les formes à côte seulement ventrale, sous le terme *'antiquata'*, ce caractère étant la seule constante que nous ayons notée.

1. **Distribution géographique actuelle** (Fig. 1)

Les deux groupes ne colonisant pas les mêmes biotopes (Carbonel 1973, Breman 1975), nous donnons leur répartition séparément.
– le groupe *'antiquata'* se rencontre sur les côtes atlantiques dans les zones infralittorale externe (80-110 m) et circalittorale (110-180/200 m) (Carbonel 1973): côtes occidentales et septentrionales des îles britanniques (Baird 1850, Brady 1868, Whatley & Wall 1969), en Manche occidentale, dans le Golfe de Gascogne (Fischer 1878, Moyes & Pachier 1968, Yassini 1969, Peypouquet 1970, Carbonel & Moyes 1971, Carbonel 1973).
  En Méditerranée, on le trouve sur les côtes occidentales de Grèce (Uliczny

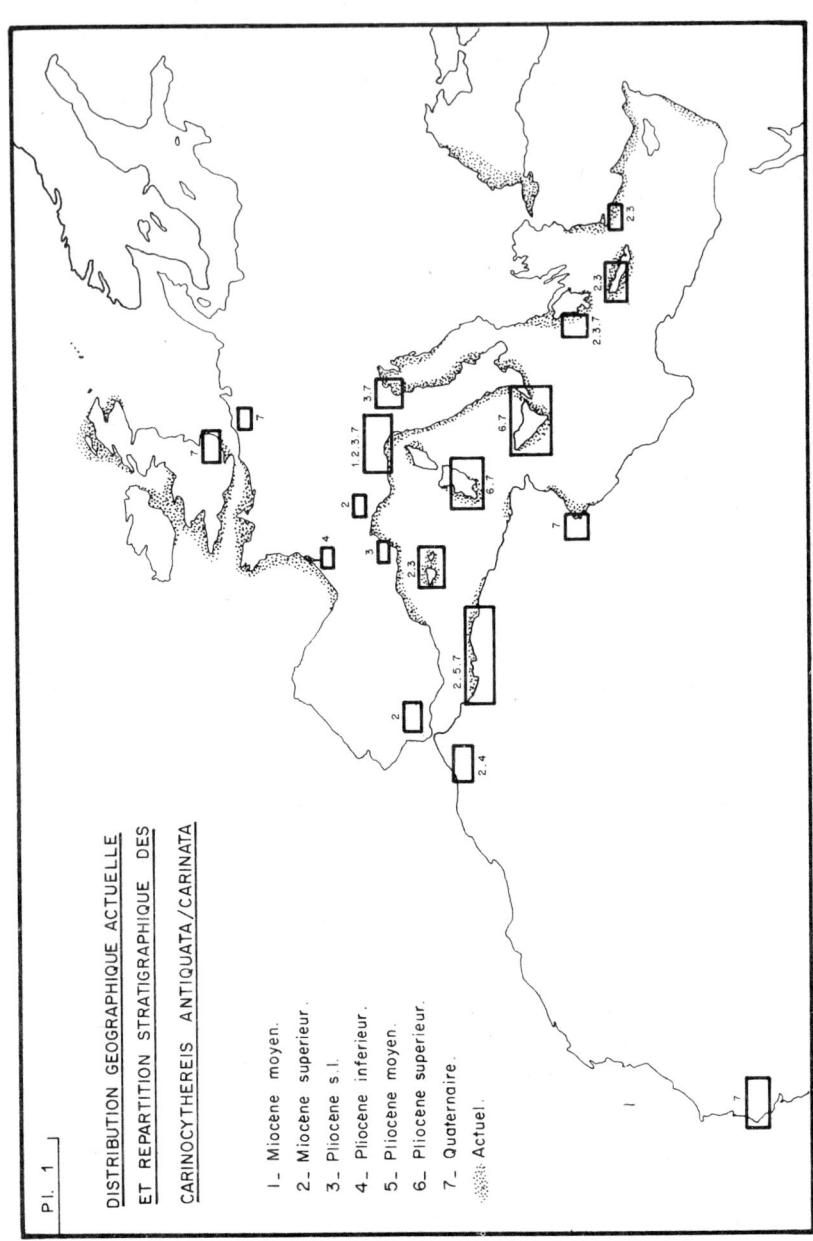

1969), en Adriatique (Uffenorde 1972, Breman 1975), probablement en baie de Naples (G. W. Mueller 1894), toujours au-delà de la zone infralittorale interne.
– le groupe *'carinata'* que l'on rencontre depuis la zone côtière jusqu'à 90 m de profondeur a une répartition géographique sensiblement analogue en domaine atlantique Cependant, il est signalé en de plus nombreux points sans doute en raison de son aire de distribution plus facile à échantillonner. Quelques points

Tabl: 1		Miocène moyen	Miocène sup.	Pliocène inf.	Pliocène moyen	Pliocène sup.	Pleistocène	Holocène
					MEDITERRANEE			
C. carinata / C. antiquata	PIEMONT - LIGURIE							
	ITALIE - SICILE							
	SUD ESPAGNE							
	ADRIATIQUE - VENETIE							
	MEDITERRANEE CENTRALE (GRECE)							
	MEDITERRANEE ORIENTALE							
	MER NOIRE							
	AFRIQUE NORD							
					ATLANTIQUE			
	AFRIQUE OCCIDENTALE							
	GOLFE de GASCOGNE							
	ILES BRITANNIQUES							

supplémentaires se situent sur les côtes bretonnes (Rouvillois & Boulanger 1962) et au débouché de la Tamise (Kilenyi 1969).

En Méditerranée, il semble par contre beaucoup plus abondant que le groupe *antiquata*. En partant des côtes espagnoles jusqu'aux côtes d'Afrique du Nord, on le signale dans la région de Valence, sur les côtes françaises (Kurc 1961, Rome 1939, 1942), sur les côtes italiennes, sardes et siciliennes (Ruggieri, pub. div., Mueller 1894, Puri et al. 1969,...), sur les côtes grecques occidentales (Uliczny 1969), en Adriatique (Uffenorde 1972, Breman 1975), sur les côtes égéennes, de la mer de Marmara, de Turquie, de Syrie (Brady 1875), en mer Noire (Marinov 1964), dans le Golfe de Gabès, en baie d'Alger (Yassini à paraître).

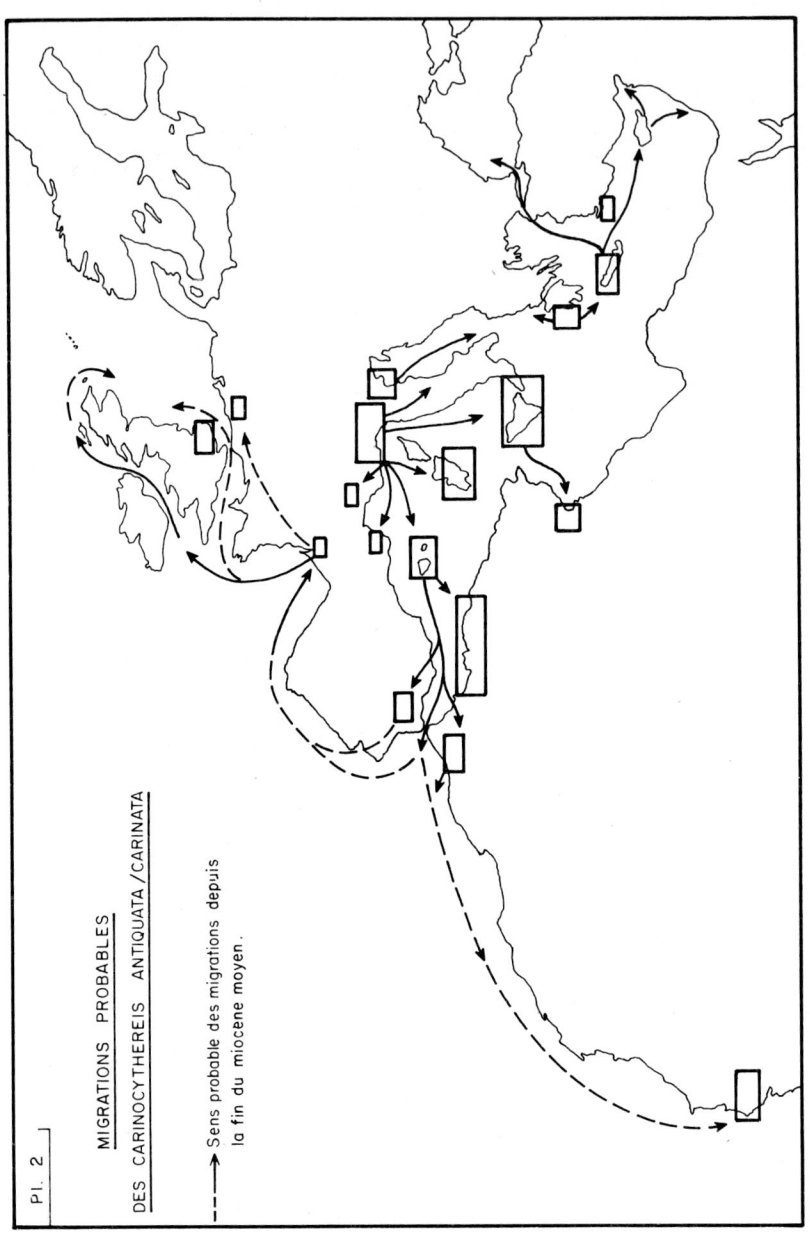

il est enfin intéressant de noter l'absense totale, à notre connaissance, des deux groupes en Manche orientale, en mer du Nord (sauf côtes britanniques) et en mer Baltique ainsi que dans une grande partie de la Méditerranée du sud (côtes d'Egypte et de Libye).

Pour Yassini (1969) qui ne sépare pas les deux groupes, il s'agit 'd'une espèce boréale de l'Atlantique Oriental et de la Méditerranée'.

2. Repartition stratigraphique (tableau)

Le groupe *antiquata* est signalé pour la première fois avec certitude dans le Pliocène moyen de l'île de Céphalonie (Grèce) (Uliczny 1969), ensuite dans le Pliocène supérieur de Rhodes et de Crète (Sissingh 1973). En domaine atlantique, ce groupe ne semble pas apparaître avant le Quaternaire.

Le groupe *carinata*, quant à lui, fait sa première apparition à la fin du Miocène moyen dans le Piémont (Capeder 1902). Dès le Tortonien, il est signalé en Méditerranée occidentale (Capeder 1902, Ruggieri pub. div., Dieci & Russo 1964), dans le bassin du Rhône (Carbonnel 1969), en Méditerranée centrale (Uliczny 1969), en Méditerranée orientale (Sissingh 1972). A la fin du Miocène le groupe se retrouve, en plus des régions déjà citées, dans le Sud de la Méditerranée (Guardia, Magne & Moyes 1971), en Espagne, à Carmona (Sissingh 1972), dans le bassin du Guadalquivir (Borragan 1964) et au Maroc septentrional, à Dar Bel Hamri. Au Pliocène, tout le domaine méditerranéen littoral est colonisé à de rares exceptions près: côtes italiennes (Roemer 1838, Capeder 1900, Namias 1900) côtes du Roussillon (Keij 1957), Algérois (Yassini l.c.), Méditerranée orientale (Terquem 1878).

C'est au Pliocène inférieur que le groupe *carinata* apparaît pour la première fois en domaine atlantique (Moyes 1965); vers la fin du Pliocène, on le trouve déjà dans le 'Scaldisien' de Belgique et dans le 'Crag' anglais (Jones 1856). Au Quaternaire, le groupe est mentionné dans presque toute la Méditerrannée (Ruggieri 1950, 1952, 1973, Masoli 1968, Uliczny 1969). En domaine atlantique, on le signale sur les côtes des îles britanniques (Brady, Crosskey & Roberton 1874); vers le sud, on le retrouve à la base de l'Holocène au niveau des côtes du Sénégal (Peypouquet 1976).

3. Origine, migration et prolifération probables du groupe

Deux faits particuliers se dégagent de cette étude:
– en premier lieu (Fig. 2), l'ensemble des données stratigraphiques montre, à notre connaissance, une origine miocène et méditerranéenne occidentale de l'ensemble *Carinocythereis carinata/antiquata* il y a environ 12 millions d'années; il est à noter que seul le groupe *carinata* apparaît. Très vite, le domaine méditerranéen est conquis dans sa quasi-totalité, Dès la fin du Miocène, la progression vers l'ouest est marquée par la présence d'individus dans les détroits nord-bétique et sud-rifain par lesquels le groupe va effectuer son passage en atlantique. Par la suite, les détroits s'étant fermés, le passage va se poursuivre par un détroit de Gibraltar plus ouvert qu'actuellement après l'effondrement du massif bético-rifain (Choubert et al. 1963).

Une fois en atlantique, le groupe va essaimer très rapidement, puisqu'on le trouve au Pliocène en Belgique et en Angleterre du sud; au Quaternaire, et jusqu'à nos jours, sa limite nord-orientale reflue vers le sud (il a disparu de Manche orientale et des côtes belges et hollandaises) alors que sa limite méridionale n'a cessé de s'étendre: on le trouve à l'Holocène sur les côtes sénégalaises.

En résumé, on peut observer deux phases dans la migration: en premier lieu, il y a conquête du domaine méditerranéen à partir d'un foyer vraisemblablement Méditerranéen occidental au Miocène moyen; à la fin du Miocène supérieur, les premiers représentants du groupe passent en domaine atlantique qu'ils colonisent ensuite pour atteindre leur extension maximale vers le nord au Quaternaire, vers le sud, peut-être progressent-ils encore actuellement.

– le second point intéressant est la diversification de l'ensemble en fonction des biotopes colonisés. A partir d'un type, *carinata* s. l., lié à un milieu côtier, voire périphytal, éthologie confirmée par l'ensemble des autres Ostracodes du même niveau, on observe une prolifération d'écotypes divers, en particulier en Méditerranée centrale, (Uliczny 1969) montrant des variations des tubercules, des côtes, de la taille, mais gardant toujours le type '*carinata*'. Toutes ces formes sont relativement stables car on les retrouve de nos jours; elles sont certainement le résultat d'une réponse de l'animal à des modifications chimiques du milieu.

En revanche, dès que le groupe gagne les domaines plus profonds du plateau continental, à partir du Pliocène moyen, le type architectural '*carinata*' se modifie pour devenir un type '*antiquata*' dépourvu de côte antérieure et au test plus convexe. Ce type est très stable et, contrairement aux formes du groupe *carinata*, ne présente que peu de variations probablement à cause de la plus grande stabilité des conditions de biotope. La différenciation des deux groupes semble donc bien liée à une différence d'environnement; nous avions donné comme raison possible l'augmentation de profondeur donc de pression subie par les tests (Carbonel 1973). Breman (1975) dans son étude des populations de la mer Adriatique observe le même phénomène et l'assujétit à la différence de température et de transparence des eaux. Des facteurs chimiques tels la matière organique ou les carbonates, dont la répartition pourraît être parallélisée à celle des *Carinocythereis* en Adriatique ne montrent aucune relation en domaine atlantique. Nous conservons donc pour le moment cette interprétation qui, pour générale qu'elle paraisse, est la seule qui concorde dans tous les cas. Sur un plan évolutif, nous pensons que le type *antiquata* est un accomodat de la forme ancestrale *carinata* adapté à un mode de vie différent; c'est pourquoi, bien qu'il soit classique de garder les deux dénominations, il semble bien qu'il s'agisse de deux écotypes d'une même espèce.

Le point important qui reste à déterminer est celui du passage des *antiquata* en Atlantique. On connaît les dates approximatives d'apparition de ces formes dans les deux domaines, mais on ignore si leur arrivée en Atlantique est consécutive à une seconde migration ou bien si l'adaptation au milieu circalittoral est une convergence identique dans les deux domaines. Dans l'état actuel de nos connaissances, nous n'avons pas les éléments nécessaires pour répondre à cette question.

En résumé, trois points ont retenu notre attention:

– l'homogénéité de l'ensemble qui en fait des formes aisées à repérer.

– son aptitude à coloniser tous les milieux de plateforme continentale, côtiers d'abord, circalittoraux ensuite, constituant très vite une proportion notable de la population.

— sa migration qui en une dizaine de millions d'années lui a permis de gagner de proche en proche des zones aussi différentes que les côtes d'Ecosse, la mer Noire ou les côtes sénégalaises.

Le devenir de ces formes montre comment, sans grandes modifications apparentes une espèce peut coloniser tous les milieux susceptibles de convenir à son mode de vie limivore.

Résumé

Les représentants de cet ensemble se rencontrent actuellement dans la plupart des sédiments marins fins des plateaux continentaux sur le pourtour de la Méditerranée et dans l'Atlantique oriental. Les deux groupes composant cet ensemble se succèdent sur le plateau continental de l'infralittoral au circalittoral.

Il n'en a pas toujours été de même depuis la fin du Miocène moyen, date à laquelle *C. carinata* est signalé pour la première fois. Cette apparition a lieu en Méditerranée. Le passage en Atlantique se fait au Pliocène inférieur. Ces premières formes ont un mode de vie limivore très côtier; peu à peu, va se différencier la forme *antiquata,* adaptée aux milieux plus profonds du plateau continental: apparue au Pliocène moyen en Méditerranée, on ne rencontre cette forme qu'au Quaternaire en Atlantique.

Bibliographie

Baird, W. 1850. The natural history of british Entomostraca. Roy. Society London, 355 pp.
Brady, G. S. 1868. A monograph of the recent british Ostracoda. Trans. Soc. Linn. London 26 (2): 353-495.
Brady, G. S., Crosskey, H. H. & Robertson, D. 1874. A monograph of the post-tertiary Entomostraca of Scotland, including species from England and Ireland. Paleont. Soc. London, vol. 4, pp. 122-232.
Brady, G. S. 1875. In: Defolin & Périer, Les Fonds de la mer. Savy Libraire Editeur, Paris.
Breman, E. 1975. The distribution of Ostracodes in the bottom sediments of the Adriatic sea. Thèse, Amsterdam, 165 pp.
Capeder, G.1900. Contribuzione allo studio degli Entomostraci dei terreni Pliocenici des Piemonte e della Liguria. Atti R. Acc. Sci. Torino 35 (1a): 60-73.
Capeder, G. 1902. Contribuzione allo studio degli Ostracodi fossili della terreni Miocenini del Piemonte. Atti R. Acc. Sci. Torino 35: 5-18.
Carbonel, P. & Moyes, J. 1971. A propos du groupe spécifique Carinocythereis gr. carinata (Roemer). Rev. Esp. Micropal. III:147-154.
Carbonel, P. 1973. Répartition des Carinocythereis dans le Golfe de Gascogne. Rev. Esp. Micropal V, 3:431-433.
Carbonnel, G. 1969. Les Ostracodes du Miocène rhodanien. Doc. Labo. Geol. Fac. Sci. Lyon, n° 32, 469 pp.
Dieci & Russo, 1964. Ostracodi tortoniani delli Appennino settentrionale. Boll. Soc. Pal. Ital. 3:38-88.
Fischer, P. 1878. Crustacés, Ostracodes marins des côtes du Sud-Ouest de la France. Act. Soc. Linn. Bordeaux 31 (4° ser): 241-249.
Guardia, P., Magne, J. & Moyes, J. 1971. Aperçu sur le Néogène autochtone de l'ouest oranais in Vème Congrès du Néogène Méditerranéen–Lyon 1971. Mém. B.R.G.M. 78 (III): 691-703.

Jones, A. J. 1957. A monograph of the tertiary Entomostraca of England. Paleont. Soc. London 1856, 63 pp.
Keij, A. J. 1956. Eocène and Oligocène Ostacode of Belgium. Mém. Instit. Sci. Nat. Belgique, 136:1-210.
Kilenyi, T. I. 1969. The problems of Ostracod ecology in the Thames estuary in the taxonomy and ecology of recent Ostracoda. Ed. J. W. Neale. Oliver and Boyd London, pp. 251-265.
Kurc, G. 1961. Foraminifères et Ostracodes de l'étang de Thau. Rev. Trav. I.S.T.P.M. 25 (2): 135-247.
Marinov, T., 1964. Untersuchungen über die Ostracodenfauna des schwarzen Meeres. Kiel. Meeresf. 20 (1): 82-91.
Masoli, M. 1968. Ostracodi recente dell Adriatico settentrionale Venezia e Triste. Mem. Mus. Tridentino Sci. Nat. XXXI – XXXII 17 (1):1-100.
Moyes, J. 1965. Les Ostracodes du Miocène aquitain: essai de paléoécologie stratigraphique et de paléogéographie. Thèse Doc. Sci. Nat. Bordeaux Imp. Drouillerd 339 pp.
Moyes, J. & Pachier, E. 1968. Répartition de quelques associations fauniques d'Ostracodes dans le Golfe de Gascogne. Bull. Inst. Geol. Aquit. 5:84-86.
Mueller, G. W. 1894. Ostracoden des Golfes von Neapel und der angrenzenden Meeresabschnitt. Fauna und Flora des Golfes von Neapel 21, 403 pp.
Peypouquet, J. P. 1970. Les Ostracodes de la région de Cap-Breton. Intérêt écologique et paléoécologique. Thèse 3è cycle Univ. Bordeaux, 266 pp.
Peypouquet, J. P. 1976. Les Ostracodes, indicateurs paléoclimatiques et paléogéographiques du Quaternaire terminal (Holocène) sur le plateau continental sénégalais. (à paraître).
Puri, H. S., Bonaduce, C. & Gervasio, A. M. 1969. Distribution of Ostracoda in the Mediterr sea. Edit. J. W. Neale, Oliver and Boyd, Edinburgh, pp. 356-409.
Roemer. 1838. Die Cytherinen des Molasse-Gebirges. Neues jahrb. f. Min. Geogr. Geol. Petrof. Stuttgart fasc. 36:514-519.
Rouvillois, A. & Boulanger, D. 1962. Un sable calcaire sur la côte Nord de la Bretagne à Primel-Tregastel (Finistère). C.R.A.S. Paris 254 (10): 1848-1849.
Ruggieri, G. 1950. Gli Ostracodi delle sabbie grigie quaternarie (Milazziano) di Imola. Giorn. di Geologia, ser. 2, 21° 1-57.
Ruggieri, G. 1952. Gli Ostracodi delle sabbie grigie di Imola. Extr. Gior. Geol. Bologna, ser. 2, 22:59-115.
Ruggieri, G. 1958. Alcuni Ostracodi del Neogene italiano. Atti soc. Ital. Sc. Nat. Mus. Civ. St. Nat. Milan WCVII (2):127-146.
Sissingh, W. 1971. Tricostate Trachyleberidinae (Ostracoda) from Neogen-Recent deposits of Europa. Proc. Kon. Ned. Akad. Wetensch. ser. B 74 (2):195-205.
Sissingh, W. 1972. Late Cenozoïc Ostracoda of the south Aegean Island arc. Utrecht Micropal. Bull. 6, 187 pp.
Sissingh, W. 1973. The Ostracode Fauna of the type Calabrian Deposits at St. Maria di Catanzaro (S. Italy) Newsl. Strat. Leiden 3 CD:25-44.
Terquem, O. 1878. Les foraminifères et les Entomostracés du Pliocène supérieur de l'île de Rhodes. Mém. Soc. Geol. France, 3 ser., 1, (1878):1-136.
Uffenorde, H. 1972. Oekologie und jahreszeitliche Verteilung rezenter benthonischer Ostracoden des Limski Kanal bei Rovinj. Gotting. Arb. Geol. Pal. 13, 121pp.
Uliczny, F. 1969. Hemicytheridae und Trachyleberididae (Ostracoda) aus Pliozän der Insel Kephallinia (Westgriechenland). Druck und Verlag, Typo-Druck-Dienst Scheffel 8 München, pp. 1-152.
Whatley, R. C. & Wall, D. R. 1969. A preliminary account of the ecology and distribution of recent Ostracoda in the southern Irish sea. In: The taxonomy and ecology of recent Ostracoda. Ed. by J. W. Neale, Olivier and Boyd, London, pp. 268-298.
Yassini, I. 1969. Ecologie des associations d'Ostracodes du Bassin d'Arcachon et du littoral atlantique. Application à l'interprétation de quelques populations du Tertiaire aquitain. Bull. Inst. Geol. Bass. Aquit. 7 (1969):288 pp.

Adresse de l'auteur:

Laboratoire de Géologie et Océanographie, Université de Bordeaux I, Avenue des Facultés, 33405 Talence, France.

Discussion

Carbonnel: Etes-vous convaincu de la détermination de *Carinocythereis carinata* s.s. au Miocène, ce qui semble en contradiction avec les travaux d'Uliczny en Grèce? Celui-ci a en effet créé de nombreuses espèces ayant pour lui valeur stratigraphique (basée sur les Foraminifères planctoniques). Celles-ci sont en effet très importantes pour nous en Micropaléontologie stratigraphique.

Carbonel: J'en suis s'autant plus convaincu que parmi les cinq espèces de *Carinocythereis* (dont deux nouvelles) que cite Uliczny, *C. carinata* s.s. est la seule qu'il trouve dès le Tortonien (Uliczny 1969, tabl. 2, p. 114-115). D'autre part, je ne sais pas trop quelle valeur stratigraphique il peut donner à ces diverses 'espèces'; en effet, si l'on excepte *C. carinata* s.s. et *C. bairdii* qui sont très proches, les trois autres formes apparaissant au Pliocène moyen et les cinq subsistent de nos jours. Il ne me semble donc pas y avoir contradiction entre mon hypothèse et les travaux d'Uliczny dans la mesure où je note l'apparition de *C. carinata* dès la fin du Miocène moyen et l'apparition de *C. antiquata* au Pliocène moyen faisant référence entre autres, précisément à cette publication.

Carbonnel: Comment sont obtenues vos données concernant *Carinocythereis carinata* en Europe mésogéenne au Mio-Pliocène: informations bibliographiques, ou nouvelles observations des individus déterminés comme tels?

Carbonel: Je n'ai pu voir, bien entendu, toutes les formes dont je parle ici. Je n'ai pu observer en particulier les faunes de Méditerranée centrale et orientale ainsi que celles de la Mer Noire (Uliczny-Sissingh-Marinov-Terquem...) Je me suis donc appuyé sur des photographies, des dessins ou des descriptions. En revanche, j'ai pu observer directement la plupart des faunes du domaine mésogéen occidental connues (Italie du Nord – Eapagne – Algérie) ou nouvelles (Maroc – Sénégal); ces dernières étant plus directement en relation avec le domaine atlantique qu'avec le domaine mésogéen.

Colin: J'ai trouvé *Carinocythereis carinata* dans du Pliocène au Portugal.

Carbonel: La présence de *C. carinata* dans cette région est un jalon supplémentaire montrant la migration du domaine mésogéen au domaine atlantique.

Uffenorde: Do you have an explanation why *Carinocythereis quata* lives in the Adriatic sea in shallower habitats than in the Atlantic Océan?

Carbonel: Il n'existe pas actuellement d'explication bien démontrée à ce phénomène. Cependant, on peut remarquer que tout au long de la côte adriatique occidentale, la succession *C. carinata – antiquata + carinata – antiquata* existe et au large de la région de Bari, elle se développe aux mêmes profondeurs qu'en Atlantique (Breman, fig. 81). On assiste donc dans la partie septentrionale de l'Adriatique à une remontée de la faune de *Carinocythereis* puisque dans cette région il n'y a pas d'étage circalittoral (100-200 m). On pourrait peut-être lier ce fait à la présence 'd'upwellings' de printemps que signale Breman (p. 7 et fig. 3). D'ap-

rès la carte de Breman (fig. 3), ces 'upwellings' n'auraient pas lieu au large de Bari.
Uffenorde: Couldn't this be an example for the 'Relatively of the bionomic depth zones' in the sens of Schmidt H. (1935)?
Carbonel: Je n'ai pas eu connaissance des travaux de Schmidt H., mais ayant lu ce que vous en citez dans votre thèse (p. 3-33-34) il est fort possible que ce soit le cas ici. De telles remontées de faune circalittorale ou profonde dans les étages supérieurs existent ailleurs (Golfe de Gascogne en particulier).

Sixth Intern. Ostracod Symposium, Saalfelden

LA FAUNE D'OSTRACODES DES DEPOTS TERTIAIRES DU PLATEAU CONTINENTAL DANS LA PARTIE CENTRALE DU GOLFE DE GASCOGNE: INTERET PALEOECOLOGIQUE – RELATIONS AVEC LE CONTINENT

O. DUCASSE

1. **Introduction**

Les coupes géologiques utilisées dans cette étude appartiennent à huit forages réalisés pour la recherche pétrolière par la société ESSO – REP. Ils se situent sur la plateforme océanique des Landes de Gascogne, à une dizaine de kilomètres de la côte, dans un domaine compris entre la latitude du Bassin d'Arcachon au Nord et celle de Mimizan au Sud (Fig. 1). Ils sont implantés suivant un alignement de

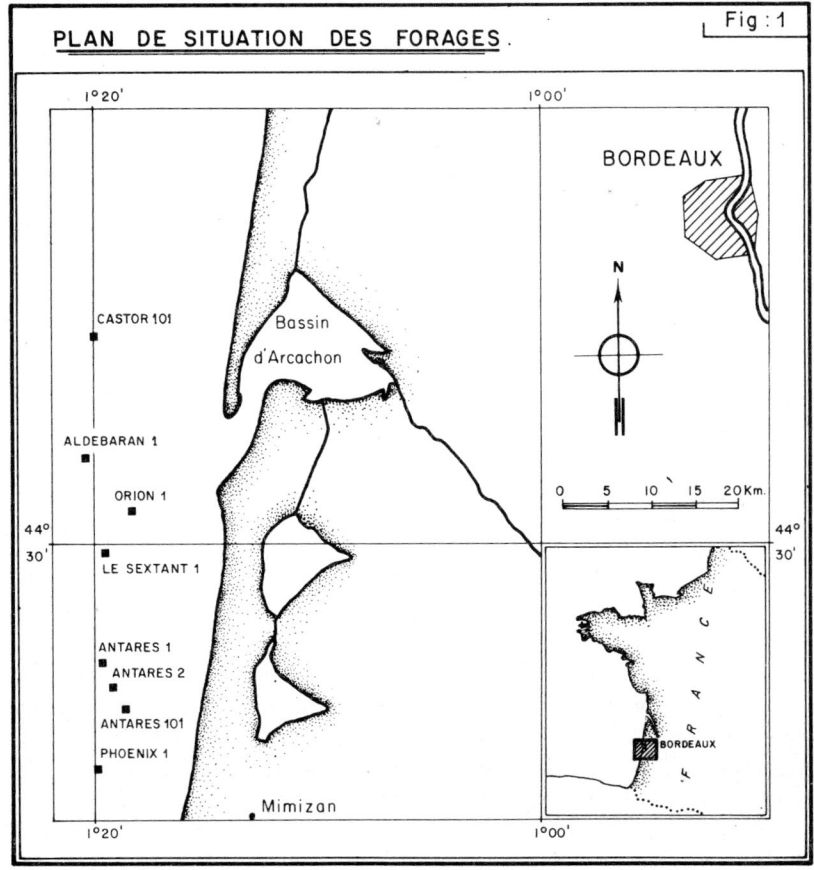

direction approximativement Nord-Sud, par des profondeurs d'eau n'excédant pas 100 m.

Les échantillons examinés sont des cuttings, conservés au Service Géologique Régional Aquitaine du B.R.G.M.; malheureusement ces 'déblais' sont parfois peu abondants et échantillonnés, le plus souvent, de façon assez lâche.

L'intérêt que présente cette étude est de nous permettre de parfaire notre connaissance du Tertiaire dans une région située à l'Ouest de l'actuel Bassin d'Aquitaine, en milieu plus marin, et de mieux suivre l'évolution qui est susceptible de se produire à l'intérieur d'un bassin, de la partie centrale vers les zones de bordure.

2. Résultats

Dans l'ensemble de ces coupes, nous avons examiné la microfaune d'Ostracodes en provenance de 224 niveaux et nous avons recueilli 132 espèces distribuées entre 57 genres.

Sur le plan stratigraphique, trois entités ont pu être reconnues: le Miocène, l'Oligocène et l'Eocène représenté en grande partie par des couches d'âge Eocène supérieur.

2.1. *Caractéristiques générales de la faune au Miocène*

Cette période est caractérisée par une association où sont principalement représentées les *Ruggieria*: *R. tetraptera, tetraptera* Seguenza, *R. tetraptera bicostata, R. nuda* Moyes, *Cytherella, Loxoconcha*: *L. parvula* Moyes, *Aurila*: *A. punctata* (Münster), *A. oblonga* Moyes, *Cytheretta*: *C. pulchra* Moyes, *Falunia*: *F. plicatula* (Reuss), *Bosquetina*: *B. carinella* (Reuss), *Bythocypris, Henryhowella*: *H. asperrima* (Reuss), *Eucytherura, Costa*: *C. punctatissima* Ruggieri, *Xestoleberis*: *X. glabrescens* (Reuss), *Echinocythereis*: *E. scabra* (Münster), *Microcytherura*: *M. angulosa* (Seguenza).

A ces formes se surajoutent quelques *Pterygocythereis, Cushmanidea, Cytherura* (s.l), *Buntonia, Krithe, Leguminocythereis, Bairdoppilata, Callistocythere, Kangarina, Neocytherideis, Occultocythereis*.

Une telle association caractérise la zone circalittorale de la plateforme continentale et présente de fortes affinités avec celles des séries connues dans la partie la plus occidentale de l'Aquitaine (Moyes 1965) (fig. 2).

2.2. *A l'Oligocène*

La faune est très homogène, essentiellement constituée de *Cytherella*: *C. consueta* Deltel, *C. transversa* Speyer, *Krithe* à incision postérieure: *K. luyensis, K. parvula, K. angusta* Deltel et *K*. cf. *caudata* Bold in Deltel, *Protoargilloecia*: *P. angulata* Deltel, *P.* sp. 2 Moyes, *Henryhowella*: *H. asperrima* (Reuss), *Bairdia*: *B. crebra* Deltel, *Pontocyprella*: *P.* sp. 1 Moyes, *Echinocythereis*: *E. scraba* (Münster). Ces formes représentent l'ossature de la faune.

Parmi les autres espèces associées, nous citerons des *Costa*: *C. tricostata* (Reuss), *Xestoleberis*: *X.* sp. 1 Ducasse, *Krausella*: *K. fulgens* Deltel, des *Krithe* dé-

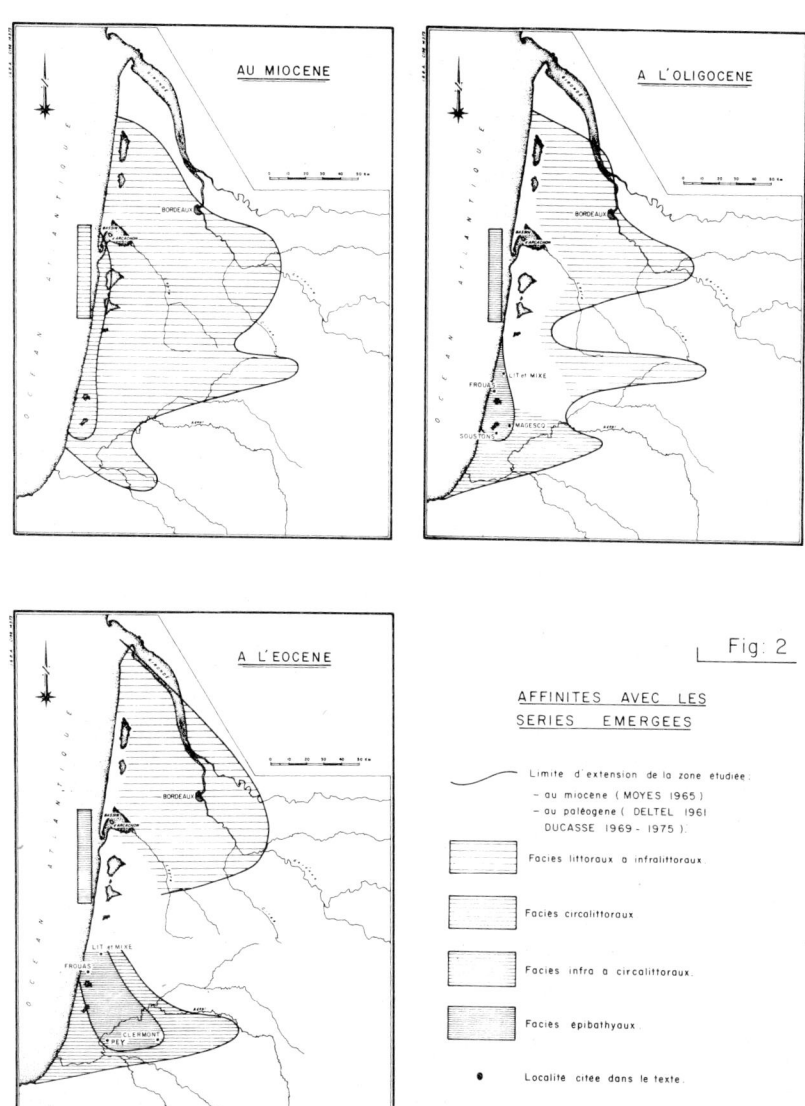

Fig: 2

AFFINITES AVEC LES SERIES EMERGEES

pourvues d'incision postérieure : *K* sp. 3 Moyes, des *Trachyleberidea* : *T. prestwichiana* (Jones & Sherborn), *Hermanites* : *H. paijenborchiana* Keij, *Agrenocythere* : *A. ordinata* (Deltel), *Cytheropteron* : *C. alveiformis* Deltel, *C.* cf. *tricorne* Bornemann in Deltel, *Loxoconcha* : *L. delemontensis* Oertli.

Nous mentionnerons la présence de nombreuses formes, souvent très rares, représentant des apports en provenance des zones sublittorales du plateau continental et l'apparition des *Buntonia* dans le Paléogène aquitain. Elles sont représentées par une seule espèce indéterminée. Il s'agit d'une petite forme, grossiè-

419

rement ponctuée, qui rappelle *B. sublatissima sublatissima* (Neviani) signalée par G. Ruggieri dans le Pliocène et le Quaternaire (?) d'Italie, par J. Moyes dans le Miocène sud-aquitain et par J. P. Peypouquet en quelques stations de la région de Capbreton, entre –200 et –700 m. Elle en diffère par l'absence de côtes saillantes dans la région médio-ventrale.

L'ensemble faunique témoigne de faciès épibathyaux, comparables à ceux que nous rencontrons, à la même époque, en Aquitaine méridionale, à Frouas, Lit-et-Mixe, Magescq et Soustons (Ducasse 1974) (Fig. 2).

2.3. A l'Eocène

Nous retrouvons sensiblement le même type d'association faunique qu'à l'Oligocène. Toutefois, si nous considérons la distribution des formes d'origine profonde au Paléogène, nous remarquons que la plupart d'entre elles sont mieux développées à l'Eocène qu'à l'Oligocène et il semble également que ce soit dans le secteur sud (Phoenix 1, Antarès 101) qu'elles sont toujours le mieux représentées (Fig. 3).
– 17 espèces existent à l'Eocène et l'Oligocène dans toutes les coupes étudiées.
– Une dizaine d'autres ont une distribution géographique plus restreinte et d'autant plus restreinte que l'on s'élève dans l'échelle stratigraphique.

A l'Eocène, quelques *Pontocyprella aturica* Deltel et *Cytherella* sp. 1 Deltel se retrouvent dans le secteur sud de Phoenix 1, Antarès 101 et Antarès 2. De rares *Saida* ont été signalées, un peu plus au Nord, jusque dans le sondage d'Orion 1.

Des *Bairdia cymbula* Deltel, *Agrenocythere ordinata* (Deltel) et *Trachyleberidea prestwichiana* (Jones & Sherborn) sont représentées jusqu'à la latitude d'Aldebaran 1 où a été trouvé un *Abyssocythere*. Des *Protoargilloecia* sp. 2 Moyes et des *Macrocypris wrightii* Jones & Hinde in Deltel sont connues dans toutes les coupes.

Trois de ces formes persistent seulement à l'Oligocène. Leur développement, en direction du Nord. est limité à la latitude du Sextant. Ce sont : *Agrenocythere ordinata* (Deltel) localisée, désormais, dans le secteur sud de Phoenix 1 et Antarès 101. *Trachyleberidea prestwichiana* (Jones & Sherborn) et *Protoargilloecia* sp. 2 Moyes représentées jusque dans le sondage du Sextant.

Nous avons noté la présence d'un *Abyssocythere* dans la coupe d'Aldebaran 1.Ces formes qui sont connues actuellement à de grandes profondeurs, à -2000 et souvent à -3000 m, n'ont été signalées jusqu'ici, en Aquitaine, qu'à l'Eocène, en deux points situés respectivement:
– dans le Sud du Bassin, dans l'axe du gouf de Capbreton, à Pey,
– et sur la plateforme continentale, en direction du Nord-Ouest, dans l'axe du canyon du Cap-Ferret, à Aldebaran.

Nous pensons:
– qu'elles ont peuplé l'Atlantique nord à l'époque considérée,
– qu'elles ont vécu dans un domaine situé à l'Ouest de celui que nous avons jusqu'ici prospecté et plus profond, mésobathyal à abyssal,
– et que les rares individus recueillis ont pu pénétrer accidentellement vers l'Est à la faveur des deux grandes brêches que représentent dans le talus continental le gouf de Capbreton et le canyon du Cap-Ferret.

Ce type d'assemblage faunique se retrouve en Aquitaine méridionale à l'Eocène supérieur à Frouas et Lit-et-Mixe et à l'Eocène moyen à Pey et Clermont (Ducasse 1975) (Fig. 2).

3. Conclusions

De cette simple étude de la faune d'Ostracodes, il ressort que la région étudiée devait se situer au Paléogène moyen et superieur sur la pente continentale aquitaine, prolongeant ainsi, en direction du Nord-Ouest, le domaine épibathyal que nous avons reconnu en Aquitaine méridionale. La profondeur semble avoir été légèrement plus faible à l'Oligocène qu'à l'Eocène et toujours un peu plus importante dans la zone sud de Phoenix et Antarès. Partout ailleurs, dans le Bassin d'Aquitaine, s'étendait une vaste plateforme continentale, sur laquelle se différenciaient des milieux littoraux à infralittoraux dans le Nord et sur la bordure Est du bassin, des milieux infra à circalittoraux dans le Sud (Deltel 1961, Ducasse 1969, 1974, 1975) (Fig. 2).

A partir du Miocène, cette région, ainsi qu'une partie du domaine ouest-aquitain, se retrouvent sur la partie externe du plateau continental, tandis que vers l'Est s'implantent des faciès plus littoraux et saumâtres (Moyes 1965).

Au toit du Paléogène, nous assistons donc à un brusque refoulement vers l'Ouest du domaine de permanence océanique. La mise en place, à l'Oligocène, de l'anomalie structurale signalée par J. Alvinerie, L. Pratviel, M. Veillon & M. Vigneaux (1974): individualisation en position médiane d'une structure en relief (secteurs d'Aldebaran 1, Orion 1, le Sextant 1) séparant deux bassins adjacents, l'un au Nord (secteur de Castor 101) l'autre au Sud (secteurs d'Antarès 1, 2, 101, Phoenix 1), pourrait être à l'origine de ce phénomène.

Résumé

Les coupes géologiques utilisées dans cette étude appartiennent à 8 forages offshore implantés à une dizaine de kilomètres des côtes landaises, sous 100 mètres d'eau environ.

A l'Eocène (Eocène supérieur) et l'Oligocène, la faune d'ostracodes est très homogène et témoigne de faciès épibathyaux comparables à ceux que nous connaissons dans le Paléogène sud-aquitain. A l'époque considérée, cette région devait se situer sur la pente continentale aquitaine, prolongeant ainsi, en direction du Nord-Ouest, le domaine de permanence océanique individualisé sur le continent. La profondeur semble avoir été légèrement plus faible à l'Oligocène qu'à l'Eocène et toujours un peu plus importante dans le secteur sud (Phoenix, Antarès).

Deux particularités de la faune méritent d'être signalées:
– la présence de *Buntonia*, signalées pour la première fois dans le Paléogène aquitain,
– la présence d'*Abyssocythere,* connues à l'Eocène, en Aquitaine, en deux points situés au débouché de deux canyons sous-marins.

A partir du Miocène, cette région se retrouve sur la partie externe du plateau continental (domaine circalittoral). L'association faunique présente de fortes affinités avec celles des séries connues dans la partie occidentale du Bassin d'Aquitaine.

La mise en place, à l'Oligocène, d'une structure en relief dans la zone médiane (Aldebaran, Orion, Le Sextant) serait à l'origine du brusque refoulement vers l'Ouest du domaine de permanence océanique.

Bibliographie

Alvinerie, J., Barrier, J., Caralp, M., Ittel, D., Klingebiel, A., Magne, J. & Moyes, J. 1968. Reconnaissance des fonds marins et des séries superficielles de la plateforme continentale au large de la côte landaise (golfe de Gascogne, France). Actes Soc. Linn. Bordeaux, vol. spec. Congrès A.F.A.S. 1967, pp. 121-136.

Alvinerie, J., Pratviel, L. Veillon, M. & Vigneaux, M. 1974. Le Cenozoïque profond de la plateforme continentale dans la partie centrale du golfe de Gascogne. 2ème Coll. intern. sur l'Exploitation des Océans, Bordeaux, 1974, Vol. 2, 8 pp.

Benson, R. H. 1971. A new Cenozoic Deep-Sea genus Abyssocythere (Crustacea: Ostracoda: Trachyleberididae) with descriptions of five new species. Smiths. Contr. Paleobiol. 7:25 pp.

Deltel, B. 1961. Les Ostracodes du Paléogène moyen et supérieur d'Aquitaine méridionale. Thèse 3e cycle Ens. sup., Bordeaux, No. 95 (ronéotypée), 215 pp.

Ducasse, O. 1969. Etude micropaléontologique (Ostracodes) de l'Eocène nord-aquitain. Interprétation biostratigraphique et paléogéographique. Thèse Doct. Sc. Nat., Bordeaux, No. 240 (ronéotypée), 381 pp.

Ducasse, O. 1974. La faune d'Ostracodes des différents domaines marins de l'Oligocène en Aquitaine méridionale. C. R. Somm. S.G.F. 1:7-9.

Ducasse, O. 1975. Les associations fauniques de l'Eocène moyen et supérieur dans le Sud du Bassin d'Aquitaine. Distribution schématique et valeur paléoécologique. Bull. Inst. Géol. Bass. Aquitaine, Bordeaux 17: 17-26.

Moyes, J. 1965. Les Ostracodes du Miocène aquitain. Essai de paléoécologie stratigraphique et de paléogéographie. Thèse Doct. Sc.Nat., Bordeaux, Drouillard imp., 339 pp.

Anonyme. Fiches synthétiques des forages marins. Golfe de Gascogne. Service de conservation des gisements d'hydrocarbures.

Adresse de l'auteur:

Université de Bordeaux I, Dépt de Géologie et Océanographie, Avenue des Facultés, 33405 Talence, France

Sixth Intern. Ostracod Symposium, Saalfelden

DISTRIBUTION OF OSTRACODA IN THE EAST CHINA SEA – A JUSTIFICATION FOR THE EXISTENCE OF THE PALEO-KUROSHIO CURRENT IN THE LATE CENOZOIC

KUNIHIRO ISHIZAKI

Abstract

In the shelf areas of the East China Sea, principal component factor analysis processed for station association (Q-matrix) and multiple regression analysis have shown that bottom salinity, referred to as the effect of the Kuroshio Current, can be predicted by the equation

$$S = 33.292 + 0.043 \, SA_3$$

In the light of this, features of the paleo-Kuroshio Current are suggested in the upper Cenozoic deposits around the area.

Samples

Most of the East China Sea, separated on the south-east by complex ridges and trough topography from the Pacific, is a shallow marginal area of the Asian continent (Fig. 1). The continental shelf ranges in water depth from an average of about 55 m on the border of the Yellow Sea to about 120 m on the outer edge of the shelf. In this area the Kuroshio current causes temperature and salinity gradation, at least in winter, as the waters of the region are characterized predominantly by southwestward-flowing countercurrents from the Yellow Sea, accompanied by discharge from the Yangtze River, and the northward flow of the warm and saline Kuroshio Current (Fig. 2).

Sediments on the shelf show a simple pattern of silt on the border of the Yellow Sea and sand in the outer areas. The 26 bottom samples studied in this work were collected by Mr. Seiji Higashikawa mostly from the shelf area of the East China Sea (Fig. 1) using a Phleger gravity core sampler aboard the training ship 'Kagoshima-maru' of the Faculty of Fisheries, Kagoshima University. The top 1 cm of the sediment cores, previously studied by Higashisaka (1970) for an analysis of grain size, were offered to the writer for the study of ostracodes.

Forty samples from the Upper Pliocene Ananai Formation were collected at five isolated regions from Ananai, Ioki, Tonohama, Ono, and Senpuku along the eastern coast of Tosa Bay, and were made available by Prof. Yokichi Takayanagi, one of the cooperative authors of the study done by Katto, Nakamura & Takayanagi (1953). Samples from the Upper Pliocene-Pleistocene Shinzato Formation in Okinawa were supplied by Dr. Hiroshi Noda, Associate Professor of Tsukuba University.

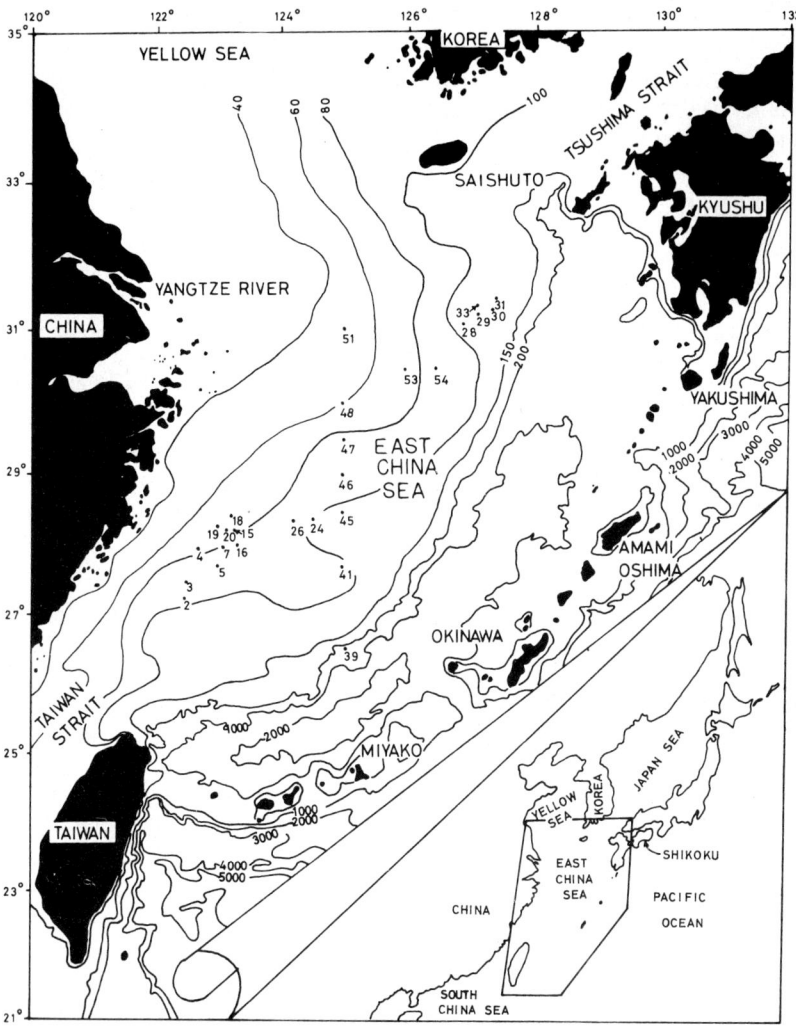

Fig. 1. Station locations in the East China Sea with bathymetry.

Analyses of Ostracoda

The 26 bottom samples, taken mostly from the shelf area of the East China Sea, were studied for ostracode analysis (Fig. 1). From the materials (except the one from St. 39 on the much deeper western slope of the trough) 66 species distributed among 41 genera are recognized to date.

Twenty-three samples, which yielded more than 115 ostracode specimens, were examined by principal component factor analysis for the Q-mode technique after Imbrie & van Andel (1964) (Tables 1, 2). Subsequently, the first three factor

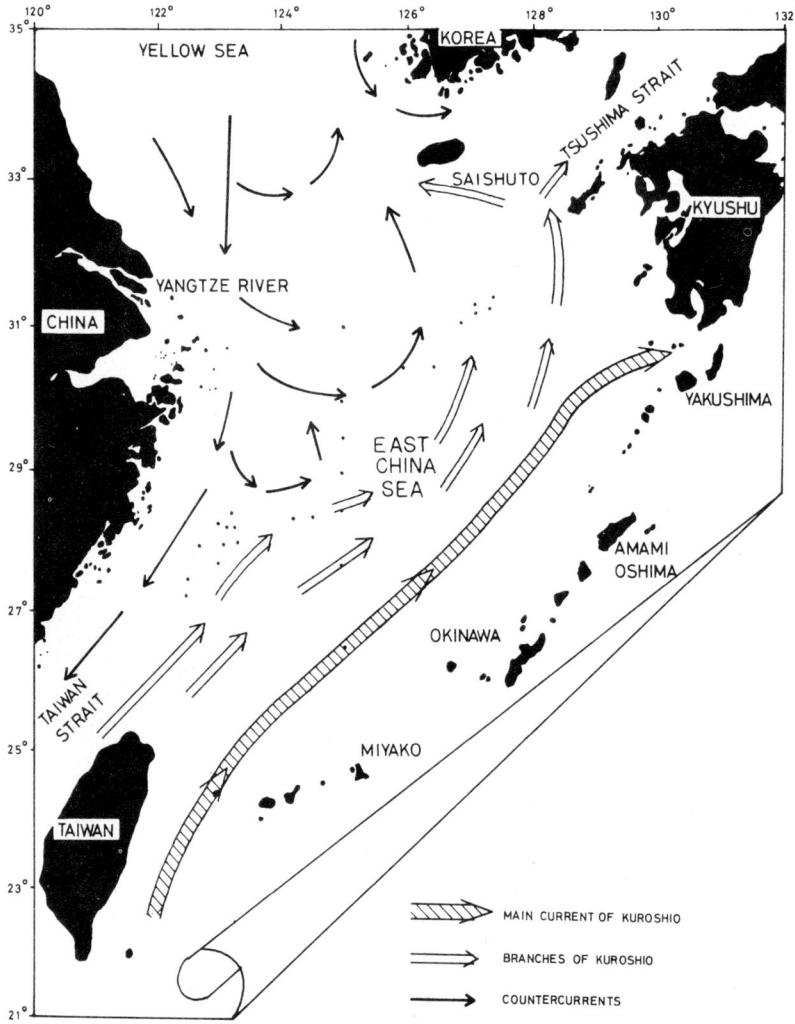

Fig. 2. Main current system in the East China Sea (after Niino & Emery 1961).

axes were rotated following the standard varimax method (Table 3). The sum of the eigenvalues of the first three factors is 87.36 percent of the total, being reasonably enough for delineating the general distribution pattern of ostracode fauna in the region (Table 1, Fig. 3). This figure briefly represents distribution of the three main elements of ostracode faunas, defined by the first three factor loadings. Each lies in a particular area, and may be subjected to a different factor.

In order to inspect this pattern based on the relative frequency of a particular ostracode species, the factor scores were calculated (Table 4). The following six species associations were extracted by factor scores of more than 1.28, which

Table 1. Eigenvalue and cumulative percentage.

Factor	Eigenvalue	Cumulative percentage
1	17.413	75.71
2	1.786	83.48
3	0.894	87.36
4	0.690	90.37
5	0.408	92.14
6	0.396	93.86
7	0.334	95.31
8	0.211	96.23
9	0.175	96.99
10	0.126	97.53

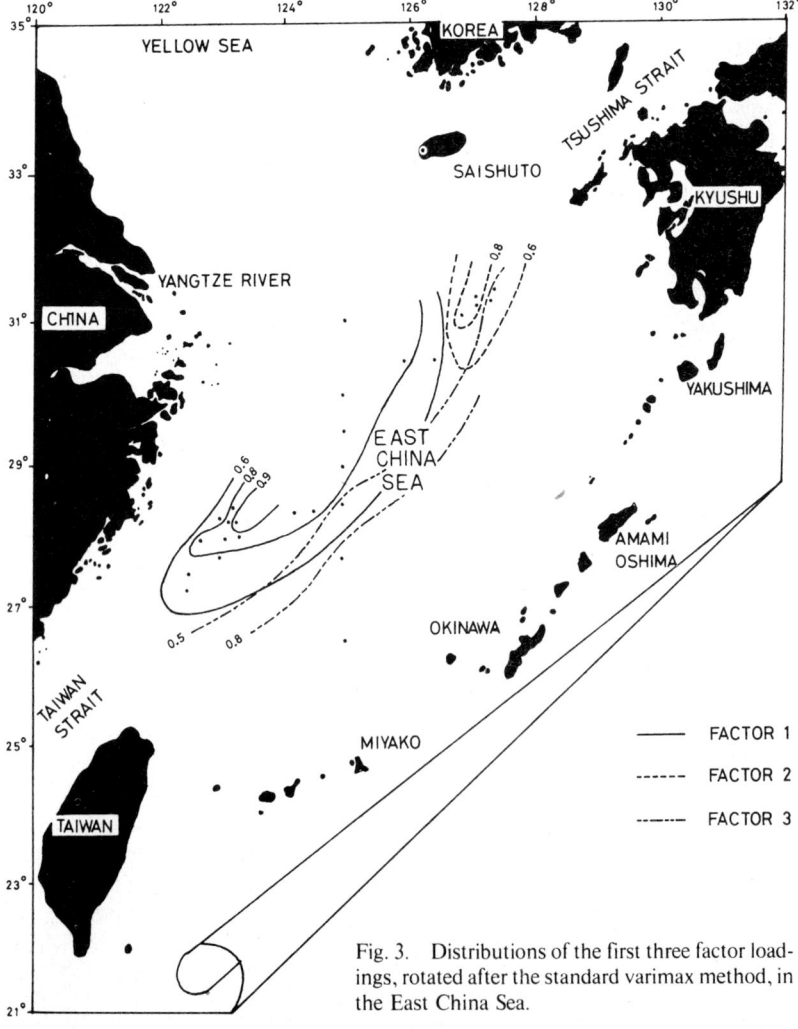

Fig. 3. Distributions of the first three factor loadings, rotated after the standard varimax method, in the East China Sea.

Table 2. Principal component factor loading matrix

Station	Communality	Factor									
		1	2	3	4	5	6	7	8	9	10
2	0.9727	0.9035	-0.0431	0.0492	0.2012	0.1853	-0.0955	0.0903	0.2142	-0.1081	0.0502
3	0.9656	0.8423	-0.0389	0.0745	0.4158	0.2201	0.0342	-0.1554	0.0027	-0.0231	0.0434
4	0.9637	0.9515	-0.1456	-0.0814	0.0118	-0.0919	-0.1370	-0.0162	-0.0210	0.0035	0.0486
5	0.9567	0.8912	-0.0642	0.0284	0.3168	0.0966	-0.0624	-0.1612	-0.0424	0.1267	0.0110
7	0.9593	0.9364	-0.0682	-0.1053	0.1260	-0.0491	-0.1871	-0.0147	-0.0016	-0.1000	-0.0558
15	0.9798	0.9263	-0.2646	-0.0848	-0.0757	0.0868	-0.0892	-0.0064	-0.0820	-0.0714	-0.1074
16	0.9846	0.9496	-0.0386	-0.1296	0.0645	-0.1790	-0.0442	-0.0370	-0.0888	-0.0432	-0.1236
18	0.9900	0.9093	-0.0723	-0.1521	-0.1612	0.2644	0.0232	0.1835	-0.0483	-0.0188	-0.0436
19	0.9874	0.8916	-0.0069	-0.2633	-0.1283	0.1499	0.1245	0.2045	-0.1281	-0.0520	0.0877
24	0.9744	0.9407	-0.1512	0.0445	0.0882	-0.0973	0.0689	0.0361	-0.2007	0.0253	0.0208
26	0.9735	0.9247	-0.2020	-0.0795	-0.1285	-0.0697	0.1070	-0.0262	0.0743	0.1246	0.1291
28	0.9674	0.6849	0.6031	-0.1974	0.0869	-0.2252	-0.0957	0.0611	0.0116	-0.1218	0.0971
29	0.9748	0.7674	0.5595	-0.1778	0.0477	-0.0093	0.0432	0.0204	0.0248	0.1856	-0.0394
30	0.9855	0.7324	0.4257	0.3053	-0.2518	0.1742	0.1515	-0.2233	-0.0123	-0.0694	-0.0550
31	0.9780	0.8272	0.3677	0.2558	-0.2012	-0.0584	-0.0383	-0.1406	-0.0684	-0.1321	0.0759
33	0.9744	0.7521	0.5965	-0.1487	0.0504	-0.0069	0.0737	0.0523	0.0877	0.0834	-0.0740
41	0.9874	0.6772	0.0468	0.6440	0.0382	-0.0594	-0.0662	0.3133	0.0505	0.0403	-0.0130
45	0.9712	0.9130	-0.1008	0.2652	0.1045	-0.1033	0.0720	0.0258	-0.1532	0.0754	0.0236
46	0.9861	0.8620	-0.2532	0.0049	0.0253	-0.1310	0.3657	-0.0103	0.1144	-0.0502	-0.1083
47	0.9672	0.9392	-0.1696	-0.0777	0.0407	-0.0651	0.1699	0.0306	0.0948	-0.0702	0.0249
48	0.9813	0.8592	-0.3772	-0.0019	-0.1803	-0.1727	-0.0732	-0.1177	0.1379	0.0117	0.0128
53	0.9774	0.9212	-0.1248	-0.0248	-0.2743	0.0827	-0.0613	-0.0758	0.0311	0.0928	0.1067
54	0.9741	0.9168	-0.0218	0.0139	-0.2125	0.0371	-0.2523	0.0028	0.0496	0.0895	-0.1109

Table 3. Principal component factor loading rotated after standard varimax method

Station	Communality	Factor 1	Factor 2	Factor 3
2	0.8206	0.7369	0.3810	0.3638
3	0.7165	0.6791	0.3485	0.3660
4	0.9333	0.8635	0.3541	0.2494
5	0.7992	0.7436	0.3633	0.3381
7	0.8927	0.8155	0.4182	0.2296
15	0.9352	0.9084	0.2436	0.2252
16	0.9201	0.8162	0.4559	0.2147
18	0.8553	0.8077	0.4146	0.1760
19	0.8644	0.7859	0.4913	0.0731
24	0.9097	0.8263	0.3098	0.3618
26	0.9022	0.8720	0.2938	0.2359
28	0.8718	0.2737	0.8843	0.1222
29	0.9336	0.3587	0.8818	0.1655
30	0.8109	0.2821	0.6210	0.5880
31	0.8849	0.4019	0.6313	0.5700
33	0.9435	0.3191	0.8974	0.1907
41	0.8755	0.3578	0.1853	0.8445
45	0.9140	0.7217	0.2782	0.5620
46	0.8071	0.8281	0.1980	0.2866
47	0.9170	0.8657	0.3273	0.2460
48	0.8946	0.8946	0.0949	0.2669
53	0.8648	0.8137	0.3416	0.2932
54	0.8411	0.7448	0.4149	0.3380

Table 4. Factor score matrix of the first three factors

Species name	Factor score		
	Factor 1	Factor 2	Factor 3
Acanthocythereis sp.	−0.010	0.070	0.054
Actinocythereis sp.	1.980	−0.857	−0.406
Alocopocythere goujoni (Brady)	0.643	−0.013	−0.427
Ambocythere sp. H.	0.170	0.569	−0.134
Ambocythere sp. P	0.082	−0.008	0.020
Argilloecia sp.	0.136	6.390	−0.599
Aurila cymba (Brady)	0.084	0.163	−0.101
Bradleya japonica Benson	0.012	0.050	−0.016
Bradleya sp.	−0.668	0.558	1.575
Bythoceratina sp. D	−0.024	−0.297	0.813
Bythoceratina cf. *orientalis* (Brady)	−0.138	−0.244	0.943
Bythoceratina utilazea Hornibrook	−0.140	−0.135	0.587
Callistocythere japonica Hanai	0.007	0.067	0.112
Callistocythere cf. *reticulata* Hanai	0.124	−0.127	0.108

Table 4 (cont.)

Campylocythere sp.	0.021	0.002	−0.002
Celtia sp. J	0.007	−0.484	2.024
Celtia sp. S	0.037	−0.022	−0.012
Cluthia sp. M	−0.042	0.192	0.114
Cushmanidea cf. *japonica* Hanai	−0.035	0.119	0.145
'*Cythere*' *euplectella* Brady	0.416	0.110	−0.510
Cytherelloidea sabahensis Keij	1.322	−0.663	1.511
Cytherelloidea sp.	0.793	−0.361	0.014
Cytherois? sp.	0.005	−0.100	0.280
Cyhteropteron cf. *miurense* Hanai	5.075	0.221	2.336
Cytheropteron uchioi Hanai	1.975	1.915	1.729
Cytheropteron sp. A	−0.050	0.152	0.266
Cytheropteron sp. B	−0.952	1.573	1.883
Cytherura sp.	0.968	0.361	0.155
Cytheruidae gen. S, sp. J	−0.261	0.709	−0.079
Echinocythereis bradyformis Ishizaki	0.279	−0.133	0.070
Echinocythereis sp. H	0.330	0.042	0.475
Eucythere subobalis Hornibrook	−0.028	−0.027	0.117
Eucythere? sp. A	−0.052	0.168	0.118
Eucythere sp. B	−0.056	−0.054	0.235
Eucythere? sp.	0.099	0.384	−0.049
Eucytherura sp. J	0.009	−0.161	0.754
Eucytherura sp.	0.094	0.137	1.280
Hemicytherura cuneata Hanai	−0.109	0.158	0.445
Hemicytherura radiata Hornibrook	−0.154	0.014	0.445
Hemicytherura sp.	0.173	−0.282	1.030
Incongruellina sp.	0.034	1.230	−0.374
Kobayashiina hyalinosa Hanai	0.397	2.362	−0.779
Krithe producta Brady	−0.443	0.960	0.199
Loxoconcha optima Ishizaki	−0.019	0.035	0.086
Loxoconcha sinensis Brady	0.044	−0.725	3.560
Macrocypris decora (Brady)	−0.081	0.013	0.248
Macrocypris maculata (Brady)	−0.031	1.023	0.030
Munseyella cf. *japonica* (Hanai)	4.379	0.224	−1.607
Neonesidea villosa (Brady)	−0.761	0.002	2.760
Nipponocythere sp. H	0.499	1.040	−0.347
Paijenborchella iocosa Kingma	0.192	0.523	0.252
Paijenborchella spinosa Hanai	−0.124	0.238	1.048
Phlyctocythere sp.	0.228	0.324	0.148
Pseudocytherura? sp. J	0.157	0.366	−0.044
Pterygocythereis scalaris (Brady)	−0.139	0.426	0.713
Saida torresii (Brady)	−0.009	−0.061	0.168
Schizocythere kishinouyei (Kajiyama)	1.006	−0.337	−0.078
Semicytherura miurensis (Hanai)	0.203	−0.164	0.401
Semicytherura sp. M	0.299	0.087	0.513
Thalmannia? sp.	0.309	−0.060	0.147
Trachyleberis niitsumai Ishizaki	0.943	−0.590	0.887
Trachyleberis scabrocuneata (Brady)	0.265	−0.307	1.099
Trachyleberis sp. S	1.475	−0.169	−0.601
Trachyleberis sp.	0.087	0.044	0.131
Xestoleberis foveolata Brady	−0.844	0.363	2.707
Xestoleberis cf. *variegata* Brady	−1.064	1.803	1.519

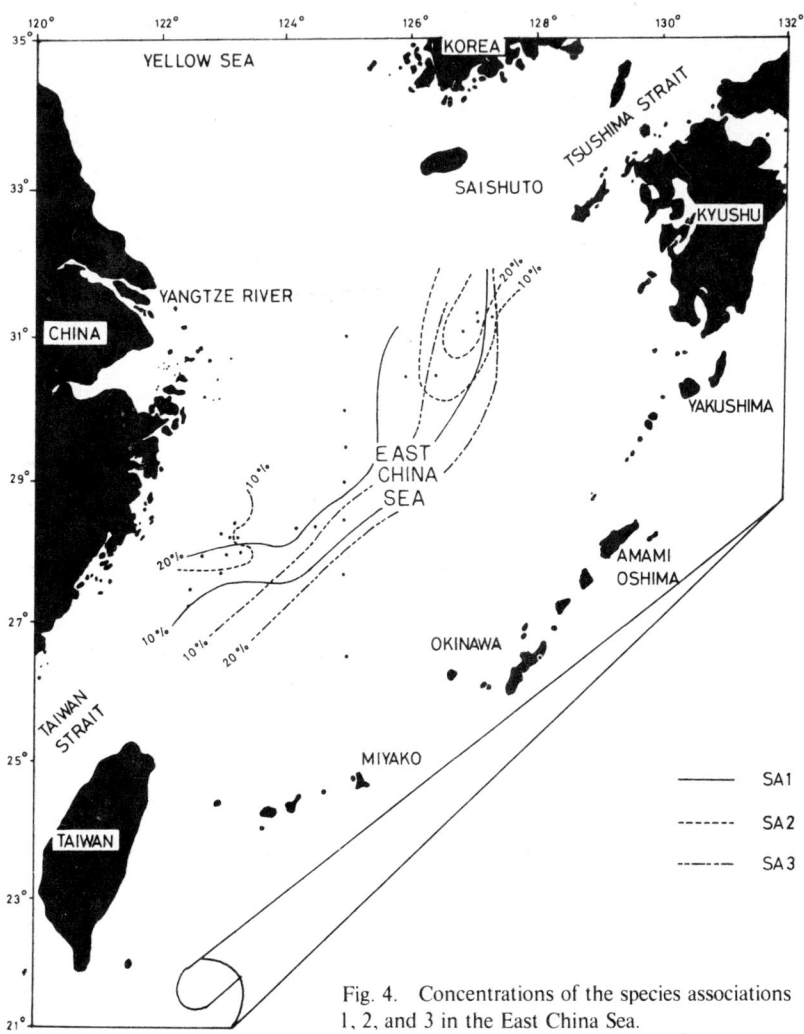

Fig. 4. Concentrations of the species associations 1, 2, and 3 in the East China Sea.

stands at ten percent in the standardized normal distribution:

Species association 1 (by the first factor): Actinocythereis sp., *Munseyella* cf. *japonica* (Hanai), 1957, and *Trachyleberis* sp. S

Species association 2 (by the second factor): *Argilloecia* sp. and *Kobayashiina hyalinosa* Hanai, 1957

Species association 3 (by the third factor): *Bradleya* sp., *Celtia* sp. J, *Loxoconcha sinensis* (Brady) 1880, *Neonesidea villosa* (Brady) 1880, and *Xestoleberis foveolata* Brady, 1880

Species association 13 (by the first and third factors): *Cytherelloidea sabahensis* Keij, 1964 and *Cytheropteron* cf. *miurense* Hanai, 1957

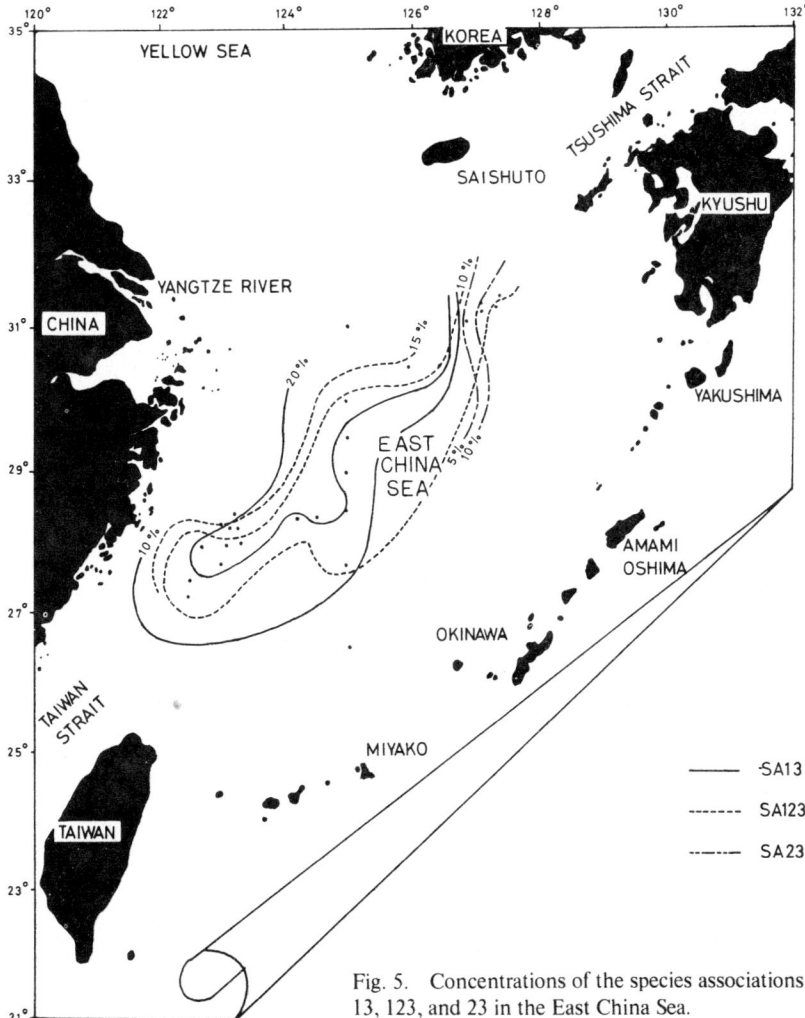

Fig. 5. Concentrations of the species associations 13, 123, and 23 in the East China Sea.

Species association 23 (by the second and third factors):*Cytheropteron* sp. B and *Xestoleberis* cf. *variegata* Brady, 1880

Species association 123 (by the first three factors): *Cytheropteron uchioi* Hanai, 1957.

The distribution patterns of these species associations are more complicated than those of the first three factor loadings in the region (Figs. 4, 5). Certain environmental data is pertinent here. After a study of grain size, Phi scale median size and mud content of the bottom sediments were listed by Higashikawa (1970), together with the water depths. Data on temperature, water colour, and the chlorinity of the surface water are illustrated by Wageman, Hilde & Emery (1970), and those on temperature and the salinity of bottom water by Polski (1959).

In order to explain the relationship between these environmental variables and the frequency of the six species associations mentioned above, multiple regression was analyzed. The results indicate that three species associations of the six have an apparent correlation with a particular environmental variable at the probability of 0.01 (Table 5). It is always important, however, in analyzing faunal assemblage and assuming the resultant environmental condition, to inquire into whether or not regression equation concerned is appropriate for satisfactorily pre-

Table 5. Results from the multiple regression analysis

Environmental variables	F value*	Contribution	Regression equation
Bottom salinity	109.851	0.902	$S=33.292+0.043SA_3$
Bottom temperature	22.486	0.652	$T=12.328+0.113SA_3$
Surface water temperature	68.316	0.851	$T=21.249+0.313SA_{23}$
Depth	18.222	0.603	$D=81.080+2.926SA_{23}$
Mud content	20.013	0.625	$M=4.187+1.777SA_2$
Md$\emptyset$	11.480	0.489	$M=2.160+0.064SA_2$

* For (12,1) degrees of freedom, F=9.33 at the prabability of 0.01

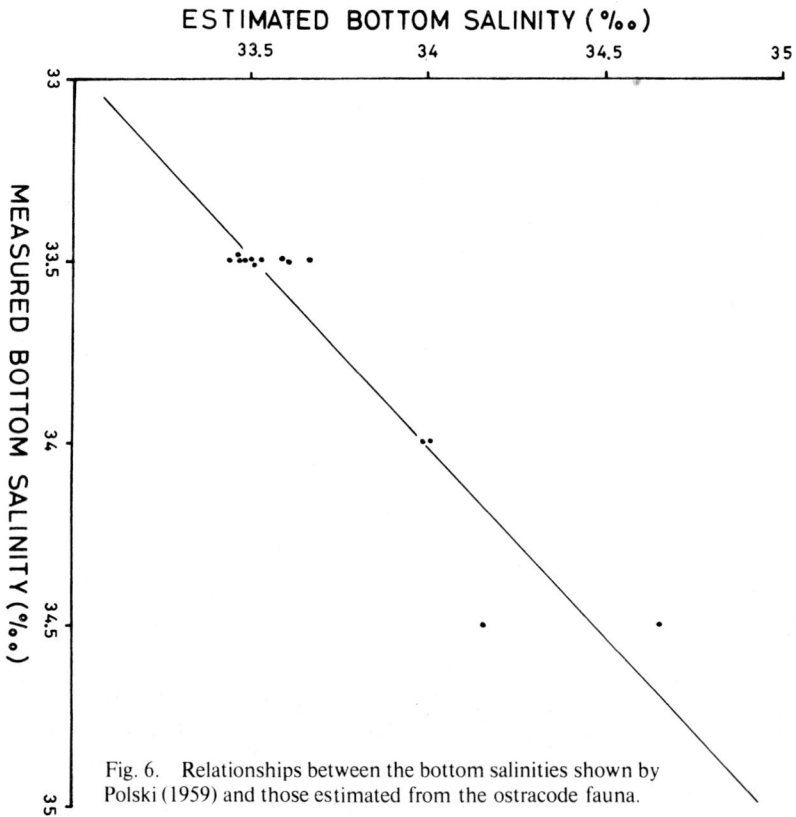

Fig. 6. Relationships between the bottom salinities shown by Polski (1959) and those estimated from the ostracode fauna.

dicting the environmental condition. In this regard, the best that can be done at present is to examine these regression equations making use of the contribution magnitude, expressed by the squared correlation coefficient. Species association 3 accounts for more than 90 percent of bottom salinity, which indicates that it is reasonably suitable for a standard statistical approach (Fig. 6).

The distribution pattern of the bottom salinity in the region, estimated from the equation, is illustrated in Fig. 7. Except for the southern part north of Taiwan, the general pattern agrees closely with those shown by Polski (1959) and corresponds with the effect deduced from the pattern of the Kuroshio Current in winter, illustrated by Niino & Emery (1961) (Fig. 2).

Loxoconcha sinensis, a member of the species association 3, was first proposed by Brady (1869) based on material from off Hong Kong and is subsequently re-

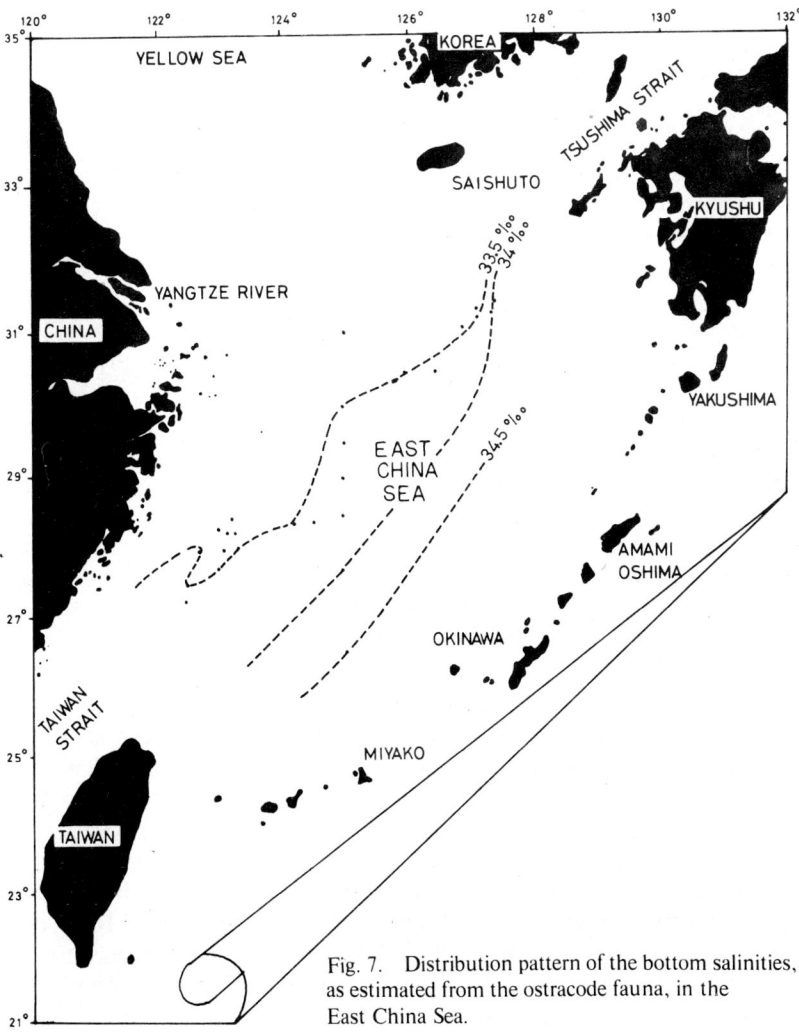

Fig. 7. Distribution pattern of the bottom salinities, as estimated from the ostracode fauna, in the East China Sea.

435

ported by Brady (1880) from Hong Kong Harbour, and from the Inland Sea (Setonaikai), Japan, at a depth of 27.5 m. *Xestoleberis foveolata* was reported by Brady (1880) off Booby Island, north of Australia, at depths between 11 m and 14.6 m, and is recorded by Hornibrook (1952) from the northern part (Aupourian Province) of New Zealand. *Neonesidea villosa* appears to be a somewhat deeper water species and to be cosmopolitan as noted by Benson (1964), judged from the record [Brady 1880: the southern oceans of the Atlantic (Nightingale Island, 183–274.5 m; Prince Edward Island, 91.5-274.5 m), the Indian (Kerguelen Island, 36.6-91.5 m; 219.6 m), and Australasia (off East Moncoeur, Bass' Strait, 69.6-73.2 m)]. Judging from the available data on geographic distribution of these species, species association 3 may represent an element, brought up by the Kuroshio Current from the southern Ocean. This fauna might have been dispersed both north- and southward from the Indo-Pacific realm, although they are not thoroughly understood to date and, on the way, equatorial currents, and eastward-flowing countercurrents might have played an important role in forming a close connection between the ostracode fauna of the Nova Zealand, Indo-Pacific, and Japonic realms.

With the provision that the doctrine of uniformitarianism is valid for such a relatively short period dating from the late Pliocene with a refrained ecogenesis of species association, resulting from a limited range of variation in the environmental conditions around the region, such an estimation may be of value in suggesting the effect of the Kuroshio Current on the Upper Pliocene-Pleistocene Shinzato Formation in Okinawa and the Upper Pliocene Ananai Formation in Shikoku. This is reasoned since, for the greatest part, both formations yield a rich fauna of ostracodes, and comprise species association 3, defined in the shelf area of the East China Sea.

Along the sequence of the Shinzato Formation in the southern part of Okinawa (Shinzato-Kudeken region), the bottom salinity generally remains low, but it increases somewhat above the pumice tuff described by MacNeil (1960) as the base of the formation, and at the lower section of the Chinen Sandstone (Fig. 8). It also remains almost unchanged through the sequence of the central part of Okinawa (Yakena region), but increases in the Ryukyu Limestone. Thus, the fluctuation of the estimated bottom salinity may show the effect of the paleo-Kuroshio Current, having been constant for most of the Shinzato Formation and becoming stronger at the time of the Chinen Sandstone-Ryukyu Limestone.

As for the Ananai Formation, Katto, Nakamura, and Takayanagi (1953) have analyzed the fossil foraminefera and allocated them to two biofacies of Ioki and Senpuku; the former, in the range of the Mesoneritic (20-30 m to 50-60 m deep) to subneritic zones (50-60 m to 100-120 m deep), comprises the two regions of Ananai and Ioki, and the latter, being at depths greater than the subneritic zone, the three regions of Tonohama, Ono, and Senpuku. In the Ananai-Ioki region, bottom salinity decreases gradually toward the middle section of the sequence, and then increases toward the top (Fig. 9). On the other hand, in the Tonohama-Ono region it shows a remarkable fluctuation throughout most of the sequence. In the Senpuku region, three short sequences are presented from different outcrops. The sequence of G is somewhat longer and shows bottom salinity greatly

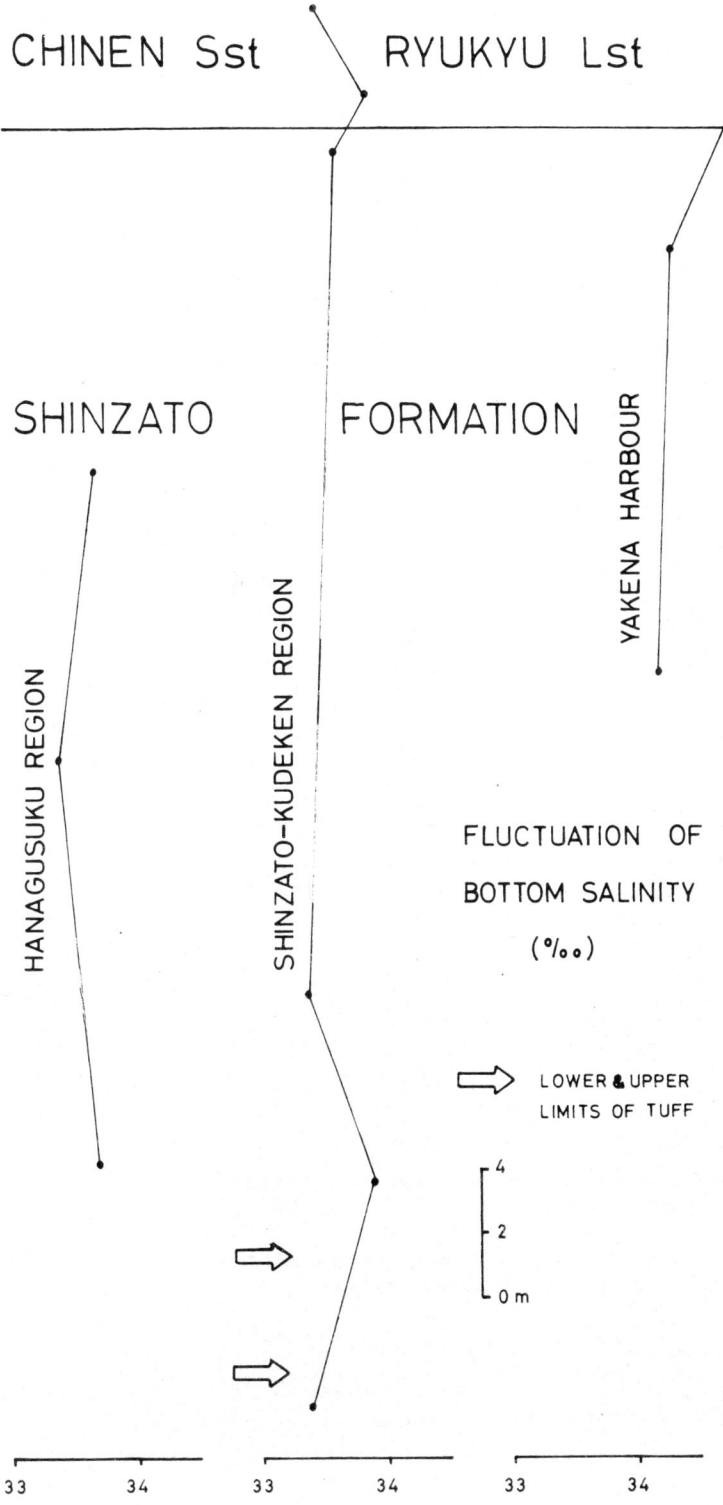

Fig. 8. Fluctuations of the bottom salinities, as estimated from the ostracode fauna, along the sequences of the Shinzato Formation, Chinen Sandstone, and Ryukyu Limestone in Okinawa.

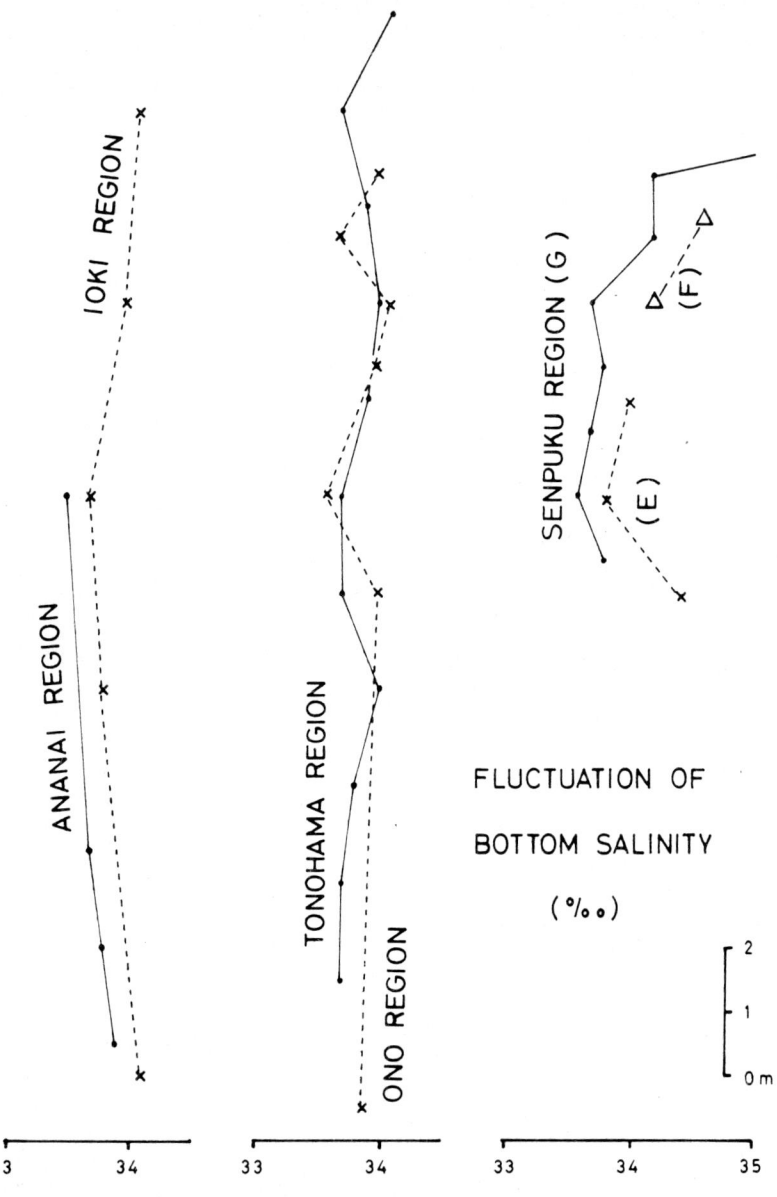

Fig. 9. Fluctuations of the bottom salinities, as estimated from the ostracode fauna, along the sequences of the Ananai Formation in Shikoku.

increasing toward the top away from the middle. Therefore, at least at the upper sections, it appears that the effect of the paleo-Kuroshio Current on the Ananai Sedimentary Basin became stronger as time advanced. This effect might have been more pronounced in the Senpuku region and more problematical west of it.

Remarks

From data established in the East China Sea and the upper Pliocene-Pleistocene in Okinawa and Shikoku, it appears that, as is the case in the East China Sea area at the present time, both the Shinzato and Ananai formations were under the influence of the Kuroshio Current at the time of their deposition and that this influence had become even stronger with time.

Acknowledgment

The writer wishes to express his sincere gratitude to the following researchers: Professor Yokichi Takayanagi of Tohoku University for his kind scrutinizing of the manuscript and the donation of samples; Dr. Hiroshi Noda, now Associate Professor of Tsukuba University, and Lecturer Seiji Higashikawa of Kagoshima University for their kind provision of many samples.

References

Benson, R. H. 1964. Recent marine Podocopid and Platycopid Ostracodes of the Pacific. Pubbl. Staz. Zool. Napoli 33 suppl.:387-420.
Brady, G. S. 1869. Les fonds de la mer, vol. 1, pt. 1, 176 pp.
Brady, G. S. 1880. Report on the Ostracoda dredged by H.M.S. Challenger during the years 1873-1876. Challenger Exped. 1873-1876, Rept., Zool., vol. 1, pt. 3, 184 pp.
Higashikawa, S. 1970. Analyses of bottom sediments of the East China Sea. Mem. Fac. Fish., Kagoshima Univ. 19:91-102 (in Japanese with English abstract).
Hornibrook, N. de B. 1952. Tertiary and Recent marine Ostracoda of New Zealand. New Zealand Geol. Surv., Paleont. Bull. 18:82 pp.
Imbrie, J. & van Andel, T. H. 1964. Vector analysis of heavy-mineral data. Geol. Soc. Amer., Bull. 75 (11):1131-1155.
Katto, J., Nakamura, J. & Takayanagi, Y. 1953. Stratigraphical and paleontological studies of the Tonohama Group, Kochi Prefecture, Japan. Res. Rept., Kochi Univ., vol. 2, no. 32, 15 pp. (in Japanese with English summary).
MacNeil, F. S. 1960. Tertiary and Quaternary Gastropoda of Okinawa. U.S.G.S., Prof. Paper, 339, 148 pp.
Niino. H. & Emery, K. O. 1961. Sediments of shallow portions of East China Sea and South China Sea. Geol. Soc. Amer., Bull. 72:731-762.
Polski, W. 1959. Foraminiferal biofacies off the North Asiatic Coast. Jour. Paleont. 33 (4): 569-587.

Author's address:

Institute of Geology and Paleontology, Faculty of Science, Thoku University, Sendai, Japan

Discussion

Carbonnel: Are the data used in your analysis original (real) data or binary ones?
Ishizaki: The equation, $R = AA'$, shows that factor analysis is to be done through the matrix of correlation coefficients, which should better be represented by quantitative coefficients than binary ones. I have calculated correlation coefficients from the frequencies (proportions) of each species, because the number of ostracode specimens in every sample is not equal, ranging from 115 to 217 (being about 200 specimens in most case).

Phlyctocythere? *globulata* D., *Protocythere revili* D., *P. divisa* O., *P. helvitica* O., *pseudoprotocythere aubersonensis* O., *Schuleridea mediocaudata* D., *Schuleridea* aff. *praethorenensis* B. et B., *Xestoleberis alta* D., *X. dimorpha* D.
2) dans les faciès hémipélagiques: *Bairdia major* D., *Cytherella dissimilis* D., *Oligocythereis bogis* D., *Procytheridea tuberculata* D.

b) Les *faunes berriasiennes d'Afrique du Nord*

Dans l'autochtone berriasien d'Algérie et de Tunisie, l'ostracofaune diffère profondément de celle qui caractérise l'ensemble précédent. Les développements évolutifs ne s'y opèrent pas de la même manière, de telle sorte qu'il existe de grandes divergences du point de vue spécifique. En outre, la composition générique est assez différente: d'une part, on y trouve des genres nouveaux, inconnus en Europe à cette époque, d'autre part, manquent les genres *Cythereis, Euriyticythere, Pseudobythocythere, Kentrodictyocythere* etc. Les genres communs correspondent pour la plupart des cas à des espèces différentes. Par exemple, les *Macrodentina* de la bordure septentrionale de la Téthys se rapprochent de l'espèce *mediostricta* S.-B.; ceux d'Afrique du Nord sont différents, et paraissent plus proches d'espèces jurassiques de l'Europe continentale (Pl., fig. 5). Le genre *Cytherelloidea* existe bien de part et d'autre de la Téthys; mais les populations en sont tout à fait différentes du point de vue spécifique. Les spécimens d'Afrique du Nord sont à rapprocher de formes du moyen-Orient, inconnues en Europe de l'Ouest (Pl., fig. 3-4).

Il est vrai qu'il existe cependant quelques formes assez proches de part et d'autre. On peut citer en particulier: *Bairdia* aff. *major* D., *Cytherella* aff. *turgida* D., *C.* aff. *dissimilis* D., *Protocythere* aff. *mazenoti* D., *P.* aff. *revili* D., *Oligocythereis* aff. *bogis* D. Mais même dans ce cas, on peut hésiter quant à l'identification spécifique, d'abord à cause des différences morphologiques apparentes, mais surtout parce que ces formes n'évoluent pas parallèlement dans les deux domaines, de telle sorte qu'on ne peut pas écarter l'hypothèse de simples phénomènes de convergence.

Soit par exemple le cas de *Cytherella* aff. *dissimilis* D., Cette forme nord-africaine a bien des caractères communs avec *C. dissimilis* du Bassin vocontien. Mais en Afrique du Nord, elle s'incrit dans une série évolutive dont des représentants se manifestent déjà au Tithonique (Pl., fig. 1-2), alors que dans le Sud-Est de la France, elle apparaît brusquement, sans précurseur connu, à la limite Berriasien-Valanginien. Il est alors légitime d'admettre, au moins à titre d'hypothèse, que les dispositions paléogéographiques devenues plus favorables à cette époque ont favorisé la migration de l'espèce vers le Nord, avec un certain retard en rapport avec la distance. De même pour les formes *Oligocythereis* aff. *bogis* D. (Pl., fig. 6). On en trouve de nombreux représentants dans le Berriasien algéro-tunisien, préférentiellement dans les milieux hémipélagiques. Mais alors que les formes du Sud-Est de la France et d'Espagne sont rigoureusement identiques, celles d'Afrique du Nord s'en différencient par des caractères morphologiques importants, qui pourraient justifier la création d'au moins une sous-espèce. De plus, comme précédemment, il n'y a pas en Europe d'ascendants connus anté-berri-

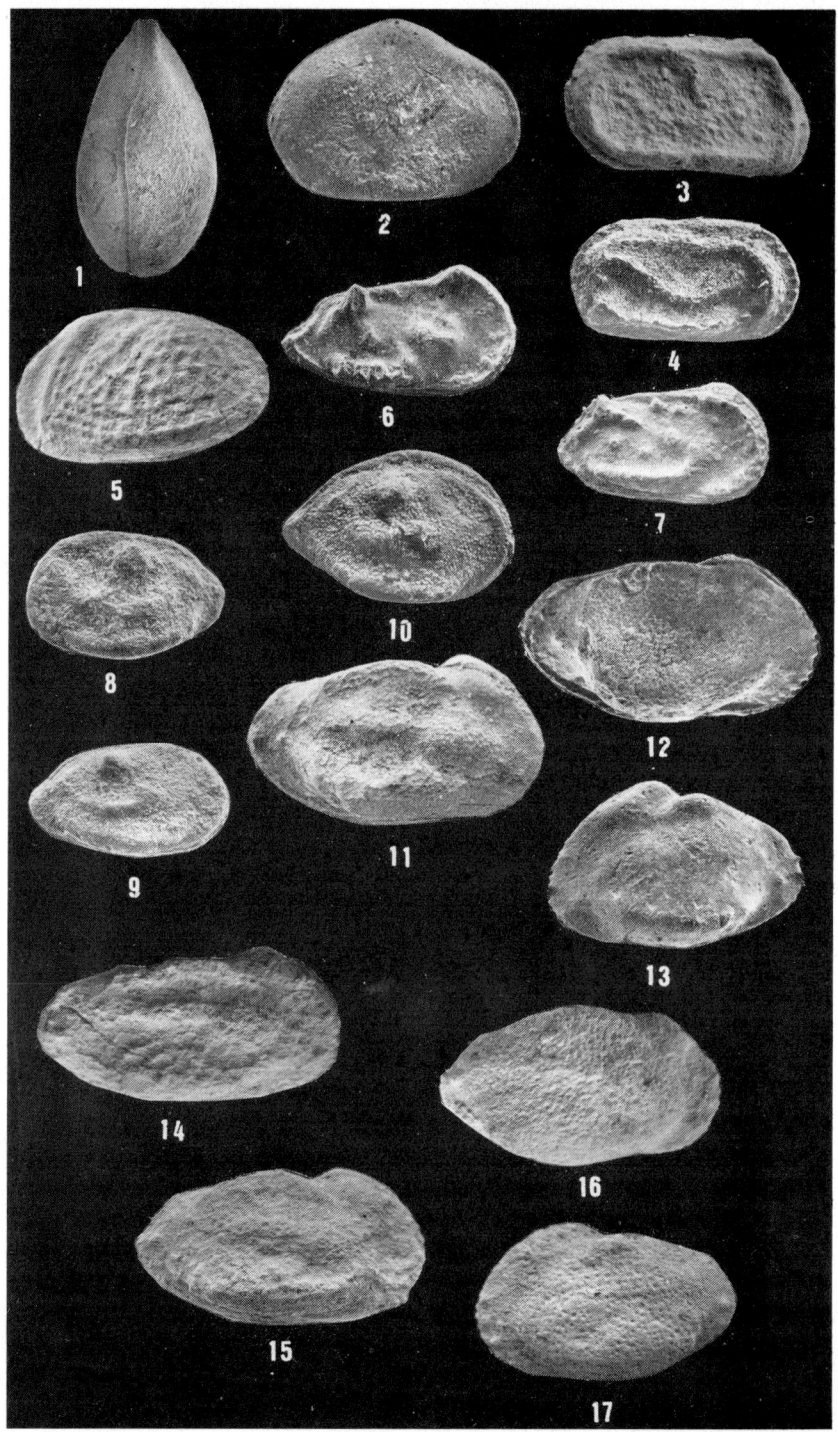

◄ Planche.

Fig. 1-2. *Cytherella* sp. B, Donze 1974. (× 54)
1 : C, vue dorsale, no. 157010. Limite Tithonique-Berriasien, 'Ravin Bleu', Monts de Batna, Algérie.
2: VD, vue externe, no. 157009. Id.
Fig. 3. *Cytherelloïdea* sp. VG, vue externe (×68). Berriasien sup., El Rhoraf, Monts de Tlemcen, Algérie.
Fig. 4. *Cytherelloïdea* sp. VD, vue externe (×53), no. 157011. Berriasien moyen, Tarhit, Constantinois, Algérie.
Fig. 5. *Macrodentina* sp. C, de gauche (×57). Valanginien basal, El Rhoraf, Algérie.
Fig. 6. *Oligocythereis* aff. *bogis*, Donze 1964. C, de droite (×58), no. 157040. Berriasien moyen, Djebel Nara, Tunisie centrale.
Fig. 7. *Oligocythereis* aff. *bogis*, Donze 1964. C, de droite (×58), no. 157100. Valanginien basal, Barret-le-Bas, Htes-Alpes.
Fig. 8. *Procytheridea* sp. A, Donze 1974. VG, vue externe (×50), no. 157029. Tithonique supérieur, Oued Soubella, Monts du Hodna, Algérie.
Fig. 9. *Procytheridea* sp. A, Donze 1974. C, de droite (×50), no. 157031. Id.
Fig. 10. *Procytheridea tuberculata*, Donze 1965. C, de droite (×51), no. 157029. Berriasien moyen, Oued Soubella, Algérie.
Fig. 11. *Protocythere mazenoti*, Donze 1973. C, de droite, mâle, holotype (×33), no. 59649. Berriasien supérieur, Entremont-le-Vieux Savoie.
Fig. 12. *Protocythere paquieri*, Donze 1967. C, de droite, mâle (×34), no. 59656. Valanginien basal, id.
Fig. 13. *Protocythere paquieri*, Donze 1967. VG, vue externe, femelle, (×40), no. 59657. Valanginien basal, Col du Granier, Savoie.
Fig. 14. *Protocythere* aff. *mazenoti*, Donze 1973. C, de droite, mâle (×45), morphotype (forme primitive). Berriasien moyen, El Rorhaf, Monts de Tlemcen, Algérie.
Fig. 15. *Protocythere* aff. *mazenoti*, Donze 1973. C, de droite, mâle (×48), morphotype B. Base du Berriasien supérieur, Lamoricière, Monts de Tlemcen, Algérie.
Fig. 16. *Protocythere* aff. *mazenoti*, Donze 1973. VD, Vue externe, mâle présumé (×48), morphotype C. Berriasien terminal, El Rhoraf, Algérie.
Fig. 17. *Protocythere* aff. *mazenoti*, Donze 1973. VG, vue externe, femelle présumée (×46), morphotype C. Id.

asiens de cette espèce*, tandis que de l'autre coté, elle paraît dériver de formes jurassiques.

Le cas de la forme africaine *Protocythere* aff. *mazenoti* D. présente encore plus d'intérêt. Elle se manifeste au Berriasien moyen au sein d'un phyllum qui s'était déjà différencié au Berriasien inférieur. Dans le Bassin vocontien, ou l'espèce a été décrite (Donze 1973), on ne trouve pas de précurseurs, et les représentants n'appasaissent qu'au Berriasien supérieur. S'il s'agit bien de la même espèce, il ne peut s'agir que de la manifestation d'un phénomène migratoire en direction du Nord, le retard étant mis en rapport avec les distances parcourues. Mais d'autre part, lorsqu'on étudie le développement phylogénétique de l'espèce, on peut constater que les évolutions ne sont pas identiques dans les deux domaines. En Afrique du Nord (Pl., fig. 14-17) les formes primitives du Berriasien moyen mor-

---

* On doit cependant tenir compte de ce que les Ostracodes du Tithonique européen sont encore mal connus, les faciès calcaires se prêtant mal à leur étude.

photype A) se différencient en un morphotype B au Berriasien supérieur, puis en un morphotype C au Berriasien terminal-Valanginien basal. Ces morphotypes B et C n'ont pas été rencontrés dans le Sud-Est de la France. L'espèce s'éteint à la base du Valanginien, où lui succède une forme assez différente, *Protocythere paquieri* D. (Pl., fig. 12-13), qui paraît en dériver.

Citons encore l'exemple de *Procytheridea tuberculata* D. (Pl., fig. 10) décrit dans le Berriasien terminal-Valanginien basal de faciès hémipélagique du Sud-Est de la France. On rencontre cette espèce avec une grande fréquence dans les mêmes faciès du Berriasien autochtone algéro-tunisien. Là encore, on ne lui connaît pas d'ascendants directs en Europe, alors qu'en Afrique du Nord, elle peut s'intégrer dans un phyllum représenté déjà au Jurassique supérieur (Pl., fig. 8-9).

Conclusions

Comme on le voit, la répartition des populations berriasiennes d'Ostracodes de part et d'autre de la Téthys est loin d'évoquer une proximité géographique, a fortiori un contact, entre la Péninsule ibérique et l'Afrique du Nord comme le suggèrent de nombreuses reconstitutions paléogéographique de l'Ouest téthysien à cette époque. Un continuité géographique aurait du se traduire par une continuité paléobiologique, ce qui n'est pas le cas. Si les territoires qui bordaient le Nord de la Téthys, c.à.d. le Sud-Est de la France et la bordure orientale de l'Espagne, forment une unité faunistiquement parlant, il existe un véritable hiatus avec les faunes d'Afrique du Nord. Celui-ci ne peut trouver son explication que si l'on admet une vaste séparation géographique, non seulement capable de faire obstacle à la facile propagation des individus, mais aussi susceptible d'engendrer de part et d'autre des conditions de milieu nettement différentes, capable d'induire de grandes divergences dans l'évolution des espèces.

Cette disparité des conditions peut encore apparaître sur la base d'autres constatations. En ce qui concerne notamment le facteur température, des indications peuvent être fournies par la considération du genre *Cytherelloïdea*. On sait que la répartition et la fréquence de ce genre sont assez étroitement conditionnées par la température du milieu (Kornicker 1963). A ce point de vue, le Sud-Est de la France et l'Espagne ne présentaient pas des différences bien significatives. Par contre, la fréquence anormalement élevée du genre en Afrique du Nord peut être mise en rapport avec une nette augmentation de la température du milieu marin, ce qui, compte-tenu de la diminution de la température avec l'abaissement du fond, lui permettait de fréquenter des eaux plus profondes.

Des résultats dans le même sens résultent de l'étude des formations d'émersions berriasiennes en Europe du Sud-Est et sur la plate-forme saharienne. Alors que dans le Sud-Est de la France et en Espagne ces dépôts sont riches en fossiles d'eau douce (Charophytes, Ostracodes .. etc.), leur rareté dans les formations contemporaines d'Afrique du Nord est un critère de bien plus grande aridité. Bref, toutes ces constatations amènent à penser que les différences de latitudes entre le Sud de l'Espagne actuelle et la bordure algéro-marocaine devaient être alors relativement grandes, ce qui rend évidemment sans justification les essais d'emboitement qui ont été tentés (fig. 1).

Mais d'autre part, les espèces qui réussirent à se propager vers le Nord ne sont pas des formes psychrosphériques, ce qui exclut des conditions océaniques et font pencher pour une mer de type épicontinental. On peut alors imaginer que la plateforme saharienne se prolongeait vers le Nord et la Meseta ibérique vers le Sud par d'importantes marges cratoniques, qui auraient constitué le soubassement de cette mer, et qui auraient été incluses par la suite dans l'édifice betico-rifain. Mais dans la théorie des plaques, on peut hésiter à envisager de telles largeurs de marges sans faire intervenir une partie océanique. L'existence d'une plaque de petite taille intercalée entre les plaques africaine et européenne au niveau de la mer d'Alboran ('sous-plaque d'Alboran'), comme l'hypothèse en a été récemment formulée pour expliquer la structure de l'arc de Gibraltar (Andrieux et al. 1971), pourait constituer un élément de solution.

Bibliographie

Andrieux, J., Fontboté, J. M. & Mattauer, M. 1971. Sur un modèle explicatif de l'Arc de Gibraltar. Earth and Planet. Sci.Let. 12 (2):191-198.
Bartenstein, H. 1959. Feinstratigraphisch wichtige Ostracoden aus dem nordwestdeutschen Valendis. Paläont. Z. 33:224-226.
Bullard, E., Everett, J. E. & Smith, A. G. 1965. The fit of the Continent around the Atlantic. A Symposium on Continental Drift. Ph.Tr.: 1088.
Dietz, R. & Holden, J. 1970. Reconstruction of Pangaea: Breakup and dispersion of continents, Permian to Present. J. Geophys. Res. 75 (26):4939-4956.
Donze, P. 1964. Ostracodes berriasiens des massifs subalpins septentrionaux (Bauges et Chartreuse). Trav. Lab. Géol. Fac. Sc. Lyon, N.S. 11:103-158.
Donze, P. 1965. Espèces nouvelles d'Ostracodes des couches de base du Valanginien de Berrias (Ardèche). Trav. Lab. Géol. Fac. Sc. Lyon, N.S. 12:87-107.
Donze, P. 1967. Deux nouvelles espèces d'Ostracodes du Crétacé inférieur vocontien. Trav. Lab. Géol. Fac. Sc. Lyon, N.S. 14:63-67.
Donze, P. 1973. Corrélations stratigraphiques dans le Berriasien-Valanginien inférieur du Sud-Est de la France, sur la base de nouveaux Trachyleberidinae (Ostracodes). Remarques paléoécologiques. Doc. Lab. Géol. Univ. Lyon, pp. 1-13.
Donze, P. 1974. Ostracodes des niveaux de la limite Jurassique-Crétacé dans le Sud-Ouest constantinois (Algérie). VIème Congrès africain de Micropaléontologie, Tunis (à paraître).
Donze, P. 1975. Espèces nouvelles d'Ostracodes du genre Protocythere Triebel, 1938, dans le Berriasien du Sud-Est de la France. Répartition stratigraphique. Rev. Espan. Micropal., num espec. pp. 97-106.
Drake, C. L. & Nafe, J. E. 1968. Geophysics of the North Atlantic region. Symposium on continental drift emphasizing the history of South Atlantic area. Montevideo, Uruguay. Can. J. Earth Sci 5:993-1010.
Kornicker, L. 1963. Ecology and classification of Bahamian Cytherellidae (Ostracoda). Micropaleontology 9 (1):60-70.
Neale, J. W. 1967. Ostracodes from the type Berriasian (Cretaceous) of Berrias (Ardèche, France) and their significance. Essays in Paleontology and Stratigraphy, Raymond C-Moore Commem. Vol., University of Kansas, pp. 539-569.
Oertli, H. J. 1966. Die Gattung Protocythere (Ostracoda) und verwandte Formen im Valanginien des zentralen Schweizer Jura. Ecl. Geol. Helv. 59:87-127.

Adresse de l'auteur:

Centre de Paleontologie stratigraphique associé au CNRS, Dépt des Sciences de la Terre, Université Claude Bernard (Lyon I), 15-43 Bd 11 Novembre 1918, 69621 Villeurbanne, France

Sixth Intern. Ostracod Symposium, Saalfelden

OSTRACOD BIOFACIES IN PALEOZOIC AND MESOZOIC LAKE SEDIMENTS OF THE USSR

I. Y. NEUSTRUEVA

Abstract

The dependence of peculiarities of ostracod thanatocoenoses in continental deposits on their facial belonging is determined. Ostracod thanatocoenoses in Paleozoic and Mesozoic lake deposits of the USSR are fixed in the following facies: littoral shallow-water (zone of wave disturbances); subaquatic parts of river deltas and temporary influxes into the lake; lake waters with quiet hydrodynamic regime (sheltered parts of the littoral); remote offshore, deeper (central) part of the lake; stagnant boggy or swampy water basins. The characteristics of thanatocoenoses connected with these facies is given. The above cited facies can be distinguished in lake sediments of different ages and various genetic and hydrochemical types. The most characteristic ostracod associations for certain types of Late-Paleozoic and Mesozoic lakes of some regions of the USSR (Minusinsk, Kuznetsk, Tunguska, Kansk-Achinsk coal fields of Siberia; Zabaikalja, Middle Asia, Kaspian depression and Russian platform) are listed.

Data on the characteristics of ostracod thanatocoenoses and their related facies given in this report are based on the investigation of large quantities of material from the continental Upper Palaezoic sediments of the Minusinsk, Kuznetsk and Tunguska Basins; from Lower Triassic sediments of the Kuznetsk Basin and from the Jurassic sediments of Middle Asia and the Kansk-Achinsk Basin in Siberia (Neustrueva 1961, 1966a,b, 1971, 1974, 1975a,b). Comparison of the author's material with data on the distribution of different ostracod genera in other regions of the U.S.S.R. is based on information available in the published literature (Belousova 1956, 1961, 1962, Kashervarova 1956, 1958, 1960, 1961, Kukhtinov 1971, Lipatova & Starozhilova 1968, Ljubimova 1956, Ljubimova, Kazmina & Reshetnikova 1960, Mandelshtam 1947, 1956, Mishina 1965, 1966, Sinitsa 1969a,b, Skoblo 1961a,b, 1975, Shneider 1948, 1960 et al.).

In investigating ostracod fossilisation, lithological-facies and taphonomic methods have been used. All characteristics – rock composition (in the form of granulometric curves), their lamination and facies, structure and composition of ostracod fossils (thickness of intercalated layers with ostracods, quantity and location of shells, form of preservation, taxonomic composition), presence of remains of other groups of fauna and plants – are represented graphically (Fig. 1). Such representation of lithological and palaeontological characteristics affords an opportunity of seeing the relationship of ostracod fossilisation peculiarities to certain facies.

In our studies of continental Palaeozoic and Mesozoic sediments these methods have shown that ostracod fossilisation was connected chiefly with limnetic facies. The ostracod remains were not found in river channel facies. Among lake

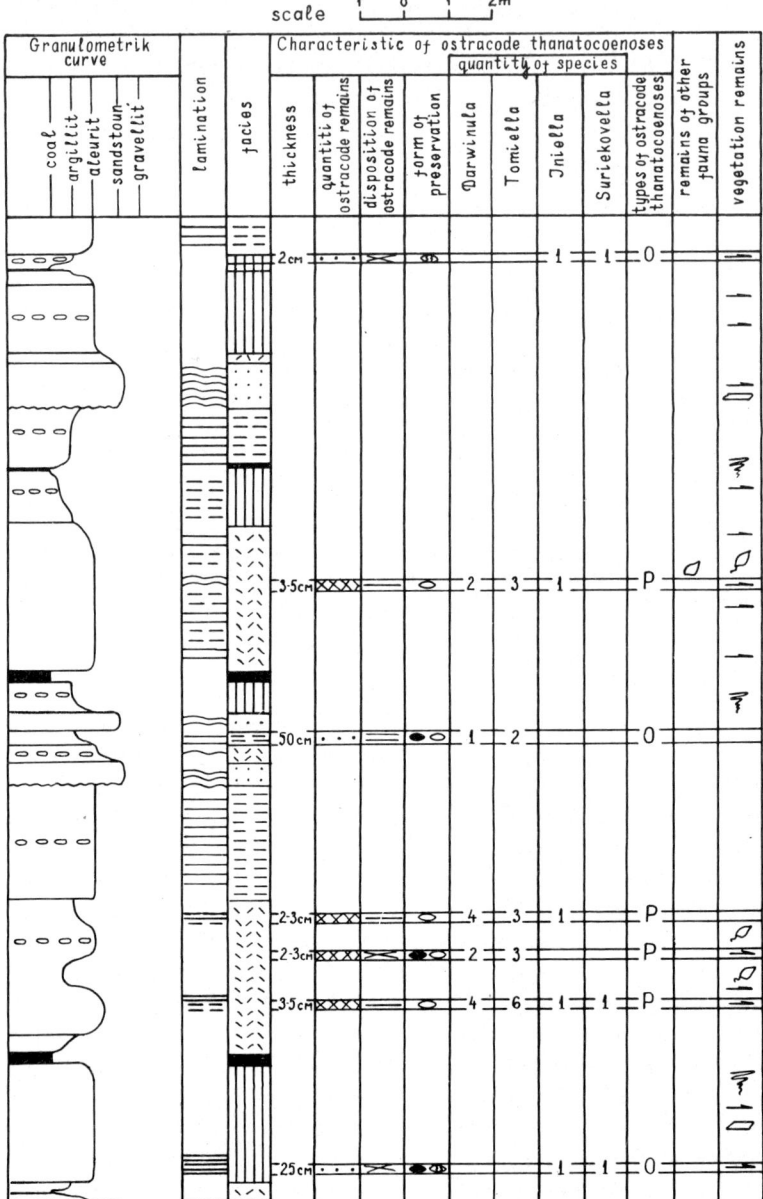

Fig. 1. Typical scheme of registration lithologic-facial and taphonomic observations (part of section of Upper-Pamian deposits by the village of Abasheva, Kuznetsk Basin).

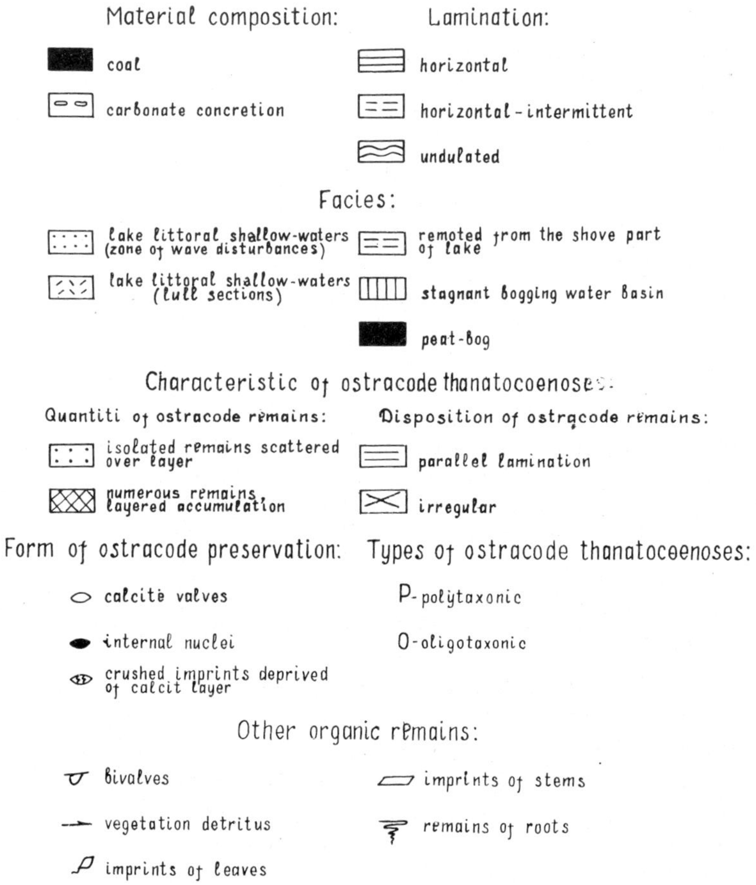

Material composition:
- ■ coal
- ⊂⊃ carbonate concretion

Lamination:
- ≡ horizontal
- = = horizontal-intermittent
- ∼ undulated

Facies:
- ⋮⋮ lake littoral shallow-waters (zone of wave disturbances)
- ∠∨∠ lake littoral shallow-waters (lull sections)
- == remoted from the shove part of lake
- ‖‖‖ stagnant bogging water basin
- ■ peat-bog

Characteristic of ostracode thanatocoenoses:

Quantiti of ostracode remains:
- ⋮⋮ isolated remains scattered over layer
- ⋈ numerous remains layered accumulation

Disposition of ostracode remains:
- ≡ parallel lamination
- ⋈ irregular

Form of ostracode preservation:
- ○ calcite valves
- ● internal nuclei
- ⌾ crushed imprints deprived of calcit layer

Types of ostracode thanatocoenoses:
- P - polytaxonic
- O - oligotaxonic

Other organic remains:
- ⌴ bivalves
- → vegetation detritus
- ♪ imprints of leaves
- ⌯ imprints of stems
- ⍭ remains of roots

Key to Fig. 1.

deposits the following facies containing ostracod remains have been distinguished: littoral shallow-water (zone of wave disturbance); subaquatic part of river deltas and temporary influxes into the lake; lake waters with quiet hydrodynamic regime (sheltered parts of the littoral); the off-shore, deeper (central) part of the lake? stagnant, boggy or swampy basins.

Ostracod thanatocoenoses in the above facies are distinguished by structure and composition, and so their characteristics are related to facies as shown below.

1. Littoral shallow-water (zone of wave disturbance)

Two types of ostracod thanocoenose are distinguished:
a. A polytaxonic thanatocoenoses consists of up to 10 species, connected with the aggradation of shelly detrital material along the shore in the beach zone. Numerous shells and debris of bivalves and ostracods form the interleaved layers

of shell deposits with thicknesses of from 3 to 10 cm. Among the ostracods large forms prevail due to the destruction (reworking) of the small and thin valves. Larval shells are absent. This sort of thanatocoenose is typical of lakes with a semi-arid climate (Early Triassic period in the Kuznetsk Basin, Middle Jurassic – Lower Cretaceous of Middle Asia).

b. An oligotaxonic thanatocoenoses is represented by isolated scattered valves and is found in sandstones with undulating lamination formed in the zone of breaking waves. This type can be found in the lake deposits of different climatic zones of various epochs.

These two types of taphocoenoses are allochthonous.

2. Subaquatic part of river deltas and temporary influxes into the lake

These ostracod thanatocoenoses are oligotaxonic, mainly allochthonous, and are represented by separated valves, sometimes by entire shells, in fine-grained sandstones with gently inclined lamination; they are found in deposits of different lake types belonging to diverse epochs.

3. Lake waters with quiet hydrodynamic regime (sheltered parts of the littoral).

This is the most favourable facies for colonisation and fossilisation of ostracods. Polytaxonic and autochtonous thanatocoenoses are represented by large quantities of complete carapaces and entire separate valves, belonging to the different instars. Genus and species composition is diverse (up to 10 species in one thanatocoeneoses). The ostracod shells form disorientated agglomerations on the bedding plane surface from 0.5 to 5 cm thick. Interleaved horizons rich in ostracod remains may be found repeatedly in one section. Ostracod thanatocoenoses of this facies are connected with aleurites, argillites (often calcareous), marls with horizontal lamination or unlaminated ones. Besides ostracods, one also usually finds here bivalve, gastropod and conchostracan shells, insect impressions, gyrogonia of charophyte algae, plant fragments and other organic remains. Such thanatocoenoses are distributed in lake sediments of different ages and hydrochemical regimes (rich in humic acid and weakly saline lakes of temperate, humid climatic zones; moderately and significantly saline lakes of semi-arid and arid climatic zones, etc.).

4. Off-shore (central) part of the lake

Oligotaxonic ostracod thanatocoeneses (1 to 3 species), mainly allochthonous, are represented by a small number of separated valves or entire carapaces, which occur throughout the profile, in horizontally laminated (sometimes unlaminated) well-sorted 'pure' argillites and clays. As a rule, larval shells are absent. These thanatocoenoses are distributed in lake deposits of different hydrochemical regimes and age.

5. Stagnant boggy or swampy water basins

Thanatoceocenoses of ostracods are oligotaxonic to monotaxonic and autochthonus. They consist of small numbers of separate valves or impressions, usually decalcified, scattered in aleurites and argillites saturated with coal bearing matter. Larval shells are absent. They are typical of lakes in a humid climate and are often found in coal-bearing deposits.

The faunal composition of ostracods in the facies considered depends on the age, hydrochemical and genetic type of the lakes. The material studied suggests that the following generic associations of ostracods are typical of certain types of lakes of one or other age.

a. The genera *Tomiella, Iniella, Suriekovella* and *Darwinula* are typical of shallow-water, humus-poor lakes of the temperate humid climatic zone of the Late-Paleozoic (Kuznetsk, Tunguska, Minusinsk Basins).

b. The genera *Inielly, Suriekovella* are typical stagnant boggy or swampy water basins with a considerable humus content such as the Late-Paleozoic of the Tunguska biogeographical region (Kuznetsk, Tunguska, Minusinsk Basins).

c. The genera *Darwinula* and *Darwinuloides* are typical of shallow-water, warm, moderately-mineralized lakes with sufficient content of calcium carbonate (Late-Permian – Early Triassic of Russian platform and Caspian depression; Early Triassic of Kuznetsk Basin).

Darwinula and *Timiriasevia* are typical of humus rich and moderately-mineralized, warm, lakes of the Jurassic epoch. They are know both from coal-bearing and from non-coal-bearing deposits of various Asian regions (USSR – Mangyshlak, Fergana, Kansk-Achinsk Basin, West-Siberian Lowland, Zabaikalje, Mongolia, China) and Europe (England).

d. *Suchonella, Gerdalia, Darwinula* are characteristic of considerably mineralized lakees of semi-arid and arid climatic zones of the Late Permian and Early Triassic of the Russian Platform and Caspian depression and the Early Triassic of Siberia.

Representatives of the genera *Cypridea* and *Theriosynoecum* are almost universal in the mineralized lakes of the Late Jurassic-Cretaceous (Europe, Asia, Africa, North and South America), (Anderson 1967, Bischoff 1963, Branson1966, Christensen 1963, Grekoff 1957, Grekoff & Krömmelbein1967, Harper & Sutton 1935, Helmdach 1971, Hou 1958, Jones 1885, Krömmelbein 1962, Krömmelbein& Weber 1971, Loranger 1951, Martin 1940, Oertli 1963, Peck 1951, Pinto & Sanguinetti 1962, Swain 1946, Wicher 1959, Wolburg 1959, and many others). It must be emphasized that, in addition to the above-mentioned generic associations of ostracods typical of lakes of different age and hydrochemical pattern, representatives of other genera are found in each individual region. The latter, however, have narrower areas of distribution and apparently reflect on the one hand the local geograpical differences, and on the other the specifity of water basins of one or another topographic zone.

On the basin of these investigations the following conclusions can be reached:
1) the structure of ostracod thanatocoenoses (i.e., the thickness of interleaved

layers containing ostracods, location of specimens in the layer, their quantity and qualitative variety) depends upon the facies of the deposits. In general outline these characteristics are irrespective of the age, of the deposits and the type of water basin.

2) the taxonomic composition of ostracod thanatocoenoses and the form of preservation of their shells depend, on the one hand, on age, and, on the other, on the specific character of the genetic and hydrochemical type of the water basins, where they have lived and are buried.

References

Anderson, F,. W. 1967. Ostracodes from the Weald Clay of England. Bull. of the Geol. Surv. of Great Britain 17.

Bischoff, G. 1963. Ostracoden Studien im Libanon: Die Gattung Cypridea im Aptien inférieur. Senck. Leth. Bd. 44, No. 4.

Branson, C. C. 1966. Fresh-water ostracode genus Theriosynoecum. Oklahoma Geol. Notes. Norman. 26, No. 4.

Christensen, O. B. 1963. Ostracodes from the Purbeck-Wealden Beds in Bornholm. Geol. Survey of Denmark, II ser. No. 86.

Grekoff, N. 1957. Ostracodes du bassin du Congo. I. Jurassique supérieur du nord du bassin. Ann. Mus. Roy. Congo Belge. ser. in-8°, Sci. Geol. 19.

Grekoff, N. & Krömmelbein, K. 1967. Etude comparée des ostracodes mesozoiques continentaux des bassins atlantiques de Cocobeach, Gabon et série de Bahia, Bresile. Rev. Inst. Français du Pétrole. 22, No. 9.

Harper, F. & Sutton, A. M. 1935. Ostracodes of the Morrison formation from the Black Hills, South Dakota. Journ. Pal. 9, No. 8.

Helmdach, F. F. Zur Gliederung limnich-brakischer Sedimente des portugiesischen Oberjura (ober Callovien-Kimmeridge) mit Hilfe von Ostracoden. N. Jb. Geol. Paläont. Mh. 11.

Hou, I. T. 1958. Jurassic and Cretaceous nonmarine ostracodes of the subfamily Cyprideinae from North-Western and North-Eastern regions of China. Mem. Inst. Paleontol. Acad. Sinica 1.

Jones, T. R. 1885. On the Ostracode of the Purbeck Formation; with Notes on the Wealden Species. Quart. Journ. Geol. Soc. 41.

Krömmelbein, K. 1962. Zür Taxonomie und Biochronologie stratigraphisch wichtiger Ostracoden-Arten aus der ober-Jurassisch? – unterkretazischen Bahia-Serie (Wealden-Facies) NE – Brasiliens. Senck. Leth. Bd. 43, No. 6.

Krömmelbein, K. & Weber, R. 1972. Ostracoden des Nordost Brasilianischen Wealden. Beih. Geol. Jhrbuch. Hf. 115.

Loranger, D. M. 1951. Useful Blairmore microfossil zone in Central and Southern Alberta, Canada. Bull. Amer. Assoc. Petroleum Geologist. 35, No. 11.

Martin, G. P. R. 1940. Ostracoden des norddeutschen Purbeck und Wealden. Senckenbergiana Bd. 22, No. 5/6.

Oertli, H. J. 1963. Ostracodes du "Purbeckien" du bassin Parisien. Rév. Inst. Français du Pétrole XVIII, No. 1.

Peck, R. E. 1951. Nonmarine Ostracodes the subfamily Cyprideinae in the Rocky Mountain area. Journ. Palaeont. 25, No. 3.

Pinto, I. D. & Sanguinetti, J. T. 1962. A complete revision of the genera Bisulcocypris and Theriosynoecum (Ostracoda) with the World geographical and stratigraphical distribution (including Metacypris, Elpidium, Gomphocythere and Cytheridella). Publ. Esp. Esc. Geol. P. Alegre 4.

Swain, F. M. 1946. Middle mesozoic nonmarine Ostracoda from Brazil and New Mexico. Journ. Palaeontol. 20, No. 6.

Wicher, C. A. 1959. Ein Beitrag zur Altersdeufung des Recôncavo, Bahia (Brasilien) Geol. Jb. No. 77.

Wolburg, J. 1959. Die Cyprideen des NW-deutschen Wealden. Senckenb. Leth. Bd. 40, No. 3/4.

Белоусова З. Д. 1956. Остракоды из разреза верхнепермских отложений. Вопросы стратиграфии, палеонтологии и литологии палеозоя и мезозоя районов Европейской части СССР. Тр. ВНИГНИ, вып. VII.

Белоусова З. Д. 1961. Остракоды нижнего триаса. БМОИП, отд. геол., т. XXXVI, № 1.

Белоусова З. Д. 1962. Остракоды верхнепермских и нижнетриасовых отложений центральных и северо-восточных районов Русской платформы. Тр. ВНИГНИ, вып. 7 (11).

Кашеварова Н. П. 1956. Остракоды татарского яруса Бугуруслано-Куйбышевской нефтеносной области. Авторефераты научных трудов ВНИГРИ (работы, выполненные в 1953г.), вып. 15.

Кашеварова Н. П. 1958. Новые виды остракод из верхнепермских отложений Южного Тимана и Волго-Уральской области. Тр. ВНИГРИ, н.с., вып. 115.

Кашеварова Н. П. 1960. Остракоды верхнепермских континентальных отложений СВ Европейской части СССР. Междунар. геол. конгресс, XXI сессия. Доклады советских геологов.

Кашеварова Н. П. 1961. Остракоды континентальной фации казанского яруса восточного склона Среднего Тимана и полуострова Канин. Тр. ВНИГРИ, вып. 179.

Кухтинов Д. А. 1971. Географическая зональность в распределении раннетриасовых (индских) остракод Европейской части СССР. Тезисы докл. на III Всесоюзном коллокв. по остракодам. Таллин.

Липатова В. В., Старожилова Н. Н. 1968. Стратиграфия и остракоды триасовых отложений Саратовского Заволжья. Изд. Саратовского Университета.

Любимова П. С. 1956. Остракоды меловых отложений восточной части Монгольской Народной Республики. Тр. ВНИГРИ, н.с. вып. 93.

Любимова П. С., Казьмина Т. А., Решетникова М. А. 1960. Остракоды мезозойских и кайнозойских отложений Западно-Сибирской низменности. Тр. ВНИГРИ, вып. 160.

Мандельштам М. И. 1947. Ostracoda из отложений средней юры полуострова Мангышлака. В кн.: Микрофауна нефтеносных месторождений Кавказа, Эмбы и Средней Азии.

Мандельштам М. И. 1956. Описания новых родов остракод. В кн.: „Материалы по палеонтологии (новые семейства и роды)".

Мишина Е. М. 1965. Расчленение татарских отложений Костромской области по остракодам. Сб. статей по геологии и гидрогеологии МГ СССР. 2-е Гидрогеологическое управление, вып. 4.

Мишина Е. М. 1966. Детальная стратиграфия отложений ветлужской серии нижнего триаса по остракодам. Изв. АН СССР. сер. геол., № 12.

Неуструева И. Ю. 1961. О возрасте континентальных пестроцветных отложений Канско-Тасеевской депрессии. В кн.: Вопросы геологии угленосных отложений Азиатской части СССР.

Неуструева И. Ю. 1966а. Остракоды верхнепалеозойских отложений Минусинского угольного бассейна. В кн.: Континентальный верхний палеозой и мезозой Сибири и Центрального Казахстана.

Неуструева И. Ю. 1966b. Верхнепермские остракоды Кузнецкого бассейна. В кн.: Континентальный верхний палеозой и мезозой Сибири и Центрального Казахстана.

Неуструева И. Ю. 1971. О палеоэкологии позднепермских и раннетриасовых остракод Кузнецкого каменноугольного бассейна. Тезисы докл. на III Всесоюзном коллокв. по остракодам. Таллин.

Неуструева И. Ю. 1974а. Остракоды юрских озер Ферганы и их палеоэкологическая характеристика. В кн.: Проблемы исследования древних озер Евразии.

Неуструева И. Ю. 1974b. Некоторые виды остракод из юрских и нижнемеловых отложений Монголии. В кн.: Фауна и биостратиграфия мезозоя и кайнозоя Монголии. Тр. СМПЭ, вып. 1.

Неуструева И. Ю. 1975а. Реконструкция условий обитания пресноводных остракод на основе изучения их захоронений. Тр. Ин-та геол. и геофиз. СО АН СССР, вып. 333.

Неуструева И. Ю. 2975b. История озер и развитие остракод раннего триаса Кузнецкого бассейна. Тезисы докл. на IV Всесоюзном симпозиуме по истории озер. т.1.

Синица С. М. 1969а. Биостратиграфия верхнего мезозоя Центрального Забайкалья по остракодам. Изв. Забайкальского филиала геогр. о-ва СССР, т.V, вып. 2.

Синица С. М. 1969b. Биостратиграфия верхнего мезозоя Восточного Забайкалья по остракодам. Изв. Забайкальского филиала геогр. о-ва СССР, т.V, вып. 4.

Скобло В. М. 1961а Новые виды рода Cypridea из нижнего мела района Гусиного озера. В кн.: Материалы по геологии и полезным ископаемым Бурятской АССР, вып. VII. Улан-Удэ.

Скобло В. М. 1961b. Новые виды остракод родов Zejaina, Limnocypridea u Darwinula из отложений гусиноозерской серии. Материалы по геологии и полезным ископаемым Бурятской АССР. вып. VII. Улан-Удэ.

Скобло В. М. 1975. Условия обитания остракод и некоторые биофации мезозоя Западного Забайкалья. Тр. Ин-та геол. и геофиз. СО АН СССР, вып. 333.

Шнейдер Г. Ф. 1948. Фауна остракод верхнепермских отложений (татарский и казанский ярусы) нефтеносных районов СССР. Тр. ВНИГРИ, н.с. вып. 31.

Шнейдер Г. Ф. 1960. Фауна остракод нижнетриасовых отложений Прикаспийской низменности. Тр. КЮГЭ, вып. V.

Author's address:

Institute of Limnology of the Academy of Sciences, Petrovskaja nab. 4, Leningrad 197046, USSR

Sixth Intern. Ostracod Symposium, Saalfelden

THURINGIAN OSTRACODS FROM THE FAMENNIAN OF THE CANTABRIAN MOUNTAINS (UPPER DEVONIAN, N. SPAIN)*

GERHARD BECKER

Abstract

Silicified ostracod faunas derived from nodular limestones of the Montó beds (Famennian) in the eastern part of the Cantabrian Mountains (N. Spain) are described. The faunas represent a mixed type, yielding benthonic assemblages characterised by spinous podocopids ('Thüringer Ökotyp') and pelagic entomozoids ('Entomozoen-Ökotyp').

The ostracod faunas show a close relationship to ostracod communities known from the Thuringian Upper Devonian – not only morphologically but also on a taxonomic level. Most of the benthonic species are described from the E. Thuringian Slate Mountains, only a few species (represented under open nomenclature) appear to be endemic to the Cantabrian Mountains. In this region, the upper part of zone 6 and the zone 7 of the ostracod zonation proposed by Blumenstengel (1965) for the Thuringian Upper Devonian can be recognised with some certainty.

The entomozoids (already identified from Thuringia and other localities in Western and Central Europe) demonstrate the *hemisphaerica-dichotoma*-Zone (= *Clymenia-Wocklumeria*-Stufe), the elder of the two faunas described, with one species of *Richterina* (*Fossirichterina*?), most probably belongs to the lower part of this entomozoid zone (= lower part of the *Clymenia*-Stufe).

The 'Thüringer Ökotyp' is indicative of calm and probably deeper, marine environments. The entomozoids are considered to be pelagic. Criticisms of this mode of life are discussed on the basis of silification of the carapace.

Introduction

For the argillo-calcareous Upper Devonian of the Eastern Thuringian Slate Mountains, Blumenstengel (1965) proposed a parachronology based on silicified benthic ostracods. As a calibration standard for his eight ostracod zones, some of which were subdivided, the author employed conodonts, entomozoids, and, in addition, a number of orthochronologically verified profiles of the Saalfeld area (Blumenstengel, 1965:9).

In searches for conodonts in the Upper Devonian nodular limestones of the Eastern Cantabrian Mountains, silicified ostracods have been discovered at several places (Budinger, oral communication; Adrichem Boogaert, 1967). Faunal descriptions are presented by Bless & Michel (1967).

Benthic ostracodes are very specific to a particular facies. Thus, it was considered that, within the framework of an interdisciplinary investigation concerned with the changes in space and time in marine biocoenoses (see Kullmann & Schönenberg, 1975), it should also be attempted to determine: first, the degree

---

* Fossil-Vergesellschaftungen, No. 46.–No. 45: Fürsich, F. T., Palaeogeography, Palaeoclimatology, Palaeoecology, 20:235-256, Amsterdam 1976.

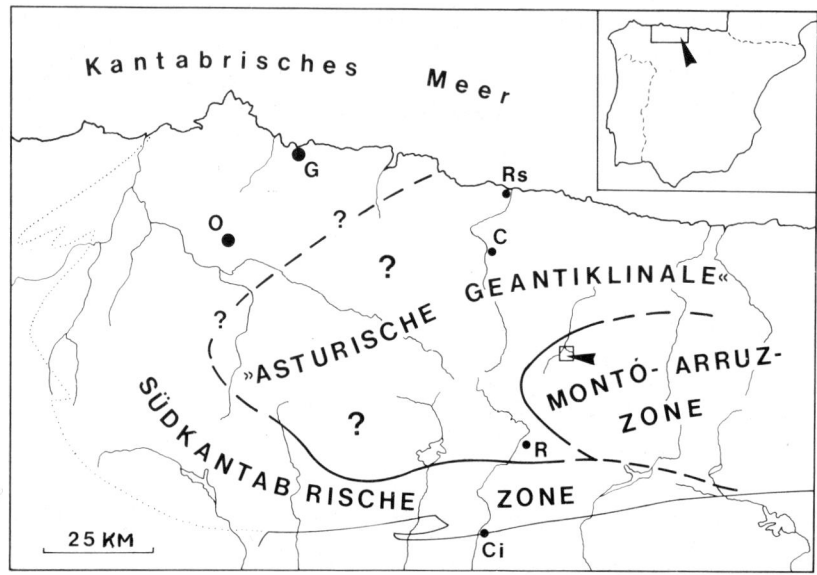

Fig. 1. Ranges of facies in the Devonian of the Cantabrian Mountains. Arrow=map section of Fig. 2. C=Cangas de Onis, Ci=Cistierna, G=Gijón, O=Oviedo, R=Riaño, Rs=Ribadesella. Adapted from Adrichem Boogaert (1967: fig. 56).

to which the Upper Devonian of the Montó-Arruz Zone (see Fig. 1), which resembles the Thuringian facies, is characterised by Thuringian faunal elements; and secondly, the extent to which the zonation proposed by Blumenstengel is of significance for other regions outside of Thuringia.

Examinations of the samples of nodular limestones and nodule-bearing shales from the Montó beds (Famennian) have revealed, at one locality (Loc. 110/1) the presence of ostracod faunas, composed of both silicified species of the Thuringian habit (mainly spinous Podocopida) and entomozoids. This was the first record of the Entomozoidae from the Cantabrian Upper Devonian. The entomozoids were similarly silicified, an extremely rare and less favourable mode of preservation for the group.

The first contribution is concerned with these ostracod associations, which are of particular interest from diverse (biostratigraphical, regional, biofacial and paleoecological) aspects.

Stratigraphy

In the Devonian of the Cantabrian Mountains, two distinct facies areas are developed (see Fig. 1): the South Cantabrian zone respectively Asturoleonese facies in the south and west of the area under discussion and the Montó-Arruz Zone or Palentine facies in the east (Kullmann, 1965; Adrichem Boogaert, 1967). In the South Cantabrian Zone, the shallow water sediments (sandstones, detrital limestones, biostromes) contain rich benthic faunas (brachiopods, crinoids, stromatoporids, colonial rugous and tabular corals, large growing solitary corals of the Zaphrentacea). In contrast, in the Montó-Arruz Zone, the predominant deposits (shales, nodular limestones, nodule-bearing marls) are poor in fauna (cephalopods, tentaculites, small solitary corals of the Cyathaxoniacea), indicative of a calm polluted environment at relatively great depth (Kullmann, 1965:43).

This facies differentiation, which was already in existence in the Lower Devonian, and which became more pronounced until the Upper Devonian, resulted from the separation of the distributive sedimentary area during the Devonian, into basins and swells, under general subsidence (Kullmann & Schönenberg, 1975:156). Adrichem Boogaert (1967:169) thought that he could recognise a subcentral geanticline which simultaneously moved upwards and extended to the north and northeast as far as the present Cantabrian coast. In his opinion, this 'Asturian Geanticline' appeared most intensively in the lower Famennian.

Accordingly, in the south Cantabrian Zone, the Famennian should have been incomplete at first (Pre-Ermita hiatus), and then, later, transgressive (Ermita sandstone). Recent observations (Kullmann, personal communication) appear to raise some doubts about this supposition.

At the same era, in the Montó-Arruz Zone, above the basal quartzites, a relatively short, continuous, stratigraphic sequence of nodular limestone and nodule-bearing shales, containing cephalopods, occurs (Montó beds; Kullmann, 1960). Based on the ammonite contents, a succession from the upper *Cheiloceras*-Stufe (to IIβ) to the *Wocklumeria-Kalloclymenia*-Stufe (toVI) can be established (Kullmann, 1960:462-473; 1964: 167-169; 1965:49-50). The conodont parachronology (Budinger & Kullmann, 1964; Adrichem Boogaert, 1967) conforms well with the ammonite orthochronology. The geological span from the *quadrantinodosa* zone to the upper *costatus* zone can be verified by the conodonts.

Sampling locality and faunas

The locality 110/1 (Mapa National de España, Hoja 80, Burón; $l=1°14'40''$, $h=43°07'10''$) is situated between Pico Gildar and Río Montó (Gildar-Montó area, Kullmann) on the western slope of the Collado de Anzo (see Fig. 2). The nearly 40 m thick stratigraphic succession, which, in the opinion of Professor Kullmann, accords the upper Famennian (presumably from toIV onwards), strikes almost perpendicular to the slope and dips at an angle of approximately 45° towards the southeast. The underlying stratum is covered and is exposed once again on the eastern slope of the Collado (Loc. 110/2). Sandy black shales constitute the hanging layers.

Yellowish brown nodule-bearing shales (sample 740916/2a) from the lower section of the Montó beds at Loc. 110/1. The pelite content diminishes towards the top. Above these shales, first succeeds a reddish nodular limestone (sample 740916/2b) then a light gray nodular limestone (sample 740916/2c).

The degree of silification also appears to decrease towards the top part of the section. Whereas the limestone nodules of the sample 'a' yielded a very rich ostracod fauna (refer to pls. 1-3), in the reddish limestone of sample 'b', only few speci-

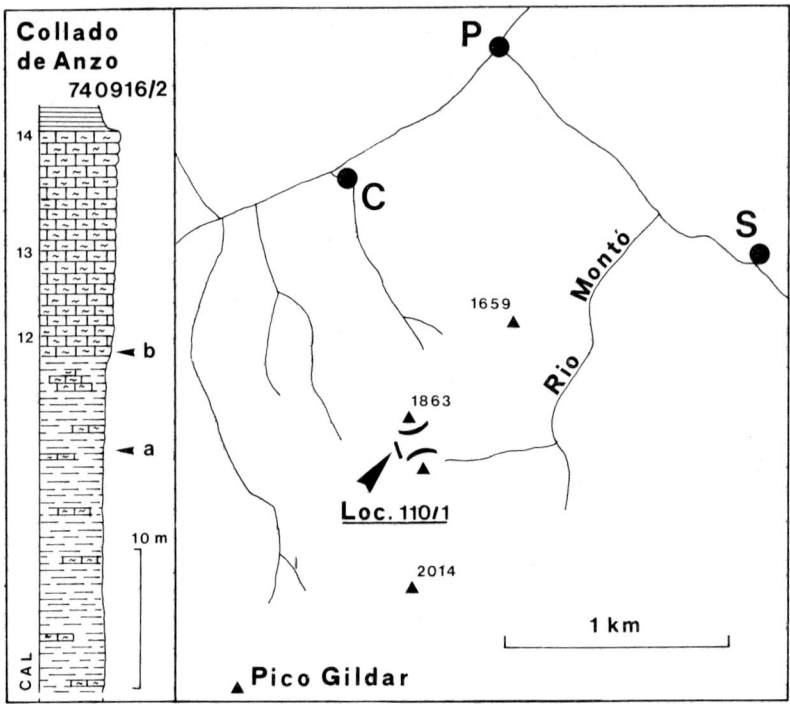

Fig. 2. Gildar-Montó area with ostracod locality (Loc. 110/1) and ostracod layers (for a description of the section, see text). C=Caldevilla, P=Posada de Valdeón, S=Santa Marina de Valdeón. 740916/2a-b=Ostracod samples, CAL 12-14=Conodont samples of Adrichem Boogaert. General map after Kullmann (1969: fig. 4), columnar section adapted from Adrichem Boogaert (1967: fig. 27).

mens (from actually a large quantity of material) could be identified. The light nodular limestone (sample 'c') in the underlying beds of black shales lacked silicified remains. Detailed identifications were made:

Sample 740916/2b: *Amphissites* aff. *irinae* Egorov, 1953 sensu Blumenstengel, 1965; *Marginohealdia blumenstengeli* n.sp.; *Marginohealdia ventrocostata* Blumenstengel, 1965; *Tricornina (Tricornina) communis* Blumenstengel, 1965; *Rectonaria muelleri* Gründel, 1961; *Rectoplacera elongata* Blumenstengel, 1965; *Triplacera triquetra* Gründel, 1961; *Bairdia (Bairdia)* spp.; *Acratia* sp.; Podocopida spp.; *Maternella dichotoma* (Paeckelmann, 1913).

The presence of conodonts could not be demonstrated correspondingly. However, in a sample (CAL 12), taken from the same sequence, (see Fig. 2), the following conodont species were described by Adrichem Boogaert (1967: encl. 3): *Palmatolepis gracilis gracilis* Branson & Mehl, 1934; *Spathognathodus stabilis* (Branson & Mehl, 1934); *Spathognathodus strigosus* (Branson & Mehl, 1934); *Spathognathodus inornatus* (Branson & Mehl, 1934); *Spathognathodus jugosus* (Branson & Mehl, 1934).

Sample 740916/2a: *Selebratina* sp.; *Amphissites* aff. *irinae* Egorov, 1953 sensu Blumenstengel, 1965; *Neochilina* aff. *capulinoda* Blumenstengel, 1965; *Marginohealdia blumenstengeli* n.sp.; *Tricornina (Tricornina) communis* Blumenstengel, 1965; *Tricornina (Bohemina) paragracilis* (Blumenstengel, 1965); *Rectonaria muelleri* Gründel, 1961; *Rectonaria inclinata* Gründel, 1961; *Orthonaria rectagona* Gründel, 1961; *Rectoplacera elongata* Blumenstengel, 1965; *Bairdia (Bairdia)* spp.; *Bairdia (Rectobairdia)* sp.; *Processobairdia posterocerata* ssp. A; *Processobairdia spinomarginata* Blumenstengel, 1965?; *Processobairdia spinaterocerata* Bless & Michel, 1967; *Processobairdia* cf. *nodocerata* Blumenstengel, 1965; *Acratia* sp.; *Ceratacratia cerata* Blumenstengel, 1965; Podocopida spp.; *Maternella dichotoma* (Paeckelmann, 1913); *Richterina (Richterina) costata* (Rh. Richter, 1869); *Richterina (Fossirichterina?)* cf. *intercostata* (Matern, 1929).

Species description

All ostracods were separated by etching with dilute formic acid. Only sample 740916/2a yielded adequately silicified and thus well preserved specimens of 'Thuringian species'. Remains of the entomozoids consisted of broken and incomplete pieces of exclusively single valves.

In the following contribution, only those few species which are of importance for the present more zoogeographically orientated study will be discussed. Presentation of detailed description of the complete material will be reserved for a later more extensive publication (Becker, in preparation). In view of this, comprehensive synonym lists, diagnosis etc. will not be included for the species dealt with in this paper (for further information see Rabien, 1954; Blumenstengel, 1959, 1965; Bless & Michel, 1967; Gooday, 1973).

The material documented is deposited in the Senckenberg-Museum, Frankfurt am Main (SMF) and in the Geological-Paleontological Institute of the University Frankfurt am Main (GPIF).

Abbreviations: C=carapace, V=valve(s), LV=left valve(s), RV=right valve(s), RI=rib intervals.

Neochilina aff. *capulinoda* Blumenstengel, 1965
Pl. 1, fig. 2

The species present in sample 740916/2a differs from *N. capulinoda* Blumenstengel from the upper toV of the E. Thuringian Slate Mountains in the comparatively small and symmetrical central knob. Two RV present; one valve (pl. 1, fig. 2, SMF Xe 10399) is amplete in outline, the other one (SMF Xe 10400) preplete in outline similar to Blumenstengel's material. This could be a new species.

Marginohealdia blumenstengeli n.sp.
Pl. 1, fig. 9; pl. 3, fig. 8

1965 *Marginohealdia* n.sp. Blumenstengel: 45-46, pl. 6, fig. 10; pl. 22, figs. 5-6.

Name: In honour of Dr. Horst Blumenstengel, Berlin, who described the species for the first time (under open nomenclature).

◄ Plate 1. (All specimens coated with magnesium oxide; phot. magnification 50×.) W Collado de Anzo, Provincia de León, N. Spain; Upper Devonian, Montó-beds, lower *hemisphaerica-dichotoma*-Zone (sample 740916/2a).
Orthonaria rectagona Gründel, 1961.

Fig. 1. C, SMF Xe 10398; a) view from right, b) view from left, c) dorsal view.
Neochilina aff. *capulinoda* Blumenstengel, 1965.
Fig. 2. RV, SMF Xe 10399; a) external view, b) ventral view.
Processobairdia posterocerata ssp. A.
Fig. 3. LV, SMF Xe 10401; a) external view, b) ventral view.
Ceratacratia cerata Blumenstengel, 1965.
Fig. 4. RV, SMF Xe 10403; external view.
Rectonaria inclinata Gründel, 1961.
Fig. 5. RV, SMF Xe 10404; external view.
Fig. 6. LV, SMF Xe 10405; external view.
Rectonaria muelleri Gründel, 1961.
Fig. 7. LV, SMF Xe 10406; external view.
Fig. 8. RV, SMF Xe 10407; external view.
Marginohealdia blumenstengeli n.sp.
Fig. 9. LV, holotype, SMF Xe 10408; a) external view, b) dorsal view.

Holotype: LV, pl. 1, fig. 9, pl. 3, fig. 8, SMF Xe 10408 (sample 740916/2a). – Type locality: Loc. 110/1; Collado de Anzo, Provincia de Léon, N. Spain. – Type stratum: Upper Devonian, Montó beds, probably lower *hemisphaerica-dichotoma*-Zone (lower toV).

Paratypes: 2 LV, 3 RV, SMF Xe 10409; 1 RV, GPIF Cr 12/1; topotypical (sample 740916/2a). 1 LV, SMF Xe 10410; Loc. 110/1 (sample 740916/2b); presumably a transitional zone, lower/upper *hemisphaerica-dichotoma*-Zone (toV).

Diagnosis: A relatively large *Marginohealdia* species with the following peculiarities: Posterior shoulder reduced into two bucklings, posterodorsal bulge knoblike, posteroventral bulge extending into a posteriorly directed spine.

Description (supplementary to Blumenstengel, 1965): Posterior shoulder characteristic for the genus, divided into two bulges: upper (posterodorsal) bulge knoblike, rounded bluntly at the posterior; lower (posteroventral) bulge more widely rounded at the posterior, low on which, a stout, cylindrical spine, directed posteriorly, is inserted.

Dimensions of the holotype: LV, SMF Xe 10408; length=1.29 mm, height=0.77 mm.

Relationships: Refer to Blumenstengel (1965:46).

Occurrence: *hemisphaerica-dichotoma*-Zone, presumably only in the lower part; toV, Upper Devonian; E. Thuringian Slate Mountains and E. Cantabrian Mountains.

Processobairdia posterocerata ssp.A.
Pl. 1, fig. 3

The material present in the sample 740916/2a markedly resembles *P. posterocerata* Blumenstengel from the toIIIβ (*veliver*-Zone, middle Hemberg-Stufe) of the E. Thuringian Slate Mountains. However, low on the dorsoposterior margin

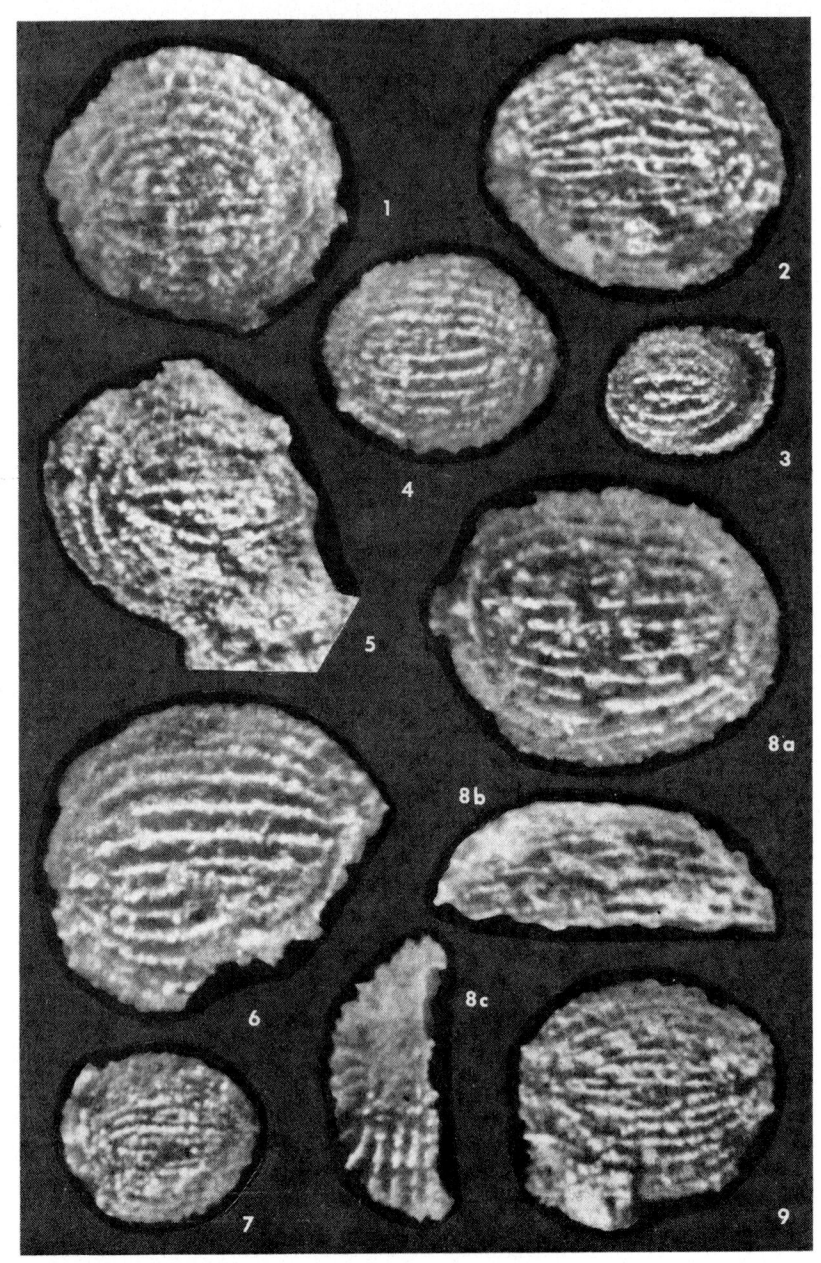

◄ Plate 2. (All specimens coated with magnesium-oxide; phot. magnification 50×.) W Collado de Anzo, Provincia de León, N. Spain; Upper Devonian, Montó-beds, lower *hemisphaerica-dichotoma*-Zone (sample 740916/2a).
Maternella dichotoma (Paeckelmann, 1913).
Fig. 1. V, SMF Xe 10411; external view.
Fig. 2. V, SMF Xe 10412; external view.
Fig. 3. V, SMF Xe 10413, external view.
Fig. 4. V, destroyed during embedding, external view.
Fig. 5. V, SMF Xe 10414; external view.
Fig. 6. V, SMF Xe 10415; external view.
Fig. 7. V, SMF Xe 10416; external view.
Fig. 8. V, SMF Xe 10417; a) external view, b) dorsal or ventral view, c) anterior or posterior view.
Fig. 9. V, SMF Xe 10418; external view.

of the larger LV, *P. posterocerata* ssp. A has an additional stout spine, which is directed inwards i.e. towards the posterior edge of the smaller RV. The species could be a subspecies of *P. posterocerata*.

Processobairdia cf. *nodocerata* Blumenstengel, 1965

In sample 740916/2a one RV of a *Processobairdia* with two processes on the lateral surface of the carapace, was present. Unfortunately, this specimen was destroyed during embedding. As far as can be remembered, it closely resembled *P. nodocerata* from the upper toV of the E. Thuringian Slate Mountains, except that the anterior knob appeared to lie more ventrally. New topotypical material should be examined to check these findings.

Maternella dichotoma (Paeckelmann, 1913)
Pl. 2, figs. 1-9; pl. 3, figs. 4-7

The material found in sample 740916/2a (approximately 30 V, SMF Xe 10411-10422, approximately 15 V, GPIF Cr 12/2) and sample 740916/2b (approximately 10 V, SMF Xe 10421) lies well within the range of variation presented by Gooday (1973, fig. 6.30) for *M. dichotoma*. The most similar feature is the pattern of the series h-j-k: the outer rib runs in spirals, from which the ribs of the median field shear off bluntly. In addition to specimens with broad striae (RI=0.08 mm) (refer to pl. 2, figs. 3, 6, 8), a 'fine-ribbed variety' also occurs (RI=0.04 mm) (refer to pl. 2, figs. 4, 7, 9). Both of these 'varieties' are apparently linked through transitional stages (see remaining figures of pl. 2). In most cases, the adductor muscle scar can be distinguished. Whereas specimens of the *Maternella* species preserved in shales always have external and internal molds with identical ornamentation (see Rabien, 1954:17, 38), generally the silicified specimens present lack sculpture on the internal surface of the valve.

Occurrences: toV-VI (Dasberg-Stufe until Wocklum-Stufe), Upper Devonian; Rhenish and E. Thuringian Slate Mountains, Harz, N. Holy Cross Mts and Devonshire, now also E. Cantabrian Mountains.

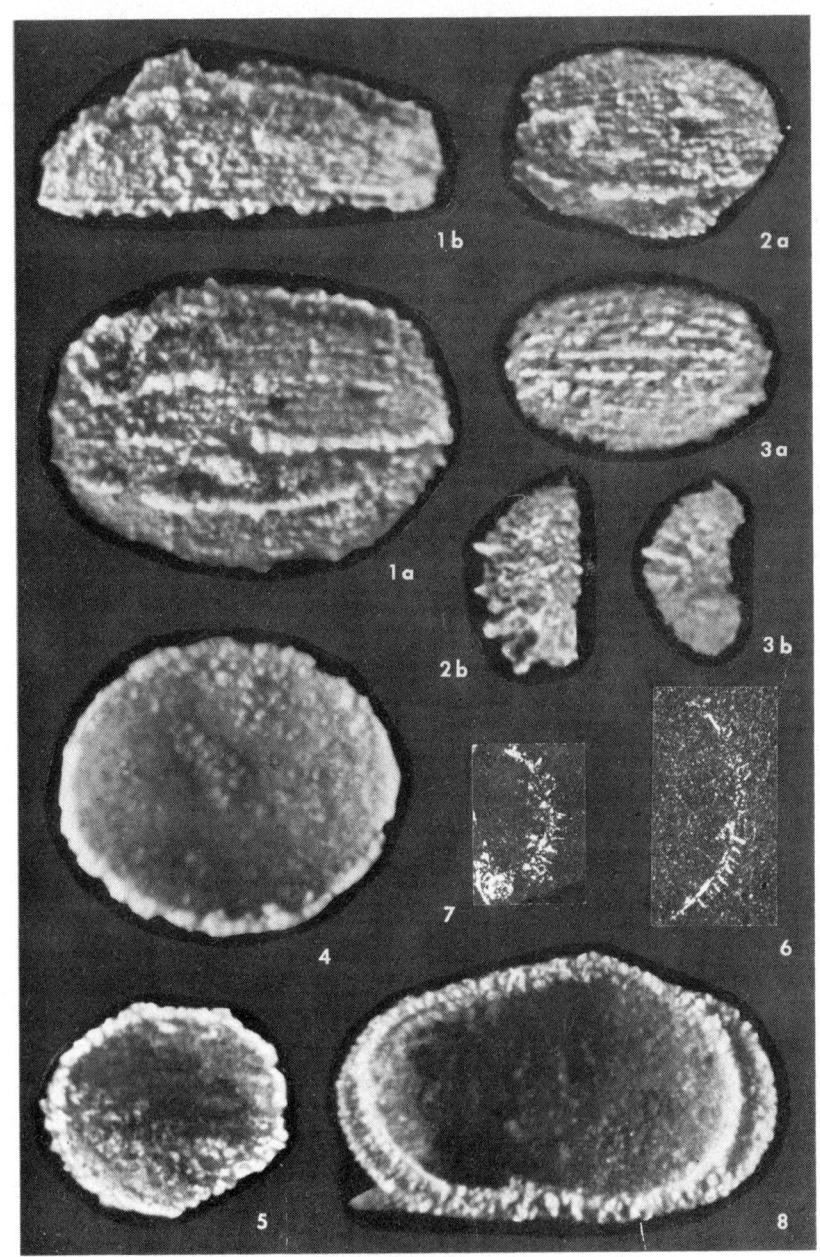

◄ Plate 3. (Fig. 1-5, 8 coated with magnesium oxide; phot. magnification 50×. Fig. 6-7, thin sections; phot. Dr. P. Benedek using Zeissphotomicroscope under crossed polarisers, 40×.)
W Collado de Anzo, Provincia de León, N. Spain; Upper Devonian, Montó beds, lower *hemisphaerica-dichotoma*-Zone (sample 740916/2a).

Richterina (Richterina) costata (Rh. Richter 1869)
Fig. 1. V, SMF Xe 10424; a) external view, b) dorsal or ventral view.
Fig. 2. V, SMF Xe 10425; a) external view, b) anterior or posterior view.
Richterina (Fossirichterina?) cf. *intercostata* (Matern, 1929).
Fig. 3. V, SMF Xe 10427; a) external view, b) anterior or posterior view.
Maternella dichotoma (Paeckelmann, 1913).
Fig. 4. V, SMF Xe 10417; interior view.
Fig. 5. V, SMF Xe 10419; interior view.
Fig. 6. V, thin sections, SMF Xe 10420; transverse section.
Fig. 7. V, thin sections, SMF Xe 10421; transverse section.
Marginohealdia blumenstengeli n.sp.
Fig. 8. LV, holotype, SMF Xe 10408; interior view.

Richterina (Richterina) costata (Rh. Richter, 1869)
Pl. 3, figs. 1-2

The species was definitely identified only in sample 740916/2a (9V, SMF Xe 10424-10426; 2V, GPIF Cr 12/3): Outline (as far as is known) widely elliptical, RI of the main ribs up to 0.19 mm, RI of the interpositioned secondary ribs up to 0.04 mm; main ribs flangelike, remarkably wide; on some valves (SMF Xe 10424, GPIF Cr 12/3) slight remains of terminal spines can be distinguished.

Occurrence: toIV-toVI (Hemberg-Stufe to Wocklum-Stufe), Upper Devonian; E. Thuringian and Rhenish Slate Mountains, N. Holy Cross Mountains, Harz, Armoricanian Massif and Devonshire, now also known from the E. Cantabrian Mountains.

Richterina (Fossirichterina?) cf. *intercostata* (Matern, 1929)
Pl. 3, fig. 3.

The specimens present (4 V, SMF Xe 10427-10428) in sample 740916/2a have more than 10 moderately high main ribs (RI=0.07 mm), between each of which a low secondary rib apparently inserts. An adductor muscle scar cannot distinguished with certainty. *R. (Fossirichterina?)* cf. *intercostata*, recorded by Rabien (1954:129; 1970: 180, 194-195, 215), from the Rhenish Slate Mountains (toIV, lower toV) is very similar to this *Richterina* species. At present it cannot be decided if they are the same species.

Relationships and age

The benthic ostracodes of the Gildar-Montó area exhibit very close relationships to those occurring in the upper part of the Upper Devonian of E. Thuringia (see Blumenstengel, 1965). More distant affinities exist with the faunas of Moravia and the Harz (see Blumenstengel, 1968: 194, 196). Probably, relationships can also be established to ostracod faunas of the Rhenish Slate Mountains (see also Gross-

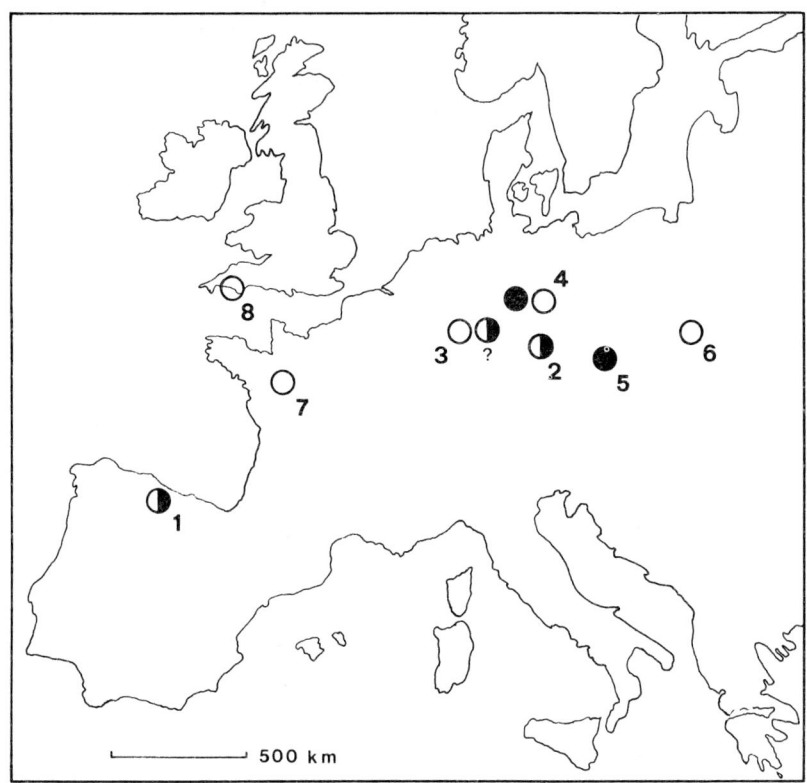

Fig. 3. Geographical distribution of the ostracod faunas from the Dasberg–Stufe (toV) mentioned in the present work. ●=Thuringian ecotype, ◐=Entomozoan ecotype, ○= mixed fauna; 1=E. Cantabrian Mountains, 2=E. Thuringian Slate Mountains, 3=Rhenish Slate Mountains, 4=Harz, 5=Moravia, 6=Holy Cross Mountains, 7=Armoricanian Massif, 8=SW. England.

Uffenorde & Uffenorde, 1974:68). The entomozoids are represented essentially by species known from both the E. Thuringian and the Rhenish Slate Mountains (for examples refer to Rabien, 1954), also from the Harz (Müller-Steffen, 1966, tab. 3), from the Holy Cross Mountains (Kosćielniakowska, 1967:81-82), from the Armoricanian Massif (Péneau, 1928:175) and from Southwest England (Gooday, 1973; 1974:60) (see Fig. 3).

With few exceptions, the benthic species have already been described fom the Thuringian Slate Mountains. Only *Neochilina* aff. *capulinoda, Processobairdia posterocerata* spp. A and *P. spinanterocerata* and a few unidentified species (=Podocopida spp. of the faunal lists) are endemic to the Cantabrian Mountains.

The assemblage in sample 740916/2a compares well with the ostracod association of Blumenstengel's zone 6. The co-occurrence of the species *Marginohealdia blumenstengeli, Rectonaria inclinata,* the common *R. muelleri* and the more rare *Cerateratia cerata,* in association with *Neochilina* aff. *capulinoda* and *Processo-*

bairdia cf. *nodocerata*, indicates a classification in the upper part of this zone.

Sample 740916/2b is more difficult to judge due to the inadequate preservation of the material. The presence of *Marginohealdia ventrocostata* indicates Blumenstengel's zone 7; *M. blumenstengeli,* which extends from zone 6, presumably indicates the lowest part of zone 7, respectively the transitional belt between the two zones.

These classifications can be checked from the entomozoids (respectively from the conodont discoveries of Adrichem Boogaert). No ammonites were found at locality 110/1.

The longest stratigraphic range is exhibited by the species *Richterina (Richterina) costata*, which extends from the Hemberg-Stufe (toIV) to the upper boundary line of the Wocklum- Stufe (toVI). *Maternella dichotoma* is restricted to the Dasberg-Stufe (toV) and to the Wocklum-Stufe (with exception of the uppermost part) (=*hemisphaerica-dichotoma*-Zeit).

Deposits of the *hemisphaerica-dichotoma*-Zone outside of Western and Central Europe (see above) form a narrow belt, which originates in the Holy Cross Mountains and extents far to the east, across Russia (Tschigova, 1971; Rozhdestvenskaja, 1973:161) and down into the south of China (Hou, 1955).

Within the *hemisphaerica-dichotoma*-Zeit, *R. (Fossirichterina)* and Rabien's *R. (Fossirichterina?)* mark the lower zone (=lower toV). A classification of the sample 740916/2a in the lower *hemisphaerica-dichotoma*-Zeit is probably valid, based on the occurrence of *R. (Fossirichterina?)* cf. *intercostata* and further, in that Blumenstengel (1965:64) also paralleled the upper part of zone 6 with this entomozoid subzone.

The conodonts of sample CAL 12, described by Adrichem Boogaert (1967: encl. 3), belong to the upper *stryriacus-*/lower *costatus*-Zone, which accords the transitional belt lower/upper *hemisphaerica-dichotoma*-Zone. Blumenstengel (1965:64) paralleled his zone 7 with the upper part of the *hemisphaerica-dichotoma*-Zone.

Paleoecological discussion

The benthos of the Thuringian ecotype is characteristic of deposits originating in calm and, probably, also deeper, off-shore waters (see Becker, in: Bandel & Becker, 1975:61). The results of the macropaleontological investigation also prove that these were essentially the conditions of the Montó beds, although Kullmann (1965:43) did not necessarily exclude the possibility of near-shore conditions. Becker (in: Bandel & Becker, 1975:59) considered that the long lateral spines functioned in flotation and stabilisation. According to Bless & Michel (1967:269), these actually enabled occasional longer swimming excursions.

Generally, the entomozoids are regarded as reliable indicators of deep water conditions. However, it is sometimes doubted if the group is exclusively pelagic. Whereas Blumenstengel (1965:9) initially hypothesised that only isolated species had adapted to the benthic environment, he (1973:67) later concluded that all Entomozoidae were marine benthic forms. Also, Kozur (1972:642) states that the entomozoids are restricted to abyssal zones.

Blumenstengel (1965) based his hypothesis on the silification of the valves of the entomozoids. From the secondary silification, he concluded that they must have been highly calcified; and that, therefore, the species in question must have been benthic forms. This conclusion is not necessarily valid.

It can be demonstrated that entomozoids are always less well preserved than other species, as has already been pointed out by Blumenstengel. This can obviously be attributed to differences in degree of silification. Further, the present author observed that the valves of *Maternella* species were generally smooth on the inside (see pl. 3, fig. 4). Only in exceptional cases of extremely thin valves, tracings of sculpture were present (see pl. 3, fig. 5). In thin sections (pl. 3, figs. 6-7), small raised ribs can be distinguished on the exterior of the shell and, at most, irregularly arranged clumps on the interior surface.

Presumably the carapaces of the entomozoids were lamellated as are those of other ostracods. It is conceivable that the internal layer of chitin was very thick and that it formed the interior sculpture in this extinct group of uncertain systematic position (Myodocopida?). This is generally not the case in other ostracods. However, a counter-example could be the (much more delicate) sculpture of the epicuticle of some recent polycopid species (see Langer, 1973:32). Calcification of the external layers was reduced as indicated by the small degree of silification. It has merely been speculated (see Langer, 1973:25) that amorphous calcium carbonate could have been involved, due to the fact that the Entomozoacea are classified with the Myodocopida.

After an early diagenetical onset of chitinolysis (see Langer, 1973:17), a smooth internal shell surface remains. Presumably, the secondary silification occurred late. It is conceivable (see the discussion in Bandel, 1972:92) that the SiO_2 required for the process originated from the transformation of the unstable montmorillonite, which is the primary form present always in the marine environment, in illite. Support for this assumption is possibly the observation, that, in the profile from the Collado de Anzo, a direct relationship exists between the pelite content of the sediment and the degree of silification of the ostracods.

An extensive discussion on the complicated subject of the structure and taphonomy of the entomozoid shell is not possible in the present paper. However, it is apparent that (the rare and inadequate) silification alone is insufficient to dismiss a pelagic mode of life. Further 'pro' arguments are presented by Langer (1973:25) and Becker (1976:216). In the case of the mixed fauna discussed, it appears to the present author, that different modes of life are quite possible. Consequently, solely thanatocoenoses are in question.

Acknowledgements

This study was incorporated in the research project 'Fossil-Vergesellschaftungen' of the Sonderforschungs-Bereich 53, 'Palökologie' (Tübingen), to whom I would like to express my thanks for financial support. Special thanks are also expressed to all members of the working group and particularly to Professor Kullmann, who conducted the field excursions.

I am also indebted to: Dr. Helga Groos-Uffenorde (Göttingen), Dr. G. Eickhoff (Hannover) and Dr. W. Struve (Frankfurt am Main) for helpfull discussions; the technicians H. Funk and E. Gottwald (both from Frankfurt am Main) for careful preparations of the photos and the thin sections; Dr. Benedek (Frankfurt am Main) for photographing the preparations (pl. 3, figs. 6–7).
The manuscript was translated by Susan Powell (Vienna).

References

Adrichem, Boogaert, H. A. van. 1974. Devonian and Lower Carboniferous conodonts of the Cantabrian Mountains (Spain) and their stratigraphic application. Leidse geol. Meded. 39:129-192.
Bandel, K. 1972. Palökologie und Paläogeographie im Devon und Unterkarbon der Zentralen Karnischen Alpen. Palaeontographica, Abt. A 141, 1-4:1-117.
Bandel, K. & G. Becker. 1975. Ostracoden aus paläozoischen pelagischen Kalken der Karnischen Alpen (Silurium bis Unterkarbon). Senckenbergiana leth. 56 (1):1-83.
Becker, G. 1976. Oberkarbonische Entomozoidae (Ostracoda) im Kantabrischen Gebirge (N. Spanien). Senckenbergiana leth. 57 (2-3):201-223.
Bless, M. J. M. & M. P. Michel. 1967. An ostracode fauna from the Upper Devonian of the Gildar-Montó Region (NW Spain). Leidse geol. Meded. 39:269-271.
Blumenstengel, H. 1959. Über oberdevonische Ostracoden und ihre stratigraphische Verbreitung im Gebiet zwischen Saalfeld und dem Kamm des Thüringer Waldes. Freiberger Forsch.-H. C72:53-107.
Blumenstengel, H. 1965. Zur Taxionomie und Biostratigraphie verkieselter Ostracoden aus dem Thüringer Oberdevon. Freiberger Forsch.-H. C183:1-127.
Blumenstengel, H. 1968. Die oberdevonischen Ostracoden Thüringens und ihre Beziehungen zu gleichaltrigen Ostracodenfaunen anderer Gebiete. Ber. deutsch. Ges. Geol. Wiss. A, Geol. Paläont. 13 (2):191-198.
Blumenstengel, H. 1973. Zur stratigraphischen und faziellen Bedeutung der Ostracoden im Unter- und Mittelharz. Z. geol. Wiss., Themenh. 1:67-79.
Budinger, P. & J. Kullmann. 1964. Zur Frage von Sedimentationsunterbrechungen im Goniatiten- und Conodonten-führenden Oberdevon und Karbon des Kantabrischen Gebirges (Nordspanien). N. Jb. Geol. Paläont. Mh. 1964 (7):441-429.
Gooday, A. J. 1973. Taxonomic and stratigraphic studies on the Upper Devonian and Lower Carboniferous Entomozoidae and Rhomboentomozoidae (Ostracoda, ? Myodocopida) from Southwest England. Unpubl. Doctoral Thesis, Univ. Exeter, 260 pp., 37 plates, 2 maps.
Gooday, A. J. 1974. Ostracod ages from the Upper Devonian purple and green slates around Plymouth. Proc. Ussher Soc. 3 (1):55-62.
Groos-Uffenorde, H. & H. Uffenorde. 1974. Zur Mikrofauna im höchsten Oberdevon und tiefen Unterkarbon im nördlichen Sauerland (Conodonta, Ostracoda, Rheinisches Schiefergebirge). Notizbl. hess. L.-Amt. Bodenforsch. 102:58-87.
Hou, Y. T. 1955. On some new ostracods from Kwangsi. Acta Palaeont. Sinica 3 (4):313-316.
Kościelniakowska, O. 1967. Devon górny w północnej części Gór Świętokryskich. Buil. geol. Uniw. Warszawa 8:54-118.
Kozur, H. 1972. Die Bedeutung triassischer Ostracoden für stratigraphische und paläoökologische Untersuchungen. Mitt. Ges. Geol. Bergbaustud. 21:623-660.
Kullmann, J. 1960. Die Ammonoidea des Devons im Kantabrischen Gebirge (Nordspanien). Abh. Akad. Wiss. Lit., math.-naturwiss. Kl. 1960 (7):457-559.
Kullmann, J. 1964. Las series devónicas y del Carbonífero inferior con ammonoideos de la Cordillera Cantábrica. Estud. geol. 19:161-191.
Kullmann, J. 1965. Rugose Korallen der Cephalopodenfazies und ihre Verbreitung im De-

von des südöstlichen Kantabrischen Gebirges (Nordspanien). Abh. Akad. Wiss. Lit., math.-naturwis. Kl. 1965 (2):35-168.

Kullmann, J. & R. Schönenberg. 1975. Geodynamische und paläökologische Entwicklung im Kantabrischen Variszikum (Nordspanien). N. Jb. Geol. Paläont. Mh. 1975 (3):151-166.

Langer, W. 1973. Zur Ultrastruktur, Mikromorphologie und Taphonomie des Ostracoda-Carapax. Palaeontographica, Abt. A 144 (1-3):1-54.

Müller- Steffen, K. 1966. Das Oberdevon des nördlichen Oberharzes im Lichte der Ostracoden-Chronologie. Geol. Jb. 82:785-846.

Péneau, J. 1928. Études stratigraphiques et paléontologiques dans le sud-est du Massif Armoricain (Synclinal de Saint-Julien-de-Vouvantes). Bull. Soc. Sci. Nat. Oust. 1928, 4°ser. 8:3-300.

Rabien, A. 1954. Zur Taxionomie und Chronologie der Oberdevonischen Ostracoden. Abh. hess. L.-Amt. Bodenforsch. 9:1-268.

Rabien, A. 1970. Oberes Oberdevon: Nehden- bis Wocklum-Stufe. Erl. geol. Kte. Hessen 1:25000, Bl. Dillenburg (5215):141-235.

Rozhdestvenskaja, A. A. 1973. Ostrakody verchnego devona Baschkirii. Akad. nauk SSSR, Baschkirskii Fil., Inst. Geol.: 3-192.

Tschigova, V. A. 1971. Geographical distribution of ostracodes in the European sea basin at Famennian time. Bull. Centre Rech. Pau – SNPA 5 suppl.: 755-761.

Author's address:

Geol.-Paläont. Institut, Senckenberg-Anlage 32-34, 6000 Frankfurt a.M., B.R.D.

PALEOECOLOGIC CONTROLS OF THE OSTRACODE COMMUNITIES IN THE TONOLOWAY LIMESTONE (SILURIAN; PRIDOLI) OF THE CENTRAL APPALACHIANS*

STEVEN M. WARSHAUER & RICHARD SMOSNA

Abstract

Analysis of the ostracode genera from the classic Tonoloway section at Pinto, Maryland, by binary Q- and R-mode cluster analyses successfully defined three ostracode communities: (1) *Leperditia*, (2) *Welleria-Dizygopleura* and (3) *Zygobeyrichia-Halliella*. Environmental preferences of each community were interpreted by analysis of diversity trends and carbonate petrography.

The *Leperditia* community was lowest in diversity and occurred in the pelmicrite microfacies. Paleoenvironmental interpretations place this community in an intertidal mudflat environment. The *Welleria-Dizygopleura* community displays an intermediate diversity, an affinity for the biopelsparite microfacies, and represents a deeper subtidal environment, though above wave base, influenced by waves of moderate energy. The *Zygobeyrichia-Halliella* community lived below wave base in the deepest subtidal environment, as evidenced by its very high diversity and biomicrite microfacies. Current energy below wave base was quite low. A fourth environment, the shallow subtidal, is characterized by an intrasparite microfacies and a conspicuous lack of ostracodes. High depositional energy from breaking waves barred ostracodes from this environment. The *Leperditia* community is restricted to the lower and upper members of the Tonoloway and the *Zygobeyrichia-Halliella* community, to the middle member. The *Welleria-Dizygopleura* community is most prevalent in the middle. The stratigraphic restrictions on the communities reflect the environmental changes, i.e., transgression and regression, responsible for differentiating the Tonoloway into three members. A general trend in the three ostracode communities is for increasing generic diversity in an offshore direction and the distribution of ostracode communities in the Tonoloway is felt to be mainly controlled by environmental energy level and only secondarily by water depth.

Introduction

The Upper Silurian (a Pridoli age is given by Berry & Boucot 1970, p. 230) Tonoloway Formation of the central Appalachians was originally named by Ulrich (1911, pl. 28) simply by including it on a correlation chart. Stose & Swartz (1912, p. 7) later described the unit and established a type section along Tonoloway Ridge in Morgan County, West Virginia. However, structural complications and poor exposure render the type section unsuitable for detailed stratigraphic studies. For this reason the relatively complete and structurally uncomplicated Tonoloway outcrop at Pinto, Maryland, was then chosen as a reference section (Swartz 1923a, p. 45).

In outcrop the Tonoloway ranges from 90 to 180 meters in thickness and con-

---

*Published with the permission of the Director and State Geologist, West Virginia Geological and Economic Survey.

sists mostly of limestone with minor beds of dolomite. Carbonate rock type, bedding character, sedimentary structures, and fossils vary considerably within the unit, enabling it to be informally subdivided into lower, middle and upper members. This tripartite subdivision of the Tonoloway was originally recognized by Woodward (1941) and later used by Smosna & Warshauer (1975) and Smosna et al. (in press) who found that the members are paleoecologically controlled. The lower and upper members represent a supratidal-intertidal mudflat environment with only a few incursions of subtidal sediments. On the other hand, the middle member was deposited mainly under subtidal conditions with minor occurrences of the supratidal-intertidal mudflat environment. Subsurface investigations indicate that the Tonoloway grades westward into the Salina evaporites of the Appalachian basin (Smosna et al. in press).

In 1923 Swartz et al. published the now-classic Silurian volume of the Maryland Geologic Survey. Included in this work is a comprehensive study of the entire Mid-Appalachian Silurian fauna, including a plethora of ostracodes (228 species and varieties of ostracodes, most of which were new). By contemporary standards, the ostracode fauna seems to be oversplit at the subgeneric level (Lundin 1971, p. 860) but, nevertheless, is surprisingly well illustrated and documented. For some of the studied localities, outcrop measurement and description are supplied in exacting detail, enabling one to place the reported fossils in an almost bed-by-bed manner. As a result, the ostracode genera identified by Swartz (1923b, pp. 114-120) from the Pinto section are the basis for this report. Additional information was gathered by recent field and laboratory studies, in particular thin-section and sedimentary-structure data.

The study was supported by the West Virginia Geological Survey, and travel funds for presentation of the results at the International Symposium on Ecology and Zoogeography of Recent and Fossil Ostracoda were made available by a grant from West Virginia University.

Methods

Analysis of the ostracode genera (9 genera, 42 samples) from the Pinto section was accomplished by the use of binary Q- and R-mode Cluster Analyses. Clustering was by the unweighted pair-group method with arithmetic averages and was based on similarities calculated with the Jaccard Coefficient (Sneath & Sokal 1973).

Characterization of the clusters in the ensuing Q-mode phenogram (Fig. 1) was done by use of constancy (C) and fidelity (F) measures, two parameters originally devised by Hazel (1970, 1971) for biostratigraphic purposes (Fidelity is termed Biostratigraphic Fidelity by Hazel) which we feel can be equally applied to paleoecologic studies. Simply stated, constancy is a measure of the percent occurrence of a taxon within a cluster and fidelity is a measure of how restricted a taxon is to a cluster. Both measures range from 0-10. Values of 10 for both constancy and fidelity indicate that a genus is found in all samples of that cluster (constancy) and is restricted to only that cluster (fidelity). Community names can then be

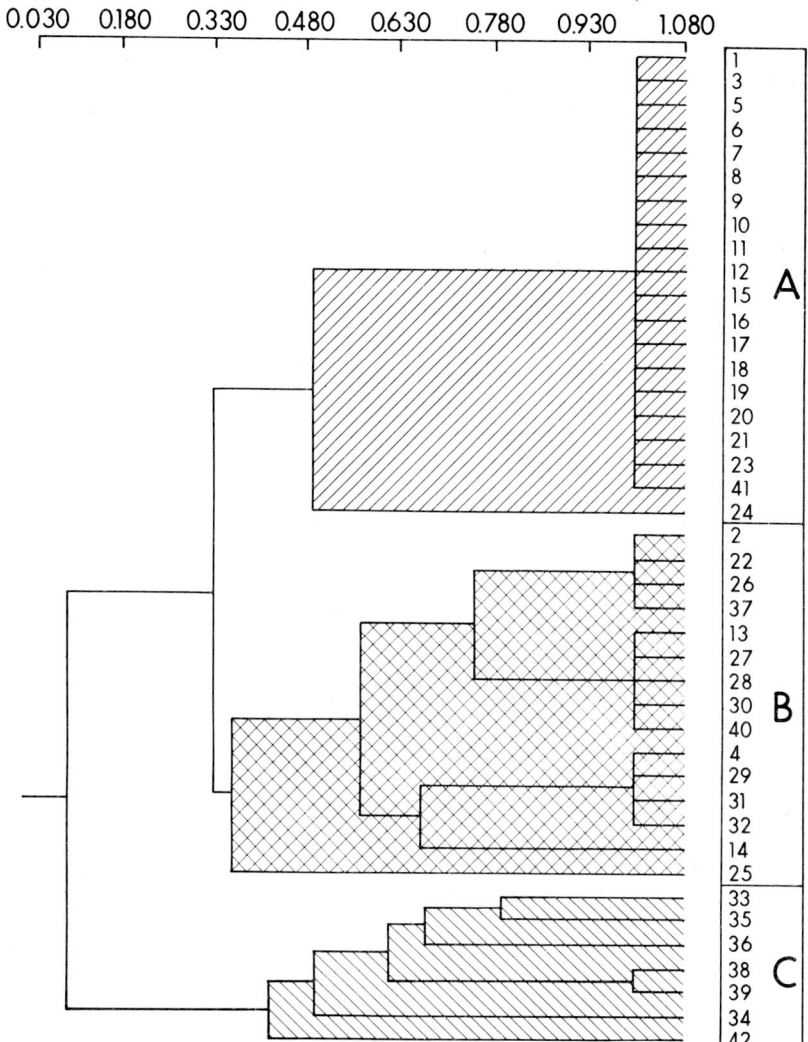

Fig. 1. Q-mode phenogram displaying relationships of the 42 ostracode samples from Pinto, Maryland. A = *Leperditia* community, B = *Welleria-Dizygopleura* community, C = *Zygobeyrichia-Halliella* community.

quantitatively chosen by scanning C and F values of each cluster and choosing those taxa with the highest values, as they best characterize the cluster.

Due to the binary nature of the data set, ostracode diversity could be calculated only by using simple diversity, the number of different genera in each sample. Ostracode diversity and total-fauna generic diversity were then logged against the stratigraphic section. In addition, average sample diversity in each cluster and total ostracode diversity (the total number of different genera within a cluster) were both calculated for each of the Q-mode clusters.

Thin sections for petrographic analysis were made from samples collected as close as possible to the ostracode-bearing beds. Standard microscopic analysis of the thin sections was then initiated for the purpose of describing the carbonate microfacies associated with each ostracode community and interpreting the ancient carbonate environments.

Results

Examination of the Q-mode phenogram (Fig. 1) reveals the existence of three major clusters.

Cluster A

This cluster is characterized (using C and F values in Fig. 2) by the almost sole occurrence of *Leperditia* and is hence termed the *Leperditia* community. Average ostracode diversity per sample and total ostracode diversity for the cluster are extremely low, the former being 1.1. and the latter 2.0. Other non-ostracode genera occurring in some of the samples of this community are '*Camarotoechia*', *Hindella, Modiolopsis, Hormotoma* and *Loxonema*. A striking association exists between this community and the occurrence of algal stromatolites. Bedding planes between large stromatolite domes are quite often covered by profuse *Leperditia*. This suggests a feeding relationship in which the *Leperditia* crop the blue-green algae that constructed the stromatolite domes.

In the field the rocks of this cluster display numerous features characteristic of a carbonate tidal flat: laminated calcilutite, algal stromatolites, a sparse fossil content (dominantly *Leperditia* but with rare gastropods and brachiopods), intraclasts that originated as desiccation chips, mud cracks, bird's-eye structure, and gypsum molds. The rocks are mainly limestone but some dolomitic limestone is present, and several beds are noticeably argillaceous.

In thin section these samples are of a pelmicrite-pelletal micrite microfacies and contain abundant (fecal?) pellets, rare intraclasts (desiccation chips of tidal-flat carbonate mud), and superficial ooids (transported from the adjacent high-energy subtidal environment). A few leperditiid ostracodes and uncommon brachiopods are the only body fossils observed, although traces of small, vertical burrows are present. Scattered, euhedral dolomite rhombs average 10 percent of each rock. Gypsum molds (now filled with very finely crystalline calcite) and mud cracks further attest to the paleoenvironment: the littoral zone, above the level of low tide.

Cluster B

Unlike the previous cluster, this group is best characterized by the occurrence of two ostracode genera, *Welleria* and *Dizygopleura* (Fig. 2), and is therefore named

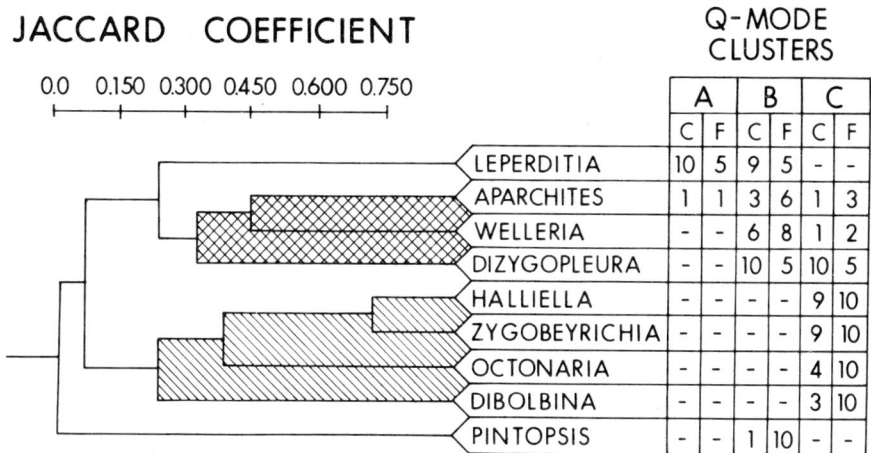

Fig. 2. R-mode phenogram and table of constancy (C) and fidelity (F) values for the Q-mode clusters.

the *Welleria-Dizygopleura* Community. Diversity is higher, with a mean ostracode diversity of 2.9 per sample and a total ostracode diversity of 5.0 for the cluster. The generic diversity of the non-ostracode fauna also increases slighty with the occurrence of *Hindella, 'Camarotoechia', Rhynchospira, Tentaculites, Hormatoma* and *Loxonema*.

In thin section the rocks of this cluster are of a biopelsparite microfacies, dominated by abundant pellets and fossils. Fossils other than the quite common ostracodes include some echinoderms, brachiopods, and rare tentaculitids, stylolinids, gastropods, and calcareous algae. Their condition ranges from whole and articulated, to disarticulated and broken but angular, to broken and rounded. Many animals possess micritic envelopes or have been completely micritized by boring algae Intraclasts are rare (though numerous in two samples); they are mostly 'rip-ups' of micrite and biomicrite limestones. On the whole, the sediments were only poorly washed, and sorting goes from poor to good. Some horizontal burrows were observed. In addition to the main rock type, biopelsparite, other microfacies containing the *Welleria-Dizygopleura* ostracode community are biosparite, biopelmicrite, and micrite – in order of decreasing abundance. On outcrop these samples are medium-bedded, fossiliferous (ostracodes and brachiopods) calcarenite.

The subtidal environment of deposition of these rocks and hence the habitat of the *Welleria-Dizygopleura* community is as follows. Water depth was relatively shallow, being greater than that of the surf zone, and the environment extended perhaps down to the level of wave base. These were normal-marine waters. The spar/micrite ratio (approximately 1) indicates that current and wave agitation was only intermittent, current strength was usually moderate, or too much micrite was produced in this environment to be washed away by currents (Folk 1962, p. 76). The rarity of intraclasts (and ooids) plus the generally angular nature of fossil fragments all point to moderately weak currents and wave action. A somewhat diverse benthic community, including browsers, filter feeders, plants, predators and/or scavengers as well as an infauna, established itself on the sea floor in this environment.

Cluster C

The final Q-mode cluster (Fig. 2) displays a high constancy and perfect fidelity for two genera, thereby becoming the *Zygoberichia-Hallella* Community. This community has the highest mean diversity, 3.7, and the highest total ostracode diversity, 7.0. In addition, the non-ostracode generic diversity of the samples within the cluster is also the highest. Non-ostracode faunal elements associated with this community are *Schuchertella, 'Camarotoechia', Rhynchospira, Hindella, Orbiculoidea, Hormatoma* and other suprageneric taxa seen only in thin section.

In the field the rocks containing the *Zygobeyrichia-Halliella* ostracode community are very fossiliferous calcisiltite and calcarenite. Petrographically, these samples are seen as numerous fossils embedded in a matrix of micrite and therefore belong to a biomicrite microfacies. In addition to the quite common ostracodes are some brachiopods and pelmatozoans (cystoids), rare calcareous algae and gastropods, and exceedingly scarce tentaculitids, pelecypods, bryozoans, and calcareous worm tubes. Most fossils are disarticulated but unfragmented, and many are partially micritized by boring algae. Horizontal burrowers thoroughly churned the sediment. Pellets are present but intraclasts, very rare. The carbonate grains typically are very poorly sorted. Beside biomicrite, less frequently occurring microfacies are micrite and biopelsparite.

The habitat of the *Zygobeyrichia-Halliella* ostracode community, based on interpretation from limestone structures, textures, and constituents, was a relatively deep, subtidal environment. The sea floor was at or near the level of wave base, but water depth was only slightly greater than that of the biopelsparite facies/*Welleria-Dizygopleura* community. Energy level of the currents in the biomicrite facies was quite low. Browsers, filter feeders, and burrowers dominated the ecosystem of the sea floor.

Another facies, representing deposition in a very shallow subtidal environment, is documented by field and petrographic study. No fossils, however, were identified from these samples by Swartz (1923), and hence the ostracodes of these rocks do not appear in the data of this report. This facies is distinguished in the field by the presence of intraclasts (often chertified), edgewise limestone conglomerate, and thin to medium bedding.

Petrographically, the rocks are intrasparite. Abundant intraclasts and pellets, rare ooids (although a 1-foot bed of oosparite occurs near the base of the upper member), and a few ostracodes – all cemented by void-filling calcite cement – are the constituent grains. The intraclasts are of micrite and pelmicrite limestone, 'rip-ups' of the carbonate mud flat and formed during storms and/or minor transgressions. Ooids are both superficial and true; many have been cracked or fragmented; a few are partly micritized by boring algae. Leperditiid and other ostracodes, brachiopods, echinoderms, and gastropods constitute the low to medium diversity faunal assemblage, but, on the whole, fossils are scarce. Factors pointing to moderately strong currents and wave action include: (1) fair sorting of grains, (2) well washed to poorly washed sediments (spar/micrite ratios greater than 1), (3) minor scour and cross bedding along bedding planes, (4) inclined to edgewise intraclasts, and (5) lack of rounded fossils. Burrows are conspicuously absent.

The occurrence of intraclasts, ooids, and sparry calcite cement argue for a high-energy environment, but not high enough to round fossils nor to wash away micrite completely. The rocks of this facies are interpreted to be very shallow, subtidal sediments on the basis of the moderate faunal diversity and the relatively high degree of environmental agitation plus the close association with intertidal sediments.

The R-mode phenogram (Fig. 2) also grouped the genera into community-related clusters, with the most important elements of each community closely associated with one another. *Pintopsis,* due to its single occurrence, was the only genus that failed to cluster. It is interesting to note that while the R-mode cluster analysis did show the most important associations of each community, it failed to delineate the total ostracode assemblage of each community. This is due to a peculiarity of cluster analysis in that each variable in an R-mode analysis is forced to cluster within only one group, even if it occurs in all groups. However, by calculating constancy and fidelity measures for each of the variables contained in each of the Q-mode clusters, this inherent difficulty can be alleviated and community compositions can be easily observed (Fig. 2).

By scanning the log of the community occurrences (Fig. 3), it is quite easy to determine the vertical distribution of the three ostracode associations. The *Leperditia* community is found in the lower and upper members but not in the middle, while the *Welleria-Dizygopleura* community is found in the lower and middle members but not, thus far, in the upper. This absence of the *Welleria-Dizygopleura* community from the upper Tonoloway at Pinto is most probably a result of the large (40 meters) covered interval and may not be a true reflection of the community distribution. Occurrences of the *Zygobeyrichia-Halliella* Community, with the exception of one bed directly below the overlying Keyser Formation, are restricted to the middle member.

Perusal of the diversity logs (Fig. 3) reveals an increase in the total-fauna (ostracodes plus non-ostracodes) generic diversity in the middle member, coincident with the *Zygobeyrichia-Halliella* community.

The upper and lower members of the Tonoloway were deposited on intertidal mud flats, and the middle member was laid down in environments where water depth fluctuated from intertidal to shallow subtidal to deeper subtidal. Energy level in the shallow subtidal environment was generally moderate and in the deeper subtidal, low. A widespread transgression was responsible for the more normal-marine conditions of the middle member.

Summary and conclusions

Three ostracode communities occur in the Tonoloway Limestone of the central

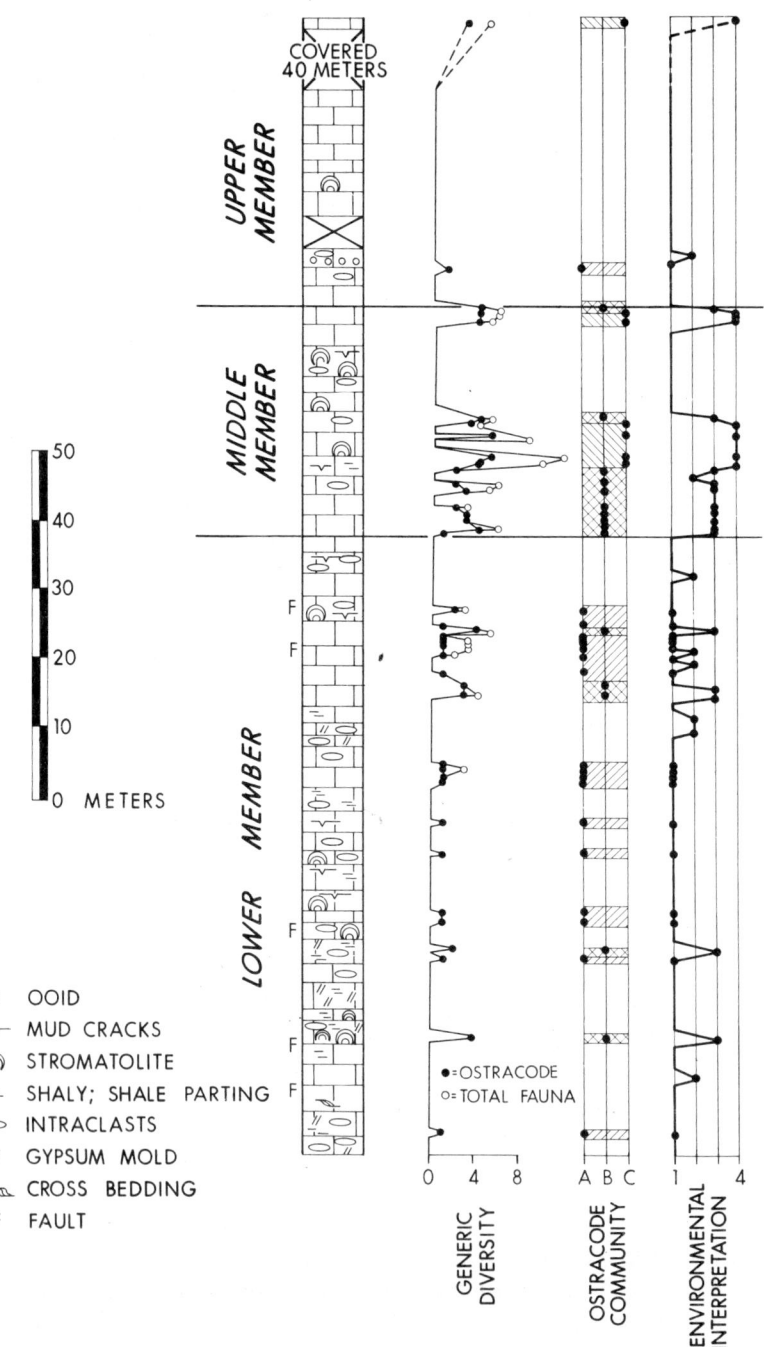

Fig. 3. Diagrammatic stratigraphic section for the Tonoloway Limestone at Pinto. When the generic diversity of the ostracodes equals the generic diversity of the total fauna, only the ostracode value is shown. Ostracode community A = *Leperditia,* B = *Welleria-Dizygopleura,* C = *Zygobeyrichia-Halliella.* Environmental interpretation 1 = intertidal mudflat, 2 = shallow subtidal, 3 = deeper subtidal, 4 = deepest subtidal.

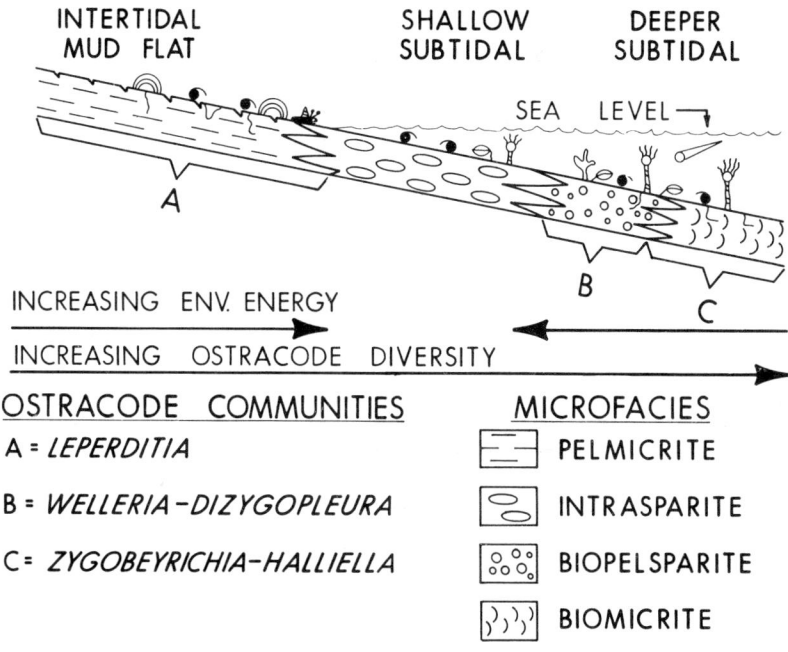

Fig. 4. Diagrammatic interpretation of the lateral relationships of both the ostracode communities and carbonate microfacies.

Appalachians. The *Leperditia* community has the lowest diversity, is associated with a pelmicrite microfacies, and is found in the intertidal-mudflat environment. In this habitat the environmental energy level was extremely low, while the environmental stress, due to the tide-induced alternation of wet and dry conditions and high salinities, was extremely high. Stress in an environment has been found to be an effective mechanism for keeping diversity low (Sanders, 1969) and is most probably the underlying cause for the low diversity of this community. Our interpretation of an intertidal-mudflat environment for the genus *Leperditia* is in close agreement with previous studies (Berdan 1968, Walker & Laporte 1970).

The *Welleria-Dizygopleura* Community, on the other hand, has the middle diversity of the three communities, is associated with the biopelsparite microfacies and occurs in a deeper subtidal environment seaward of the *Leperditia* community. Environmental energy level was moderate due to the location of this community above wave base. Environmental stress would be low as this community enjoyed normal-marine salinities and was buffered from sudden atmospheric changes by deeper water.

Highest diversity is displayed by the *Zygobeyrichia-Halliella* Community which is associated with the biomicrite microfacies and is therefore felt to occur in waters even deeper than those for the *Welleria-Dizygopleura* Community.

Due to fragmentation, the sparse ostracodes of the intrasparite microfacies were, for the most part, unidentifiable. Physical parameters, however, point to a high-energy, shallow-subtidal environment for the accumulation of these rocks. A general lack of ostracodes in this type of environment can be easily understood as the depositional energy would have been too high for free living, micro-crustaceans to be able to survive. Ostracode densities, and hence environmental preferences, increase on either side of this microfacies where energy conditions were lower (Fig. 4).

The ostracode communities and carbonate microfacies occur in a 1:1 relationship because both are controlled by the same underlying mechanism, environmental energy. That is, each different community is restricted to a particular microfacies. Water depth is a secondary control mechanism as it helps to define the energy-distribution pattern of an environment.

Three major conclusions can be formulated from this study:

1. The main factor controlling the ostracode community distribution in the Tonoloway Limestone was the level of environmental energy.

2. Ostracode communities in the Tonoloway follow the general trend of increasing diversity offshore as found for other Paleozoic benthic invertebrates (Anderson 1971).

3. The widespread transgression postulated (Smosna & Warshauer 1975, Smosna et al. in press) for the middle member of the Tonoloway is reflected by the distribution of the ostracode communities.

References

Anderson, E. J. 1971. Environmental models for Paleozoic Communities. Lethaia 4:287–302.

Berdan, J. 1968. Possible paleoecological significance of Leperditiid ostracodes. Geol. Soc. Amer., Prog, Annu. Mtg., Northeastern Sect. Washington, D. C., p. 17 (Abstr.).

Berry, W. B. N. & A. J. Boucot. 1970. Correlations of the North American Silurian rocks. Geol. Soc. Amer. Spec. Paper 102, 289 pp.

Folk, R. L. 1962. Spectral subdivision of limestone types. In: W. E. Ham (ed.), Classification of carbonate rocks. Amer. Assoc. Petroleum Geologists, Memoir 1, pp. 62–84.

Hazel, J. E. 1970. Binary coefficients and clustering in biostratigraphy. Geol. Soc. Amer. Bull. 81:3237–3252.

Hazel, J. E. 1971. Ostracode Biostratigraphy of the Yorktown Formation (upper Miocene and lower Pliocene) of Virginia and North Carolina, U.S. Geol. Surv. Prof. Paper 704, 13 pp.

Lundin, R. F. 1971. Possible paleoecological significance of Silurian and Early Devonian ostracode faunas from mid-continental and north-eastern North America. In: Paleoecologie de Ostracodes, H. J. Oertli (Ed.), Pau, France, pp. 853–868.

Smosna, R. & S. M. Warshauer. 1975. Depositional environments of the upper Silurian Tonoloway Formation. Geol. Soc. Amer., Prog., Annu. Mtg., Southeastern Sect., Memphis, Tennessee, p. 536 (Abstr.).

Smosna, R., D. G. Patchen, S. M. Warshauer & W. J. Perry, Jr., in press. Relationships between Tonoloway Limestone environments of deposition and the distribution of Salina evaporites in West Virginia. In: Reefs and evaporites. Amer. Assoc. Petroleum Geologists, Studies in Geology 6.

Sneath, P. H. A. & R. R. Sokal. 1973. Numerical Taxonomy. W. H. Freeman and Company, San Francisco, 573 pp.

Stose, G. W. & C. K. Swartz. 1912. Pawpaw-Hancock Folio, U.S. Geol. Surv. No. 179.
Swartz, C. K. 1923a. Stratigraphy and paleontologic relations of the Silurian strata in Maryland. Maryland Geol. Surv., Silurian Vol., pp. 25–50.
Swartz, C. K. 1923b. Sections of the Wills Creek and Tonoloway formations. Maryland Geol. Surv., Silurian Vol., pp. 105–181.
Ulrich, E. O. 1911. Revisions of the Paleozoic System. Geol. Soc. Amer. Bull. 22:281–680.
Walker, K. R. & L. F. Laporte. 1970. Congruent fossil communities from Ordovician and Devonian Carbonates of New York. J. Paleontology 44:928–944.
Woodward, H. P., 1941. Silurian System of West Virginia. West Virginia Geol. Surv., Vol. 14, 326 pp.

Authors' address:

S. M. Warshauer, Dept of Geology and Geography, West Virginia University & Co-operating Geologist, West Virginia Geological and Economic Survey, Morgantown, West Virginia, USA.
R. Smosna, West Virginia Geological and Economic Survey, Morgantown, West Virginia, USA.

Discussion

Kornicker: In one of your slides you show diversity increasing seaward from your shallowest to deepest water environments. I would think that diversity would be lowest (in a living environment) at the place where you found no unbroken specimens (high energy environment) and, therefore, diversity would increase both landward and seaward from that place.
Warshauer: Yes, Dr. Kornicker, you are correct in your observation. However, while the highest energy deposits do contain broken, and therefore unidentifiable, ostracode shells, the thin section analysis reveals these broken shells to belong to several different groups. You will recall that the intertidal-mudflat environment was almost exclusively inhabited by *Leperditia,* causing the high energy shallow subtidal environment, since it included several different types of ostracodes, to have a slightly higher diversity. However, we are not sure of how many of the ostracodes were either living in or washed into the shallow subtidal environment. The arrow showing an increase in diversity in an offshore direction was, therefore, simply meant to display the general diversity trend.

Sixth Intern. Ostracod Symposium, Saalfelden

SOME PALEOECOLOGICAL AND TAXONOMIC PROBLEMS IN CONNECTION WITH GROWTH SERIES OF THE OSTRACOD GENERA BAIRDIA AND PARAPARCHITES FROM A SCOTTISH LOWER CARBONIFEROUS SHALE

LAING FERGUSON

Abstract

Several thousand specimens of *Bairdia* and *Paraparchites* were recovered from the 2.8 m shale below the Second Abden Limestone (Visean) near Kinghorn, Fife, Scotland. Scatter diagrams of their Length and height measurements suggest that only one species in each genus is present and that previous records of numerous species of each form are spurious.

The L:H measurements of *Bairdia* form relatively good clusters representing the various instars and possibly the male dimorph. The measurements of *Paraparchites* do not cluster well but form a more or less continuous plot. *Bairdia* occurs only in the upper, more truly marine part of the shale while *Paraparchites* occurs principally in the lower, nearshore, part of the shale. It is thought that the development of good instar clusters in *Bairdia* is the result of growth in more stable offshore conditions while the lack of any distinct clusters in *Paraparchites* is the result of their growth in nearshore and highly variable conditions.

The measurements of specimens housed in the museums of Britain (and which have been assigned to those species of *Bairdia* previousy recorded from the Abden locality), plot up in such a way as to suggest that individual instars and also the male dimorphs were given different specific names by previous workers. Many of these species are thus considered invalid.

The measurements of museum specimens of species of *Paraparchites* recorded from the Abden locality also suggest that small, medium and large forms of only one species may have been given different specific names and that certain of these species are invalid.

Author's address:

Dept of Geology, Mount Allison University, Sackville, New Brunswick, Canada E0A 3CO.

Discussion

Maddocks: There appears to be much need for such work on Paleozoic *Bairdia* in the USA. Based on my experience of Post Paleozoic *Bairdias* one does not need to depend on just size to identify juveniles but can use the width of the inner lamellae.
Ferguson: Clusters in *Bairdia* were quite distinct (better than the illustrated slide) so I could see juvenile instars quite clearly. As yet I have not studied the interiors although I have some available and will certainly examine these carefully in later stages of the study.
Sohn: I have not had much success with instar clusters over the last thirty years

suspect that any two successive instars will overlap. In *Paraparchites* did you only only measure Length and Height, or did you plot up Length to Width measurements too? I suspect that dimorphism might be apparent in the L:W plots although it is not apparent in the L:H plots.

Ferguson: No, I have not measured the width of the specimens but agree that this might well be something worth measuring and plotting up.

Kornicker: In the slide showing the distribution of *Bairdia* and *Paraparchites* within the shale there are some levels where both occur together, i.e., before *Bairdia* completely replaces *Paraparchites*. If your idea is true (that lack of clustering in *Paraparchites* is due to highly variable near-shore conditions and that the good clustering in *Bairdia* is the result of growth in more stable offshore conditions) would you not get good clustering in the *Paraparchites* which occur along with the *Bairdia*? In other words, where conditions are relatively stable for *Bairdia*, would not the *Paraparchites* also show good clusters?

Ferguson: The bed was examined in layers only two centimetres thick so it might well be possible to examine the plots of the specimens from the layers where both *Bairdia* and *Paraparchites* occur and to test this idea. Some complications might arise, however, due to sorting, etc., and one would have to be sure that both forms were indigenous at these horizons before drawing any conclusions.

Warshauer: I am worried by the non-alignment of the instar clusters in Kelvingrove Museum material.

Ferguson: As I mentioned, it was a slide of an atypical collection, i.e., a collection not even just from one locality but from several – and also possibly several different horizons at each locality. It might be a series of different sized species overlapping, each with a male dimorph cluster of slightly different size. However, if they are of one species it might be indicative of a situation such as Whatley described in this symposium, i.e., with males and females having different alignments.

Whatley (written question): Anyone who attempts to reduce the number of taxa should be congratulated. I feel, however, that it is necessary to consider more than just length/height relationship in attempting to achieve this. The author should carefully consider which of his specimens of *Bairdia* exhibit juvenile and adult carapace characters, respectively.

Ferguson: In reading Jones & Kirkby's original papers on the species in question, one finds that in many cases they claimed that the species were naming were in fact identical in all respects except height and length. It thus seems logical to sort them out again on to the basis of H:L.

Despite this fact, I will endeavour to look for adult characteristics (such as a well developed inner lamellae) in the various growth stages during later phases of this study.

V. ABSTRACTS OF THE DISCUSSION SESSIONS

DISCUSSION ON SAMPLING METHODS

MARTIN V. ANGEL

Our ideas and theories concerning the nature of world ecosystems both past and present are based on our experiences gained by sampling. Our sampling methods, that is both the samplers we use and our procedures in deploying them, inevitably distort what we see. So the problems of interpretation are inextricably involved in the limitations imposed by these methods. Thus I believe that many of us run the danger of reading too much into our results, and may even write results into our observations by our sampling regimes. Conversely I also believe that many of the indications within our samples about the prevailing ecological conditions are missed because we have not learnt to read or interpret the signs.

As a biological oceanographer I will concentrate on the problems of sampling present day midwater communities (see also Angel 1977) but many of these have analogues in palaeontology. Perhaps the most basic question is what species occur at any particular depth stratum at a location in the ocean. By using a sampler that takes a contiguous series of subsamples, it is possible to look at the accumulation rate of species in the samples. Arithmetic plots of the number of species against sample size give a hyperbolic curve that tends towards an asymptote. This plot can be used to predict the number of species to be expected in a sample of as given size, and will also indicate what size of sample is needed to identify all the important components of the community. Jumps in the accumulation curve indicate the crossing of a community boundary. The slope of log plots of the same parameters which will give a straight line can be used as an estimate of species diversity. Thus by looking at small replicated subsamples will provide more information than studying the equivalent simple large sample.

The next question that is asked is how many. Quantitative estimates are much more difficult to make because of the patchiness of the populations. Geologists are much less interested in patchiness for its own sake than biologists, but as a result tend to ignore its effects on their sampling. Using sequences of subsamples the minimum detectable patch size is twice the minimum subsample size, and the maximum is about an eighth to a quarter the size of the total sampling length. Samples of one that are randomly taken, and so can not be statistically blocked up into progressively larger sample sizes, only give information about patchiness of the dimension of the individual samples. The type and degree of patchiness will vary with sample size. Thus a sampling programme may only allow study a very narrow band or window within the spectrum of variability. Beyond the bounds of this window of discrimination random noise irretrievably fogs the true patterns and may even stochastically generate artefacts. Far too many studies endeavour to investigate problems outside the bounds of the sampling windows. One obvious example of such a window is with a net sampler (or sieve). A lower size limit of what is retained is determined by the mesh size employed and the upper limit by the ability of organisms to avoid capture by the net. These limits will vary between even quite closely related species.

Recent studies in oceanography (e.g. Fasham & Pugh 1976) have suggested that the causal factors of patchiness vary with the dimensions of the patches concerned. Automated sampling techniques using pump samples allow the study of a wide band of patch size scales of fluorescence (i.e. chlorophyll concentrations) to be examined in the near-surface layers of the sea. By using statistical analytical methods of spectral and coherence analysis, it was demonstrated that at patch sizes of less than 50 m, there is no clear relationship between chlorophyll distributions and the physical mixing processes. These are the scales at which biological interactions such as predator/prey relationships are dominant in determining distribution patterns. However, diffusion and turbulent mixing result patterns only having longevity of a few minutes which imposes an insuperable time limit on the duration of series of observations that can be meaningfully interpreted. At patch sizes of 50 m–1 km there is a close coherence between the distribution of temperature microstructure and chlorophyll. These patches are generated either by the mechanical effects of internal wave trains or by turbulent mixing. These patches are also continually being eroded by diffusion.

At even larger scales surprisingly little is known about biological patchiness. Theoretical studies imply important factors are likely to be the growth parameters of the populations. The growth parameters are effected by the physical environment; the size and energy of physical features are awe-inspiring. Rings spawned off the Gulf Stream are 200 km in diameter and can persist for five years. 99% of the ocean's dynamic energy at these scales is involved in such eddies which are analogous to atmospheric disturbances; the remaining 1% being the energy of the major ocean currents.

The eddies which exist below the permanent thermocline are less energetic but move faster. Bottom topography has a dramatic effect on these eddy systems, inthe North Atlantic small 200 m high features in the middle of a flat abyssal plain have been show shown to influence the physical structure of the water over 500 m above the features. We are almost totally ignorant of the biological influence of these sizeable phenomena.

Perhaps much more needs to be researched on the observation and analysis of time series. In midwater of the ocean in situ growth rates of long lived plankton can not be measured as diffusion and turbulence disperses a population too rapidly. Much the same classes of problems will be encountered when analysing cores. Investigations of variations at different scales (e.g. time scales) may give different impressions of the dominant ecological factors regulating the fossil populations. It is essential to look at the within sample variability otherwise figures are meaningless and cannot be compared. The use of modern analytical techniques such as factor, principal components and clustering analyses, provide us with powerful tools. As the CLIMAP programme has shown that if sampling procedures are standardised so that intercore comparisons can be sensibly made, much more of the information locked in the samples can be discerned and interpreted. Many of these methods are best used initially in situations which we understand. Kaesler's contribution on the influence of the Antarctic convergence on ostracod distributions was an excellent example of this. It told us little we did not know before, and doubters will therefore ask why do it. Its main value is in providing a feel for the method and confidence in its results. It can now be used to analyse more complex situations where the patterns are not at all obvious prior to analysis.

To sum up we should be much more aware of the limitations of our sampling procedures. Our samplers are our eyes on the natural world. Each sampler even when deployed to best advantage is limited in the range of its discrimination and its depth of focus. It provides us with a limited window into the spectrum of confusion. Sampling programmes must be conducted within the limits of these windows otherwise the results will be blurred into meaninglessness by random noise. Even within these windows, random variation obscures much of the information in which we are interested. Powerful techniques of statistical analysis are now beginning to prove invaluable in unscrambling the information from the noise.

References

Angel, M. V. 1977. Windows into a sea of confusion; sampling limitations to the measurement of ecological parameters in oceanic midwater environments. Pp. 217–248. In: Oceanic sound-scattering prediction N. R. Andersen & B. J. Zahuranec, eds., Marine Science Vol. 5. Plenum Press.

Fasham, M. J. R. & P. R. Pugh, 1976. Observations on the horizontal coherence of chlorophyll a and temperature. Deep-Sea Res. 23:527–538.

Author's address:

Institute of Oceanographic Sciences Wormley, Godalming, Surrey GU8 5UB, England.

Comment

Kaesler: I have five points to make:

1. A sample must be a sample of something. We must be able to define the population (i.e. the statistical universe) from which samples are being collected. Spatial heterogeneity (patchiness) is very important here.
2. samples must be collected with a specific test of a hypothesis in mind. For this reason, we may find that many samples do not contribute to the solution of the problem in hand. Similarly, if we decide on a particular statistical test, we must collect our samples with this analytical method in mind.
3. Variability among samples must be studied by using replicated samples.
4. Can we afford to collect and process enough samples to answer the question being asked?
5. The statistician must be consulted before the sampling regime is designed, so that with his help the chances of answering the problem in hand can be optimized.

Sixth Intern. Ostracod Symposium, Saalfelden

DISCUSSION ON OSTRACOD TERMINOLOGY

JOHN W. NEALE

The Chairman introduced the disussion by referring to the confusion that existed in reference to the nomenclature of the ostracod appendages and suggested that some standardisation of terminology would be advantageous. A lively discussion established that not only was there uncertainty of the division between cephalic and thoracic limbs, but in description there was also a tendency to confuse homology with function and to use a mixture of terms. Professor Hartmann was requested to supply a clear statement of the situation and has kindly written as follows:

'The reduction of the outer segmentation of the ostracod body caused by the development of the carapace, which envelops the whole body and provides protection and an anchor for the musculature, makes an outer skeleton of the body itself unnecessary and makes the recognition of the homologues of the ostracod body segments with those of other crustacea very complicated. The same is true of the Phyllopoda. If the homologies of the segments are not clear the homologies of the appendages belonging to certain segments of the body are also obscure.

Indications of the homologues of segments of the ostracod body may be found on the one hand in the structure of inner organs of the body such as segmental organs (nephridia) and the nervous system, and on the other hand in the ontogenetic development. In general, crustaceans have five limbs which belong to the head: Antennula (= first antenna), Antenna (= second antenna), Mandible, Maxillula (= first maxilla) and Maxilla (= second maxilla). The number of thoracic appendages varies, as does the number of the appendages of the abdomen. The situation becomes difficult to resolve when a part of the thorax fuses with the head to form a 'cephalothorax'. In this case the function of the appendages of the part fused with the head changes since they become appendages which help with feeding, the so-called maxillipeds. Anatomically they remain appendages of the thorax; only their function has changed.

In the ostracods we normally have seven appendages. As the outer segmentation cannot indicate if they belong to the head or to the thorax and abdomen, we have to look for other indications. There is no doubt that two antennae are present (now named following international usage the antennula and the antenna). There is also no doubt that a mandible is present, and that the fourth appendage is the first maxilla (= maxillula). A lot of trouble has arisen, however, with the fifth appendage. This appendage varies in shape, and hence in function, in different groups of ostracods. In the Myodocopida it is mainly shaped as a feeding limb and, like the maxillula, often carries a respiratory plate. The Halocypriformes have a similarly shaped fifth limb as have the Platycopa, the Metacopa and the Cypridacea and Darwinulacea among the Podocopida. Only the Bairdiacea and the Cytheracea have a fifth limb which is leg-like, and although the respiratory plate is still fully present in the Bairdiacea, in the Cytheracea it is completely absent. Consequently, the fifth limb was mostly called the second maxilla or maxilliped in those groups where it functioned as a feeding limb; whereas in the Bairdiacea and Cytheracea it was called a thoracic leg. As it is fixed to the endoskeleton of the thorax in these groups, scientists believed that it was a thoracic leg, and that the second maxilla was reduced in these two ostracod superfamilies. Moreover, G. W. Müller (1894) stated that during the larval development the segment of the fifth limb was formed after a pause in the development and took this to indicate a reduced maxilla. Weygoldt (1960) showed clearly that there is no pause but that the fifth appendage is formed during the third moulting, directly after the formation of the first maxilla.

Crustaceans have two so-called segmental organs. These are glands which are homologous with the metanephridia of the Annelida. In Crustacea these segmental organs are situated in the segments of the antenna (= second antenna) and maxilla (= second maxilla). In ostracods, in the Myodocopida we have only the segmental organ in the antennal segment; that of the maxilla segment is missing. In the Podocopida we have both organs developed during the larval stages. In adults only the segmental organ of the maxilla segment is present. It was Cannon (1925) who clearly demonstrated that in the Cyprididae the segmental organ belongs to the segment of the fifth limb and hence that this appendage must be regarded as the maxilla (= second maxilla). It is interesting, and a pity, that this paper was overlooked for so long. Later Weygoldt (1960) found the segmental organ present in the segment of the fifth limb in the Cytheridae (*Cyprideis*). He showed clearly that the fifth limb is formed during the ontogeny of *Cyprideis* like a maxilla (= second maxilla) in other crustacean groups. Unfortunately the Platycopa have not yet been studied.

After Cannon and Weygoldt there cannot be the slightest doubt that the fifth limb of the Bairdiacea and Cytheracea is morphologically homologous with the maxilla (second maxilla) of the other crustacean groups, in spite of its different function. We have seen above that the function of appendages may vary, even thoracic limbs may become feeding limbs. The opposite occurs in some ostracods where a feeding limb may become a walking leg. This has to be seen in the context of the general reduction of the number of appendages and segments in Ostracoda (carapace development). Thus the Polycopidae have only five limbs, that is only the appendages of the head. Nearly all other ostracods possess the head appendages and one or two thoracic limbs. There are no abdominal appendages left. The function of the fifth limb varies. It may be a feeding limb like the maxillula, a maxilliped which helps feeding and locomotion, or a walking leg. Nevertheless, the fifth limb is always homologous with the maxilla of other crustacean groups and must therefore be called the 'maxilla'.

Crustacean workers are agreed on the nomenclature of the appendages, and ostracod workers should not hesitate to follow this nomenclature:

Head: Antennula, Antenna, Mandible, Maxillula, Maxilla.
Thorax: First and second Thoracic Limbs (or appendages).

Description can then take the following form: 'the maxilla is a walking leg', or 'the maxilla is a feeding limb', or 'the maxilla is a maxilliped', and 'the second thoracic limb is a cleaning limb', or 'is a walking leg' etc., etc."

The discussion then moved on to consider the terminology of the hinge. Some workers thought that the number of the terms introduced for this structure was excessive and tended to obscure rather than clarify the situation. As was pointed out, however, it was important to differentiate between terms used to describe the various parts of the hinge structure and the terms introduced to cover a complete hinge structural type. It was universally agreed that the former, which included terms such as 'anterior hinge plate', 'denticulate median bar', 'rounded postjacent socket' etc., was essential. It was generally agreed that if space was no problem a full description of the hinge structure using a limited number of simple, unambiguous terms was the ideal at which to aim. Opinion on the value of a single term to describe a complete hinge structure varied. With a limited number of terms which were easily remembered and clearly defined such terminology served as a useful and convenient shorthand. However, some workers felt that the increasing numbers of such terms made them difficult to remember so that constant reference back to the literature was necessary. This detracted from their original conception of being self-explanatory and time saving. In addition some terms had been used in different senses by different authors. After valuable exchanges of views on these problems the feeling of the meeting was that the use or otherwise of existing terminology was a matter of personal preference and whilst every effort should be made to make quite clear the sense in which a term was used, it was inappropriate at this time to formulate a set of madatory instructions.

Author's address:

Dept of Geology, University of Hull, Hull, England

Sixth Intern. Ostracod Symposium, Saalfelden

DISCUSSION ON PALEOZOIC OSTRACODA

I. G. SOHN

More than an hour and twenty minutes was devoted to an informative discussion convened by I. G. Sohn and F. Adamczak on the relation of soft parts to the valve structures of Paleozoic Ostracoda. The discussion was taped by Dr. R. F. Maddocks. By arrangement, Dr. J. M. Berdan projected slides of the Ordovician genus *Ceratopsis* Ulrich, 1894 as a starting point for the discussion in which many interesting observations were made.

Because of lack of space, a verbatim transcript of the discussion has not been published. The following 20 participants contributed to the discussion: Drs. F. Adamczak, G. Becker, J. M. Berdan, W. K. Braun, D. L. Danielopol, P. De Decker, G. Hartmann, K. Ishizaki, L. S. Kornicker, E. Kristan-Tollmann, K. Krömmelbein, A. Liebau, R. F. Maddocks, I. D. Pinto, R. Schallreuter, D. A. Siveter, I. G. Sohn, F. M. Swain, P. C. Sylvester-Bradley, and S. M. Warshauer.

I wish to thank Dr. Maddocks for copies of her cassettes, without which this report could not have been prepared. Drs. J. M. Berdan and L. S. Kornicker checked my transcription of the tapes.

Author's address:

Room E-214, Nat. Museum Nat. History, Washington, D.C. 20560, USA

GENERAL INDEX

Authors

Absolon, A., 298
Adrichem Boogaert H. A. van, 459, 460, 461, 462, 471
Agalarova, D. A., 400, 402, 403
Agalarova, D. A., Djafarov, D. I. & Halilov, D. M., 402
Agalarova, D. A., Kadyrova, Z. K. & Kulieva, H. M., 403
Agassiz, L., 356
Ahmad, M., 186
Alvinerie, J., Pratviel, L., Veillon, M. & Vigneaux, M., 422
Anderson, E. J., 484
Anderson, F. W., 455
Andrieux J., Fontbote, J. M. & Mattauer, M., 449
Angel, M. V., 51, 54, 256, 268, 491
Angel, M. V. & Fasham, M. J. R., 46, 48, 51, 256
Apstein, C., 271
Archibald, C. P., 250
Ascoli, F., 23, 397
Assemien, P. et al., 392

Baird, W., 407
Baker, A. de C., Clarke, M. R. & Harris, M. J., 48
Baker, F. C., 289
Baker, J. H., 165, 167, 245, 246, 248, 251, 252, 254
Baker, J. H. & Hulings, N. C., 176, 179
Bandel, J., 472
Bandel, K. & Becker, G., 471
Barrington, B., 356
Bassler, R. S. & Kellett, B., 10
Bate, R. H., 74, 219
Becker, G., 472
Begin, Z. B., Ehrlich, A. & Nathan, Y., 58, 59, 63
Belousova, Z. D., 451
Benson, R. H., 16, 17, 18, 19, 20, 33, 35, 94, 107, 161, 216, 337, 436
Benson, R. H. & Coleman, G. L., 176

Benson, R. H. & MacDonald, H. C., 327, 336
Benson, R. H. & Kaesler, R. L., 216, 218, 336
Bentor, L. K., 56
Benzecri, J. P., 127
Berdan, J., 483
Berger, F., 324
Berry, E. W., 356
Berry, W. B. N. & Boucot, J., 475
Bertels, A., 94
Bischoff, G., 455
Björnberg, T. K. S., 256
Bless, M. J. M. & Michel, M. P., 459, 463, 471
Blondeau, M.-A., 23
Blumenstengel, H., 459, 463, 465, 469, 471, 472
Bottger, O., 353, 356
Bold, W. A. van den, 177, 179, 186, 207, 216, 217, 218
Boltovskoy, E., 359
Bonaduce, G., Ciampo, G. & Masoli, M., 223
Borragan, 411
Bowman, T. E. & Kornicker, L. S., 250
Brady, G. S., 176, 271, 232, 399, 407, 409
Brady, G. S., Grosskey, H. W. & Robertson, D., 332, 333, 411, 435, 436
Brandhorst, W., 256
Branson, C. C., 455
Brelie, G. van der, 64
Breman, E., 180, 183, 223, 408, 409, 412, 415
Brezanu, G. H. et al., 299
Brillouin, L., 34, 35
Brockmann, Chr. von, 63
Broecker, W. S. & Kaufmann, A. 311, 312
Broecker, W. S. & Orr, P. C., 311, 312
Bronstein, Z. S., 399
Brooks, J. L. & Dodson, S. I., 250
Brown, S. G., 336
Brown, S. G. & Shumann, H. H., 336
Buch, A., 64

Budinger, P. & Kullmann, J., 461
Bullard, E., Everett, J. E. & Smith, A. G., 441, 442
Butzer, K. W., Isaac, G. L., Richardson, J. L. & Washbourn-Kamau, 392
Buzas, M. A. & Gibson, T. G., 161

Camacho, H., 94
Cameron, S. P., 337
Camp, W. H., 359
Cannon, G., 266, 496
Capeder, G., 411
Carbonel, P., 183, 407, 412, 415
Carbonel, P. & Moyes, J., 407
Carbonel, P., Moyes, J., Peypoquet, J. P., 369
Carbonnel, G., 66, 114, 125, 130, 391, 400, 411, 415, 440
Chase, J. S. & Hunt, A. S., 327, 330
Childress, J. J., 269
Choubert et al., 411
Christensen, O. B., 455
Cita, M. B., 18
Clements, G., 118
Coates, D. R., 336
Colin, J. P., 31, 202
Conrad, T. A., 353, 356
Cooper, J. R., 336
Cooper, W. E., 250
Coryell, H. N. & Fields, S., 216
Curry, D., 27
Curtis, D. M., 216

Daday, E. von, 271
Dall, W. H., 356
Damon, P. E., Long, A. & Sigalove, J. J., 340
Danielopol, D. L., 295, 296, 297, 298, 300, 301, 302, 304
Darby, D. G., 164, 167
DeDeckker, P., 293, 294
Delamare-Deboutteville, C. L., 297
Delibrias, G., 391
Delorme, L. D., 327, 328, 336, 337, 345, 346
Deltel, B., 422
Den Hartog, C., 208, 213
Devoto, G., 402
Dickinson, K. A. & Swain, F. M., 318, 337
Dieci & Russo, 411
Dietz, R. & Holden, J., 441, 444
Dittmer, E., 64
Dobzhansky, T. H., 303
Donze, P. 441, 447
Drake, C. L. & Nafe, J. E., 441, 443
Ducasse, O., 420, 421, 422

Edmondson, W. T., 245, 250
Ehrlich, A., 57
Ekman, S., 218
Elofson, O., 85, 87, 121, 124, 203, 247, 248, 249, 250
Emery, K. O., 246
Engel, P. L. & Swain, F. N.M., 217
Etheridge, R., 353

Fagetti, E., 256
Fagetti, E. & Fischer, W., 256
Fasham, M. J. R. & Pugh, P. R., 492
Faure, H., 387
Faure, H. & Elouard, P., 387, 391
Felber, J. E., 5
Ferguson, E., 273
Ferguson, L., 487, 488
Fillon, R. H. & Hunt, S. A., 327
Findenegg, I., 324
Fischer, A. G., 166
Fischer, P., 407
Folk, R. L., 479
Fowler, G. H., 45, 48
Fox, H. M., 280
Fredoux & Tastett, 394
Freed, W. K., Hunt, A. S. & Fillon, R. H., 327
Frith, H. J., 289, 290, 291
Frith, H. J., Braithwaite, L. W. & McLean, J. L., 289, 290
Furtos N. C., 336

Gabb, W., 353, 356
Gardner, J., 355, 356
Gellert, J. F., 197
Ghetti, P. F., 280
Gooday, A. J., 52, 463, 470
Gooday, A. J. & Angel, M. V., 52
Gottsche, C., 58
Grahle, H. O., 64
Gramann, F., 399, 402
Gramm, M. N., 134
Gramm, M. N. & Gründel, J., 139
Green, J., 280
Green, P. E. & Carmone, F. J., 238
Grekoff, N., 455
Grekoff, N. & Molinari, V., 403
Grekoff, N. & Krömmelbein, K., 455
Greve, L. de, 353, 356
Grochmalicki, J., 275
Groos-Uffenorde, H. & Uffenorde, H., 470
Gründel, J., 8, 134
Guardia, P., Magne, J. & Moynes, J., 411
Guber, A. L., 70, 71
Gunther, E. R., 256
Gunther, F. J. & Hunt, A. S., 327, 329, 332

Gurney, R., 271
Gutendag, E. D. & Benson, R. H., 336

Hall, D. J., 250
Hanganu, E., 403
Hardy, A., 268
Harper, F. & Sutton, A. M., 455
Harten, D. van, 66
Hartl, H. & Sampl, H., 321
Hartmann, G., 4, 189, 203, 213, 216, 218, 256, 297, 495
Hartmann, G. & Puri, H. S., 200
Hartmann-Schröder, G. & Hartmann, G., 188, 189, 190, 256, 304
Hartt, Ch. Fr., 356
Hazel, J. E., 29, 125, 164, 328, 476
Heck, H. L. & Brockmann C., 64
Hedgpeth, J. W., 168
Helmdach, F. F., 455
Henningsmoen, G., 8
Herrig, E., 115
Hessler, R. R., 161
Hevly, R. & Martin, P. S., 336
Higashikawa, S., 425, 433
Hillman, N. S., 45, 48
Höff, C. C. 87, 336
Hornibrook, N. de B., 436
Hou, I. T., 455, 471
Howe, H. V., 5
Howe, H. V. & Bold, W. A. van den, 178
Hulings, N. C., 178, 250
Hulings, N. C. & Puri, H. S., 211
Hutchinson, G. H., 50, 309, 311

Ihering, 359
Inbrie, J., Andel, T. H. van, 426
Ishizaki, K., 440
Ishizaki, K. & Gunther, F. J., 216, 218
Ivanova, V. A., 8, 10

Jaanusson, V., 70
Jäger, P., 324
Jones, A. J., 411
Jones, M. E., 165
Jones, R. S. & Cushman, R. L., 336
Jones, T. R., 455
Jones, T. R. & Kirkby, J. W., 488
Juday, O., 216

Kaesler, R. L., 33, 42, 43, 54, 125, 327
Kaesler, R. L., Crossman, J. S., Toole, T. D. & Urban, R. D., 237
Kashervarova, N. P., 451
Katto, J., Nakamura, J. & Takayanagi, Y., 425, 436
Katzer, Fr., 356, 359

Keen, M. Ch., 31
Keij, A. J., 411
Kempf, K., 283
Kesling, R. V., 45, 247
Keyser, D., 207
Kilenyi, T. I., 42, 43, 142, 332, 409
King, C. E. & Kornicker, L. S., 207, 213, 216, 218
Klein, L. N., 403
Klie, W., 166, 200, 216, 217, 275, 289, 297, 299
Knox, G. A., 169, 192, 194
König, D., 64
Kontrovitz, M., 176
Kornicker, L. S., 69, 70, 87, 162, 164, 166, 167, 236, 239, 241, 247, 249, 250, 252, 269, 293, 448, 484, 488
Kornicker, L. S. & Caraion, F. E., 167
Kornicker, L. S. & Sohn, I. G., 8, 289, 290
Kornicker, L., Wirsing, S. & McManus, 266
Kornicker, L. S. & Wise, Ch. D., 165, 213
Koscielniakowska, O., 470
Kozur, E., 471
Kozur, H., 8, 147
Kramp, P., 256
Kristan-Tollmann, E., 139, 140, 142
Krömmelbein, K., 455
Krömmelbein, K. & Weber, R., 455
Krstic, N., 31, 302, 399, 402
Kruskal, J. B., 238
Krutak, P. R., 176
Kukhtinov, D. A., 451
Kullmann, J., 460, 461, 471
Kullmann, J. & Schönenberg, R., 459, 461
Kurc, G., 409

Lafrenz, H. R. von, 58, 64
Lange, W., 63
Langer, W., 115, 472
Latreille, P. A., 10
Lerner-Segev, R., 57
Levy, Y., 57
Lie, U., 165, 245, 246, 249
Liebau, A., 94, 95, 96, 97, 107, 108, 109, 111, 113, 114, 115, 116
Lipatova, W. W. & Starozhilova, N. N., 451
Lister, K. H., 327
Ljubimova, Kazmina, T. A. & Reshetnikova, M. A., 451
Loczy, L. de, 359
Löffler, H., 293, 294, 297, 301, 302, 321, 322, 324, 325, 327
Löffler, H. & Leibetseder, J., 293
Long, A., 336, 340, 346, 347
Loranger, D. M., 455
Loyd, M., Inger, R. F. & King, F. W., 35

Lucas, V. Z., 165, 245
Lundin, R. F., 476

MacArthur, R. H., 50
MacDonald, A. G., 268
MacDonald, J. D., 291
MacNeil, F. S., 436
Maddocks, R. F., 53, 125, 393, 487
Malz, H., 5, 10
Malz, H. & Moyedpur, 302
Mandelstam, M. I., 451
Mandelstam, M. I., Markova, L. Rozyeva, T. R. & Stephanaitys, N. E., 403
Mandelstam, M. I. & Schneider, G. F., 302
Manton, S. M., 10
Margalef, D. R., 34
Marinov, T., 409, 415
Marshall, W. B., 356
Marshall, W. B. & Bowles, E., 356
Martens, J. M., 256
Martin, G. P. R., 455
Martin, P. S., 335, 336, 339, 340, 435, 347, 349
Martin, P. S. & Moismann, J. E., 336
Martinsson, A., 70, 71, 89
Masoli, M., 411
Maury, C. J., 353, 355, 356
McKenzie, K. G., 166, 289
McKenzie, K. G. & Hussainy S. U., 291
McKenzie, K. G. & Swain F. M., 216
Meinzer, O. E. & Kelton, F. C., 336
Mendes, J. C., 359
Mishina, E. M., 451
Moore, R. C., 8
Morales, F. G. A., 176, 207, 213, 216, 218
Morkhoven, F. P. C. M. van, 69, 70, 176, 178, 179, 180, 357, 393
Moroni, A., 280
Moyes, J., 411, 418, 420, 422
Moyes, J. & Pachier, E., 407
Müller, G. W., 55, 87, 166, 248, 255, 408, 409, 496
Müller-Steffen, K., 470

Naidina, N. N., 397
Neal, R. H., Johns & Erickson, R., 176
Neale, J. W., 4, 5, 16, 31, 203, 271, 283, 293, 328, 441, 495
Neale, J. W. & Howe, H. V., 28, 29, 328
Neale, J. W. & Victor, R., 277
Neev, D. & Emery, K. O., 58
Neustrueva, I. Y., 451
Niino, H. & Emery, K. O., 427, 435
Niitsuma, N., 33
Noodt, W., 297
Norman, F. I., 289

Oertli, H. J., 4, 20, 66, 111, 304, 369, 455
Okubo, I., 271
Oliveria, A. I. & Carvalho, P. F., 356
Omatsola, M. E., 55, 393
Orton, J., 353, 356

Parker, R. H., 163
Pearse, A. S., 10
Peck, R. E., 455
Peneau, J., 470
Petkovski, T., 303
Peypouquet, J. P., 28, 367, 393, 394, 407, 411
Pielou, E. C., 34, 35
Pine, G. L., 336
Pinto, I. D. & Sanguinetti, J. T., 455
Pokorny, V., 107, 216
Polski, W., 434, 435
Pipikin, B. W., 336, 347
Por, F. D., 56, 57
Poulsen, E. M., 45, 48, 50, 70, 245, 255
Powell, S., 322
Proctor, V. W., 289, 290
Proctor, V. W. & Malone, C. R., 289, 290
Puri, H. S., 4, 55, 176, 177
Puri, H. S., Bonaduce, G., Gervasio, A. M., 409
Puri, H. S. & Dickau, B. E., 55
Purper, I., 369

Rabien, A., 463
Rafle, M. A., 176
Reid, J. L., 257
Remane, A., 191
Remane, A., Storch, V. & Welsch, U., 107
Reuss, A. E., 332
Reys, S., 201
Robinson, R. C., 336
Rohlf, F. J., 238
Rome, D. R., 409
Rosenfeld, A. & Vesper, B., 66
Rothwell, W. T. Jr., 179
Rouvilois, A. & Boulanger, D., 409
Roxo, G. de Oliveira, 353, 356
Rozhdestvenskaja, A. A., 471
Rudjakov, A., 45
Ruggieri, G., 18, 400, 409, 411, 420
Russel, I. C., 309, 310
Ryan, W. F. B., 18
Ryan, W. B. F. & Hsu, K. J. et al., 18

Sandberg, Ph. A., 176, 313, 357
Sandberg, Ph. A. & Hay, W. W., 55
Sandberg, Ph. A. & Plusquellec, P. K., 55, 289, 290
Sanders, H. L., 161, 483

Sanders, H. L. & Hessler, R. R., 161
Sars, G. O., 200, 327, 400
Schäfer, H., 297
Schattner, J., 57
Schmalfuss, H., 107
Schmidt, H., 416
Schminke, K., 304
Schreiber, J. F. et al., 335, 339
Schweyer, A. V., 403
Servant, M. 392
Shannon, C. E. & Weaver, W., 35
Sherov, A. G., 10
Shneider, G. F., 451
Shornikov, I., 399
Siddiqui, Q. A. & Grigg, U.M., 162
Siegel, S., 35
Sievers, H., 255
Simpson, G. G., 356
Sinegub, V. V., 400, 402, 403
Sinitsa, S. M., 451
Sissingh, W., 411, 415
Skoblo, W. M., 451
Skogsberg, T., 48, 50, 248
Smosna, R. & Warshauer, S. M., 476, 484
Smosna, R., Tarchen, D. G. & Perry, W. Y., 476, 484
Sneath, P. H. A. & Sokal, R. R., 237, 238, 476
Sohn, I. G., 3, 54, 66, 487
Sokac, A., 223, 227, 302
Sokal, R. R. & Sneath, P. H. A., 125, 237
Spjednaes, N., 71
Stancheva, M., 400, 401, 403
Staplin, F. L., 336, 347
Steimann, G., 356
Stephenson, T. A. & Stephenson, A., 197
Stevanovic, P., 399
Stose, G. W. & Swartz, C. K., 475
Suzin, A. V., 399, 403
Sverdrup, H. U., Johnson, W. M. & Fleming, R. H., 235
Swain, F. M., 176, 216, 217, 318, 336, 445
Swain, F. M., Becker, J. & Dickinson, D. A., 318, 337, 345
Swain, F. M. & Gilby, J. M., 216
Swain, F. M., Kornicker, L. A. & Lundin, R. F., 4
Swain, F. M. & Meader, R. W., 309, 311
Swartz, C. K., 475, 476, 480
Sylvester-Bradley, P. C., 20
Sylvester-Bradley, P. C. & Benson, R. H., 107, 114, 115, 116, 118
Szczechura, J., 85, 86

Tastet, J. P., 394
Teeter, J. W., 179, 336
Terquem, O., 411, 415
Theisen, B. F., 129
Triebel, E., 216, 218
Tschigova, V. A., 471
Turner, H. W., 309

Uffenorde, H., 223, 408, 409
Uliczny, F., 407, 409, 411, 412, 415
Ulrich, E. O., 475

Valentine, J. W., 26
Vandel, A., 297, 300
Van Dolah, R. F., Shapiro, L. E. & Rees, C. P., 250
Vavra, W., 279
Vekua, M. L., 403
Vesper, B., 55, 56, 313
Voorthuysen, J. H. van, 64

Wagner, C. W., 313
Walker, K. R. & Laporte, L. F., 483
Wall, D. R., 85
Waller, R. A., 165
Walton, W. R., 33
Warshauer, S. M., 485, 488
Wass, M. L., 164
Weygoldt, P., 496
Whatley, R., 31, 254, 294, 304, 329, 325, 488
Whatley, R. C. & Kaye, P., 332
Whatley, R. C. & Wall, D. R., 201, 202, 407
Whittaker, R. H., 169
Wicher, C. A., 455
Wieser, W., 200, 202
Williams, D. & Hynes, 304
Williams, R., 202
Williams, W. T., 237
Wilt, J. C., 336
Wise, C. D., 3, 5
Wolburg, J., 455
Wolf, J. P., 297
Woodward, H., 353, 356, 476
Wooster, W. S. & Gilmartin, M., 256
Woszidlo, H., 63
Wyrtki, K., 256

Yassini, I. L., 407, 409, 410, 411
Yen, T.-Ch., 289

Subjects

Abundance (ostracods) 344–350
Antarctic/Subantarctic regions 235, 239, 240, 243
Antarctic convergence 235, 236, 243, 244

and distribution of podocopids 235, 236, 243
and distribution of myodocopids 235, 239, 243
Antarctic Intermediate Water 239, 343, 257
Atlantic (ostracods) 29, 409, 410, 411, 412, 413 (see also Coast Atlantic)
Assemblages (ostracods) 330, 331
   Candona zone 332
   Cytheropteron zone 332
   Estuarine 373
   Euryhaline 215, 219
   and factor scores 427
   Freshwater 332
   Lagoonar 372, 373, 386
   and lake types 455
   Limnic-Oligohaline 213, 219
   Mangrove 207, 219
   Marine 332
   Marine littoral (phytal) 369
   Monoceratina 379
   Myodocopids 242
   Oligo-Mesohaline 215, 219
   Nearshore 346, 350
   Poly-Euhaline 216, 219
   Sand – clayey 228
      – coarse 233
      – fine 228, 231
      – medium 231
   and mineral composition of sediment 233

Bar-3, 233
Behaviour (ostracods) 267, 268
   Movement 268, 299, 300
Beriassian (ostracods) 445
Biocenose/Thanatocenose 369, 371, 372
Biodynamics (ostracods) 17
Biofacies
   Skogsbergiella – Empoulsenis 241
Biogeographic provinces (ostracods) 23, 26, 218 (see also Zoogeographic provinces and zonation)
   Anglo Gallic 24, 26, 27
   Aquitaine – North 24, 26, 27
            – South 24, 26, 27
   Brittanic (northern) 28
   Celtic 26, 28, 29
   Gasconian (southern) 28
   Virginian 29
   Tethyan 26
Basin
   Aquitanian 31
   Hampshire 27
   Paris 27, 31

Brackish water (ostracods) 207, 310, 311, 399
   Currents 213, 219
   pH 213, 219
   Substrate 211, 219
   Temperature 211, 213, 219
   Turbidity 213, 219
   Water depth 213, 219
Budva 2, fine sand 228, 231
Budva 3, medium sand 231, 233

Carapace morphology (ostracods)
   Carinocythereis 412
   Macrocyprididae 148
   Tyrrhenocythere 396, 398, 399
Carapace ornamentation 357
   Advanced Cytheracea 108
   Australicythere 94
   Celation 111
   Conation 107, 108, 116, 117, 118
   Costulation 107, 108, 115, 118
   Ecologic distribution 186
   Evolution and functional significance 107
   Limnocytherinae 118, 119
   Loxoconchidae 129, 130
   Primitive Cytheracea 117
   Post-Triassic Bythocytheridae 118, 119
   Reticulation 107, 108, 111, 114, 115, 118
Carapace pores
   Normal type 55
   Sieve type 55
   Sieve pores and salt concentration 55-62
Carapace sexual dimorphism 49, 50, 51
   Precocious 69, 70, 71, 72, 89
Carapace size 252
   Depth 156, 184
   Environment factors 45, 50
   Inner lamella 487, 488
   Latudinal zonation 156
   Post maturation molt 45
   Seasonal variation 85, 86
Carson Sink 309
Champlain Lake 327
Change (ostracods with water-mass) 17
Chaetognatha 266
Circulation water (gyroidal) 31
Coast Atlantic 162
   Gulf 164
   North American 162
   Pacific 165
Cochise Lake
   Pleistocene 335, 336, 337, 339
   Ostracods 340-347
Coralline Crag (faunas) 31
Community structure

Leperditia 478, 483
Ostracods 33
Organisms 43
Welleria-Dizygopleura 479, 481, 483
Zygoberichia-Haillella 480, 481, 483
Competition (ostracods) 51, 53

Deep sea faunas (ostracods) 15, 20
Development (ostracods)
 methods of instar identification 247, 256
 Euophilomedes productus 247, 250, 252
Displacement 180, 332, 346, 347
Distribution (ostracods) (see also geographic distribution)
 Air exposure 201, 202, 203, 204
 Algae 201, 202, 203, 204
 Alkaline waters 309
 Altitudes high 292
 Amazon basin, upper 353, 354, 357
 Asia minor 280
 Asia South East 271, 277
 Australian Coast 187, 188, 195
 Calcium sediment (lacustrine) 316
 Circumpolar 327, 328
 Coast of Montenegro 223
 Current Kuroshio 436, 439
 Depth 179, 180, 184
 Endemic to Australia 258
 European subteranean forms 295, 304
 Giant forms 285
 Halocyprididae 255
 Horizontal 200, 256-261
 India 277
 Indonesia 277, 280, 282
 Lakes, quaternary pluvial 307
 Littoral 453, 454
 Mytilocypridini 285-288, 293, 311
 Passive 285 (see also passive distribution)
 Rocky shore (Helgoland) 197, 202-204
 Running waters 298
 Saliné water 288
 Salinity 208-210, 213, 328, 329
 Seasonal 246, 249, 250, 251, 252
 Sri Lanka 271, 277, 282
 Substrate 201, 211, 213, 215
 Swampy water basins 455
 Temperature 328, 329
 Thermal spring 298
 Tropical 189, 190, 191, 195
 Vertical 263, 264, 265
 Warm antiboreal zone 191
 Water masses 239, 243, 256-261
Diversity (ostracods) 160, 295, 300, 302, 332, 477, 478, 479, 481, 483, 484, 485
Assemblages 33

Australian Mytilocypridini 293
Bays 166
Bathymetry of the ocean 41, 161, 169, 170, 171
Benthic myodocopids 159
Brillouin's index 34, 37, 41, 43
Carbonate banks and coral reefs 167
Continental shelves 167
Generic 36, 39, 41
Geologic time 40, 41
Methods of sampling 160
Southern oceans 167
Species 33, 35, 41
Stability of the environment 171
Taxonomic hierarchy 38, 40, 42, 43
Zoogeographic areas 162-169

Eggs (ostracods)
 Desiccation 288, 289
 Dispersal 289, 290, 292
 Transport 289
Eh sediment measurements 347
Entomozoidae
 Ecology 471
 Systematics 463-470
Euhedra conditions 347
Evolution (ostracods)
 Berriassian 444, 447
 Candoninae 304, 305
 Cypridacea 300
 Subterranean forms 300, 302

Faunas (ostracods)
 Caribbean 178, 179, 180, 186
 Psychrospheric 18
 St. Erth in Cornwall 29, 31
Formation (geologic)
 Ananai 436, 439
 Esmeralda 316
 Gilda-Morito beds 460, 461
 Monte Leon 93, 94
 Santa Julian 93, 94
 Shinzato 436
 Tonoloway 475, 476, 481
Fossilisation (ostracods) 451, 472

Geographic distribution (ostracods)
 Caribean, Gulf of Mexico 175, 178, 179, 180, 186, 207, 216, 217, 218, 327, 328, 329 (see also distribution)
 Goggausee 324
 Gunther current 257
 Hydrochemical condition 455
 Macrocyprididae 147, 153, 154, 155, 156
 Mediterranean basin 400, 407, 409, 412, 415

507

Pacific 207, 217, 218
Vocontian and Iberian realms 443

Helgoland Island
　Ecologic characters 200
　Geographic description 197, 198
Holocene/Pleistocene (ostracods) 29, 30, 31, 307-316, 321-325, 327, 329, 335-337
Hypogean (Ostracods) 295, 297, 298, 304

Interstitial (ostracods) 202, 295, 297, 303, 304

Juveniles (ostracods)
　Growth factor 45, 48, 49
　sampling methods 54

Kleinsee 321, 322, 324
Klopeiner See 321
Kotor-3, Clayey sand 228

Lahonton Lake 307
Längsee 324
Life History (ostracods) 245

Mediterranean Sea 18, 19
Meromixis 321, 323, 324
Mesozoic (ostracods) 7
Migration (ostracods) 411
Miocene (ostracods) 19
Morphology (ostracods) 267, 268, 297, 298, 299, 300, 301, 305 (see also carapace morphology)
　Eye tubercles 17
　Heteromorphs 254
　Loxoconchidae (morphology groups) 125
　Morphocline (Macrocyprididae) 154-156
　Morphotypes 54
　Muscle scar patterns 133, 134, 135, 136, 139, 148
　Polymorphism 74
　Preservation 290
　Standardization of limbs terminology 495-497
　Terminology of the hinge 497
Molting (ostracods)
　Postadult 69, 87, 88, 252

Neogene (ostracods) 17, 399-403
Nutrition (ostracods) 413
　Gut contents 266, 268, 269
　Stomach contents 263

Origin (ostracods)
　Freshwater 297, 300
　Marine 297

Paleclimatic indicators (ostracods) 391, 392
Paleoecologic indicators (ostracods)
　Alkaline mesothrophic lake 345
　Benthal 472
　Brackish water 357, 359
　Deep water 471
　Depth 412
　Desiccation 346
　Environmental enegy 373, 484
　Estuarine environment 377
　Freshwater 357
　Lagoonar conditions 373, 377, 381, 383, 384, 386
　Litoral (infra) 381
　Marine transgression 484
　Pelagial 472
　Phytal 377, 381
　Pluvial interpluvial conditions 345, 346, 350
　Salinity 372, 373, 382
　Shallow waters 346
　Shore (near) highly variable 487
　Shore stable conditions 487, 488
　Temperature 448, 449
Paleogeographic indicators (ostracods) 357, 391
　Gascogne basin 421, 422
　Nouakchottian transgression 369
Paleozoic (post.) ostracods 5, 7, 499
Parthenogenesis (ostracods) 288
Passive distribution (ostracods) (see also distribution)
　Birds 280, 285, 289, 291, 292, 293, 294
　Fishes 289, 290, 292
　Rice plants 280, 282, 293
　Ships 280
　Wind 289, 292
Peru current system 255, 256, 257
Phylogeny (ostracods)
　Costa 18
　Cytherelidae, Healdidae 140
　Platycopina 141
Phytal (ostracods) 202, 203, 205
Planktonic halocyprids 45
Plate tectonics
　Alboran subplate 449
Pollenanalysis 339, 340, 347-350
Population analysis (ostracods) 250, 251, 252
Predation (on ostracods) 251, 252, 289
Predators (ostracods) 269
Pyramid lake 309, 310, 311, 313

Radiometric chronology 340
Reconstruction (paleogeographic/paleoecologic)

Europe (eocene) 25-27, 31
Lagoonar environment 387-391
Litoral zone 478, 480
Mangrove conditions 388, 389
Marine transgression 388
Subtidal zone 479
Remane's brackish water rule 192
Reproduction period (ostracods) 249, 250, 252
Rice-fields (ostracods) 271, 277, 280

Salinity crisis 18, 400, 403
Sampling methods
  Accumulation rate of species in samples 491
  Biological patchiness 491, 492
  Subsampling program 491, 493
San Diego, California 246
SEM (ostracods) 7, 16
Senegal, continental shelf 369
Sex ratios (ostracods) 89, 90, 247, 248, 250, 264, 288
Shallow water (ostracods) 399
Silicified (ostracods) 459, 471, 472
Stratigraphy (ostracods) 344, 347-356, 411
Subantarctic surface water 257
Sulphur springs valley 336, 339
Statistics methods
  CLIMAP programme 492
  Cluster analysis 33, 237
  Correspondence analysis 127, 128, 129
  Coefficient of cophenetic correlation 237, 238
  Jaccard's coefficient 24, 25, 237
  Nonmetric multidimensional scaling 237, 238, 242
  Ordination techniques 238
  Q and R-mode analysis 237, 238, 239, 242, 425, 476, 480, 481
  Recurrent groups analysis 245
  Regression (multiple) 434, 435
  Sperman's correlations coefficient 36
  Wilcoxon's signet rank test 248, 252
Sympatric species (ostracods) 54
Systematics (ostracods)
  Coquimbids 101
  Cythereidinae n.s. 116
  Hemicytheridae 93
  Macrocyprididae 154, 155
  Nomenclatorial notes 186

Tertiary faunas 29, 31
Tertiary ostracods 7, 23, 30
Tethys ocean 15, 18, 19, 361, 441, 442, 448
Tethys/Paratethys realms 399-403, 444, 448
Thermal barrier 17, 19

Thermal gradient 17, 26
Triassic ostracods 133
Troglobitic/troglophilic ostracods 295
Tropical fauna 218, 219

Ulcinj-4, fine sand 228, 231
Upper Amazon basin sediment 355-357
  paleogeography 359

Water mass structure of the ocean 19
Willcox Playa 335
Williston's law 107

Zoogeographic provinces (see also biogeographic provinces)
  Antillean 177
  Central American 177
  Eocene recent 36
  Gulph of Mexico, South Florida 177
  South American 177
  Venezuela 177
  Virginian 328
Zoogeographical zonation (see also biogeographic provinces)
  Australian Coast 188
  Monto-Arruzo zone 461
  South America 188
  South Cantabrian zone 460, 4461
  Subtropical tropical transition zone 195
  Suptropical Zone 194
  Tropical fauna 189 (see also faunas)
  Warm antiboreal zone 194
  West coast of Africa 188

Genera and species

Abyssocythere 184, 420, 421, 422
Acantocythereis 225
 – histrix 224, 231
 – sp. 430
Acetabularia 211
Acratia sp. 462, 463
Acratina 147
Acrocythere 444
 – constricta 444
 – diversa 444
Actinocythereis 177
 – scutigera 189
 – subquadrata 208, 209, 210, 211, 212, 214, 216, 217
 – triangularis 216
 – sp. 430, 432
Acuticythereis sp. A 217
 – sp. B 217

Aglaiocypris 181
- eulitoralis (f. floridensis) 208, 209, 210, 212, 214, 216
- sp. 218

Agrenocythere 184
- ordinata 419, 421

Alocopocythere goujoni 430
Alteratrachyleberis 109
Alteratrachyleberis? striatopunctata 113
Alveolina 26
Ambocythere 180, 181, 184
- sp. H 430
- sp. P 430

Ambostracon 93, 94, 96, 104
- - (Ambostracon) 96, 100
- - costatum 97, 98
- - diegoensis 98
- - flabellicostata 99
- - glaucum 98
- - hulingsi 99
- - vermilloensis 99
- - sp. 99
- - sp. 1 94, 97, 98, 101
- - sp. D 99, 101
- - sp. E 98, 99
- - sp. G 98, 99
- - sp. J 98, 99
- - sp. K 98, 99
- - sp. L 98
- - sp. M 98, 99
- - sp. O 98, 99
- (Patagonacythere) 96, 100, 101
- - japonicus 99
- - longiductus antarcticum 98
- - pumila 99
- - tricostata 96, 98, 99, 101, 102
- - wyvillethomsoni 98, 99
- - sp. 2 94, 98, 101, 102
- - sp. 3 94, 98, 101, 103
- - sp. A 99

Amphissites aff. irinae 462, 463
Amphycythere cf. vonvalensis 444
Anathron 70
- dithrix 235, 237, 240, 241, 242
- reticulata 256

Anchistrocheles? sp. aff. A? angulata 182
Aquitaniella transversa 27
Archasterope 70
Argilloecia 181
- cylindrica 328, 329, 330, 331
- sp. 182, 430, 432

Asciocythere circumdata 444
- montis 444

Astenocypris sp. 218
Aurila 101, 104, 180, 181, 192, 225, 233, 371, 372
- amygdala 208, 209, 210, 212, 213, 214, 216, 217
- convexa 104, 224, 228, 233
- conradi 213
- cymba 430
- interpetis 224, 231
- oblonga 418
- prasina 224, 231, 233
- punctata 418
- speyeri 224, 228, 231, 233
- sp. 224, 231
- sp. 1 98, 101, 104, 369
- sp. 2 98, 103, 104, 371, 372
- 3 98, 103, 104
- (Hemicytheria) 398
- - josephinae 398

Australicythere 93, 94, 100, 104
- californicum 94, 95, 99, 101
- devexa 99, 101
- megalodiscus 99
- polylyca 94, 95, 99, 101
- sp. 1 94, 98, 99, 102
- sp. 2 94, 98, 99, 102
- sp. 3 94, 98, 99, 102
- sp. 4 94, 98, 99, 102
- sp. 5 94, 98, 99, 102

Australocypris 285
- hypersalina 285
- insularis 286
- robusta 286, 289

A? rectangularis n. sp. 286
Avicennia 195
- marina 189

Bairdia 23, 27, 181, 184, 225, 233, 487, 488
- amygdaloidea 183
- antillea 182
- bradyi 217
- crebra 27, 418, 420
- cymbula 27, 421
- dimorpha 182
- longisetosa 182, 183
- longivaginata 224, 231, 233
- major 445
- aff. B fusca 182
- succinata 27
- tenuis 27
- victrix 182
- aff. victrix 183
- sp. 183, 462, 463
- sp. 2 182
- sp. 4 182
- sp. 5 182

Bairdoppilata 418
- gliberti 28

Basslerites 180, 181

- minutus 183, 217
Bathyconchoecia 70
Bathyvargula 70
Beyrichia 70, 71
Bisulcocypris 357
- sp. 354, 365
Bosmina 323
- cf. coregoni 321
Bosquetina 225
- carinella 418
- pectinata 224, 228
Bradleya 26, 101, 181, 184
- dictyon 19
- japonica 430
- kaasschieteri 28
- normani 98, 101, 103
- oertlii 28
- ordinata 27
- sp. 430, 432
Buntonia 418, 419, 422
- sublatissima sublatissima 420
- sp. 2 420
Bythoceratina 184
- utilazea 430
- cf. orientalis 430
- sp. D 430
- sp. 111, 182
Bythocypris 181, 184, 370, 371, 372, 418
- prozoma 150
Bythocythere 181, 184
- bilobatus 328, 329, 330, 331
- constricta 330

Callistocythere 180, 181, 191, 193, 225, 369, 418
- adriatica 224, 231
- japonica 430
- cf. reticulata 430
Camarotoechia 478, 479, 480
Camdenidea 147
Campylocythere 180, 181
- laevissima 217
- perieri 175, 177, 179
- sp. 431
Campylocythereis 370
Candona 225, 295, 296, 302, 303, 321, 323, 324, 332, 343, 371
- acuminata 313, 315
- annae 208, 209, 210, 212, 213, 214
- bretzi 312, 315
- devexoides 338, 342, 343, 349
- neglecta 224, 228, 231
- patzcuaro 338, 342, 343, 349
- aff. patzcuaro 312, 315
- pratensis 342, 343, 349
- protzi 312, 315

- rawsoni 327, 328, 330, 331, 332
- rostrata 324
- simpsoni 338, 342, 343, 347, 349
- subtriangulata 327, 328, 330, 331, 332
- trunctata 338, 342, 343, 349
- wanlessi 338, 342, 343, 347, 349
- cf. crogmaniana 312, 315
- cf. ohioensis 313, 315
- sp. 224, 313
C.? balatonica 208, 209, 210, 212, 213, 214
Candonocypris 296
Candonopsis 295, 296, 305
Cardobairdia 184
Caribella 178
C.? yoni 182
Carinocythereis 412, 415
- antiqua 182
- antiquata 224, 228, 231, 407, 408, 409, 410, 411, 412, 413, 415
- bairdii 415
- carinata 371, 407, 408, 409, 410, 411, 412, 413, 415
Carinocytherideis 225, 231, 233
Cativella 178, 179, 180, 181
- dispar 217
- sp. 182
Caudites 93, 104, 180, 181
- angulatus 183
- howei 182, 183
- japonicus 99
- nipeensis 182, 183
- sp. 98, 101
Cavellina 135, 140
Celtia sp. J 431, 432
- sp. S 431
Centrocypris viridis 273, 275, 277, 279
Ceratacratia cerata 463, 465, 470
Ceriodaphnia laticaudata 310
Chaoborus 322, 323, 324
Chara sp. 310
Chasmites cujus 310
Chrysocythere 370
Clausocalanus furcatus 267
Clavoflabella 70, 71, 118
Cletocythereis 109
Clithrcytheridea 27
- faboides 27
Cluthia sp. M 431
Cnestocythere lamellicostata 186
Cobanocythere subterranea 217
- labiata 217
Conchoecia alata minor 45
- ametra 49, 51
- bispinosa 45, 51
- edentata 52
- elegans 45, 46, 47, 48, 52, 53

- giesbrechto giesbrechti 255, 260, 261
- haddoni 46, 49, 50, 51
- aff. haddoni 255, 258, 261
- hyalophyllum 46, 49, 50, 51
- imbricata 46, 48
- incisa 52
- inermis 46, 49, 50, 51
- lophura 46, 49, 51, 255, 261
- loricata 46, 49, 51
- macroprocera 51
- magna 46, 51
- microprocera 51
- oblonga 54, 255
- obtusata 46, 48, 49, 51
- parthenoda 51
- porrecta 45, 51, 52
- aff. porrecta 255, 261
- procera 46, 51
- pseudoparthenoda 51
- rhynchena 46, 49, 51
- rotunda 255
- rotundata 46
- secenenda 45, 51
- serrulata 45, 48, 255, 259, 261
- spinifera 46, 49, 51
- spinirostris 45, 51, 52
- stigmatica 46, 49, 50, 51
- striata 255
- striola striola 255
- teretivalvata 46, 49
- sp. 267

Congeria 302
Coquimba 93, 101
- bicostata 101
- piscicula 101
- sp. 1 98
- sp. 2 103
- sp. 182

C.? sp. 2 98
Cordocythere 295
Cornucoquimba 93, 101
- angulosa 101
- sp. 1 98, 103
- sp. 2 98, 101, 103

Costa 18, 179, 225
- bellipulex 179
- edwardsii 224, 228
- laticostata 181
- maquayensis 179
- punctatissima 418
- recticostata 181
- tricostata 418, 420
- variabilocostata 179, 180
- – laticostata 180
- – recticostata 180

Cryptocandona 295, 296, 303, 305

Cursina 116
Cushmanidea 181, 418
- cf. japonica 431
- sp. 182, 183

Cyamocytheridea 27
- hevertiana 28

Cyclocypris 296, 323
Cycloleberis 70
Cylindroleberis 70
Cymbicopia 70
Cypretta 211, 296
- brevisaepta 208, 209, 210, 212, 213, 214
- globosa 271, 275, 279
- globula 273, 275, 277, 278, 280

Cypria 295, 296, 323, 324, 357
- ophthalmica 299, 321, 323
- pseudocrenulata 208, 209, 210, 212, 214, 215, 219

C.? sp. 301, 354, 365
Cypricercus 296
C.? sp. 313, 315
Cypridea 357, 455
- sp. 354, 365

Cyprideis 18, 66, 181, 191, 211, 225, 296, 313, 318, 319, 357, 372, 496
- beaveni 208, 209, 210, 211, 212, 214, 215, 217, 218, 219
- bensoni 217
- castus 217
- mexicana 217
- pacifica 217
- salebrosa 208, 209, 210, 211, 212, 213, 214, 217, 219
- sp. nov. A 354, 363
- sp. nov. B 354, 363
- stenophora 218
- torosa 55, 60, 61, 62, 64, 217, 224, 231, 313, 317, 318
- cf. baconensis 313, 317, 318

Cypridinodes 70
Cypridopsis 211, 295, 296, 323
- okeechobei 208, 209, 210, 212, 213, 214
- subterranea 297
- vidua 313, 317, 318, 338, 342, 343, 349
- cf. glaucus 313, 315
- sp. 313, 315, 321, 323
- sp. C.44 273
- sp. 45 273

Cyprinotus glaucus 338, 342, 343, 349
- (H.) incongruens 86

Cypris 296
- subglobosa 273, 275, 277, 279, 283

Cyprois 296
Cythere amnicola 395, 399

- devexa 94
- diegoensis 99
- euplectella 431
- lutea 205, 329, 330, 331
- pseudoconvexa 395
- pumila 99
- (Bairdia) reussi 186
- sicula 395, 399

Cythereis 114, 116, 296, 445
- azerbaidjanica 395
- bailovi 395
- (s.l.) ciplensis 113
- devexa 94
- donetziensis 395
- flabellicostata 99
- glaucum 97
- longiductus 99
- matura 444
- megalodiscus 94
- pontica 395
- praeazerbaidjanica 395
- truncata 395
- wyvillethomsoni 99

Cytherella 23, 135, 140, 181, 184, 225, 371, 372, 418
- alvearium 224, 231
- consueta 27, 418, 420
- dissimilis 445
- mejanguerensis 217
- pandora 182
- pustulosa 28
- transversa 418, 420
- aff. harpago 217
- aff. turgida 445
- sp. 224, 231, 364
- sp. 1 421
- sp. B 447

Cytherelloidea 180, 181, 225, 445, 448
- dameriacensis 27
- keiji 189
- precipua 182
- sabahensis 431, 432
- sorida 224, 233
- aff. rehburgensis 444
- sp. 182, 183, 431, 447

Cytheretta 26, 177, 225, 231
- costellata 28
- crassivenia 28
- eocaenica 27
- forticosta 28
- pulchra 418
- pumicosa 182, 183
- sculpta 28
- subradiosa 224, 228, 231
- vulgaris 28
- sp. 224, 228, 231

Cytheridea 27, 225, 231, 233, 357
- atlantica 28
- neapolitana 224, 228, 231
- rigida 28
- sp. nov. A 354, 361
- sp. nov. B 354, 361
- sp. nov. C 354, 363
- sp. nov. D 354, 363
- sp. nov. E 354, 363

Cytheridella 357
- alosa 208, 209, 210, 211, 212, 213, 214, 217
- sp. nov. A 354, 365

Cytherissa 296, 322
- lacustris 322, 323, 327, 328, 330, 331

Cytherois 203, 296
- fischeri 329, 331
- vitrea 329, 330, 331

C.? sp. 431

Cytheromorpha 119, 296
- macchesneyi 329, 331
- paracastanea 208, 209, 210, 212, 214, 216, 217
- warneri 216, 217
- sp. 183

Cytheropteron 181, 225, 331, 332, 371, 372
- alveiformis 419
- arcuatum 329, 331, 332
- alatum 330
- inflatum 329, 331
- montrosiense 329, 331, 333
- paralatissimum 329, 330, 331
- rotundatum 224, 231
- uchioi 431, 433
- vespertilio 328, 329, 330, 331, 332
- cf. miurense 431, 432
- cf. tricorne 419
- sp. 182, 369
- sp. A 431
- sp. B 431, 433
- sp. 1 331
- sp. 2 331
- sp. 3 331

Cytherura 23, 181, 192, 211, 215, 418
- elongata 208, 209, 210, 212, 214, 215, 216, 217
- osticola 217
- palacii 217
- radialirata 217
- sandbergi 208, 209, 210, 212, 214, 215, 216, 217, 218, 220
- striata 329, 331
- swaini 217
- aff. forulata 208, 209, 210, 212, 214, 215, 216, 217
- aff. johnsoni 182

513

- sp. 182, 183, 217, 431
- sp. 1 369

Daphina 322, 323
- pulex paralex 310
Darwinula 141, 225, 296, 298, 357, 455
- boteai 295, 297, 301
- furcabdominis 208, 209, 210, 212, 213, 214
- stevensoni 208, 209, 210, 212, 214, 224, 228, 323
- sp. 354, 365
Darwinuloides 455
Diasterope 70
Dizygopleura 478
Dolerocypria fastigata 208, 209, 210, 212, 214, 215
Dolerocypris 296
Doloria 70
- pectinata 235, 237, 242, 243
Dolocythere sp. 444

Echinocythereis 116
- bradyformis 431
- margaritea 183
- multicostata 28
- scabra 418, 420
- septentrionalis 28
- scabropapulosa 28
- sp. H 431
Ecklonia 192
- radiata 192
Elkocythereis 318
- sp. 313, 317
Elofsonia 121
- baltica 123
Elofsonella concinna 200, 201
- salvadoriana 217
Empoulsenia 241
- pentathrix 235, 237, 241, 242
Enteromorpha 199, 201, 202, 203
Eocytheropteron 27
- sherborni 28
Erpetocypris 322
- cf. reptans 321, 323
Euchaeta sp. 267
Euchirella rostratum 267
Eucopia grimaldi 267
- unguiculata 267
Eucypris 295, 296, 322
sp. 321, 323
Eucythere subovalis 431
- sp. A 431
- sp. B 431
E.? sp. 431
Eucytherura 181, 418

- sp. 431
- sp. J 431
Eucytheridea declivis 329, 331
- macrolaminata 329, 331
- punctillata 200, 201, 202, 329, 330, 331
Euphilomedes 70
- producta 245, 246, 248, 249, 250, 251, 252
- rhabdion 70
Euprimites 70
Euriyticythere 445
- subtilis 444
Exophtalmocythere insignis 444

Fabaeformiscandona 295, 296, 305
Falunia 370, 371
- plicatula 418
Fucus 199, 201, 203
- platycarpus 203
- serratus 203
- vesiculosus 203

Gaidius cf. tenuispinus 267
Galeolaria caespitosa 193
Gammarus palustris 250
Gangamocytheridea 181
Gerdalia 455
Ghardaglaia setigera 150
Gigantocypris 264, 266, 268
- muelleri 263, 264, 265, 266, 268, 269
Glabellacythere dolabra 72, 83, 84, 85
Glyptobairdia 181
- coronata 182
Glyptocythere 80, 90
- bindosa 72, 76, 77, 78, 86, 89
- guembeliana 72, 73, 74, 75, 78, 86, 89
- oscillum 72, 79, 80
Grantia sp. 310

Hadracypridina 70
Halocypria globosa 45
Hammatocythere sp. 113
Haplocytheridea 178, 211
- bradyi 178, 217
- setipunctata 208, 209, 210, 211, 212, 214, 215, 216, 217, 329, 331
- sp. (gr. setipunctata) 177
H.? setipunctata 183
Healdia 133, 135, 139, 140
- simplex 136
- sp. 136
Hemicyprideis 178
Hemicypris pyxidata 271, 275, 277, 279, 281
- sp. B 273
Hemicythere 296
- villosa 200, 201, 202

Hemicytheria filipescui 395
Hemicytherura 180, 181
- cellulosa 200, 201, 202
- cranekeyensis 182, 217
- cuneata 431
- radiata 431
- sp. 183, 431
Henryhowella asperrima 418, 420
Hermanites 93, 104, 178, 180, 181
- hornibrooki 182
- paijenborchiana 27, 419
- cf. dohmi 98, 103, 104
Hermiella 139, 140
Herpetocypris 296
H.? sp. 313, 315
Heterocyprideis sorbyana 329, 330, 331
Heterocypris 296
- dentatomarginatus 273, 275, 277, 278
- punctata 208, 209, 210, 211, 212, 214, 215, 219
Heterocythereis 225
- albomaculata 200, 201, 202, 203, 224, 231
Heterorhabdus 266
Hiltermannicythere 225
- turbida 224, 228, 231, 233
Hindella 478, 479, 480
Hirschmannia 121
- tamarindus 123, 124
- viridis 200, 201, 202
Hormotoma 478, 479, 480
Hornibrookella 93
Hulingsina 178
- americana 329, 331
- ashermani 217
- sandersi 217
- aff. rugipustulosa 217
- sp. 183
Hungarella 133, 136, 140, 142, 143
- limbuta 136
- lunata 136
Hungarocypris 296

Ilyodromus 295, 296, 299
Ilyocypris 225, 296, 322
- bradyi 224, 228
- taprobanensis 227
- cf. lacustris 323
- sp. 273, 321
Incongruellina sp. 431
Indiacypris luxata 273, 277
Idiocythere aquitanica 27
Iniella 455
Isocypris 296

Jugosocythereis 178, 180, 181
- pannosa 182, 183

Kangarina 181, 418
- tridens 27
Kentrodyctyocythere 445
- typica 444
Kikliocythere sp. (gr. labyrinthica) 117
Kliella 295, 296
Kobayashiina hyalinosa 431, 432
Kovalevskiella 295, 296, 297, 302
Krausella fulgens 418, 420
Krithe 23, 180, 181, 225, 418
- angusta 418, 420
- dolichodeira 182
- luyensis 27, 418, 420
- parvula 418, 420
- producta 431
- rutoti 28
- cf. caudata 418, 420
- sp. 224, 228
- sp. 5 420

Laminaria 199, 200, 202, 202, 203, 204, 205
- digitata 203
- saccharina 203
Leguminocythereis 23, 26, 418
- angulatopora 28
- inflata 28
- magna 28
- pertusa 28
- striatopunctata 28
Leperditia 477, 478, 482, 483, 485
Leptocythere 203, 225, 296
- darbyi 208, 209, 210, 212, 214, 216, 217, 219
- nikraveshae 217
- pellucida 86
- ramosa 224, 231
Lessonia nigrescens 192
- corrugata 192
- flavicans 192
- variegata 192
Leucocythere 296
Leviella 139, 140
- raibliana 136, 141
- sp. 136
Limburgina 108, 109, 111, 113
- ornata 109, 113
Limnocythere 118, 225, 295, 296, 340
- ceriotuberosa 338, 340, 341, 342, 343, 344, 345, 346, 349
- floridensis 208, 209, 210, 212, 214, 215, 217, 218
- friabilis 327, 328, 330, 331
- herricki 338, 342, 343, 349

515

- inopinata 111, 119, 224, 228, 318, 323, 324
- pterygoventrata 338, 340, 341, 342, 343, 344, 345, 346, 349
- robusta 338, 342, 343, 344, 345, 346, 349
- sanctipatricii 217
- staplini 338, 340, 341, 342, 343, 344, 345, 346, 349
- cf. posterolimbata 313, 315
- cf. ceriotuberosa 313, 315, 317
- sp. aff. varia 313, 315, 317, 318
- sp. aff. verrucosa 313, 317
- sp. 217, 218, 313

Loculicytheretta sp. 371
Lophocythere fulgurata 72, 83
- multicostata 86
- scraba scraba 72, 80, 81, 82

Loxoconcha 121, 127, 132, 178, 180, 181, 186, 189, 203, 211, 225, 231, 233, 296
- agilis 224, 228, 231
- alata 123
- atlantica 121, 123
- bulgarica 123
- delemontensis 419
- dorsotuberculata 182, 183
- elliptica 123, 129
- fischeri 180, 182, 183
- fragilis 123
- gallilea 123
- gibberosa 224, 231
- granulata 123
- griganensis 129, 130
- hattorii 122, 123
- hadonica 129, 130
- impressa 123
- japonica 123
- kattoi 123
- kitani 123
- laetea 123
- lapidiscola 175, 177, 179, 217, 218
- levis 123
- longipes 123
- magna 182, 183
- matagordensis 208, 209, 210, 211, 212, 214, 215, 216, 217
- mediterranea 122, 123
- modesta 124
- multipunctata 124
- nana 124
- napolitana 224, 231
- optima 122, 124, 431
- parallela 224, 233
- parvula 418
- postdorsalata 183
- pulchra 124

- rhomboidea 86, 205, 224, 231, 233
- rubritincta 224, 231
- rugosa 182
- schusterae 217
- semistriata 124
- separata 124
- sinensis 431, 432, 435
- stollifera 124, 224, 228
- subrhomboidea 124
- subulata 122, 124
- tosaensis 122, 124
- tumida 224, 228, 231
- uranonchensis 124
- valerii 122, 124
- versicolor 124
- viva 124
- warneri 124
- wilberti 182, 183
- zamia 124
- aff. sarasotana 217
- sp. 182, 183, 205, 224, 228, 231, 370
- sp. 1 371
- sp. 3 371, 372
- sp. A 123
- sp. B 123
- sp. OC 124
- sp. OE 124
- sp. OF 124

Loxocorniculum 178, 186
Loxonema 478, 479
Lycopterocypris? sabaudiae 444

Macrocyprina 148, 154, 180, 181
- africana 148
- angusta 154
- decora 148
- dispar 148
- gracilis 148
- inaequalis 148
- maculata 148
- pacifica 148
- propinqua 148, 150
- schmitti 148
- succinea 148
- tensa 148
- turbida 148, 150
- vargata 148
- aff. maculata 182
- sp. 180, 183

Macrocypris 53, 147, 148, 149, 153, 154, 156, 193
- adriatica 148, 155, 156
- angusta 148
- bathyalensis 148, 149, 151, 155, 156
- canariensis 148, 150
- decora 431

- gracilis 148
- maculata 431
- minna 148, 153, 154, 156
- sapeloensis 148, 151, 153
- siliquosa 148, 149, 150, 156
- similis 148, 150, 151, 152, 155, 156
- tenuicauda 148, 156
- trigona 148
- wrightii 420, 421
- sp. 148, 182

Macrocyprissa 154
- cylindracea 154

Macrodentina mediostricta 444
- sp. C 447

Marginohealdia 465
- blumenstengeli 462, 463, 465, 469, 470
- ventrocostata 462

Maternella 467, 472
- dichotoma 462, 463, 467, 469, 471

Matronella 116
Mauritsina 116
Megacythere johnsoni 217, 218
Mercierella 195
- enigmatica 190

Metacypris 296, 324
- cordata 323, 324

Metridia lucens 267
Microcharon 297
Microcythere 181, 370
Microcytherura angulosa 418
Mixtacandona 295, 296, 303, 304, 305
Modiolopsis 478
Monoceratina 225
- striata 27
- sp. 224, 228

M.? sp. 183
Monsomirabilia oblonga 28
- subovata 28

Moorites 118
Morkhovenia 178, 180, 181
- inconspicua 182, 183

Mosaeleberis 109, 111, 113
Mungava marthapuriae 208, 209, 210, 212, 213, 214
Munseyella 180, 181
- cf. japonica 431, 432

Mytilocypris 285
- ambiguosa 287, 288
- henricae 286, 288, 289
- minuta 287, 288
- mytiloides 286, 288
- praenuncia 287, 288
- splendida 287, 288, 290, 292
- tasmanica 287, 288
- sp. 287, 288

Nannocandona 296
nannocoquimba 101
- apiata 101
- sp. 1 98, 101, 102

Nannokliella 295, 296
Neocaudites 180, 181, 217
- nevianii 182, 208, 209, 210, 212, 214, 216, 217

Neochilina ff. capulinoda 463, 465, 470
Neocyprideis 370
Neocytherideis 225, 231, 370, 418
- fascinata 224, 228, 231, 233
- subspiralis 224
- sp. 224, 228, 231

Neomonoceratina 180, 181, 371, 372
- koenigswaldi 189
- sp. 111

Neonesidea villosa 431, 432, 436
Neopomatus 195
- ushakovi 190, 195

Nereina? sp. 395
Nipponocythere sp. H 431
Notodromas 296
Novopris eocaenica 28

Oblitacythereis 17, 20
Occultocythereis 180, 181, 372, 418
- angusta 182

Ocypode 189
Oepikella 70
Oertiella 109, 113
- aculeata 28
- sp. 113

Ogmoconcha 133, 136, 140, 142, 143
- amalthei rotunda 139

Oligocythereis bogis 445, 447
Opimocythere gr. elonga 114
Orionina 93, 178, 179, 180, 181
- bradyi 179, 183, 217
- serrulata 179, 182

Orbitolites 26
Osphranticum sp. 310
Orthoconchoecia aff. haddoni 258
Orthonaria rectagona 463
Orthonotacythere aff. favulata 444

Paijenborchella 23
- iocosa 431
- mediterranea 217
- spinosa 431

Paijenborchellina 370
Palaciosa vandenboldi 217, 218
Palmatolepsis gracilis gracillis 462
Paracandona 296
Paracaudites sp. 117
Paracypris 372

517

- arctutilis 444
- contracta 28
- n.sp. 150
- sp. 183

Paracytheridea 180, 181, 192, 225
- remanei 189
- troglodyta 217
- tschoppi 182
- vandenboldi 217
- aff. altila 182
- aff. hispida 182
- sp. 182, 183, 224, 231

P.? sp. 1 354, 365
P.? sp. 2 354, 365

Paracytheroma 180, 181
- costata 217
- magna 217
- stephensoni 208, 209, 210, 211, 212, 214, 216, 217
- undulimarginata 217

Paradoloria 70

Paradoxostoma 181, 189, 192, 202, 225
- salvadorianus 217
- simile 224, 228, 231
- variabile 200, 201, 202
- sp. 182, 183, 200, 201, 202

P.? sp. 217

Parakrithe 180, 181
- sp. 182

Parakrithella 181

Paranesidea bradyi 182
- tuberculata 182
- aff. tuberculata 182
- aff. fortificata 182
- sp. 182, 183

Paraparchites 487, 488

Paraphalyx effusa 311

Parapontoparta 211

Parapontoparta subcaerulea 208, 209, 210, 212, 214, 215
- sp. 218

Parasterope 70
- pollex 250

Paravargula 70

Parexophtalmocythere aff. berriasensis 444

Parvocythere dentata 217

Parvocythereis 116

Patagonacythere 94, 104
- longiductus antarticus 99
- tricostata 94, 99

Pellicistoma 178
- magniventra 217
- sp. 183

Pericythere foveata 217

Periphylla 266

Perissocytheridea 180, 181, 211, 217

- brachiforma forma inferior 208, 209, 210, 212, 214, 215, 216, 217, 219
- cribrosa 208, 209, 210, 212, 214, 215, 216, 217, 219
- dentatomarginala 217
- excavara 217, 218
- meyerabichii 217
- punctata 217, 218
- rugata 217, 218
- swaini 217
- sp. nov. A 354, 365

Philomedes 70
- globosa 247, 249, 250
- orbicularis 235, 237, 241, 242

Phlyctocythere sp. 431
P.? globulata 445

Phyloctenophora zeelandica 189

Phreatocandona 295, 296

Physocypria 296
- pustulosa 338, 342, 343, 349

Pintopsis 481

Pleurammia robusta 266, 267

Pokornyella 27, 104
- longicostis 28
- talencis 28
- ventricosa 28

Polycope 225
- reticulata 224, 228
- sp. 224, 228

Polyleberis 70

Pontocyprella aturica 27, 421
- sp. 1 418, 420

Pontocythere 225, 231, 233
- elongata 224, 228, 231, 233

Pontoparta hartmanni 208, 209, 210, 212, 214, 215

Potamocypris 295, 296
- granulosa 338, 342, 343, 349
- humilis 280
- smaragdina 218
- sp. 218, 321
- sp. B 218

Praemacrocypris 147

Praezerbaidjanica 396

Polydora 193

Prionocypris? sp. 313, 315

Prionotoleberis 70

Processobairdia 467
- cf. nodocerata 463, 467, 471
- posterocerata ssp. A 463, 465, 467, 470
- spinaterocerata 463, 470
- spinomarginata 463

Procythereis 192

Procytheridea tuberculata 445, 447, 448
- sp. A 447

Propontocypris 181, 225

- setosa 224, 228, 231
- sp. 183
- sp. 1 182
Protoargilloecia angulata 418, 420
- sp. 2 418, 421
Protocythere divisa 445
- helvetica 445
- paquieri 447, 448
- revili 445
- tricostata 116
- aff. mazenoti 445, 447
Protocytheretta 177
Pseudocandona 295, 296, 303, 305
- bilobata 298
- bilobatoides 298
- brevicornis 298
- n.sp.gr. eremita 301
- spelaea 298
Pseudocertina 181
- droogeri 182
Pseudoconchoecia 259
Pseudocypretta maculata 271, 275, 277, 278
Pseudocythere sp. 182
Pseudocytherura 225
- calcarata 224, 231
- sp. 224, 231
P.? sp. J 431
Pseudohealdia 135, 140
Pseudolimnocythere 296
- hypogea 295
Pseudophilomedes 70
Pseudoprocythere aubersonensis 445
Pterygocythereis 27, 180, 181, 225, 233, 418
- aquitanica 28
- ceratoptera 224, 231
- cornuta 28
- jonesi 224, 228, 231
- scalaris 431
- tuberosa 28
- sp. 114, 183
Pumilocytheridea 180, 181
- ayalai 217
Puriana 180, 181, 372
- sp. 369
- aff. rugipunctata 182
Pyrgulopsis nevadensis 311

Quadracythere 93, 178, 180, 181
- lamarckiana 28
- lankfordi 182, 183
- orbignyana 28
- producta 180, 182, 183

Radimella 93, 181
- confragosa 182
- floridana littorala 208, 209, 210, 211, 212, 213, 214, 215, 216, 217
- wantlandi 180, 182, 183
Rectonaria inclinata 463, 465, 470
- muelleri 462, 463, 465, 470
Rectoplacera elongata 462, 463
Recytellina 135
Reticulocythereis 211, 215
- floridana 208, 209, 210, 212, 214, 215, 217, 218
- multicarinata 217, 218
- purii 208, 209, 210, 212, 214, 215, 217
- sp. 183
Rehacythereis 116
- sp. 113, 116
Reubenella 135, 139, 140
Reussicythere howei 175, 177, 179, 186
- reussi 186
Rhynchospira 479, 480
Richterina 469
- (Richterina) costata 463, 469, 471
- (Fossirichterina?) cf. intercostata 463, 469, 471
Roundstonia globulifera 329, 331
Ruggieria 180, 181
- nuda 418
- tetraptera 4128
- sp. 370
Ruppia 345
Rutiderma 70

Saida 420
- torresii 431
Sarsiella 70
Schizocythere 23
- appendiculata 28
- kishinouyei 431
- tessellata 28
Schuleridea mediocaudata 445
- perforata 28
- aff. praethorensis 445
Sclerochilus 181
- centroamericanus 217
- contortus 329, 330, 331
- sp. 183
Scleroconcha gallardoi 235, 237, 241, 242
Schleroconcha 70
Schuchertella 480
Scottia 296
Selebratina sp. 463
Semicytherura 225, 231, 233
- acuminata 224, 228
- affinis 329, 331
- incongruens 224, 231, 233
- inversa 224, 231
- miurensis 431
- nigrescens 85, 200, 201, 202

- sulcata 224, 228, 231
- sp. 224, 228, 369
- sp. M 431

Signohealdia 139, 140
- robusta 136, 137, 139, 141

Skogsbergia 70

Skogsbergiella 70, 241
- macrothrix 235, 237, 240, 241, 242

Soudanella 371, 372

Spathonathodes inorinatus 462
- jugosus 462
- stabilis 462
- strigosus 462

Sphaeromicola 296

Sphenocytheridea gracilis 28

Spinacopia 70
- antarctica 70
- octo 70
- sandersi 70, 247
- torus 70
- variabilis 69

Suchonella 455

Suriekovella 455

Stenocypria 296

Stenocypris major 273, 275, 277, 278, 280
- sp. 273

Strandesia 275, 296
- elongata 273
- marmorata 273, 277
- purpurescens 273, 277, 281
- striatoreticulata 275
- wierzejskii 273, 275, 277
- sp. nov. 273
- sp. 273

Synasterope 70

Syphlocandona 296

Tanella 180, 181
- gracilis 189, 217

Tentaculites 479

Tetracytherura 225
- angulosa 224, 231, 233

Tetradella 70
- quadrilirata 71
- scotti 71

Thalassocypria gesinae 208, 209, 210, 212, 214, 215
- vavrai 208, 209, 210, 211, 212, 214, 215
- sp. 218

Thalmannia? sp. 431

Theriosynoecum 118

Timiriasevia 455
- mackerrowi 118

Tomiella 455

Torohealdia 139, 140
- opistocostata 138

Trachyleberidea 184
- prestwichiana 419, 420, 421

Trachyleberis 184
- cibaria 395
- gavrilovi 395
- horrescens 28
- immediata 395
- imnadzeae 395
- niitsumai 431
- parula 27
- pontica 395
- scabrocuneata 431
- sp. S 431, 432
- sp. 395, 431

Triadohealdia 135, 140

Triangulocypris 181
- laeva 182, 183
- pachyconcha 183
- sp. 182

Tricornina (Bohemina) paragracilis 463
- communis 462, 463

Triebelina 27, 181
- gierloffi 217
- punctata 27
- serlata 182

Trigonocypris 285
- globulosa n.sp. 286
- timmsi 286
- sp. 286

Triplacera triguetra 462

Tyrrhenocythere 395, 396, 398, 399, 403
- amnicola 396, 399, 400
- azerbaidjanica 396, 399, 402, 403
- bailovi 396
- ballesioi 395, 396
- cavernosa 396, 399
- ciberia 396
- dendropora 396
- donetziensis 396, 397
- filipescui 396, 398, 399, 401, 402
- gavrilovi 396
- immediata 396
- imnadzeae 396, 399, 400, 403
- notabilis 396, 398, 399
- papilosa 402
- pignattii 395, 396, 397
- pontica 396, 399, 400, 401, 402, 403
- praezerbaidjanica 396, 397, 399, 403
- pseudoconvexa 396
- ruggierii 395, 396, 397, 402
- sicula 396
- taurica 396, 398, 399, 401, 403
- triebeli 395, 396, 397, 399, 402
- truncata 396, 399, 402
- sp. 396, 403

Undenchaeta plumosa 267
Urocythereis 101, 225
– californicum 94
– sp. 1 98, 101, 102
– sp. 2 98, 101, 102
– sp. 224, 231, 233
Uroleberis 181
– angulata 182
– striatopunctata 27
– sp. 183

Vargula 70

Welleria 478

Xenocythere cuneiformis 329, 331
Xestoleberis 181, 211, 225, 231, 233, 371, 372
– alta 445
– communis 224, 228, 231, 233
– depressa 329, 330, 331
– dimorpha 445
– dispar 224, 231, 233
– eulitoralis 217
– foveolata 431 432, 436
– glabrescens 418
– mixohalina 208, 209, 210, 211, 212, 214, 215, 216, 217
– plana 224, 233
– rigbyi 216, 217
– cf. variegata 431, 433
– sp. 1 182, 418, 420
– sp. 2 182
– sp. 3 182
– sp. 183, 216, 217

RAYMOND H. FOGLER